# DESIGN WITH THE DESERT

## CONSERVATION AND SUSTAINABLE DEVELOPMENT

# DESIGN WITH THE DESERT

## CONSERVATION AND SUSTAINABLE DEVELOPMENT

EDITED BY
RICHARD MALLOY · JOHN BROCK
ANTHONY FLOYD · MARGARET LIVINGSTON
ROBERT H. WEBB

CRC Press is an imprint of the
Taylor & Francis Group, an **informa** business

CRC Press
Taylor & Francis Group
6000 Broken Sound Parkway NW, Suite 300
Boca Raton, FL 33487-2742



Printed in the United States of America on acid-free paper
Version Date: 20120827

International Standard Book Number: 978-1-4398-8135-4 (Hardback)



**Library of Congress Cataloging-in-Publication Data**

Design with the desert : conservation and sustainable development / editors, Richard Malloy ... [et al.].
p. cm.
Includes bibliographical references and index.
ISBN 978-1-4398-8135-4
1. Desert ecology. 2. Desert ecology--Southwestern States. 3. Desert ecology--West (U.S.)
4. Sustainable development. 5. Sustainable development--Southwestern States. 6. Sustainable development--West (U.S.) I. Malloy, Richard (Richard A.)

QH541.5.D4D475 2013
577.54--dc23 2012023189

**Visit the Taylor & Francis Web site at**
**http://www.taylorandfrancis.com**

**and the CRC Press Web site at**
**http://www.crcpress.com**

# Contents

## Part III Desert Planning



## Part IV Ecology in Design of Urban Systems

## Part V Urban Sustainability

# *Foreword**

In the desert, we have no sustainable alternative to design with nature when it comes to our human environments. We have limited, long-term options for the world at large, but the desert poses special problems including the extremes of temperature and scarcity of water. The realities of the desert environment, combined with the need to make our developments more sustainable for future generations, make it obvious that we must be guided by ecological knowledge in desert regions when designing new living and working spaces or retrofitting old ones.

Our current condition requires that we reconnect with the nature of our regions instead of designing spaces under the old ethos of "conquering nature" and isolating humans from their natural environments. We need to look back at what our society has collectively learned about this seemingly harsh environment in order to move ahead.

The Roman Marcus Vitruvius Pollio wrote the first guide to architecture and dedicated *On Architecture* to his emperor, Augustus. A good architect, according to Vitruvius, was not a narrow professional but an intellectual of wide-ranging abilities. For example, Vitruvius included medicine in his extensive list of subjects of which an architect should "have some knowledge." An architect should understand medicine, "in its relation to the regions of the earth (which the Greeks call *climata*)" in order to answer questions regarding the healthiness and unhealthiness of sites. A knowledge of air ("the atmosphere") and the water supply of localities is essential, "[f]or apart from these considerations, no dwelling can be regarded as healthy."[1]

Vitruvius devotes much of his writing to site-specific, or landscape, considerations. As one classicist observed, "Vitruvius' conception of architecture is... wide, at times almost approaching what we define as urban studies."[2] Vitruvius made detailed pronouncements for planning new urban developments. The very first consideration must be salubrity. He noted, "First, the choice of the most healthy site. Now this will be high and free from clouds and hoar frost, with an aspect neither hot nor cold but temperate. Besides, in this way a marshy neighborhood shall be avoided. For when the morning breezes come with the rising sun to a town, and clouds rising from these shall be conjoined, and with their blast, shall sprinkle on the bodies of the inhabitants the poisoned breaths of marsh animals, they will make the site pestilential..."[3]

In addition to his directions for using an understanding of nature to design houses and plan cities, Vitruvius provided considerable advice for building civic structures and spaces. The Romans constructed many new communities, a good number of which continue to prosper today throughout Europe, the Middle East, and North Africa. Twenty centuries after Vitruvius, in their detailed study of architectural education for the Carnegie Foundation, Ernest Boyer and Lee Mitgang urged architects to shift their focus from designing objects to "building community."[4] Such a change requires careful consideration of what constitutes "community" and what is the relationship of communities to their physical and biological regions.

---

* Revised from Steiner, F. *The Living Landscape*, McGraw-Hill, New York, 2000 (Paperback edition Island Press, 2008); Burke, J. and Ewan, J., *Sonoran Preserve Master Plan*, Department of Parks, Recreation, and Library, Phoenix, AZ, 1998. With permission).

Community activities appear to flourish or to wane depending upon their regional context. Harkening back to Lewis Mumford's use of the term, several contemporary architects and planners advocate the notion of the "regional city" or the "city-region," including Peter Calthorpe, William Fulton, Gary Hack, and Roger Simmonds.[5] Healthy city-regions fit their natural environments and foster civil interactions. Healthy building and landscape designs, in turn, fit their city-regions and deepen human interactions.

For example, Calthorpe and Fulton contend, "the Regional City must be viewed as a cohesive unit—economically, ecologically, and socially—made up of coherent neighborhoods and communities, all of which play a vital role in creating the metropolitan region as a whole."[6] Arizona's Sun Corridor, which combines the Tucson and Phoenix metropolitan areas, forms such a regional city or, alternatively, a megaregion. Several other metropolitan regions in the arid Southwest also continue to grow, including Las Vegas–Henderson and El Paso–Juarez.

University of Texas Professor Steven Moore, a leading sustainability theorist, links regionalism to place-making. He concludes "it is politically desirable and ecologically prudent to reproduce regionalism as a practice relevant to contemporary conditions."[7] Moore builds on Kenneth Frampton's advocacy of critical regionalism. According to Frampton,

> The fundamental strategy of Critical Regionalism is to mediate the impact of universal civilization with elements derived indirectly from the peculiarities of a particular place.
> … Critical Regionalism depends upon maintaining a high level of critical self-consciousness. It may find its governing inspiration in such things as the range and quality of local light, or in a tectonic derived from a peculiar structural mode, or in the topography of a given site.[8]

Regional understanding is important for architects and landscape architects to design specific buildings and sites much the way medical doctors need to understand human anatomy in order to treat an individual patient. Such a design practice would involve the critical understanding of the region as well as the shaping of its futures.

What are the implications of this view of city-regions and their wealth for architecture, landscape architecture, and planning? Two essential needs emerge: *Think comprehensively and broadly* and *Make places matter.*

First, if one sphere of knowledge is privileged at the expense of others, then the result is deleterious and not sustainable. For example, if a building design favors aesthetics and ignores concerns about its environmental or social context or the economics of the project, it will sooner or later fail. My own discipline, planning, tended to ignore the aesthetic principles of good community design for much of the second half of the twentieth century. One does not need to look far to see the negative consequences of this oversight.

To be comprehensive, we need to heed the advice of Vitruvius and think broadly again. If an architect is a wide-ranging intellectual, then architecture should reflect a broad understanding of other fields. Such understanding certainly should encompass those fields closest to architecture, arguably landscape architecture and city planning.

Conversely, landscape architects and planners can gain much by seeking to understand architecture, rather than by leaving building design solely to architects. As reading Vitruvius reminds us, architecture has a more ancient history than related fields. That history, as well as the theories architects have promulgated for designing interior and exterior spaces and for planning cities and regions, sets the stage for those of us in sibling fields.

Such strategic thinking should be grounded in theory. The design and planning disciplines could benefit from a few more good theories. Landscape architecture is a discipline that illustrates the adage that there is nothing as practical as a good theory. Two theories have catapulted landscape architecture into greater prominence. In the mid-nineteenth century, Frederick Law Olmsted, Sr. advocated the use of public parks to address the ills of urbanization brought on by the Industrial Revolution. A century later, Ian McHarg urged us to "design with nature," publishing his immensely influential book with the same title. McHarg's theory highlights the integration of understanding the biophysical setting, ecology, planning, design, and execution of projects that reflect an understanding of people and nature.

These two theories—that public parks have social benefits and that design should be derived from environmental understanding—sustained landscape architecture for two centuries. Like jazz, landscape architecture originated as a particularly American art form. Landscape plays a central role in American culture akin to the city in Italian culture. A newer theory than McHarg's shows signs of emerging by combining concerns about urban welfare with ecologically based design. New urban ecology–based theories promise to address a range of pressing issues from environmental justice to the reclamation of postindustrial, marginalized sites. Such theories are beginning to yield new forms of landscape urbanism, like those generated by the firm West 8 in The Netherlands and James Corner Field Operations in the United States. Fresh urban theory could move landscape architecture closer to emerging ideas about the structure of cities in architecture and planning as well as new theories in urban ecology being put forth by biologists and geographers.

Theory fuels the academic engine. In the sciences, theories are tested through experimentation. In the design fields, they are explored in studios and through reflection upon projects. Ecological design and urban ecology extend outside the bounds of the traditional sciences and arts. This suggests, to advance these new theories, science education needs to learn from the creativity of studios, while designers could benefit from more fact-based education.

With Modernism, design education turned its back on history. Postmodernism embraced history, but its design applications (including some that are New Urbanist) incorporate past elements too literally and romantically. We must learn from precedent without becoming prisoners of the past. By thinking broadly, we can design several solutions based on local and regional considerations rather than looking in on a single, predetermined course of action.

A second implication for architectural and planning education within this city-region perspective is perhaps the most obvious but one often overlooked by our own culture: place matters. To remain competitive in the global economy, city-regions must offer compelling places for people to live. Architects can, and should, contribute to creating such urban places. Several institutions provide ongoing advances to our knowledge of desert environments, as well as native flora and fauna, including the Desert Botanical Garden, the Arizona-Sonora Desert Museum, the Great Basin Institute, and the Lady Bird Johnson Wildflower Center.

Arizona State University (ASU) provides a nice model in this regard. From 1989 until 2001, I directed ASU's former School of Planning and Landscape Architecture, which included several authors in this book who have contributed much to the planning and design in the Sun Valley region. Beginning in the late 1960s, then architecture dean Jim Elmore began advocating for converting the abandoned, dry Salt River bed into a linear greenway through the metropolitan region. Generations of ASU architecture, planning, and landscape architecture faculty and students followed Dean Elmore's vision, and components of the Rio Salado project are now realized in Phoenix and Tempe. Water now again occurs in the once dry river bed of Tempe, which enhances recreational and economic development opportunities. More recently, former dean John Meunier encouraged faculty and students to become engaged in the pressing issues affecting the region, especially through design and planning charrettes. ASU's influence on the design and planning of the northern—rapidly urbanizing—20% of the city of Phoenix is especially evident.[9] As a result of collaboration between ASU faculty and city staff, large areas of the north area have been set aside as desert preserves. Most of ASU's sustainability programs and projects are now wrapped into the Global Institute of Sustainability, as envisioned by President Michael Crow.

As the global population continues to grow and to become more urban, place-making possibilities expand. At the beginning of the twentieth century, 2 billion people inhabited the planet. The Earth currently has almost 7 billion inhabitants. The United Nations projects the world's population to plateau at 9.4 billion by the year 2050 and then slowly rise to 10.4 billion by 2100.[10] This translates into some 12.6 billion additional individuals appearing on the planet over the next century.[11] Half of the world's population now lives in cities, and the number of these urban inhabitants is expected to double by 2030.[12] We live in the first urban century. By 2050, two-thirds of the people in the world will be living in urban regions. Our challenge is to design healthy, sustainable, and safe city-regions for those who will be joining us on the planet.

To make place matter, designers and planners must become ecologically literate. Certainly, this was the foundation of McHarg's argument that we should "design with nature." Architect Grant Hilderand contends that such design is fundamental to our species. He writes, "some characteristics of our surroundings, natural and artificial, may bear to some of our innate survival-supportive behaviors."[13] In his exploration of architecture's biological roots, George Hersey concludes, "we build and inhabit giant plants, animals, or body parts."[14] Stephen Kellert and others call such an approach "biophilic design," which emphasizes "the necessity of maintaining, enhancing, and restoring the beneficial experience of nature in the built environment."[15]

The artifacts of design provide physical shape to cultural identity. An improved environment can provide the context for positive interactions among people. The more we know about and care for our surroundings, and the more we interact with and care about other people on that account, the greater is the potential for knowledge to thrive. Such knowledge is capital. Only with such capital can a civilization—a culture—be created for a city-region that is worthy of the grandeur of its natural surroundings. The pages that follow take on that timely objective.

**Frederick R. Steiner**

# *Preface*

When I moved to the Phoenix area in the late 1990s, there was an active public debate forming on the trajectory of growth in the Valley of the Sun in particular and state of Arizona in general. There was a group of concerned residents who wanted to see measures put in place to curb unbridled growth and development activities and to protect sensitive lands. Cory Filler comments: "Developers have been ruining our cities for too long, and our quality of life has suffered to fill their fat wallets."[15] Another group thought these measures could be initiated within the jurisdictions and with incentives rather than through regulation, warning of dire consequences to the region's economy if these measures were adopted. An *Arizona Republic* article on the growth management initiative cites analysis from Professor Peter Gordon from the University of Southern California with a predicted loss of 235,000 person-years of employment (a person-year is the effort of one worker employed for a year), assuming a two-year construction moratorium if the initiative is passed. In addition, Elliott Pollack, a Scottsdale developer and economist, forecasts a loss of 219,000 jobs if lawsuits bring development to a halt, as some attorneys believe. Otherwise, he says that relying on state population forecasts, which tend to be low, would force construction cutbacks and cost 90,000 jobs in all.[15]

As this debate gathered momentum into a formal referendum on growth, the issue appeared to me to become lost in the rhetoric of words, sound bites, and dire scenarios presented by opposing viewpoints on this important issue that could be responsible for shaping future development of the region. If and how sustainable growth and development could be implemented was not a clear and decisive matter, and even educated people were confused about where to stand in this debate. I was personally concerned about the pace of growth and development, seeing large tracts of virgin desert in my previous home in Tucson, Arizona, converted to generic tract housing; the same pattern was playing out in the Phoenix area. I began to wonder about the future of the city and whether the current system of urban development in the desert was indeed serving the good of the overall community. I told myself there must be a better way.

A vision for this possible better way came to me during my first year of graduate school at Arizona State University. During my undergraduate training in landscape development and planning, I was exposed to the work of Ian McHarg, author of *Design with Nature* and founding director of the School of Landscape Architecture and Regional Planning at the University of Pennsylvania. McHarg has been widely credited as the founder of the ecological planning profession. He developed a unique method of overlays that allowed a number of features of the landscape to be separated into layers that are used for suitability analysis of the proposed site. I built on McHarg's work with the creation of *Design with the Desert*—a manner of adopting rational planning in regions shaped through the scarcity of natural resources and defined by the climatic and hydrological regimes. This concept accepts the need to put forth options to development that feature sustainability as a fundamental component of building design and projects that minimize waste, enhance the quality of life, and make smart use of energy, food, and the hydrologic system. Applying the principles of the *Design with the Desert* concept will provide a more profound approach to working in any area of the world facing limitations to life by the natural setting.

The purpose of this book is to serve as an educational and inspirational tool for anyone concerned with conservation and development in fragile regions of the world. This is a *transdisciplinary* volume that spans the fields of science, ecology, planning, landscape development, architecture, and urban design. The area of focus for this book is defined as the geographic area of the four warm deserts of the American West, which include the Great Basin, Mojave, Sonoran, and Chihuahuan deserts. In spatial terms, this includes an area from southern California, north beyond Las Vegas, south to Mexico, and east to El Paso, Texas. This area represents the most actively developing regions in the American deserts where the interface of conservation and development has long-lasting impacts from development activities. The concepts presented in this book have specific relevance to this region, but also apply in large part to other areas of this continent or other parts of the world by incorporating the natural, historical, and cultural considerations for that region into the study. We hope that this book will create the inspiration and opportunity to apply these principles of sustainability to other parts of the country and world.

This book was written for educated readers from many backgrounds. We chose chapters that would be the most informative on the theme and easy to read. The authors of each chapter in this book are considered to be top level authorities in their field of expertise in this region; many have written several books on their own. All of the contributors have thoughtfully developed their chapters with the specific goal of providing their vast technical and professional knowledge on this region in a condensed format to provide a thumbnail understanding on each theme.

This book is designed to be read from beginning to end in sequence, although each chapter can be read without the need to review the preceding chapters. Where possible, we have cross-referenced in the text other parts in the book if the reader has a desire to explore parallel themes on a given topic. The flow of the book begins with Part I covering the physical aspects of the desert realm—the land, geology, water, climate, and related themes. Part II deals with the "living" and ecological aspects of deserts, including plants, animals, ecosystems, and restoring habitats of degraded ecosystems. This is followed by Part III on desert planning, including ecological planning, water planning, resource planning, and community development. We then move on to Part IV on ecology in the design of urban systems, that is, how you can bring nature into the built environment through the use of native plants, creation of habitats for nature in the urban setting, and design of urban building and projects that create life. Lastly, in Part V, we explore the concepts of urban sustainability—how to design urban systems that provide a secure future for community development through water security, sustainability building practices, bold architecture and community designs, and experimentation in futuristic communities and urban designs that integrate ecological and resource sustainability into every aspect of the urban community designs.

My desire is that this book serves as a source of hope for many people across the region and world to explore ways to connect development with nature without destroying the desert in the process. I hope that many people including students, developers, planners, engineers, environmentalists, community leaders, city councils, urban planners, people relocating to the desert region, and many others use this book as a point of discussion, inspiration, reference, or contemplation. Deserts are fragile regions by nature. We have limited chances to get it right on development to avoid leaving permanent scars on the landscape that can take generations or more to recover. This book provides a path to help create or enhance the connection of life from the natural to the built environment in the American Southwest.

In 2001, I organized a *Design with the Desert* conference at Arizona State University that included some of the most notable experts in science, ecology, planning, and development. Ian McHarg was slated to be the keynote speaker for this event. When I had the program for the event complete, I sent a copy to Ian to outline his role in the program prior to his trip to Arizona. He was truly impressed with the program I had put together. He told me "tell your people, you do good things." Sadly, Ian died two weeks before this event was held. However, Ian's spirit lives on through his students, apprentices, projects, books, and teachings. This book captures Ian's desire for all of us to act in manners that collaborate with nature, rather than participate in endeavors that lead to degrading the natural environment or the human condition.

**Richard A. Malloy**
*Executive Editor*

# *Acknowledgments*

This book was forged as a collaborative project designed to educate people about our desert environment and manners to sustain the region with rational solutions to conservation and development. As such, this work garnered grassroots support from many institutions, organizations, and people, without which this project would not have been as successful in engaging people and reaching the hearts and minds of many of our participants over the years.

I would like to thank the institutions that provided support in making past *Design with the Desert* events happen. First of all, the Desert Botanical Garden in Phoenix, Arizona, which helped host my first public event on this topic in 2000. The involvement of the Department of Horticulture and Desert Landscaper School and all of the students was a great help in carrying out a flawless event. In particular, I would like to thank Caroline O'Malley, Linda Raish, and Cesar Mazier for their involvement in this event. Secondly, Arizona State University (ASU) played a major role in providing a setting and framework for me to expand on this theme and reach a larger group of professionals to engage in large conferences and forums. The School of Planning and Landscape Architecture, Center for Environmental Studies, and the Department of Applied Biological Sciences were full or cosponsors of these events. I would like to personally thank Frederick Steiner, Mary Kihl, Joseph Ewan, Ignacio San Martin, Cheryl McNabb, Hemelata Dandekar, Ron McCoy, Charles Redman, Gail Reisner, Cindy Zisner, and Ward Brady. We also had support from the ASU Department of Geographical Sciences in creating maps and charts, including Barbara Trapido-Lurie, Kimberly Hall, Don Thorstenson, Derek Weatherly, and Samantha Yates. From the USGS office in Tucson, we would like to thank Diane Boyer, Peter Griffiths, and Helen Raichle.

As the *Design with the Desert* program was expanding and reaching larger audiences, there were several organizations that willingly provided financial resources to make these public events possible. These sponsors included the Arizona Municipal Water Users Association, ASI Sign Company, Inc., CF Shuler, Desierto Verde, Coy Landscaping, Inc., DMB, Inc., Grady Gammage, Jr., Jones Studio, Sierra Club, Grand Canyon Chapter, Swaback Partners, Ten Eyck Landscape Architects, and Valley Crest Landscape Company. The support provided by these sponsors helped to provide the setup and promotion of these conferences and forum.

To the authors of each of the chapters in this book, I appreciate your dedication in putting your knowledge down on paper to share with the world. I also want to thank you for your patience and efforts made in our chapter review and editorial process. In the end, it made this volume better by incorporating these comments and suggestions to your chapters.

We would like to thank several people who took the time to provide feedback for the authors in this volume. The following people have participated in the peer review of a chapter in this book, including Michael Applegarth, Ramon Arrowsmith, Nancy Selover, Mike Crimmons, John Keane, Anthony Brazel, David Edwards, Robert Patterson, Matt Johnson, David Patton, William Miller, John Alcock, Jeffrey Whitney, Eddie Alford, Grady Gammage, Jr., Laurel Arnt, Bobby Creel, Tom Bruschatske, Romella Gloriso-Moss, Sandy Bahr, Laurence Moss, David Pijawka, Christopher Boone, Nan Ellin, Frederick Steiner, Shankara Babu, James Riley, Tim Pope, Judy Milke, Ron Gass, Jeff Homan, Milton Sommerfeld, Barry Spiker, Emily Talen, Nancy Levinson, Kim Sorvig, Tom Martinson, Vernon Swaback, Pliny Fisk III, Amanda Ormond, John Motlock, Lucia Athens, and Tony Brown.

Finally, I would like to thank Robert Webb of the USGS in Tucson, Arizona. "Bob" Webb willingly took on the role as mentor for our interdisciplinary editorial group through the publishing process. His vast experience in book publishing provided the editorial team with valuable insights that made the editing of this complex and multifaceted book seem easy and straightforward, in spite of the challenges posed by working with authors from many disciplines and backgrounds. We all sincerely appreciate his professional guidance, which helped all of us as we learned the ropes of the editing and publishing of important works of environmental literature.

# *Introduction*

The deserts of the world are in trouble. The International Fund for Agricultural Development reports that 250 million people worldwide are affected by desertification and 1 billion people are at risk. In addition, over one quarter of the Earth's surface is threatened by desertification—an area of 8.9 billion acres of land.[15] Global issues of climate change, rapid population growth, and depletion of natural resources are a few of the main pressure points on the ability of these regions to follow a sustainable path for human populations, particularly in developing nations. Deserts have a limited capacity to support human populations, but have experienced increased development activities, particularly in areas that had the resources to engineer and construct projects to transport water long distances to areas once dependent on the limited annual rainfall.

Where there is no water, there is no life. Deserts are defined by the scarcity of water, and the extremes of climate, as such, will govern the areas that can support life and those areas that will remain devoid of life. Human activities to develop desert lands have the potential to have dramatic and long-term effects. Disturbed lands in the desert lack resiliency to recover from activities such as grazing, mining, mineral extraction, and the removal of the natural vegetation cover.[15] In some parts of the world, expansive oases in the desert are built using large amounts of financial resources to create communities largely dependent on water and resources not directly tied to place. These developments create the illusion of sustainable life for a while. At some point, the forces of nature will expose the fragility of these developments and their intense need to consume resources to sustain their existence. The desert will take these areas back, eventually.

## Development in the American Southwest: An Unsustainable Path

The American southwestern desert region faces these same global challenges, particularly when confronted with decisions on conservation or development. The Great Basin, Mojave, Sonoran, and Chihuahuan Deserts in the American Southwest have been the site of exponential growth over the past several decades (see Figure I.1). This geographical area includes the modern southwestern cities of Phoenix, Tucson, Las Vegas, Albuquerque, and El Paso, which occupy lands that once supported a rich desert life. Growth and development activities across this region are fueled by the low cost of land, desirable climate, and an unquenchable desire for people to own their own homes. Typical development activities in the desert often resulted in scraping these lands of an ancient living landscape and replacing it with one that is man-made and dependent on a large consumption of energy and natural resources to sustain the growing human population. Major water projects (Hoover, Glen Canyon, Roosevelt, and Elephant Butte dams) funded through the U.S. Bureau of Reclamation have dammed nearly all of the flowing rivers across this region. In addition, modern engineering has made air conditioning possible for comfortable living in a harsh environment, making the modern desert city comparable in comfort to the homes left behind. In short, urban desert residents in the Southwest are increasingly becoming removed from the harsh extremes of the desert by the nature of development and design.

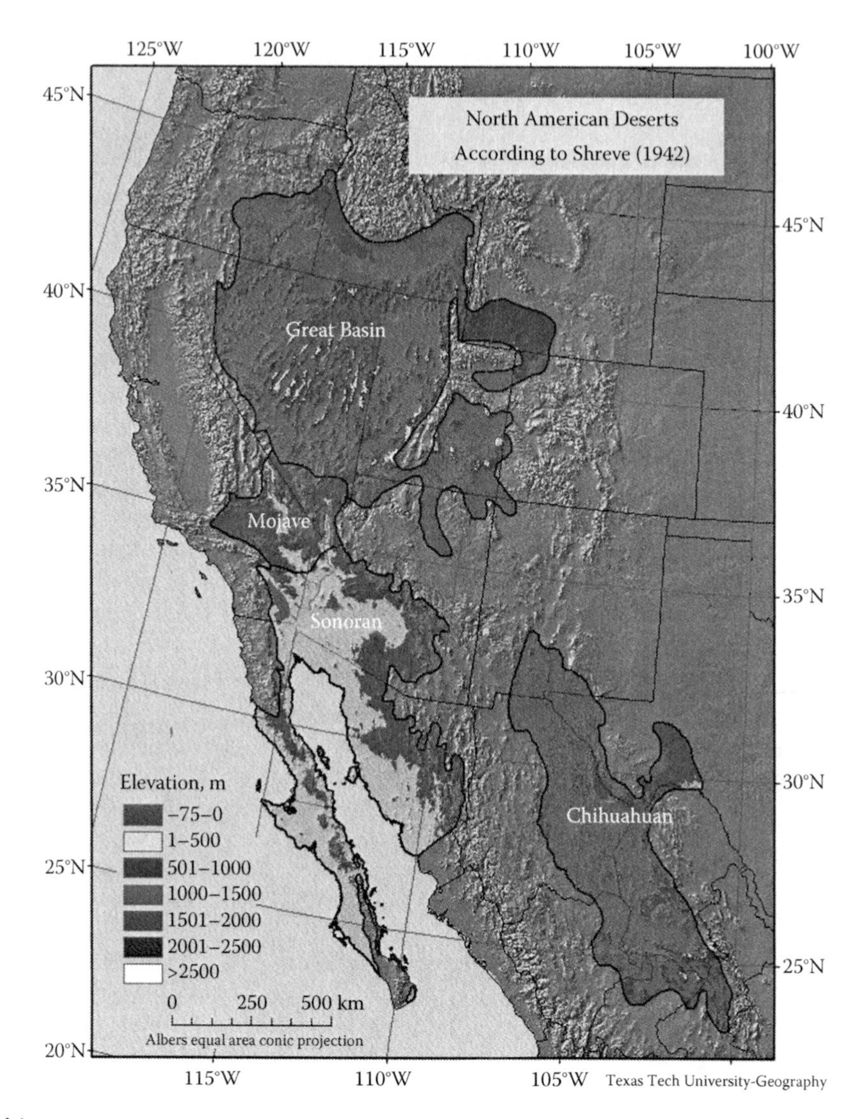


**FIGURE I.1**
Southwestern desert regions. (From Lee, C., R. Arroyo, J. Lee, M. Dimmitt, M. McGinley, and Arizona-Sonora Desert Museum, Deserts of North America, *Encyclopedia of Earth*. C.J. Cleveland, ed. (Washington, D.C.: Environmental Information Coalition, National Council for Science and the Environment). Accessed October 25, 2012. http://www.eoearth.org/article/Deserts_of_North_America. With permission.)

Problems persist across the region. Southwest cities are dependent on water from the Colorado and Rio Grande Rivers, which have experienced periods of drought and reduced annual flows on the rivers. The uncertainty of water is one of the most critical aspects to plan for the future of these cities. Moreover, the single family home, the typical feature of residential development, consumes large tracts of land that expands the spatial dimensions of the urban area. Current development trends favor development on the fringe of the urbanized area where the low cost of land will result in higher profit for the developer. Development on the fringe contributes substantially to urban sprawl, longer commute times, the need for new roads, and increases in air pollution. The once quiet and community feeling of historic desert towns can be transformed by development dependent on the automobile and residents disconnected from each other. Architect and visionary Paolo Soleri encourages society to demand development alternatives that respond to urban sustainability problems at their core, rather than "a better kind of wrongness."[15] That is, new development solutions can be just a repackaged version of the same urban problem.

## Pondering the Lessons from the Mites and the Fern: A Story of Community Collapse

Many years ago I purchased a fresh, green Boston fern for my apartment from the grocery store. I placed the fern in a prominent spot in my kitchen for all to see. Over the next several weeks, my busy schedule did not allow me to be home other than to sleep and change my clothes. I noticed the fern as beginning to look a bit yellow in color; then developed brown tips, which worsened each day. I realized I needed to stop my busy life and do a close examination on what was causing these abnormal symptoms on my plant.

When I picked up the plant and examined it closely, I was *amazed* to discover a whole world of small mites that had created an entire network made from silk-like threads. The mites covered the whole plant and moved effortlessly along a geometric maze of threads that appeared like 1000 expansion bridges spanning off in every direction imaginable. The intricate design of this network of webbing appeared to have been built by a highly skilled community designer. The mites' plump, nearly translucent bodies radiated health and vigor as they moved quickly and rapidly in all directions up and down the plant. I was truly amazed with the beauty of what they created on the plant, the building of a healthy and *vibrant* community and stunning architectural designs of the network of webbing that supported the community. One week later, the plant was brown and dead. The only thing left were whispers of the web-like threads flapping in the breeze.

I was at once stunned, angry, and confused all at the same time. I asked myself:

—What happened to this healthy insect community, the architecture, the vibrant lifestyle they were living?
—How could these insects be oblivious to the fact that they are destroying the place that provides the support for their community to function? Didn't they know that a slower development rate would give the plant and the community opportunities to survive together?

I pondered this situation for a while and had a startling revelation that transcended the situation at hand. The behaviors demonstrated by the mite community are the same as those I see in human societies. I asked myself:

—Isn't this the same thing we are doing as a society in our understanding and connection to nature and our own community development—exploiting the resources that sustain us (water, land, etc.) beyond a point that can be recovered?
—Are our development activities leading us to a point of collapse from a lack of understanding of preserving the resources required to sustain our future?
—Is this hyperconsumption that takes the colony to near extinction a trait that all living species possess?
—How can we educate ourselves to avoid the path to our own self-destruction?

This is a lot to glean from a seemingly natural or mundane reaction between two living species, but it made me realize the path to wisdom on these matters requires a conscious thought and effort on our part to protect the sustainable path between us and nature. The default thinking that the overuse or exploitation of resources is contributing to community health may someday be the mechanism that destroys the foundation of civilization.

In his book, *The Hidden Connections*, Frijof Capra states:

> As this new century unfolds, there are two developments that will have major impacts on the well-being and ways of life of humanity. Both have to do with networks, and both involve radically new technologies. One is the rise or capitalism; the other is the creation of sustainable communities based on ecological literacy and the practice of ecodesign. Whereas global capitalism is concerned with the electronic networks of financial and informational flows, ecodesign is concerned with ecological networks of energy and material flows. The goal of ecodesign [is] to maximize the sustainability of life.
>
> These two scenarios—each involving complex network and special advanced technologies—are currently on a collision course. We have seen that the current form of global capitalism is ecologically and socially unsustainable. The so-called "global market" is really a network of machines programmed according to the fundamental principle that money-making should take precedence over human rights, democracy, environmental protection, or any other value...[h]owever, human values can change, they are not natural laws.[15]

This collision appears to already have occurred in the American Southwest. The economic downturn starting in 2008 hit the Southwest particularly hard, where the economic vitality of many cities depends on growth and development activities. The economic collapse in Phoenix first hit new developments along the fringes of the urbanized area, but eventually spread in waves to the more established areas, resulting in historic high home foreclosure and unemployment rates.[15] Las Vegas, once boasting a booming and vibrant economy, is ranked near the bottom of the 150 major world metropolitan areas, with a grim outlook for positive economic recovery based on the area's dependence on tourism and construction in its economic base.[15] Speculation and greed were driving investments in these markets that appeared to have no end. Development fueled development as long as the market would bear. This era has come to an end, and it appears to be several years in the future before a sustainable path for urban development is possible.

It seems somewhere in human nature is the desire and ability to follow paths of self-destruction when disconnected from the source of continuity and sustainability of natural systems. Unsustainable developments would include those that result in short-term gain for one party but contribute to a greater loss to the community as a whole by creating undue burdens on community members due to environmental degradation, generation of hazardous wastes, depression of home values, or decrease in the quality of life. Garrett Hardin wrote on this human dilemma in a classic article entitled "The Tragedy of the Commons" in 1968. Hardin observed that rational people, acting independently, will make conscious decisions to deplete a limited natural resource based on their perspective notion of personal gain for this action.[15] Unsettling as it is, this aspect of human behavior is exhibited in other parallel areas of land development, banking, marketing, and government. This type of decision-making lacks the contemplation of the individual's highest potential and core values in this process.

## Following the Wisdom of the Native People

Several Native American tribes have inhabited the Southwest since time immemorial. The ancient *Hohokum* tribes that settled along the Salt and Verde rivers in Arizona were noted as sophisticated engineers by diverting water through a network of canals that allowed them to flourish as a sedentary agrarian culture. The *Anasazi* tribes of New Mexico, Colorado, and beyond built dwelling in the cliffs along the rugged desert mesas and made baskets and pottery from natural materials. These ancient cultures had a deep connection to nature, the cycles of the animals, the Earth, the moon, and the sky. The ancients were followed by several other native tribes that now occupy reservations in the Southwest. Many indigenous cultures honor and celebrate the forces of nature through ceremony, tradition, or belief. They also believe personal and community decisions should take into consideration seven generations.[15] This ethical principle called for restoring the balance of the Earth elements in a fair and equitable manner, and the requirement to link the economic need with environmental protection and the well-being of the community as a whole. The native tradition of long-term vision on decision-making allows for continuity and sustained community development when consideration is placed beyond the option to expend resources for immediate use. In modern times where profit is driving the course of development, many people struggle with the ability to think beyond the immediate project, land sale, or development. There are many lessons Western civilization could learn from this native wisdom, which was formed from a deep connection to the land.

## Desert Visionaries

When we begin to contemplate conservation and development, there is need for a vision or foundation to apply these concepts in the world. For this purpose, we need to turn to people of vision—leaders that have risen above the normal intellectual or pedagogical processes to present a new and different way of interacting with the environment. For change to happen in the human condition, it takes one person with a complete understanding of a system of human interaction to influence the society as a whole by sharing an enlightened understanding of this system by theory or practice. There are a few people who stand out for having profoundly shaped our understanding of the relation between man and nature in the Southwestern United States. These visionaries include John Wesley Powell, Aldo Leopold, Frank Lloyd Wright, and Ian McHarg. Their lives and legacies live through their work and their apprentices. These people had a deep and far-reaching grasp of knowledge that will guide society to create or preserve the connection between man and nature. Each of these visionaries left an indelible mark in their profession in their own ways, but by combining the wisdom of these leaders, a deeper and more holistic understanding of sustaining the desert environment may emerge for the reader.

### John Wesley Powell

John Wesley Powell is described as a soldier, explorer, and scientist and was a self-made man of determination from humble upbringings. Powell was born in New York State and served in the military during the Civil War, where he reached the rank of major. He became famous for his expeditions of the Colorado River, where he documented the natural and evolutionary history of the river canyon starting in 1869, which was later published in his book *Report on Lands of the Arid Regions of the United States*.[15] His bold experiences made him a popular speaker across the country on the natural settings of the American West.

In 1881, Powell became director of the U.S. Geological Survey, where he was able to work at the national level to help establish rational policy for natural resource use based on sound science and exploration. Powell formed a viewpoint that development in the West should be limited and targeted based on the limitation of available water. He believed the arid West was unsustainable for agricultural development, except for certain areas along existing perennial rivers. He advocated for a survey of irrigable lands of the West; larger homestead requirements for arid lands; the use of watersheds as boundaries to promote wise resource use and avoid political controversies; and the slow, rational development of western lands based on equitable sharing of natural resources, particularly water. His viewpoint was in direct conflict with the railroads, which had been granted large tracts of land to dispose of in exchange for the construction of rail lines. The political and business interests of the time formulated policies on western lands that favored capitalism and profit that were not based on rational science or planning.

Reisner states in his book, *Cadillac Desert*, on Powell's observations on Western land policy:

> Speculation. Water Monopoly. Land Monopoly. Erosion. Corruption. Catastrophe. By 1876, after several trips across the plains and the Rocky Mountain States, John Wesley Powell was pretty well convinced that those would be the fruits of western land policy based on wishful thinking, willfulness, and lousy science. And everything he predicted was happening, especially land monopoly, water monopoly, graft, and fraud.[15]

Powell's vision for a slow approach to development of the West was seen as too slow, limited, and unnecessary by those with investments in the West. He lost out to those with power and influence in shaping of western land policy. The rush to settle was met with the realities forecasted by Powell in these barren lands of the arid West, most notably the huge losses of land, crops, and livelihood of the Dust Bowl in the 1930s. Again, Reisner sums up the lessons gained in hindsight from Powell:

> What is remarkable, a hundred years later, is how little has changed. The disaster that Powell predicted—a catastrophic return to a cycle of drought—did indeed occur, not once but twice: in the 1880s and again in the 1930s. When it happened, Powell's ideas—at least the insistence that the federal irrigation program was the only salvation of the arid West—were embraced, tentatively at first, then more passionately, then with a kind of desperate insistence.[15]

### Aldo Leopold

Born in 1887, Aldo Leopold is noted as one of the foremost leaders in ecology and conservation. He is often described as an ecologist, forester, and environmentalist. He had a keen interest in the natural world from an early age with avid interests in ornithology. He attended Yale School of Forestry and was later stationed in New Mexico with the U.S. Forest Service. Among other things, Leopold was noted for the development of the first comprehensive plan for the Grand Canyon and proposed the Gila Wilderness Area, the first national wilderness area in the country.

Leopold's 1949 book, *A Sand County Almanac*, is considered one of the most profound books dealing with the relation of man to the natural world. His writing promotes the understanding of the natural world and the preservation of wildlife through conscious human actions that protect their habitats. In this book, Leopold outlined his "land ethic"—a new vision that establishes a connection of man in relation to the land and life forms that inhabit it. His land ethic is based on moral actions that result in creating maximum benefit for all people and living things on land.

Leopold wrote on land ethic:

> The land ethic simply enlarges the boundaries of the community to include soils, waters, plants, and animals, or collectively: the land…A land ethic of course cannot prevent the alteration, management, and use of these "resources," but it does affirm their right to continued existence, and, at least in spots, their continued existence in a natural state.[15]

His vision and philosophy of conservation of the natural world survives today as inspiration for all those seeking to follow a path that promotes ethical actions of man in relation to any action or project that has an impact on nature. *A Sand County Almanac* is still widely read and considered one of the most influential books dealing with man's role with nature.

### Frank Lloyd Wright

Considered one of the most prominent architects and designers in America, Frank Lloyd Wright set new standards in building design as demonstrated in over 1000 projects spanning many decades. Wright was born in Wisconsin in 1867, where he grew up to learn the foundations of the profession of architecture and building design. Wright was noted for his mastery of "organic architecture," a philosophy that attempts to promote harmony and balance between human habitation and the natural world through design in a manner that integrates the building, interior elements and surrounding with the site. Wright was considered the leader of the Prairie School of Architecture, which sought to develop a North American style of architecture that focused on developing the unique qualities of a building and site through design.

Fallingwater in Bear Run, Pennsylvania, is considered one of the most notable and famous building designs in modern times. This private home was constructed on a waterfall in a manner that makes the home appear as to have always been part of the natural setting. Broadacre City was Wright's model for suburban development. This design was the antithesis of the ubiquitous small lot tract developments that were largely dependent on the automobile. Wright's vision for this suburban community was one connected with nature carved out of large lots using natural materials from the site.

Wright established Taliesin in Spring Greens, Wisconsin, as a school to train aspiring architects in his unique approach to buildings and community designs and Taliesin West in Scottsdale, Arizona, in 1937 as a winter retreat in the Southwest. These schools are still serving to educate architectural students on the principles of organic architecture and techniques that he developed. Wright died in 1959, but his spirit and legacy lives with us through his former students, projects, and designs that serve as examples of buildings that are connected to nature and its surroundings. He stated:

> Organic buildings are the strength and lightness of the spiders' spinning, buildings qualified by light, bred by native character to environment, married to the ground.

### Ian McHarg

A Scottish native, Ian McHarg is considered by many to be the father of ecological planning. McHarg, a graduate of Harvard, went on to become the founding chair of the Department of Landscape Architecture and Regional Planning at the University of Pennsylvania. His novel approach of combining landscape architecture and regional planning helped establish a framework for multidisciplinary teaching and project development. McHarg was one of the first to develop the map "manual overlay" method, which involved creating separate layers of variables on a site that could be manually manipulated to display relevant information on a map. This method is considered the basis now used in computer-based geographic information system analytical tools. In addition, McHarg developed what is called suitability analysis, which is used to assign ratings of the elements on a site as suitable or not suitable for the objectives of the project.

McHarg's method of ecological planning is now a fundamental part of the curriculum of advanced programs in landscape architecture and planning in most of our leading universities. He believed we all need to become educated in methods that allow humanity to create or restore the Earth in a greater capacity than it can be destroyed. The biosphere must be understood as something that sustains us and our role is to avoid creating adverse harm through human activities. A fallacy of human nature, McHarg believes, is to hold on to the notion that there is an architectural or engineered solution for all problems.

On the trends of modern urban settings McHarg states:

> Today, the modern metropolis covers thousands of square miles, much of the land is sterilized and waterproofed, the original animals have long gone, as have the primeval plants, rivers are foul, the atmosphere is polluted, climate and microclimate have retrogressed to increased violence, a million acres of land are transformed annually from farmland to suburban housing and shopping centers, asphalt and concrete, parking lots and car cemeteries, yet slums accrue faster than new buildings, which seek to replace them. The epidemiologist can speak of urban epidemics—heart and arterial disease, renal disease, cancer, and, not least, neuroses and psychoses.[15]

This frank assessment by McHarg on the plight of the urban center has a degree of truth in the reality of what we are creating here and across the globe—that is, anthropogenic activities over time can result in alterations in the landscape that can have significant impacts on the quality of life in the urban setting.

## Creating a Sustainable Future

Taken together, the wisdom of these visionaries—Powell, Leopold, Wright, and McHarg—forms an integral understanding of how the human species can interact with the natural environment without the need to destroy it or alter it beyond its ability to sustain the human population without adverse effect. With the rapid pace of development in the Southwest, a bold vision for conservation is needed in design and planning for the future. To be clear, development by itself is not inherently bad or undesirable. We need development to support the natural growth of our communities, but what we also need is to create or follow a vision for our cities that respects the natural processes, protects areas of ecological and cultural importance, preserves part of the natural setting in the built environment, and creates solutions to sustainable resource use without long-term harm to the environment.

To address these concerns, we put together this book that we feel will present a unique approach to this topic by providing a logical format for understanding the natural environment of the Southwestern deserts, then expand this knowledge into how the built environment can include qualities and attributes of the natural desert. There is an emphasis on understanding ecological systems of hydrology, climate, ecology, and energy flows to create communities and places designed and built with ecological literacy and consciousness. This book features a *transdisciplinary* approach to the topic of desert sustainability, which we believe will create a bridge to better understand how the built environment is inherently interdependent on the natural environment for its future. The application of the principles outlined in this book can help desert communities work toward a sustainable future—one that leads to greater health, happiness, and quality of life for all of its residents.

**Richard A. Malloy**

## References

1. Vitruvius, *On Architecture* (trans. F. Granger, Cambridge, MA: Harvard University Press, 1931), Book I, Chapter I, p. 10.
2. Casson, L., *Everyday Life in Ancient Rome* (Baltimore, MD: Johns Hopkins University Press, 1998), p. 134.
3. Vitruvius, *op. cit.* p. 1.
4. Boyer, E.L. and L.D. Mitgang, *Building Community: A New Future for Architectural Education and Practice* (Princeton, NJ: The Carnegie Foundation for the Advancement of Teaching, 1996).
5. Calthorpe, P. and W. Fulton, *The Regional City* (Washington, DC: Island Press, 2001); Simmonds, R. and G. Hack, eds., *Global City Regions: Their Emerging Forms* (London, U.K.: Spon Press, 2000).
6. Calthorpe, P. and W. Fulton, *op. cit.* p. 10.
7. Moore, S.A., Reproducing the local, *Platform* Spring (1999): 9.
8. Frampton, K. Towards a critical regionalism: Six points for an architecture of resistance, in H. Foster, ed., *The Anti-Aesthetic: Essays on Postmodern Culture* (Port Townsend, WA: Bay Press, 1983), p. 21.
9. Steiner, F., *The Living Landscape* (New York: McGraw-Hill, 2000; paperback edition Island Press, 2008); Burke, J. and J. Ewan, *Sonoran Preserve Master Plan* (Phoenix, AZ: Department of Parks, Recreation, and Library, 1998).
10. Barrett, G.W. and E. P. Odum, The twenty-first century: The world at carrying capacity, *BioScience* 50(4) (2000): 363–368.
11. Brand, S., *The Clock of the Long Now: Time and Responsibility* (London, U.K.: Weidenfeld & Nicolson, 1999).
12. United Nations Development Programme, United Nations Environment Programme, World Bank, and World Resources Institute, *World Resources 2000–2001, People and Ecosystems, the Fraying Web of Life* (Amsterdam, the Netherlands: Elsevier, 2000).
13. Hilderbrand, G., *Origins of Architectural Pleasure* (Berkeley, CA: University of California Press, 1999), p. xvii.
14. Hersey, G., *The Monumental Impulse: Architecture's Biological Roots* (Cambridge, MA: MIT Press, 1999), p. 183.
15. Kellert, S.R., J. H. Heerwagen, and M. L. Mador, eds., *Biophilic Design* (Hoboken, NJ: John Wiley & Sons, 2008), p. vii.

# Editors

**Richard A. Malloy** is a manager for environmental projects at the School of Applied Sciences and Mathematics and the Biodesign Institute at Arizona State University. He has an AS in landscape operations from the University of Massachusetts at Amherst, a BS in landscape horticulture from Texas A&M University, and an MEP in environmental planning and an MS in applied biological sciences from Arizona State University. He currently oversees projects and facilities dedicated to advancing biotechnology and sustainable solutions to environmental challenges for Arizona State University. He has past experience working in biodiversity planning and land management on an international scale while serving as a volunteer for the U.S. Peace Corps in Ecuador, South America, and as an invited scholar to the Chinese Academy of Sciences in Beijing and the Guizhou Academy of Sciences in Guiyang, China. Malloy has worked professionally in many capacities to bring people of diverse backgrounds together to explore issues of pressing ecological and human concerns in workshops, forums, and educational events. His personal background and professional skills are uniquely focused toward helping people develop a better relationship with nature and their communities.

**John H. Brock** is professor emeritus at the School of Applied Sciences and the Mathematics Department at Arizona State University (ASU) and founder of the firm Brock Habitat Restoration & Invasive Plant Management LLC. He was awarded a BS in agriculture and an MS in botany from Fort Hays State University in 1966 and 1968, respectively, followed by a PhD in range science from Texas A&M University in 1978. Since 1977, he has conducted research and taught natural resources management courses at ASU. In 2007–2008, he served as the president of the ASU Polytechnic Academic Senate and as the chair of ASU's Academic Faculty Council. In July 2008, he retired from ASU and the department formally named Applied Biological Sciences Department after 31 years of service. Dr. Brock is an active member of the Governor's Invasive Species Advisory Council (ISAC) and is the co-lead for its control and management work group. His professional interests are habitat restoration and invasive plant management. Dr. Brock now operates a consulting firm in the area of invasive plant management and landscape restoration.

**Anthony C. Floyd**, American Institute of Architects, is a registered architect and green building program manager for the City of Scottsdale. He has a civil engineering and architecture degree from Pennsylvania State University and an MPA in public administration from Arizona State University. Floyd served as a building official for the City of Scottsdale from 1988 to 1995. In 1998, he acted as city liaison to a local citizen group, which was responsible for establishing Arizona's first Green Building Program. Floyd later served as Scottsdale's manager for this program. He is involved in the administration of green building program guidelines, which include project qualification, education, energy code compliance, and evaluation of alternative materials. He maintains the city's green building rating standards for residential and commercial development and helps to facilitate Scottsdale's LEED gold mandate for newly constructed and renovated public facilities. He is cochair of the U.S. Green Building Council (USGBC) Codes Committee and serves on the project committee for the new ASHRAE 189 commercial green building standard and the ICC Sustainable Building Technology Committee in the development of the first International Green Construction Code. Floyd also served on the National Green Standard Committee in the development of the ICC 700 Residential Green Building Standard.

**Margaret Livingston** is a professor in the School of Landscape Architecture and Planning at the University of Arizona, Tucson. She teaches a range of courses related to ecological and environmental issues in arid environments and has locally and internationally conducted lectures and workshops that focus on water conservation, wildlife habitat, and use of native plants in urban areas. As an urban ecologist, Livingston's work emphasizes the importance of evaluating and maintaining natural and seminatural ecosystems within and surrounding urban areas. In her role as a designer, she focuses on the use of native plants and design of urban wildlife spaces.

**Robert H. Webb** has worked on long-term changes in natural ecosystems of the southwestern United States since 1976. He has degrees in engineering (BS, University of Redlands, 1978), environmental earth sciences (MS, Stanford University, 1980), and geosciences (PhD, University of Arizona, 1985). Since 1985, Dr. Webb has been a research hydrologist with the U.S. Geological Survey in Tucson and an adjunct faculty member of the Departments of Geosciences and Hydrology and Water Resources at the University of Arizona. He has authored, coauthored, or edited 12 books, including *Environmental Effects of Off-Road Vehicles* (with Howard Wilshire); *Grand Canyon, A Century of Change; Floods, Droughts, and Changing Climates* (with Michael Collier); *The Changing Mile Revisited* (with Raymond Turner and others); *Cataract Canyon: A Human and Environmental History of the Rivers in Canyonlands* (with Jayne Belnap and John Weisheit); *The Ribbon of Green: Long-Term Change in Woody Riparian Vegetation in the Southwestern United States* (with Stanley Leake and Turner); *Damming the Colorado, The U.S. Geological Survey in Grand Canyon in 1923* (with Diane Boyer); *The Mojave Desert: Ecosystem Processes and Sustainability* (with five other editors); and *Repeat Photography: Methods and Applications in the Natural Sciences* (with Boyer and Turner).

# *Contributors*

**Sandy Bahr**
Sierra Club
Phoenix, Arizona

**David Berry**
Western Resource Advocates
Phoenix, Arizona

**Bob Bolin**
Department of Sociology
Arizona State University
Tempe, Arizona

**Ward W. Brady**
Department of Applied Sciences and Mathematics
Arizona State University
Tempe, Arizona

**Anthony J. Brazel**
School of Geographic Sciences and Urban Planning
Arizona State University
Tempe, Arizona

**Chad Campbell**
Arizona State House of Representatives
Phoenix, Arizona

**Carol Chambers**
School of Forestry
Northern Arizona State University
Flagstaff, Arizona

**Ron Cooke**
University College London
London, United Kingdom

**William Wallace Covington**
Ecological Restoration Institute
Northern Arizona University
Flagstaff, Arizona

**Mark A. Dimmitt**
Arizona-Sonora Desert Museum
Tucson, Arizona

**Barbara Dugelby**
Latin America Programs
Round River Conservation Studies
Blanco, Texas

**Allan Dunstan**
Desierto Verde
Tempe, Arizona

**Mark Edwards**
Morrison School of Management and Agribusiness
Arizona State University, Polytechnic Campus
Mesa, Arizona

**Nan Ellin**
City and Metropolitan Planning
University of Utah
Salt Lake City, Utah

**Pliny Fisk III**
Center for Maximum Potential Building Systems
Austin, Texas

**Anthony C. Floyd**
Office of Environmental Initiatives
Greenbuilding Program City of Scottsdale
Scottsdale, Arizona

**Dave Foreman**
The Rewilding Institute
Albuquerque, New Mexico

**Geoffrey Frasz**
Department of Social Studies
College of Southern Nevada
Las Vegas, Nevada

**Andrew Goudie**
School of Geography and the Environment
University of Oxford
Oxford, United Kingdom

**Douglas Green**
Department of Applied Sciences and Mathematics
Arizona State University, Polytechnic Campus
Mesa, Arizona

**Sara Grineski**
Sociology and Anthropology
The University of Texas at El Paso
El Paso, Texas

**Renée Guillory**
Tempe, Arizona

**Edward J. Hackett**
Department of Sociology
Arizona State University
Tempe, Arizona

**Andy Holdsworth**
Minnesota Department of Natural Resources
St. Paul, Minnesota

**Robert Howard**
New Mexico Wilderness Alliance Council
Arroyo Hondo, New Mexico

**Jack Humphrey**
The Rewilding Institute
Albuquerque, New Mexico

**Heather Kinkade**
Forgotten Rain
Phoenix, Arizona

**Stanley A. Leake**
United States Geological Survey
Tucson, Arizona

**Rurik List**
Institute de Ecologia
Universidad Nacional Autónoma de México
Coyocan, Mexico

**Margaret Livingston**
Department of Landscape Architecture
University of Arizona
Tucson, Arizona

**Richard A. Malloy**
Department of Applied Sciences and Mathematics

and

Biodesign Institute
Arizona State University
Tempe, Arizona

**Sharon B. Megdal**
Water Resources Research Center
University of Arizona
Tucson, Arizona

**Joanna B. Nadeau**
Water Resources Research Center
University of Arizona
Tucson, Arizona

**David Orr**
Environmental Studies and Politics
Oberlin College
Oberlin, Ohio

**Stephen J. Pyne**
School of Life Sciences
Arizona State University
Tempe, Arizona

**Paolo Soleri**
Arcosanti/Consanti Foundations
Scottsdale, Arizona

**Kim Sorvig**
Department of Landscape Architecture
University of New Mexico
Albuquerque, New Mexico

**William L. Stefanov**
Department of Science Applications Research and Development
Johnson Space Center
National Aeronautics and Space Administration
Houston, Texas

**Frederick R. Steiner**
College of Architecture
The University of Texas at Austin
Austin, Texas

**Brian K. Sullivan**
Division of Math and Natural Sciences
Arizona State University
Glendale, Arizona

**Vernon D. Swaback**
Swaback Partners
Scottsdale, Arizona

**Thomas R. Van Devender**
Arizona-Sonora Desert Museum
Tucson, Arizona

**David R. Van Haverbeke**
Arizona Fisheries Resources Office
United States Fish and Wildlife Service
Flagstaff, Arizona

**Andrew Warren**
Department of Geography
University College London
London, United Kingdom

**Robert H. Webb**
United States Geological Survey
Tucson, Arizona

# Part I

# Physical Aspects of the Desert Environment

**Robert H. Webb**

Some of the most rapidly growing urban areas on Earth are in desert regions. Whether on the Indian subcontinent, North Africa, western China, central Australia, or the southwestern United States, growth has steadily transformed deserts from what is perceived as a hostile, dangerous environment to a network of infrastructure and houses. Part and parcel of this growth is the joint assumption that the desert environment is unchanging, offering stable building sites; that water supplies can be found, either on the surface, in the subsurface, or through transbasin transfers from more mesic regions; and that hazards are reduced in a region of low rainfall and seasonally high temperatures. While these assumptions may hold true for short or even long time periods, eventually the reality of the desert environment sets in, creating constraints on urban design and development.

This part provides a general introduction to the physical environments of deserts worldwide and particularly in the southwestern United States, where considerable research on the characters and processes affecting the desert environment provides a wealth of information useful for urban design. We begin with a modified version of a chapter describing deserts of the world from the classic book *Desert Geomorphology* by Ronald Cooke, Andrew Warren, and Andrew Goudie. Without changing its substance, we were compelled to convert the units from metric to English and add explanatory footnotes to define technical terms that may not be well known within the community concerned with design in deserts. The scope of this chapter, as well as its discussion of evolution of deserts on six continents, makes it an invaluable contribution to this book for those who want a deeper understanding of how deserts originated and their common characteristics worldwide.

We follow the opening chapter with contributions on desert soils and climate, two closely related subjects. These chapters focus on the southwestern United States and particularly the desert regions of Arizona, which sustains several of the largest cities in the arid regions of North America. The chapter on desert geology and soils, by William L. Stefanov and Douglas Green, expands on general concepts presented in the opening chapter to give a more detailed perspective in a smaller region of the Sonoran Desert. Likewise, Anthony J. Brazel's discussion of climate uses a focus on desert cities in the Southwest and that region to provide a more expansive view of short- and long-term desert climate, particularly in light of predicted future changes expected to be caused by greenhouse-gas emissions.

These chapters are followed by two chapters that discuss the related topics of desert hydrology, water supplies, and hazards in the southwestern United States, particularly in Arizona. This region provides a microcosm of hydrology and hazards in deserts worldwide, where the details change but the processes remain the same. The water resources of surface water and groundwater are intertwined in process and supply, and other characteristics important to design where environmental concerns are paramount, especially riparian vegetation, are discussed. In particular, an understanding of the hydrologic cycle in deserts, and how that influences the supply and movement of groundwater and the availability of surface water, is presented to allow some basic understanding of the potential for water supplies in deserts and their limitations.

Hazards in the desert environment, and how they relate to urban design, are discussed in detail from the perspective of the physical environment instead of design. These hazards range from natural, such as flooding and earthquakes, to human caused, such as toxic waste dumps and land subsidence.

In Part I, we strive to provide overviews of characters and processes, providing the reader with additional resources for deeper understanding of this unique physical environment. In particular, we point to various tools that enable those interested in urban design to determine climatic characteristics, examine the spatial distribution of soil types and properties in an area of interest, estimate water resources, and determine what hazards might be of concern and what their magnitude might be. Needless to say, the physical environment of deserts is both complex and nuanced, and our introduction in this part hopefully will provide sufficient information for the reader to investigate further on issues of specific concern to urban design.

# 1

# *Deserts of the World**

**Ron Cooke, Andrew Warren, and Andrew Goudie**

**CONTENTS**



## 1.1 Introduction

To some, deserts are simply barren areas, barely capable of supporting life forms. Many places meet this criterion: Mangin's *The Desert World*, published in 1869,[1] embraced environments as diverse as the waste heaps of the china clay quarries in Cornwall, the steppes of Tartary, the Dead Sea, and the Arctic wildernesses. But most deserts are areas of aridity and they are usually defined scientifically in terms of some measure of water shortage. Such measures, indices of aridity, are commonly based on the relationships between water gained from precipitation and water lost by evaporation or transportation. There are plenty of indices to choose from, the differences between them reflecting different objectives of classification.[2]

The areas shown in Figure 1.1 constitute the warm deserts realm. Within it, there are five major regions of aridity: the deserts of North and South America, North Africa, Eurasia, southern Africa, and Australia. They cover a third of the Earth's land surface and are the context for this study of desert geomorphology.[†] There are also arid areas in the polar

* Adapted from R. Cooke, A. Warren, and A. Goudie, Deserts of the world, in *Desert Geomorphology* (London, U.K.: University of London Press, 1993), pp. 423–447.

† Definitions used in footnotes in this chapter are derived from the combination of W. R. Osterkamp and A. Allaby and M. Allaby.[3]

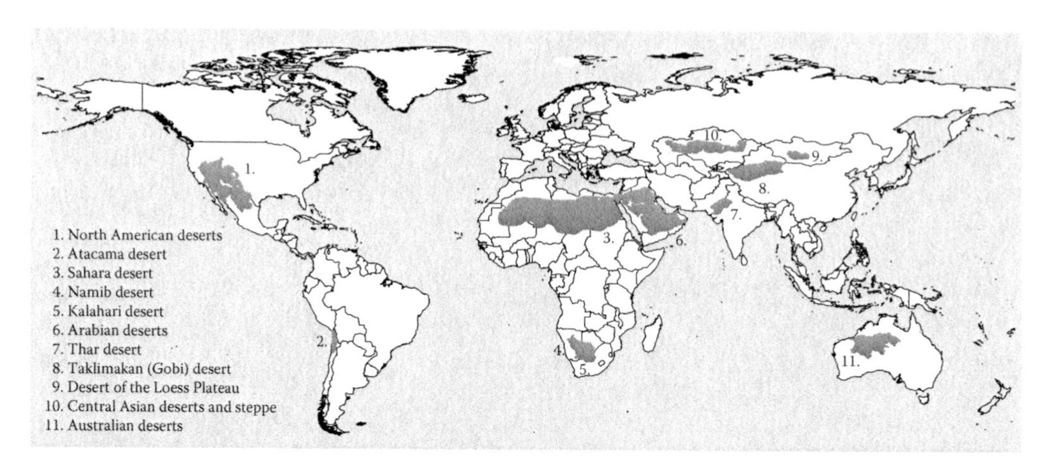


**FIGURE 1.1**
Map showing world-wide distribution of the warm deserts.

latitudes, but, geomorphologically, they are very different from the subtropical deserts and are excluded from this review.

## 1.2 The Sahara and Its Margins

The Sahara is the world's largest desert (covering c. 2.7 million miles$^2$), and the region comprising the Sahara and the Nile occupies about half of the entire African continent.[4] The greater part of the region is free of surface water and is sparsely vegetated, and, being exposed to dry, descending, northeasterly airstreams, its mean annual rainfall is less than 16 in. and over vast areas less than 4 in.[5] Temperatures are also high, and evaporation losses from free water surfaces and transpiration losses from vegetated areas are greater than anywhere else on the globe.*

The general morphology of the Sahara has been discussed by Mainguet,[7] who suggests that its most distinctive characteristic, save only the relief provided by the Hoggar and Tibesti massifs, is its flatness. This flatness is associated with great sandstone plateaus, a series of broad, closed basins (of which Chad is the most notable), and a series of wind-molded landscapes, which include deflational *regs*, *corrasional* fields of *yardangs*, and areas of sand deposition (*ergs*).† Some of the major geological, geomorphological, and climate features of the area are shown in Figure 1.2.

Like most other deserts, the Sahara shows the imprint of a long evolution. For example, the Sahara was glaciated in Ordovician times, when from palaeomagnetic evidence, it is apparent that the South Pole was located about the center of this region. Well-preserved

* General discussions of the Saharan environments are provided by E. F. Gautier, J. L. Cloudsley-Thompson, and R. Capot-Rey.[6]

† Regs are gravelly plains, typically formed by wind deflation. Yardangs are wind-sculpted landforms, and corrasion is a process akin to sand blasting. Ergs are sand seas composed of a large area of dunes. Duricrusts are exposed soil horizons cemented by various minerals and compounds, typically calcium carbonate (calcretes), silica (silcretes), or iron compounds (fericretes).

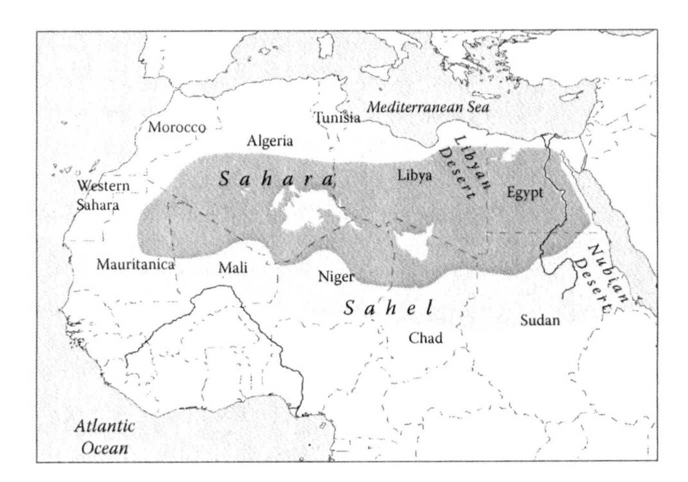


**FIGURE 1.2**
Map of North Africa showing the general distribution of the Sahara Desert and the Sahel.

striations, crescentic gouges, erratic boulders, and glacial lineations are still evident in the present landscape.[8] In the early Tertiary, there was a long interval of intense weathering, producing *duricrusts* on the southern side of the Sahara in tectonically stable lowlands,[9] while Tertiary and Quaternary uplift, with associated volcanism, produced major Saharan massifs such as the Hoggar and Tibesti. In the late Tertiary, climatic deterioration and tectonic movements between Africa and Europe led to the gradual diminution of the Tethys Sea and the formation in the so-called Messinian salinity crisis (c. 6 million year BP) of a large-closed depression in the vicinity of the present Mediterranean basin. The northern Sahara must have been arid at that time, and large spreads of evaporites formed before a marine transgression through the Straits of Gibraltar caused a reestablishment of marine conditions.[10] The Nile cut down deeply to the low base level, forming a canyon 8200 ft deep, 813 miles long, and 6–12.5 miles wide, dimensions which comfortably exceed those of the present-day Grand Canyon in Colorado (Figure 1.3).[11]

**FIGURE 1.3**
Photograph of the eastern Sahara Desert in the Sudan. (Courtesy of J. Woodward.)

In late Cenozoic times, aridity became a prominent feature of the Saharan environment,[12] probably because of the occurrence of several independent but roughly synchronous geological events[13]:

- As the African plate moved northward, there was a migration of northern Africa from wet equatorial latitudes (where the Sahara had been at the end of the Jurassic) into drier subtropical latitudes.
- During the late Tertiary and Quaternary, uplift of the Tibetan plateau had a dramatic effect on world climates, helping to create the easterly jet stream, which now brings dry subsiding air to the Ethiopian and Somali deserts.
- The progressive build-up of polar ice caps during the Cenozoic climatic decline created a steeper temperature gradient between the Equator and the Poles, and this in turn led to an increase in Trade Wind velocities and their ability to mobilize sand into dunes.
- Cooling of the ocean surface may have reduced the amount of evaporation and convection in low latitudes, thus reducing the amount of tropical and subtropical precipitation.

Thus, although the analysis of deep-sea cores in the Atlantic offshore from the Sahara indicates that some aeolian activity dates back to the early Cretaceous,[14] it was probably around 2–3 million year BP that a high level of aridity became established. From about 2.5 million year BP, the great tropical inland lakes of the Sahara began to dry out, and this is more or less contemporaneous with the onset of mid-latitude glaciation. Aeolian sands become evident in the Chad basin at this time, and such palynological work indicates substantial changes in vegetation characteristics.[15]

In the Pleistocene, a clearer pattern of climatic oscillations became apparent, with alternations of aridity and greater humidity, although dates are a matter of controversy, especially in North Africa.[16] Each cycle may have been of the order of 100,000 years in duration, with nine-tenth of the cycle involving a gradual buildup toward peak aridity, coldness, and aeolian activity, followed by a rapid but short-lived return to milder and wetter conditions.[13] During dry phases, ocean cores demonstrate that large quantities of dust were generated by the Sahara,[17] and there was an equator-ward spread of dune fields in the Sahelian-Soudano zone.[18] The situation becomes especially clear at the time of the maximum of the last glaciation (the "Ogolian" of French workers), when *aeolian turbidites** were deposited on the Atlantic continental shelf. There was a substantial increase in dust output, and fluvial inputs from rivers such as the Senegal were minimal.[19] Desiccation of lakes took place on the southern side of the Sahara between 23,000 and 16,000 year BP, and active dunes extended up to 280 miles southward into the Sahel, blocking the courses of the Senegal and Niger rivers and probably also crossing the Nile.[20]

Toward the end of the last glacial, at around 12,500 year BP, there was a major change in environmental conditions, characterized by a redevelopment of lakes in the Chad basin and eastern Africa. Extensive lakes and swamps also formed along the Blue and White Niles in the Gezira area. A peak of humidity may have occurred around 9000–8000 year BP,[21] and, if one assumes that temperatures were broadly the same as today, the rainfall increase may have been as much as 65%. There were other lake oscillations during the course of the Holocene, but from about 5000 year BP conditions began to deteriorate irregularly toward the present situation. The general trend toward aridity is reflected in

* Aeolian turbidites result from undersea mass movement of sediments deposited by wind.

a sharp drop in the influx of montane and Sudano-Guinean pollen into the Chad area between 5000 and 4000 year BP.[22]

Saharan climates have also undergone significant changes in recent centuries, based on an analysis of historical sources, meteorological records, and studies of lake and river fluctuations.[23] For example, from the early sixteenth century until the eighteenth century, the Sahel experienced increased rainfall in comparison with the present century; in the eighteenth century, there were frequent famines and droughts; there was increased precipitation in the late nineteenth century; and in the early twentieth century, there was marked aridity along the Saharan margins. The rivers and lakes showed low positions around 1912, again in the early 1940s, and in the period since the late 1960s. This last fluctuation caused a marked diminution in the area and volume of Lake Chad, and it led to a marked increase in the level of dust-storm activity in a great belt from Mauritania to Sudan.[24]

## 1.3 Southern Africa

There are three main desert areas in southern Africa. In the west, there is the coastal desert of the Namib[25] that extends some 1750 miles from south of Luanda in Angola to St. Helena bay in the Republic of South Africa (Figure 1.4). It is bounded on the east by the Great Escarpment and forms a narrow strip, generally less than about 94 miles wide. In the central and coastal parts of the Namib, hyperaridity prevails (the mean annual rainfall at Walvis Bay is around 0.91 in.), although fog and dew are of frequent occurrence. The desert

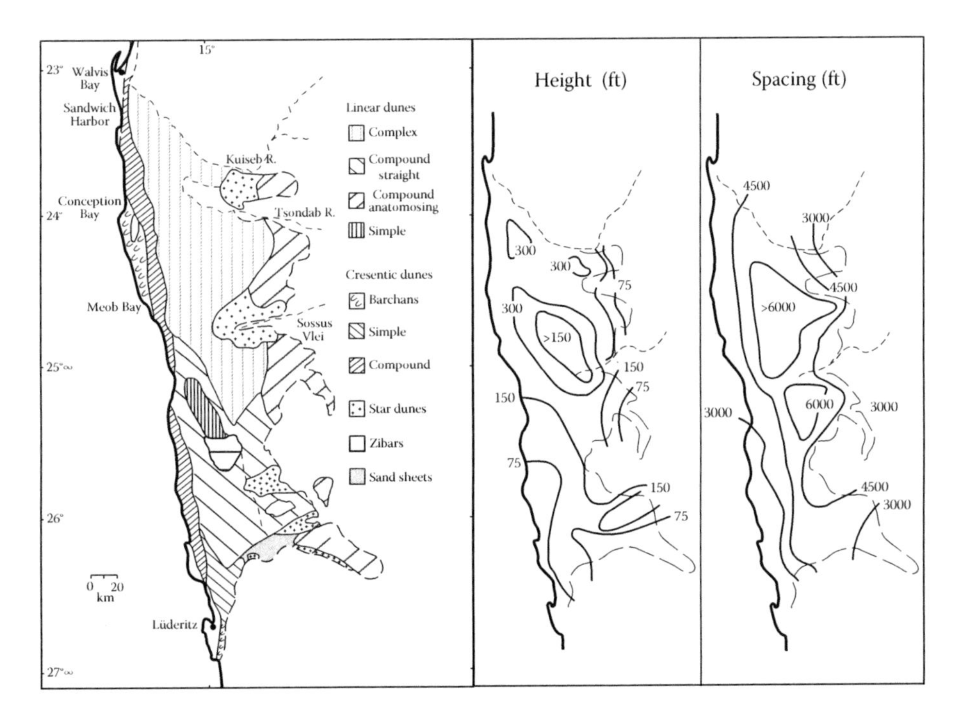


**FIGURE 1.4**
Map of the Central Namib Desert. (From previously unpublished data. With permission.)

**FIGURE 1.5**
Photograph of the Namib Desert (Soussusvlei). (Courtesy of N. Lancaster.)

contains two major dune fields separated by inselberg-studded gravel plains: the Skeleton Coast erg[26] and the great Namib erg to the south of the Kuiseb Valley.[27]

The most striking feature of the Namib is the 13,400 miles$^2$ sand sea that stretches for over 188 miles between Luderitz and the Kuiseb River (Figure 1.5). Three major dune types occur: *transverse* and *barchanoid dunes** that occur aligned normal to SSW to SW winds in a 12.5 miles wide strip along the coast; *linear dunes* on N–S to NW–SE alignments, reaching heights of 490–568 ft in the center of the erg; and *dunes of star form* that occur along the eastern margins of the erg (Figure 1.5).†

The Namib seems to be a desert of considerable antiquity, for the character of most Tertiary sediments in the Namib is suggestive of arid or semiarid conditions. The cross-bedded Tsondab Sandstone Formation represents the accumulation of a major sand sea in the central and southern Namib over a period of 20–30 million years prior to the mid to late Miocene.[29] In addition, extensive *calcrete* formation seems to have occurred at the end of the Miocene, while in the Pliocene, a climate of modern affinities was developing in the region.

Seisser's[30] investigation of offshore sediments has indicated that upwelling of cold waters intensified significantly from the late Miocene (7–10 million year BP) and that the Benguela Current developed progressively thereafter. Pollen analysis of such sediments indicates that hyperaridity occurred throughout the Pliocene and that the accumulation of the main Namib erg started at that time. For much of the Pleistocene, aridity was also the norm, and although there have been periods of increased fluvial and lacustrine activity, most of the sedimentological and faunal record suggests that moist phases were relatively short-lived and of limited intensity. However, fossil silts caused by ponding of river waters in the dune field occur in the Kuiseb Gorge, the Sossus Vlei, and the Tsondab Valley,[31] but no coherent picture emerges as yet from the few dates that are available. There are also *speleothems*‡ in the Rössing cave of the central Namib that indicate phases of greater hydrological activity in the late Pleistocene before 25,000 year BP.[32] The last glacial maximum (c. 18,000 year BP) may have been dry.[33] To the east of the Great Western Escarpment is a second major desert: the interior desert of the Kalahari,[34] a word derived from the Setswana word

* Names of sand dune types are extensive and descriptive. For example, barchan dunes tend to have a parabolic form.

† Detailed morphometric and sedimentological data on the Namib Desert are provided by N. Lancaster.[28]

‡ Speleothems (aka dripstone) are calcium carbonate accumulations in caves (e.g., stalactites, stalagmites).

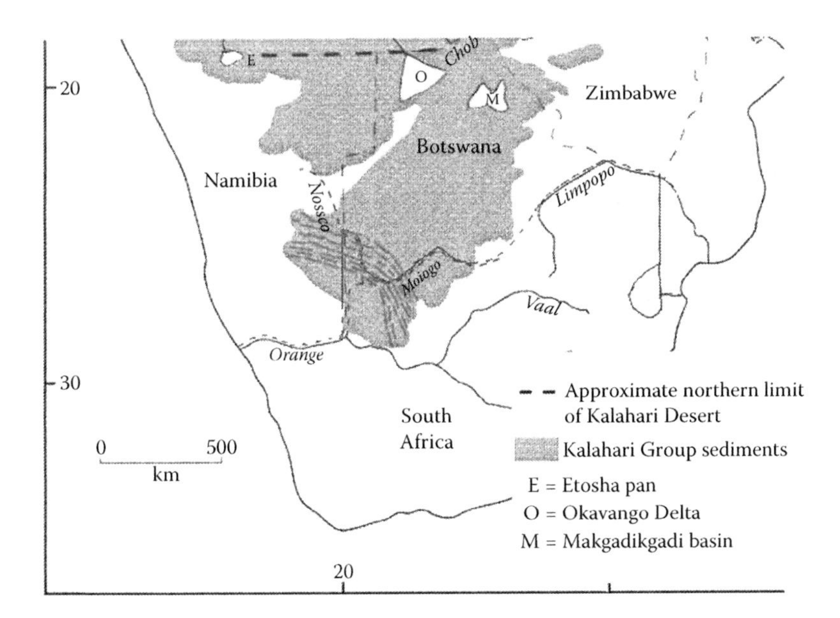


**FIGURE 1.6**
Map of the Kalahari Desert in southern Africa. (From previously unpublished data. With permission.)

"Kgalagale," which means "always dry." However, the area covered by the term is far from clear. Three main regions occur (Figure 1.6).[35]

1. The Kalahari dune desert in the arid southwest interior of Botswana and adjoining parts of Namibia and South Africa. The primarily summer rainfall is less than 8 in. per annum and is just sufficient to stabilize the plinths of a major field of dominantly linear dunes.
2. The Kalahari region (or thirstland) approximately delineated in the north by the Okavango Swamps and in the south by the Orange and Limpopo Rivers. This is an area of little or no surface drainage despite a relatively higher rainfall (c. 23.6 in. per annum). It is almost entirely covered with grass and woodland and has extraordinarily low relief.
3. The Mega-Kalahari is an extensive area consisting of a basin infilled by continental sediments of the Kalahari Beds. This extends from beyond the Congo River in Zaire to the Orange River in South Africa. Precipitation may be as high as 59 in., and vegetation may range from savanna to tropical moist forest. Nonetheless, it displays evidence for formerly more extensive aridity, in terms of both the development of ancient dune systems and of the widespread distribution of closed depressions, called pans (Figure 1.7).[36]

The Kalahari is another long-continued area of terrestrial sedimentation, with sequences of *marls*,* sands, lake deposits, calcretes, silcretes, etc., dating back to the Cretaceous. These beds, which are called the Kalahari Group, are ill exposed, and much of the information comes from borehole records. Dating evidence is still slender, but there is general agreement that the sediments accumulated under arid to semiarid environments. The Kalahari Group

* Marls are clay deposits containing large amounts of calcium carbonate that are deposited in lakes.

**FIGURE 1.7**
Photograph of Kalahari Desert dunes. (Courtesy of Robert H. Webb.)

is especially noted for the extensive development of calcretes[37] and of silcretes,[38] while the Kalahari Sands (which are either in situ or reworked aeolian materials for the most part) stretch over large tracts of country (c. 1 million miles$^2$) between the Orange River in the south and the Congo River in the north. The river systems of parts of Zambia and Zimbabwe inherit their alignments from a previously more extensive aeolian cover,[39] and throughout the Kalahani, there are extensive fossil drainage networks that are now either ephemeral or dry.[40]

The internal basin of the Kalahari, in which these sediments accumulated, was created by tectonic processes in mid-Jurassic times at the final division of Gondwanaland. There are various subbasins and graben structures within the area, and the greatest depths of Kalahari sediments occur in the Etosha Pan area of northern Namibia, locally exceeding 1200 ft in thickness).

There is abundant evidence for environmental changes in the Pleistocene within the Kalahari, with ancient dunes and palaeolakes being the most striking manifestations.[41] The dune systems of the Mega-Kalahari occur in areas where mean annual rainfall currently exceeds 31 in., and they have been mapped in the Hwange and Victoria Falls areas of Zimbabwe, Zambia, and Angola[42] and in Botswana.[43] They indicate formerly very extensive late Pleistocene aridity (with rainfalls less than 5 in. per annum), and there is also some suggestion that wind directions may have been different from those pertaining in the region today.[44] The presence of sandy and relatively clayey lunettes in association with small-closed depressions (pans) has also been used to indicate the hydrological fluctuations of the late Pleistocene (Figure 1.8).[45]

The greatest palaeolake in the area, however, is that of the Makgadikgadi Depression in northern Botswana (Figure 1.6). Strandlines extend to an altitude of 3080–3110 ft (c. 160 ft above current pan floor level), and the maximum extent of the lake was probably around 23,600 miles$^2$, about the size of today's Lake Victoria and larger than palaeolake Bonneville in the United States.[46] The palaeoclimatic significance of the Makgadikgadi lakes is, however, uncertain given the increasing body of evidence for tectonic instability in the Okavango delta region.

The third desert of southern Africa is sometimes called the Great Karoo semidesert. It occurs as a plateau at an altitude of 1970–3280 ft, tends to be underlain by horizontally bedded Palaeozoic sediments of the Beaufort Series, is bounded on the north and south

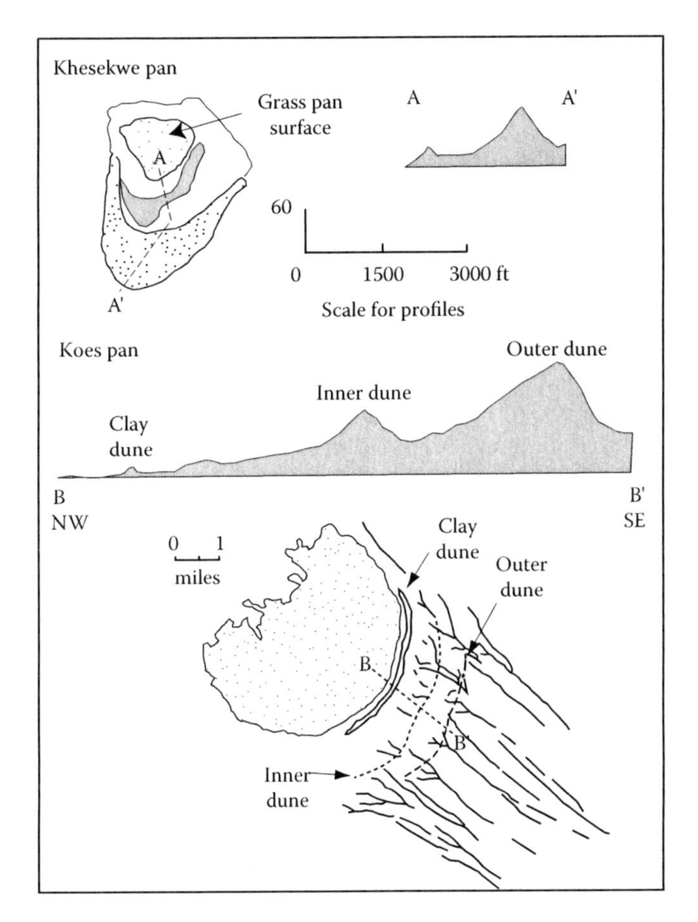


**FIGURE 1.8**
Diagrams of dunes associated with pans in the Kalahari Desert. (Modified from material supplied by N. Lancaster.)

by mountain ranges, and has a primarily winter rainfall regime that produces 5–16 in. per annum. Information on the evolution of this arid region is sparse.

The dating of the various fluctuations of climate in the southern African region is still highly uncertain. Lancaster[30] remarks that "there is no reliable chronology of events, nor any agreement upon the nature of the changes in regional climatic patterns," while Butzer[47] finds that "the Kalahari–Namib evidence is both patterned and ichoate. No distinctive interregional contrasts emerge, and different categories of data are often difficult to reconcile within one area."

## 1.4 The Great Indian Desert or Thar

The arid zone of the northwestern part of the Indian subcontinent extends from the Aravalli Range in the east to the Indus Plain and the mountains of Baluchistan in the west (Figure 1.9). In Rajasthan, it is traditionally called Marwar or "place of death," but

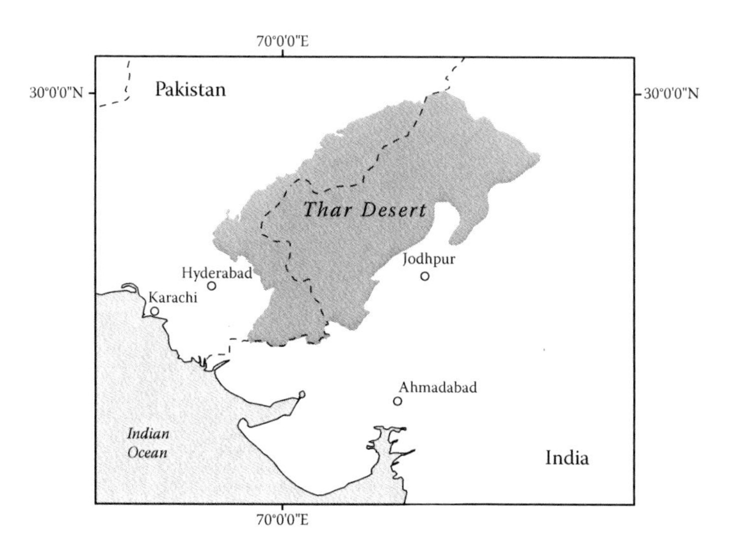


**FIGURE 1.9**
Map of the Thar Desert in India.

desert conditions are not especially extreme, for only locally does mean annual precipitation fall below 4 in. The southwest monsoon manages to produce summer rainfall in the area, and rural population densities are quite high. In the driest parts of the area, in the west, the Indus and its tributaries are through-flowing rivers that are now much used for irrigation.*

An important feature of the Great Indian Desert is the presence of the snow-fed Indus and its tributaries, both past and present. The discharge pattern is highly seasonal, and floods can be very destructive. Its average annual flow is about twice that of the Nile, and during floods, the river in the plains of Sind can be over 10 miles wide. The river carries a huge sediment load, so that the Indus Delta is thought to have extended 50 miles in the past 2000 years, and some 32 ft of aggradation has taken place in the past 5000 years.[49] However, perhaps, the most interesting hydrological features of the Thar rivers are their propensity to change course or to disappear and the history of competition between the Ganga and Indus systems in the northern Punjab. Ancient river courses of different ages have been located,[50] and the geomorphology of ancient Indus courses in the Thal *doab* (between Chenab and Indus) has been described elsewhere.[51]

Another distinctive feature of the Thar is the nature of its dunes. The coastline of the Arabian Sea, the alluvial plain of the Indus, and the weathering of widespread outcrops of sandstones and granites provide plentiful sand for aeolian reworking. The dominant sand-moving winds come from the southwest, and this accounts for both the long-distance transport of *foraminiferal tests*† from the coast and for the overall alignments of the dunes.[52] In the coastal regions of Saurashtra and Kutch, the dunes are composed of *calcareous aeolianites*, which are locally called miliolites,[53] but as one moves inland they become generally *quartzose*. The general pattern of the dunes has been mapped, and the particular importance of clustered parabolics of rake-like form in much of the desert was emphasized.[54] This may reflect the relatively high rainfall levels with the concomitantly relatively dense vegetation cover or it may be one response to the fact that, in terms of

* General background information on the Indian Deserts is provided by B. Allchin et al.[48]

† Foraminiferal tests are shells of foraminifera, a group of marine organisms. Calcareous aeolianites are wind-blown sediments cemented with calcium carbonate. Quartzose deposits mostly contain particles of quartz. Sayf dunes are longitudinal dunes that are parallel to wind direction.

**FIGURE 1.10**
Lag gravels in the Thar Desert, India. Mohangarh (Jaisalmer) playa (exposed at center) is located in Jaisalmer district of Rajasthan and is the largest (about 501 ha) gypsum quarry in India. (Courtesy of Navin Juyal; from previously unpublished data. With permission.)

wind-energy levels, the Thar is one of the least energetic of the world's deserts. Large *sayf dunes* that occur in the west of the desert, especially near Umarkot in Pakistan, have some seasonally inundated lakes in their swales (*dhands*) and may be derived as blown-out parabolics.[55]

Locally, within the Thar, there are closed basins and salt deposits (Figure 1.10). Some may be the result of the blocking of drainage systems by dunes as at Sambhar or on some of the tributaries of the Luni River,[56] while others, such as the Jaisalmer and Pokaran Ranns, may be related to faulting.[57] Coastal deposits are also important areas of salt, and the Rann of Kutch is a major sabkha area.[58]

Possibly, the most contentious issue surrounding the Thar Desert is its age and origin. Fossil evidence for pre-Pleistocene climates is scanty in Gujarat, Sind, and Rajasthan, but the records of *Dipterocarpoxylon malavii*, *Cocos*, *Mesua tertiara*, and *Garcinia borocahii* from the Tertiary beds of Kutch and Barmer and the Eocene lignite at Palna near Bikaner may be suggestive of conditions rather similar to those currently pertaining in eastern Bengal, upper Burma, and Assam.[59] However, it is far from clear when the desert became established. Many authors have maintained that the desert is only of Holocene age and is the result of postglacial progressive desiccation.*

The stratigraphy of the Rajasthan lakes at Sambhar, Didwana, and Lunkaransar has shown conclusively that a major aeolian layer predates Holocene freshwater deposits,[61] and there are also hypersaline evaporite layers that date back to the last glacial maximum.[62] There are several phases of dune formation in the late Pleistocene, and the dune fields were formerly more extensive and active than they are today.[63] Many of the dunes are now stable, vegetated, gullied by fluvial action, and overlain by slopewash deposits, and, in the case of the coastal and near-coastal miliolites, they have been strongly cemented into material used for building. The lake stratigraphy also shows evidence for fluctuations of humidity in the Holocene, and this may have influenced the fortunes of the Harrappan civilization in the Indus Valley and its margins. The late Pleistocene aridity of the area may be confirmed by the presence of loess layers in river terrace deposits dated to c. 20,000–10,000 year BP in the Allahabad region[64] and by high dust loadings in Indian Ocean cores.[65]

* See, for example, M.S. Krishnan, V.M. Meher-Homji, and B. Allchin et al.[60]

Furthermore, detailed investigations of planktonic foraminifera in the Bay of Bengal[66] indicate high levels of salinity at the time of the Last Glacial Maximum, suggesting that there was reduced runoff in the Ganga–Brahmaputra at that time, probably as a result of a less vigorous monsoonal circulation. High-global radiation receipts at around 9000 year BP caused aridity to be reduced as the vigor of the monsoonal circulation returned.[67]

## 1.5 Arabia and the Middle East

The Middle East is an area of sometimes great aridity and also of great topographic diversity. On the one hand, there are major mountain ranges: the Zagros and Elburz mountains of Iran, the Taurus Mountains of Turkey, the Asir Mountains of Arabia, and the Jebel al Akhdar of Oman. On the other hand, there are the extensive inland plains and plateaus of Arabia, with their two great sand seas, An Nafud and Rub' al Khali (Figure 1.11), and the large intermontane basins in which lie the *kavirs* (salt plains) of Iran.

This topographic diversity owes much to the tectonic history and plate-tectonic setting of the region. Much of Arabia represents the remnant of part of the ancient landmass of Gondwanaland, while the mountain ranges are associated with the interaction of three great plates: the African, the Eurasian, and the Arabian. The Red Sea and the Gulf of Aden have been formed as the result of sea-floor spreading as Arabia moved away from Africa, and there has also been about 63 miles of left-lateral movement along the Dead Sea Fault system since Miocene times, as the Arabian plate has moved northward relative to the microplate of Sinai. In Iran, the same northward movement of Arabia toward Eurasia has caused widespread overthrusting, and sediments have been folded into a series of major

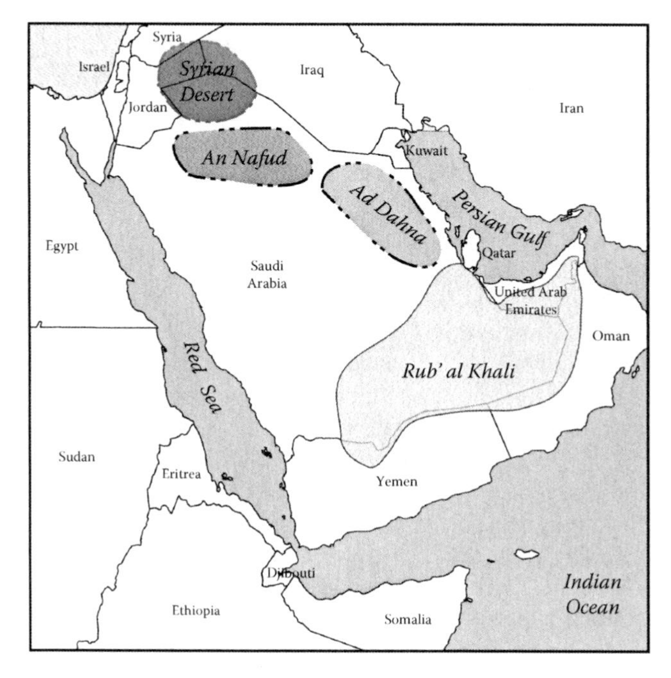


**FIGURE 1.11**
Map of the Al Rub' al-Kali (the Empty Quarter) on the Saudi Arabian peninsula.

synclines and anticlines. The underthrusting of Iran by the Arabian plate resulted in the complex folding of the Zagros Mountains. Eruptive rocks occur in zones of structural weakness, in the highland zones of Turkey and Iran, and also adjacent to the major faulting zones of the Dead Sea Lowlands and the Red Sea.

Such structural considerations thus underlie the gross morphology of the region. The detailed morphology owes much to environmental changes of the Tertiary and Pleistocene. Widespread humid conditions in the late Tertiary may have been of particular importance.[68] The evidence for this is provided by the development of erosional and weathering features on basalt lava flows of known age. Intense *lateritic weathering** is evident on a basalt flow 3.5 million year old, as is extensive fluvial dissection. By contrast, a younger flow, dated to the early Pleistocene (c. 1.3 million year old), shows no such features. Blocked-up linear drainage systems may date back to this stage, and associated gravel fans interfinger with the deposits of the regressing late Pliocene/early Pleistocene sea.

During the Quaternary, long-continued humid periods appear to have been much less significant, though the evidence for pluvials and interpluvials is well known. Indeed, it was in the Dead Sea trough that some of the first evidence for *pluvials* was identified.[69] Nonetheless, aridity has probably been a dominant feature for much of the Pleistocene (Figure 1.12).[70] Unfortunately, precisely dated, reliable palaeoenvironmental information is not readily available, especially before about 40,000 year BP, and attempts at correlation across the very varied climatic environments of the Near East have so far produced results that are confused both temporally and spatially. So, for example, an analysis of palynological data for a range of sites, for the Pleniglacial (c. 50,000–14,000 year BP), found that not only are there striking differences in vegetational and climatic history between the Levant and western Iran, but that even within the Levant the climatic history deduced for northern Israel cannot be brought into line with that of northwestern Syria.[71] For that reason, at present, it is probably prudent to provide some results from relatively well-dated situations from selected sites across the area, rather than to try to produce premature correlations.

**FIGURE 1.12**
Photograph of the Al Rub al' Kali in northern Yemen. (Courtesy of Robert H. Webb.)

* Lateritic weathering typically occurs in tropical climates and results in distinctive deposits bearing iron and aluminum oxides. Pluvials are geologic periods of increased precipitation.

Whitney[72] has undertaken an analysis of information for Saudi Arabia and, on the basis of a large number of radiocarbon dates, finds a pattern. He believed that the late Pleistocene alluvium, calcrete, and lake deposits define a clear pluvial episode between about 33,000 and 24,000 year BP, with the most intense phase of pluvial conditions probably being between 28,000 and 26,000 year BP. Quite large lakes occurred in the Rub' al Khali at this time,[73] and there seems to be a broad correspondence in climatic history with that encountered in Africa.[74] From about 19,000 to 10,000 year BP, aeolian activity prevailed through Arabia, and dune development occurred at times of low sea level on the floor of the Arabian Gulf,[75] permitting Saudi sand to enter Bahrain Island.[76] There are also extensive spreads of cemented aeolianites that extend below the present sea level in the Gulf States and Oman and that is formed at the time of low glacial sea levels.

Ocean core deposits in the Arabian Sea also indicate that large quantities of silt were being exported around 18,000 year BP.[67] A second major cluster of dates suggests that pluvial conditions began again in Arabia about 9000 year BP or slightly later, causing the deposition of *tufa*,* lake deposits in the Rub' al Khali, soil carbonates, and spreads of fine alluvial silts along major wadis. Pluvial conditions lasted until c. 5000 year BP.

The Konya Basin, in one of the more arid parts of the Anatolian Plateau, Turkey, has a series of dated shorelines that permit the reconstruction of hydrological changes in the northwestern part of the Near East. Lake-level curves have been established by Roberts,[77] with three phases of high level: Konya I prior to 30,000 year BP, Konya II between 23,000 and 17,000 year BP, and Konya III around 12,000–11,000 year BP. The important Konya II event appears to have occurred at the same time as the build-up of the last major Northern Hemisphere ice sheets, while the dramatic fall of palaeolake Konya around 17,000 year BP occurred well before the northward retreat of the Laurentide and Fennoscandian ice sheets. The early Holocene lacustral phase, noted in Saudi Arabia, is not represented in the Konya sequence. Roberts believed that the palaeolakes of Iran and Anatolia were more a product of reduced evaporation brought about by temperature depression than of changes in precipitation.

Nonetheless, when one considers another area, comprising the high plateaus of western Iran, detailed pollen analytical work from Lake Zeribar[73] indicates that vegetation conditions have changed very markedly in response to changes in precipitation levels over the past 40,000 years. Until c. 14,000 year BP, during a period broadly coincidental with the Pleniglacial of the European Würm glacial chronology, the vegetation was predominantly open, with steppe or desert-steppe in which *Artemisia*, Chenopoiaceae, and Umbellferae were important. This is seen to be the result not so much of coldness but of accentuated aridity. Conditions for tree growth improved after 14,000 year BP, but it was not until well into the Holocene that *Quercus* and *Pistacia* forests became established over wide areas.

The differences that have been observed in the pollen and lacustrine records for the late Pleistocene in the Middle East at different sites may be susceptible to a climatological explanation based on the varying importance of major systems (e.g., the summer monsoon, cyclonic westerlies, and cold Eurasian air masses) at different times and in different places. For example, during the late Pleistocene glacial, the northern shores of the Mediterranean Basin may have been under the influence of the cold, dry air masses generated by the giant ice sheets of Eurasia,[78] while the southern shores (including the southern Levant) would have received precipitation from westerly cyclonic systems that were compressed between the northern cold air and the more or less fixed subtropical high-pressure belt. Southern Arabia would have been influenced by the varying strength of the monsoons, and thus it shows a certain similarity of climatic trends to those observed in East Africa and the Thar.

* Tufas are calcium carbonate deposits caused by springs beneath lakes.

## 1.6 China and Central Asia

The deserts of China cover an area of around 0.43 million miles$^2$ and occupy about 11.6% of the total land area of the country. They are located in the temperate zone, stretching from 75°E to 125°E and from 35°N to 50°N. Within this huge area, extreme aridity characterizes the Taklimakan Desert (also known as the Taklamakan) of the Tarim basin. The locations of the main desert zones are shown in Figure 1.13. They are positioned in the great inland basins and high plateaus, with elevations generally lying between 1640 and 5000 ft, although there are some areas, such as the Turfan Depression, that lie below sea level. A distinction is often drawn between rocky and gravel deserts, termed "gobi," and the sandy deserts, termed "shamo."

The Chinese deserts appear to be very old,[79] being formed as early as the late Cretaceous and early Tertiary. At this time, the area was mostly under a subtropical high-pressure belt. Late in the Tertiary, the Tibetan Plateau was uplifted and the great Himalayan orogeny occurred. The continentality of the climate was greatly strengthened, the monsoon system became well established, and northwestern China became even more arid. Ancient lakes in the Tarim and other inland basins diminished or dried out gradually, and the Taklimakan and other sandy deserts probably enlarged considerably at that time. Continued uplift of the mountains during the Pleistocene and Holocene has further accentuated the aridity.

There is abundant evidence of climatic fluctuations during the Quaternary in the form of very complex *loess** profiles, suites of old lake shorelines (most notably around Lop Nor),

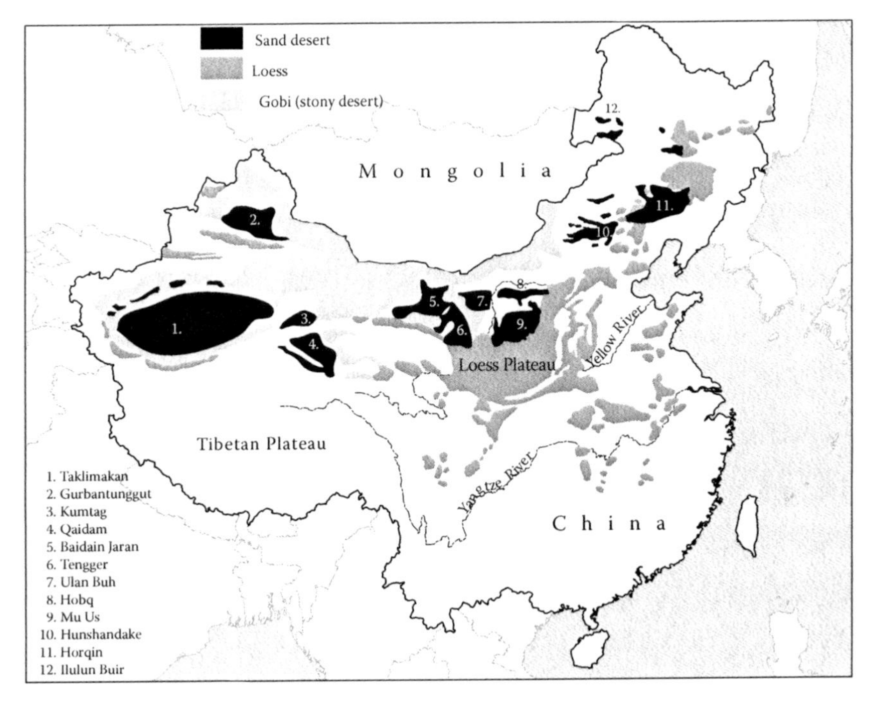


**FIGURE 1.13**
Main desert areas of China. (Modified from *Encyclopedia of Quaternary Sciences*, Sun, J. and Muhs, D.R., Mid-latitudes, Ed. S. Elias, p. 609, Figure 2, Copyright 2007, Elsevier.)

* Loess deposits, which typically are spatially extensive, are aeolian deposits associated with deglaciation.

**FIGURE 1.14**
Photograph of a part of the Gobi Desert in China. (Courtesy of Jayne Belnap.)

and miscellaneous historical and archaeological evidence.* Of these, the loess profiles give the longest record of environmental change, for maximum thicknesses of 1100 ft have been observed.[81] The materials have also proved susceptible to dating by palaeomagnetic and thermoluminescence methods. The oldest loess in the Central Loess Plateau has been dated palaeomagnetically at about 2.4 million years,[82] and this confirms that aridity has a lengthy history, for much of the silt is derived from the Gobi (Figure 1.14) and Ordos deserts.[83] Seventeen periods of loess sedimentation, separated by periods of nondeposition and soil development, have been recognized in the classic section of Luochuan for the period since 1.77 million year BP, and the depositional episodes have been correlated with glacial periods. It has been suggested that glacial maxima in Tibet, Tien Shan, and the Kun Lun Mountains were accompanied by a greater frequency of cyclonic depressions and sandstorms in the Gobi Desert and by more effective easterly transport of dust by a westerly jet stream centered north of the Tibetan anticyclone.[84]

The Turkestan desert of the (former) USSR lies between 36°N and 48°N and between 50°E and 83°E. It is bounded on the west by the Caspian Sea, on the south by the mountains bordering Iran and Afghanistan, on the east by the mountains bordering Sinkiang, and on the north by the Kirghiz Steppe. Two great ergs are included: the Kara-Kum ("black sands") and the Kyzyl-Kum ("red sands").

As in China,[85] loess deposits are both extensive and thick (up to 650 ft), and they have been dated in a similar manner. The loess record in Uzbekistan extends as far back as 2 million years, there are at least nine major soils in loess above the Brunhes/Matuyama boundary (c. 690,000 year BP), and loess deposition appears to have been relatively slight during the Holocene. Likewise, in Tajikistan, the loess record goes back to the Pliocene, and impressive sections contain more than 20 loess units with intervening palaeosols, many of which show heavy calcification.

The loess horizons themselves appear to have formed under more arid conditions than today,[86] for they contain large amounts of carbonates and soluble salts, have a xerophilous mollusc fauna, and show few indications of waterlogging. Rates of accumulation in the

* This archaeological evidence is of the type employed by E. Huntington.[80]

late Pleistocene appear to have been about four times faster than in the early Pleistocene and may have reached a rate of 4.9 ft $10^{-3}$ year$^{-1}$. Shortly before the Holocene, loess accumulation ceased almost everywhere in Central Asia. A progressive trend toward greater aridity through the Quaternary is evident from soil and pollen evidence within the profiles and may be related to progressive uplift of the Ghissar and Tien Shan Mountains.*

The Aral–Caspian basin in the western part of the Turkestan Desert shows dramatic evidence of marked hydrological fluctuations and, during glacial times, may have been occupied by the greatest known pluvial lake, covering an area of around 0.43 million miles². The highest shoreline was 250 ft above present Caspian level, and the Aral and the Caspian were united and extended some 813 miles up the Volga River from its present mouth. The largest transgressions occurred during early glacial phases, partly because reduced temperatures caused less evaporative loss, partly because of inputs of glacial meltwater, and partly because the surface over a large part of the catchment was sealed by permafrost.[88] Interglacials were times of regression.

## 1.7 Australia

With the exception of Antarctica, Australia is the driest of the continents, with a total of around 2.13 million miles² experiencing appreciable aridity (Figure 1.15). Paradoxically, however, aridity is not especially extreme in its intensity, and mean annual precipitation levels do not fall below 4–5 in. Indeed, because of another major control on its landscape evolution and its relatively long history of tectonic stability over large areas, many of the present features of its geomorphology are inherited from a great variety of climates that may go back to the Jurassic or earlier. Dunes as old as the Eocene are still preserved.[89]

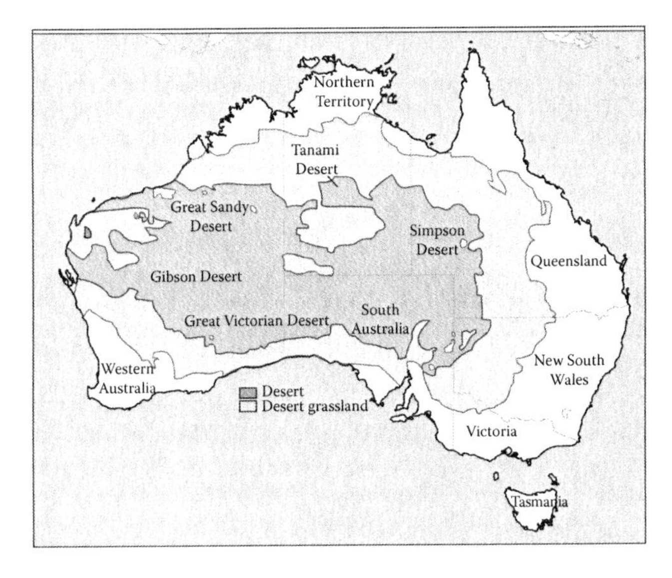


**FIGURE 1.15**
Deserts of Australia.

* Detailed analyses are provided by R.S. Davis et al.[87]

**FIGURE 1.16**
Photograph of the outback in central Australia near Alice Springs. (Courtesy of Robert H. Webb.)

The fundamental base for much of the flat or gently undulating desert landscape is Cretaceous; most of the macroforms such as plateaus and mesas and structural features such as the larger lake depressions are Tertiary; while mesoforms like sand dunes, prior stream formations, and the many small playas are Pleistocene (Figure 1.16). The Holocene has had little bearing on the deserts of today, apart from European man's contribution to the degeneration of ecosystems on some semiarid/arid margins. Useful general reviews of the distinctive landscapes of the Australian Deserts are provided by Mabbutt.[90]

The impact of post-Tertiary climates can be appreciated only within the context of plate tectonics and continental drift. Around 50–60 million year BP, Australia began to drift apart from Antarctica, and it migrated northward, drifting to within 4° of its present latitude by the early Miocene (c. 25 million year BP). Thus, not only has Australia been subjected to climatic changes resulting from changes in latitude, but it has also been subjected to the climatic effects that such movement had on the nature of oceanic and climatic circulation systems in the Southern Hemisphere and to the global climatic changes associated with the so-called Cenozoic climatic decline.

Among the important relict Tertiary features of the Australian desert are the widespread duricrusts[91] that include silcretes[92] and laterites with associated deep weathering profiles. There are also widespread relict palaeodrainage systems on vast erosion surfaces,[93] and there was a virtual inland sea in the Lake Eyre depression.

However, in the late Miocene–Pliocene, there was a gradual transition to a more arid climate. There is, however, scant evidence about the nature of climates for much of the Pleistocene, and it is only for the past 40,000–50,000 years that there is much information. The importance of major Pleistocene climatic and hydrological changes is clearly evident in the riverine plains of the east of the arid zone, which are part of the Murray system.[94] There are large spreads of alluvium deposited by ancient "prior streams," which were sinuous bedload channels indicative of coarse sediment transport in flash floods. These plains are mantled by aeolian clays, called *parna*, and the active floodplains are slightly entrenched beneath the prior stream deposits and are occupied by meandering, suspended-load channels. Cooper Creek, which flows into Lake Eyre, shows a comparably varied history in the late Pleistocene.[95]

The widespread sand deserts of Australia also display the impact of Pleistocene changes,[96] and a striking feature of the linear dune fields of the Australian deserts is that they extend well beyond the present desert areas. They are relict features of Pleistocene aridity, and some may be of late Pleistocene age. This applies to the stable dunes of the northern part of the Great Sandy Desert that extends beneath the Holocene alluvium of the Fitzroy Estuary[97] and that thus formed at a time of low sea levels. Lacustrine and lunette sediments in the Willandra Lakes area of New South Wales also afford evidence for a period of dune encroachment and lake desiccation in the late Pleistocene.[98] This glacial aridity may have become prevalent after 25,000 year BP, before which lake levels in the southern parts of Australia appear to have been relatively high. Northeastern Australia, however, was significantly less humid than today well before 25,000 year BP, for the pollen spectra from Lynch's Crater in Queensland indicate a drastically reduced rainfall (100–20 in. $year^{-1}$) in the interval between about 80,000 and 20,000 year BP. Overall, the combined evidence from marine cores on the Timor continental shelf, pollen analysis, and the study of lake and dune deposits indicates that at the time of the last glacial maximum (c. 25,000–18,000 year BP), much of Australia was drier and windier than today and was surrounded by a much broader continental shelf.[99] Soon after 17,000 year BP, temperatures, precipitation, and sea levels began to rise, and most of the desert dunes were probably becoming stabilized by c. 13,000 year BP. In the early Holocene, rainfall levels were higher, and forests became more widespread. There is some evidence for renewed dune activity in western New South Wales beginning c. 2500 year BP,[100] and this may signify the passing of the mid-Holocene humid phase and a brief return to relative aridity.

## 1.8 South America

The main desert areas of South America are inextricably associated with the Andean cordillera. The most extensive zone of aridity includes the coastal Peruvian and Atacama deserts to the west of the mountains from c. 5°S to c. 30°S; to the east of the cordillera lie the Monte and Patagonian deserts of Argentina (Figure 1.17).

The Monte and Patagonian deserts both lie essentially in the lee of the Andes. The Monte Desert, which is more or less continuous with the deserts to the west, is composed of basin-range topography, including mountain blocks, extensive piedmont surfaces, and largely internal drainage. Volcanic features are also to be found. The evolution of the region is not well understood, but Walter Penck was amongst those who have contributed toward the description of slope evolution and piedmont development. The Patagonian Desert stretches for over 313 miles between the Andes and the sea. It owes its aridity to the mountains, which block the rain-bearing winds from the west, and to the cold Falkland Current off the coast. The region is dominated by piedmont plains that slope eastward toward the Atlantic, where they are terminated by marine surfaces, by ephemeral rivers that are entrenched into them, and by enclosed drainage basins. Volcanic, glacial, and fluvial deposits occur extensively in the region. Several glacial episodes during the Quaternary in the Patagonian Andes certainly influenced the evolution of this arid area strongly, especially in feeding fluvioglacial gravels into the desert.[101]

West of the Andes, the Peru–Chile desert has several distinctive features. Climatically, the aridity is created by subtropical atmospheric subsidence reinforced by the upwelling of cold coastal waters associated with the north-flowing Peru current. As a result, it is one of the world's driest areas, although precipitation does increase eastward with elevation in the Andes (Figure 1.18). The coastal zone is characterized by fogs (*camanchaca*) that roll in

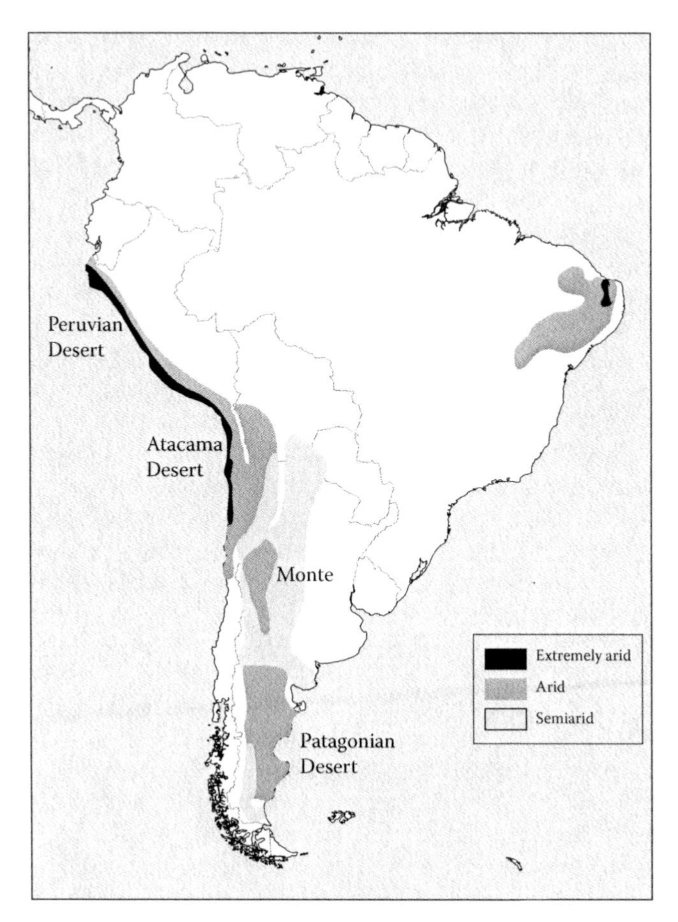


**FIGURE 1.17**
Arid lands of South America. (Modified from Meigs, P., The world distribution of arid and semiarid homoclimates, in *Reviews of Research on Arid Zone Hydrology*, UNESCO, Paris, France, 1953, pp. 203–209.)

from the Pacific on many days, providing unexpectedly high humidities. The region has many local and regional winds, but there are very few areas of sand dunes. Nevertheless, mobile barchans and yardangs are well developed in southern Peru.

Geologically, the region is dominated by the Andean cordillera (up to 22,800 ft asl) and, offshore, the Peru–Chile trench (up to 25,000 ft bsl). Both features are associated with the westward migration of the South American plate over the eastward-moving Nazca plate. The region has abundant evidence of volcanic activity, including volcanoes, calderas, enormous pyroclastic flow deposits (ignimbrites), lava flows, geysers, and solfatara.[102] Andesitic rocks are typical of the volcanic region, and contemporary activity is confined to a few moderately explosive volcanoes. Tectonic activity, in the form of extensive folding and faulting, is widespread and is responsible, inter alia, for the longitudinal differentiation of the topography.

The principal longitudinal features of the Atacama topography (Figure 1.19) are, from west to east, the Coastal Cordillera (up to c. 6500 ft asl); the Longitudinal Valley, the precordillera (an area of basin-range topography, salt domes, and slopes rising up to the Andes), and the Altiplano (an extensive, arid plateau at over 13,100 ft elevation where volcanic activity is widespread). The Andean flanks and the Longitudinal Valley are

**FIGURE 1.18**
Photograph of the Atacama Desert, Chile, showing one of many extinct wetlands (diatomaceous badlands in center) that were active during the late glacial-early Holocene in the now waterless terrain at the base of the Andes ~3000 m elevation, where the regional water table intersects the toe of a massive alluvial fan. (From Quade, J. et al., *Quat. Res.*, 69, 343, 2008; Courtesy of Julio L. Betancourt. With permission.)

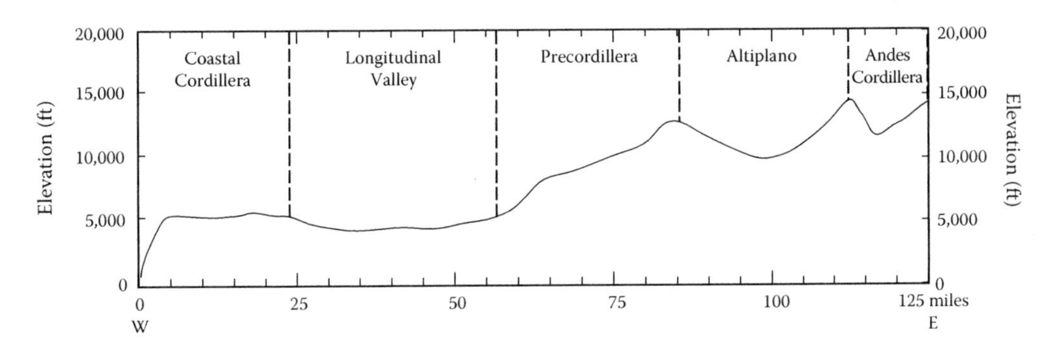


**FIGURE 1.19**
Topographic profile of the Atacama Desert (22°S) in Chile.

dominated by massive aggradation during the Oligocene and, possibly as a consequence of the onset of aridity, by subsequent extensive, bedrock-dominated pediment surfaces developed across a variety of rocks, including granite.[103] Segerstrom[104] referred to the pediment topography as a "matureland," although, in reality, it probably consists of several uplifted and dissected erosional surfaces. There are extensive and probably complementary piedmont deposits within the Longitudinal Valley, the internal drainage basins, and along the perennial rivers.

The drainage of the region, especially in Chile, is dominated by enclosed drainage basins that focus on salars. In addition, snowmelt in the Andes nourishes a few perennial streams that cross the desert to the Pacific. Only the Rio Loa crosses the driest zone of the Norte Grande, but to the south, as the precipitation rises and the snowline falls in the Norte Chico, the rios Copiapó, Huasco, and Elqui all cut across the major zones of longitudinal relief. Salts, many of which may be of volcanic origin, are widespread. They are found in many salars; nitrates and iodates are distinctively concentrated, especially on the eastern side of the coastal cordillera[105]; and the salt domes occur in the precordillera of the Norte Grande.

The evolution of the Peru–Chile deserts is still a matter of considerable speculation, but a few observations are in order. First, and contrary to common opinion, it seems likely that there has not been enormous Andean uplift during the Quaternary. One study, based on K-Ar dating of ignimbrite flows, suggested that in the high Andes of the Norte Chico, the *pediplain topography** has suffered remarkably little erosion since the Upper Miocene and that over the past 9–12 million years there has been entrenchment only in the canyons of c. 328–650 ft.[106]

Second, the Atacama Desert is probably very old. It is generally believed that it has been arid since at least the late Eocene, with hyperaridity since the middle to late Miocene.[107] The uplift of the Central Andes cordillera during the Oligocene and early Miocene was a critical palaeoclimatic factor, providing a rain-shadow effect and also stabilizing the southeastern Pacific anticyclone. However, also of great significance (and analogous to the situation in the Namib) was the development 15–13 million year BP of cold Antarctic bottom waters and the cold Humboldt current as a result of the formation of the Antarctic ice sheet.

Third, it is a widely held view that much of the region is a "core desert," where climatic change has been quite limited during the Quaternary. Certainly much of the Atacama gives this impression. Although there is some evidence of glaciation in the high Andes of the desert during the Quaternary, it was very local and only on the highest mountains. Morainic deposits showed that short glaciers extended to c. 13,800 ft. Morainic deposits and associated features have been identified in the northern Atacama, east of the Salar de Atacama.[108] Farther south, in the Norte Chico, glacial features suggest three or four glacial "periods," although their ages are uncertain.[109]

Detailed research leaves no doubt that there has been a series of roughly synchronous glacier fluctuations of similar magnitude in the Andes during the late Pleistocene and that these events were preceded by glaciations of similar magnitude during the past 3.5 million year BP.[110] Clapperton[110] recognized significant glacial episodes in the Holocene (c. 16,000–10,000 year BP), the last glaciation (c. 18,000–16,000 year BP), the penultimate glaciation (c. 170,000–140,000 year BP), the prepenultimate glaciations (<1.9 million year BP), and the pre-Pleistocene glaciations (1.9–5 million year BP). The glacial episodes undoubtedly reflect climatic changes. In the Atacama region, the glacial features were minor, but the impact of the climatic changes on the evolution of landforms is not yet clear, except in so far as there is evidence of fluctuations in lake levels in the salars.

Evidence of precipitation changes is provided by lake basins in the Altiplano. Hastenrath and Kutzbach,[111] for example, have shown that in the late Pleistocene (before 28,000 year BP and from 12,500 to 11,000 year BP) lakes in the Peruvian–Bolivian altiplano were four to six times more extensive than today and that this implies rainfall increases of around 50%–75%. By contrast, in the mid-Holocene (c. 7700–3650 year BP), Lake Titicaca was at a very low level.[112]

Along the perennial river valleys and the coast, there is also evidence of Quaternary evolution in the form of marine and fluvial terraces and their associated deposits. Because the number of these surfaces and deposits varies from sector to sector along the coast, it seems probable that the sequences may reflect differential tectonic movements as well as fluctuating sea levels. As a result, clear generalizations are difficult, but it seems likely that the highest major surface of erosion-aggradation is at least Pliocene in age and possibly

---

* Pediplain topography refers to coalesced pediments or slopes overlying bedrock created by scarp recession.

much older. Paskoff[113] has suggested that the high cliff, which is up to 2600 ft high and is so characteristic of the Chilean desert coast, probably originated as a Miocene fault scarp, which has retreated and been embroidered by an oscillating sea level ever since.

## 1.9 North America

The deserts of North America occupy much of western United States and northern Mexico between 44°N and 22°N (Figure 1.20). They extend southward from central and eastern Oregon, embracing nearly all of Nevada and Utah, into southwestern Wyoming and western Colorado, reaching westward in southern California to the eastern base of the Sierra Nevada, the San Bernardino Mountains, and the Cuyamaca Mountains. From southern Utah, the desert extends into Arizona and on into the Chihuahua Desert of Mexico. The Sonoran Desert of California extends into Baja California and to the eastern side of the Gulf of California. These deserts owe their aridity to a variety of conditions. Orographic barriers are especially important in the north and in parts of California,

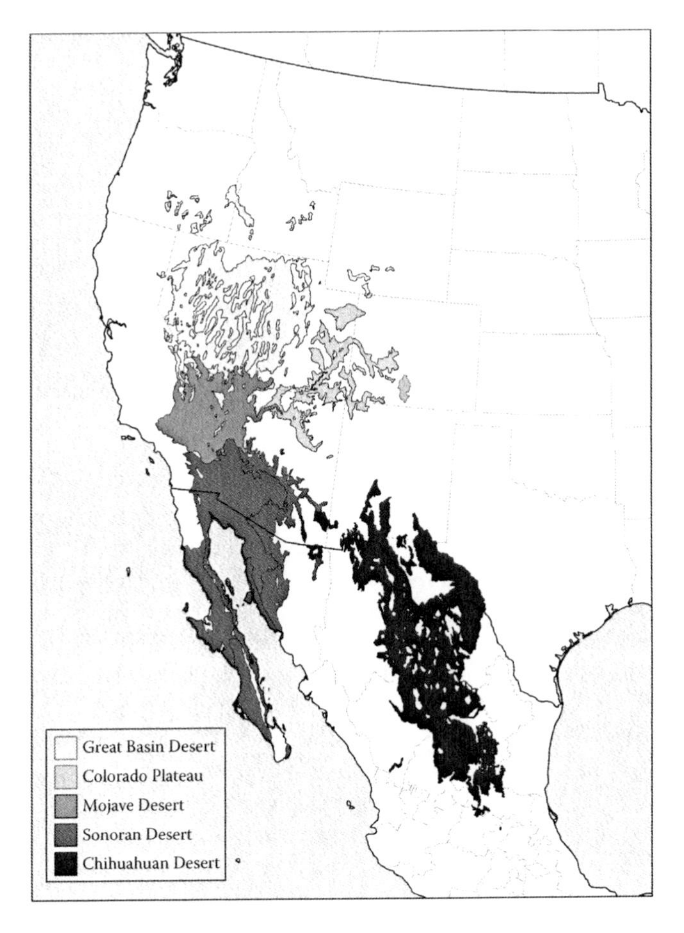


**FIGURE 1.20**
Deserts of North America.

**FIGURE 1.21**
Photograph of the Mojave Desert in California. (Courtesy of Todd C. Esque.)

**FIGURE 1.22**
Photograph of the Sonoran Desert east of Guaymas, Sonora. (Courtesy of Robert H. Webb.)

while the southern portion comes under the influence of a subtropical high-pressure cell and has a summer maximum of precipitation. Extreme aridity is relatively limited in extent, with the most arid regions lying along the Gulf of California and in the Mojave. Even in such hyperarid areas, fluvial activity is probably significant because of the close proximity of high mountain ranges (Figures 1.21 and 1.22).

A major control of the development and climatic evolution of the deserts of North America was the Cordilleran orogeny, which began at the close of the Cretaceous period and continued into the Cenozoic. This involved the intrusion of massive granite batholiths in the Sierra Nevada and Idaho and low-angle thrusting of immense slabs of rock eastward over one another along a line extending from Mexico to northwestern Canada. This thrusting ceased in Miocene times (c. 12 million year BP), but uplift and tectonic processes continued thereafter. Structural domes were uplifted, volcanism occurred, and rivers cut spectacular canyons into the uplifted masses. The Basin-Range province started to develop in the late Oligocene, creating a landscape of fault-bounded blocks and troughs. This was caused by crustal extension, either because high heat flow above the subducted plate may have caused doming and extension of the crust or because the subducting plate may have broken, causing very high heatflow and extension of the crust for a limited period.

These tectonic events provided the setting for some of the most important geomorphological features of the arid west playas, although deflation and other processes have contributed to their development and form, they are essentially products of a particular tectonic setting. That these mainly dry playa basins formerly contained large lakes was documented around a century ago.[114] It is now recognized that more than a 100 closed basins in the western United States contained lakes during the late Wisconsin, but only about 10% of them are perennial and of substantial size today. The largest of the basins was Bonneville, which had a length of around 313 miles, a maximum area of 20,300 miles$^2$, and a depth of about 1,100 ft. Lake Lahontan, the second largest lake, had an area of 9000 miles$^2$ and a 918 ft maximum depth.

The dating of hydrological changes in these basins over the past 40,000 years is relatively precise,[115] and the lake-level curves from a number of basins show a considerable degree of similarity. It is strikingly evident that from c. 24,000 to 14,000 year BP (i.e., more or less contemporaneous with the maximum of the last Wisconsin, glaciation) that lake levels were high. There was then a phase when lake levels showed marked fluctuations, before a period of drought between 10,000 and 5,000 year BP, which culminated between 6,000 and 5,000 year BP. While the temporal patterns of the fluctuations may be relatively well known, the explanation for the greatly expanded pluvial lakes is more controversial, and there has been a longstanding debate as to whether the crucial control was the diminished evapotranspirational losses brought about by late Pleistocene temperature reductions or some increase in precipitation levels.[116]

Although the pluvial legacy is widely evident in the presence of old lake beds, high lake shorelines, and formerly integrated fluvial systems, elsewhere there is abundant evidence for the formerly greater power and distribution of aeolian processes, as revealed by the presence of aligned drainage systems, yardangs, shaped depressions, and associated lunettes and, most importantly of all, of palaeodune systems. These were first recognized by Price[117] and have recently been mapped in a great belt of country in the lee of the western cordillera between the Canadian Line and the Gulf of Mexico.[118] The dating of the largely relict dune fields, of which the Nebraskan Sandhills are the largest example,[119] is controversial. Debate surrounds the relative importance of late Pleistocene and mid-Holocene (altithermal) arid phases.[120] The stratigraphy, sedimentology, and arrangement of lunettes in the lee of large deflation basins cut into palaeodrainage systems on the High Plains of Texas indicate that there were multiple phases of aeolian activity. Indeed, aeolian activity in the High Plains, as represented by the Blackwater Draw Formation, dates back to beyond 1.5 million year BP.[121]

## 1.10 Conclusion

The world's deserts show the imprint of climatic changes at many different time scales. These range from runs of a few dry or wet years, through to phases of the duration of a decade or decades (e.g., the Dust Bowl years of North America in the 1930s or the persistent drought of the Sudano-Sahel belt since the late 1960s), through to extended periods of some centuries or millennia (e.g., the dry hypsithermal phase of the American High Plains in the mid-Holocene or the intense pluvial or lacustral phase of tropical Africa in the early Holocene), through to major Pleistocene events related to the glacial and interglacials of higher latitudes (which may have had a duration of the order of 100,000 years), and through to the long-term "geological changes" of the order of millions of years associated with major shifts in the positions of the continents, major tectonic and orogenic events, and the configurations of the ocean basins and their associated circulation systems.

The short-term fluctuations are related to changes in the general atmospheric circulation, with such changes as the water temperatures of the Pacific Ocean associated with the El Niño effect or the zonality or meridionality of the Rossby waves playing a major rôle.[122] Medium-term fluctuations, such as the early Holocene pluvial of the tropics, may be related to changes in Earth geometry, and, at around 9000 year BP, theoretical insolation receipt during the Northern Hemisphere summer may have been larger than now (by about 7%), thereby leading to an intensification of the monsoonal circulation and associated precipitation.[69] At a longer timescale, related to the interglacial and glacial fluctuations of the Pleistocene, a variety of factors may have been involved. For example, during a major glacial phase, tropical aridity may have been heightened by an increased continentality of climate resulting from the withdrawal of the sea from the continental shelves as a consequence of glacial eustasy. Such a drop in sea level, together with an extension of sea ice, would result in less evaporation from the ocean surface, leading to less rain. Moreover, the worldwide cooling of the oceans would lead to less evaporation and convection and the generation of fewer tropical cyclones.

Furthermore, thermal variations provoked by the growth and decay of the great ice sheets decisively influenced the patterns of the general atmospheric circulation. For example, in theory, an increased temperature gradient resulting from the presence of the great Scandinavian and Laurentide ice sheets would result in stronger westerlies, an equatorial displacement of major circulation features, and an intensification and shrinking of the Hadley cell zone and the zone of extratropical Rossby wave circulation. The result would have been a greater degree of westerly flow and midlatitude cyclogenesis over an area such as North Africa.[123] A corresponding displacement of the subtropical high-pressure zone would have displaced the aridity maximum into West Africa. The decreased thermal contrast between the two hemispheres compared to the situation today would have had an impact. At present, the Southern Hemisphere, in comparison with the Northern, is much cooler and its temperature gradient greater. This is because of the varying amounts and distribution of ocean and land in the two hemispheres. In the Southern Hemisphere, the stronger temperature gradient produces a more intense atmospheric circulation and is probably largely responsible for the asymmetry that exists, whereby the meteorological equator lies in the Northern Hemisphere. During a glacial phase, intense continental glaciation should have led to a displacement of this meteorological equator to a position more coincident with the geographical equator (i.e., southward). This would disrupt the annual march of the monsoon and thereby reduce precipitation amounts in such areas as the Thar and southeast Arabia.

The trigger for the glacial/interglacial fluctuations themselves is now thought to lie in the Milankovitch mechanism of Earth orbital fluctuations, with amplitudes of the order of about 100,000, 43,000, and 19,000–24,000 years.[124] Spectral analysis has shown such wavelengths in ocean-floor sediments and in Chinese loess profiles.[125]

The longest scales of change, which include the establishment of the world's arid areas, are related to variations in the configurations of the continents and oceans, occasioned by global tectonics. This mechanism operated in a variety of ways. First, the drift of the continents changed their latitudinal positions and thus their position with respect to major climatic belts. Second, because of the rise of the great Cenozoic mountain systems caused by plate collision, rainshadow deserts were created on the continents, especially in Asia, northern Argentina, the southwestern United States, and the Atlas. Third, changes in the configurations of the oceans and continents caused a general world cooling: the so-called Cenozoic climatic decline. Such cooling would have affected the nature and intensity of the general atmospheric circulation and of ocean currents. The significance of this last point is that in the Southern Hemisphere, the continents all have dry zones along their western margins. Each of these deserts is to a large degree caused by an upwelling of cold water off the coast, and such cold water is mostly of Antarctic origin. Such cold currents would, therefore, not exist until Antarctica was producing large quantities of cold meltwater and sea ice. For this to happen, Antarctica had to have moved to a suitable latitudinal position following its separation from Australia.[126] These sorts of influences are seen in the history of the Atacama Desert, an area which has been arid since at least the late Eocene. The uplift of the Central Andes cordillera during the Oligocene and early Miocene was one critical palaeoclimatic factor, providing a rain shadow for precipitation from the Amazon basin, and stabilizing the southeastern Pacific anticyclone. However, it was the dramatic cooling of Antarctic bottom waters and of the Humboldt Current around 13–15 million year BP, associated with the formation of the Antarctic ice sheet, that enabled hyperaridity to be established for the first time in the area.[110]

## 1.11 Postscript

Since the writing of this chapter was completed in 1992, there have been some major changes in our knowledge of the nature and history of the world's main deserts.[127] These have been brought about by such factors as new dating techniques (e.g., optical dating) and the development of methods for high-resolution environmental reconstruction. In addition, our knowledge has been greatly expanded as a result of the availability of free remote sensing imagery of increasingly high resolution and also the growth of indigenous studies of deserts, most notably in China. We now have a clearer idea of the antiquity of deserts of the climate changes that have taken place in the Quaternary and the distribution and characteristics of such features as sand sea, pans, and yardangs. The burgeoning of quaternary sciences has indicated the degree, frequency, and abruptness of climatic changes in the world in general and in deserts in particular. We now have a far better appreciation of changes at millennial, century, and decadal time scales.

## References

1. A. Mangin, *The Desert World* (London, U.K.: Nelson, 1869).
2. P. Meigs, The world distribution of arid and semiarid homoclimates, in *Reviews of Research on Arid Zone Hydrology*, pp. 203–209 (Paris, France: UNESCO, 1953); C. C. Wallen. Aridity definitions and their applicability. *Geografiska Annaler* 49 (1967): 367–384; G. M. Howe et al., Classification of world desert areas (U.S. Army National Laboratories Technical Report 69-38-ES, 1968); M. I. Budyko, *Izmeniya Klimata*. Gidrometeoizdat, also published as: M. I. Budyko, *Climatic changes*

(transl. *Izmeniia Klimata* Leningrad: Gidrometeoizdat, Washington, DC: American Geophysical Union, 1974); R. D. Thompson, The climatology of the arid world (Reading, U.K.: University of Reading, Department of Geography, Geographical Papers 35, 1975).

3. W. R. Osterkamp, Annotated definitions of selected geomorphic terms and related terms of hydrology, sedimentology, soil science, and ecology (Reston, VA: U.S. Geological Survey Open File Report 2008-1217, 2008); A. Allaby and M. Allaby, *The Concise Oxford Dictionary of Earth Sciences* (New York: Oxford University Press, 1990).
4. M. A. J. Williams and H. Faure, Eds., *The Sahara and the Nile* (Rotterdam, the Netherlands: A. A. Balkema, 1980).
5. A. T. Grove, Geomorphic evolution of the Sahara and the Nile, in *The Sahara and the Nile*, Eds. M. A. J. Williams and H. Faure, pp. 7–16 (Rotterdam, the Netherlands: A. A. Balkema, 1980).
6. E. F. Gautier, *Sahara: The Great Desert* (New York: Columbia University Press, 1935; New York: Octagon Press, 1970); J. L. Cloudsley-Thompson, Ed., *Sahara Desert: Key Environments* (Oxford, U.K.: Pergamon Press, 1984); R. Capot-Rey, *Le Sahara Français* (Paris, France: Presses Universitaires de France, 1953).
7. M. Mainguet, Tentative mega-geomorphological study of the Sahara, in *Megageomorphology*, Eds. R. A. M. Gardner and H. Scoging, pp. 113–133 (Oxford, U.K.: Oxford University Press, 1983).
8. S. Beruf, B. Biju-Duval, O. de Charpal, P. Rognon, P. Gariel, and O. A. Bennacef, Les grès du paleozoique infèrieur au Sahara (Paris, France: Editions Technip., 1971).
9. H. Faure, Reconnaissance Géologique des Formations Post-Paléozoiques du Niger Oriental, Mémoires No. 47 (Paris, France: Bureau de Recherches Géologiques et Minieres, 1962).
10. K. J. Hsü, L. Montadert, D. Bernoulli, M. B. Cita, A. Erickson, R. E. Garrison, R. B. Kidd, F. Mélières, C. Muller, and R. Wright, History of the Mediterranean salinity crisis, *Nature* 267 (1977): 399–403.
11. R. Said, The geological evolution of the River Nile in Egypt, *Zeitschrift für Geomorphologie* 26 (1982): 305–314. The Grand Canyon actually is in Arizona, where it is carved by the Colorado River.
12. E. M. van Zinderen Bakker, Sr., Elements for the chronology of late Cainozoic African climates, in *Correlation of Quaternary Chronologies*, Ed. W. C. Mahaney, pp. 23–37 (Norwich, U.K.: Geobooks, 1984).
13. M. Williams, Geology, in *Sahara Desert: Key Environments*, Ed. J. L. Cloudsley-Thompson, pp. 31–39 (Oxford, U.K.: Pergamon Press, 1984).
14. A. Lever and I. N. McCave, Eolian components in Cretaceous and Tertiary North Atlantic sediments, *Journal of Sedimentary Petrology* 53 (1983): 811–832.
15. S. Servant-Vildary, Le Plio-Quaternaire ancien du Tchad: Evolution des associations de diatomées, stratigraphie, paléogéographie. Série Géologique 5 (Paris, France: Cahiers ORSTOM, 1973), pp. 169–217; A. Street and F. Gasse, Recent developments in research into the Quaternary climatic history of the Sahara, in *The Sahara: Ecological Change and Early Economic History*, Ed. J. A. Allan, pp. 7–28 (London, U.K.: Menas Press, 1981).
16. J. C. Fontes and F. Gasse, On the ages of humid Holocene and late Pleistocene phases in North Africa: Remarks on 'Late Quaternary climatic reconstruction for the Maghreb (North Africa)' by P. Rognon, *Palaeogeography, Paleoclimatology, Palaeoecology* 70 (1989): 393–398.
17. C. Parmenter and D. W. Folger, Eolian biogenic detritus in deep-sea sediments: A possible index of equatorial ice age aridity, *Science* 185 (1974): 695–698; D. W. Parkin and N. J. Shackleton, Trade wind and temperature correlations down a deep-sea core off the Saharan coast, *Nature* 245 (1973): 455–457.
18. A. T. Grove and A. Warren, Quaternary landforms and climate on the south side of the Sahara, *Geographical Journal* 134 (1968): 194–208; P. Michel, Les bassins des fleuves Sénégal et Gambie: études géomorphologiques, 3 vols. Mémoires No. 63. (Paris, France: ORSTOM, 1973); M. Sarnthein, Sand deserts during the last glacial maximum and climatic optimum, *Nature* 272 (1978): 43–46; J. E. Nichol, The extent of desert dunes in northern Nigeria as shown by image enhancement, *Geographical Journal* 57 (1991): 13–24.
19. M. Sarnthein and L. Diester-Haass, Eolian sand turbidites, *Journal of Sedimentary Petrology* 47 (1977): 868–890.

20. P. Rognon. Essai d'interpretation des variations climatiques au Sahara depuis 40,000 ans, *Revue Géographie Physique et Géologique Dynamique* 18 (1976): 251–282; M. A. J. Williams and D. A. Adamson, Late Pleistocene desiccation along the White Nile, *Nature* 248 (1974): 584–586.
21. H. J. Pachur and P. Hoelzmann, Palaeoclimatic implications of late Quaternary lacustrine sediments in western Nubia, *Sudan Quaternary Research* 36 (1991): 257–276.
22. J. Maley, Palaeoclimates of central Sahara during the Holocene, *Nature* 269 (1977): 573–577.
23. S. E. Nicholson, Saharan climates in historic times, in *The Sahara: Ecological Change and Early Economic History*, Ed. J. A. Allan, pp. 35–59 (London, U.K.: Menas Press, 1981).
24. N. J. Middleton, Effect of drought on dust production in the Sahel, *Nature* 316 (1985): 431–434.
25. N. Lancaster, *The Namib Sand Sea: Dune Forms, Processes and Sediments* (Rotterdam, the Netherlands: Balkema, 1989).
26. N. Lancaster, Dunes on the Skeleton Coast, Southwest Africa/Namibia: Geomorphology and grain size relationships, *Earth Surface Processes and Landforms* 7 (1982): 575–587.
27. N. Lancaster, Aspects of the morphometry of linear dunes of the Namib Desert, *South African Journal of Science* 77 (1981): 366–368; Good general reviews of the Namib Desert are provided by R. F. Logan, *The Central Namib Desert, South West Africa* (Washington, DC: National Academy of Sciences Press, 1960); G. Beaudet and P. Michel, Recherches Géomorphologiques en Namibie Centrale (Strasbourg, Austria: Association Géographique d'Alsace, 1978).
28. N. Lancaster, Linear dunes of the Namib sand sea, *Zeitshcrift für Geomorphologie, Supplementband* 45 (1983): 27–49.
29. N. Lancaster, Aridity in Southern Africa: Age, origins and expression in landforms and sediments, in *Late Cainozoic Palaeoenvironments of the Southern Hemisphere*, Ed. J. C. Vogel, pp. 433–444 (Rotterdam, the Netherlands: A. A. Balkema, 1984).
30. W. G. Seisse, Aridification of the Namib Desert: Evidence from oceanic cores, in *Antarctic Glacial History and World Palaeoenvironments*, Ed. E. M. van Zinderen Bakker, pp. 105–113 (Rotterdam, the Netherlands: A. A. Balkema, 1978); W. G. Seisser, Late Miocene origin of the Benguela upwelling system off northern Namibia, *Science* 208 (1980): 283–285.
31. K. W. Butzer, Archaeology and Quaternary environments in the interior of Southern Africa, in *Southern Africa Prehistory and Paleoenvironments*, Ed. R. G. Klein, pp. 1–64 (Rotterdam, the Netherlands: A. A. Balkema, 1984).
32. K. Heine and M. A. Geyh, Radiocarbon dating of speleothems from the Rotting Cave, Namib Desert, and palaeoclimatic implications, in *Late Cainozoic Palaeoenvironments of the Southern Hemisphere*, Ed. J. C. Vogel, pp. 465–470 (Rotterdam, the Netherlands: A. A. Balkema, 1984).
33. J. C. Vogel, Evidence of past climatic change in the Namib Desert, *Palaeogeography, Palaeoclimatology, Palaeoecology* 70 (1989): 355–366.
34. S. Passarge, *Die Kalahari*, 2 vols. (Berlin, Germany: Reimer, 1904); D. S. G. Thomas and P. A. Shaw, *The Kalahari Environment* (Cambridge, U.K.: Cambridge University Press, 1991).
35. D. S. G. Thomas, Late quaternary environmental change in Central Southern Africa with particular reference to extension of the arid zone (PhD dissertation, Oxford University, Oxford, U.K., 1984).
36. A. S. Goudie and D. S. G. Thomas, Pans in Southern Africa with particular reference to South Africa and Zimbabwe, *Zeitschrift für Geomorphologie* 29 (1985): 1–19.
37. F. Netterberg, Dating and correlation of caicretes and other pedocretes, *Geological Society of South Africa, Transactions* 81 (1979): 379–392.
38. M. A. Summerfield, Silcrete as a palaeoclimatic indicator: Evidence from southern Africa, *Palaeogeography, Palaeoclimatology, Palaeoecology* 41 (1983): 65–79.
39. D. S. G. Thomas, Geomorphic evolution and river channel orientation in northwestern Zimbabwe, *Geographical Association of Zimbabwe, Proceedings* 14 (1984): 45–55.
40. E. Boocock and O. J. Van Straten, Notes on the geology and hydrogeology of the central Kalahari, Bechuanaland Protectorate, *Geological Society of South Africa, Transactions* 65 (1962): 125–171.
41. A. T. Grove, Landforms and climatic change in the Kalahari and Ngamiland, *Geographical Journal* 135 (1969): 191–212.

42. D. S. G. Thomas, Ancient ergs of the former arid zones of Zimbabwe, Zambia, and Angola, *Institute of British Geographers, Transactions* 9 (1984): 75–88.
43. N. Lancaster, Palaeoenvironmental implication of fixed dune systems in southern Africa, *Palaeogeography, Palaeoclimatology, Palaeoecology* 33 (1981): 327–346.
44. K. Heine, The main stages of the late Quaternary evolution of the Kalahari region, southern Africa, in *South African Society for Quaternary Research, Proceedings*, Eds. J. A. Vogel, E. A. Voigt, and T. C. Partridge (Rotterdam, the Netherlands: A. A. Balkema, 1982).
45. N. Lancaster, The pans of the southern Kalahari, Botswana, *Geographical Journal* 144 (1978): 81–98.
46. H. J. Cooke, Landform evolution in the context of climatic change and neo-tectonism in the middle Kalahari of northern central Botswana, *Institute of British Geographers, Transactions* 5 (1980): 80–99.
47. K. W. Butzer, Archaeology and Quaternary environments in the interior of Southern Africa, in *Southern Africa Prehistory and Paleoenvironments*, Ed. R. G. Klein, p. 51 (Rotterdam, the Netherlands: A. A. Balkema, 1984).
48. B. Allchin, A. S. Goudie, and K. Hegde, *The Prehistory and Palaeogeography of the Great Indian Desert* (London, U.K.: Academic Press, 1978).
49. H. T. Lambrick, *Sind: A General Introduction*, 2nd edn. (Hyderabad, Pakistan: Sindhi Adabi Board, 1975).
50. H. Wilhelmy, Das urstromtal am ostrand der industene und der Sarasvati-problem, *Zeitshcrift für Geomorphologie, Supplementband* 8 (1969): 76–93.
51. G. M. Higgins, A. Mushtuq, and R. Brinkman, The Thal interfluve, Pakistan: Geomorphology and depositional history, *Geologie en Mijnbouw* 52 (1973): 147–155.
52. A. S. Goudie and C. H. B. Sperling, Long distance transport of foraminiferal tests by wind in the Thar Desert, north-west India, *Journal of Sedimentary Petrology* 47 (1977): 630–633.
53. S. K. Biswas, The miliolite rocks of Kutch and Kathiawar (western India), *Sedimentary Geology* 5 (1971): 147–164; C. H. B. Sperling and A. S. Goudie, The miliolite of western India: A discussion of the aeolian and marine hypothesis, *Sedimentary Geology* 13 (1975): 71–75; A. R. Marathe, S. N. Rajayuri, and V. S. Lele, On the problem of the origin and age of the Miliolite rocks of the Hiran Valley, Saureshtra, western India, *Sedimentary Geology* 19 (1977): 197–215.
54. C. S. Breed, S. C. Fryberger, S. Andrews, C. McCauley, F. Lennartz, D. Gebel, and K. Horstman, Regional studies of sand seas using LANDSAT (ERTS) imagery, in *A Study of Global Sand Seas*, Ed. E. D. McKee, pp. 305–397 (Reston, VA: U.S. Geological Survey Professional Paper 1052, 1979).
55. H. Th. Verstappen, On the origin of longitudinal (seif) dunes, *Zeitshcrift für Geomorphologie* 12 (1968): 200–220.
56. B. Ghose, Geomorphological aspects of the formation of salt basins in the Lower Luni Basin, in *Papers from Symposium on Problems of Indian Arid Zone*, pp. 79–83 (Cazri, Jodhpur, India: Ministry of Education, Government of India, and UNESCO, 1964).
57. S. Pandey and P. C. Chatterji, Genesis of 'Mitha Ranns,' 'Kharia Rann' and 'Kanodurala Ranns' in the Great Indian Desert, *Annals of Arid Zone* 9 (1970): 175–180.
58. K. W. Glennie and G. Evans, A reconnaissance of the recent sediments of the Ranns of Kutch, India, *Sedimentology* 23 (1976): 625–647.
59. G. Singh, Palaeobotanical features of the Thar Desert, *Annals of the Arid Zone* 8 (1969): 188–195.
60. M. S. Krishnan, Geological history of Rajasthan and its relation to present-day conditions, pp. 19–31, *Proceedings of the Symposium on the Rajputana Desert* (Delhi, India, 1952);. The arguments have been reviewed by V. M. Meher-Homji, Is the Sind-Rajasthan Desert the result of a recent climatic change? *Geoforum* 15 (1973): 47–57, and dismissed by Allchin et al.[48]
61. O. Singh, R. D. Joshi, and A. P. Singh, Stratigraphic and radiocarbon evidence for the age and development of three salt lake deposits in Rajasthan, India, *Quaternary Research* 2 (1972): 496–501.
62. R. J. Wasson, G. I. Smith, and D. P. Agrawal, Late Quaternary sediments, minerals, and inferred geochemical history of Didwana Lake, Thar Desert, India, *Palaeogeography, Palaeoclimatology, Palaeoecology* 46 (1984): 345–372.

63. A. S. Goudie, B. Allchin, and K. T. M. Hegde, The former extensions of the Great Indian Sand Desert, *Geographical Journal* 139 (1973): 243–257.
64. M. A. Williams and M. F. Clarke, Late quaternary environments in north-central India, *Nature* 308 (1984): 633–665.
65. V. Kolla and P. C. Biscaye, Distribution and origin of quartz in the sediments of the Indian Ocean, *Journal of Sedimentary Petrology* 47 (1977): 642–649.
66. J. L. Cullen, Microfossil evidence for changing salinity patterns in the Bay of Bengal over the last 20,000 years, *Palaeogeography, Palaeoclimatology, Palaeoecology* 35 (1981): 315–356.
67. J. E. Kutzbach, Monsoon climate of the early Holocene: Climatic experiment with the Earth's orbital parameters for 9,000 years ago, *Science* 214 (1981): 59–61.
68. J. Hötzl, F. Kramer, and V. Maurin, Quaternary sediments, *Quaternary Period in Saudi Arabia*, Eds. S. Al-Sayari and J. G. Zötl, pp. 264–311 (Wien, Germany: Springer, 1978).
69. L. Lartet, Sur Ia formation du bassin de la mer morte ou lac asphaltite, et stir les changements survenus dans le niveau de ce lac, *Académie des Sciences a Paris, Comptes Rendus* 60 (1865): 796–800.
70. D. Anton, Aspects of geomorphological evolution: Palaeosols and dunes in Saudi Arabia, in *The Quaternary of Saudi Arabia (vol. 2): Sedimentological, Hydrogeological, Hydrochemical, Geomorphological, Geochronological and Climatological Investigations in Western Saudi Arabia*, Eds. A. R. Jado and J. G. Zötl, pp. 275–296 (Berlin, Germany: Springer, 1984).
71. S. Bottema and W. van Zeist, Palynological evidence for the climatic history of the near East, 50,000–6,000 BP, *Colloques Internationaux du CNRS* 598 (1981): 111–132.
72. M. I. Whitney, Eolian features shaped by aerodynamic and vorticity processes, in *Eolian Sediments and Processes*, Eds. M. E. Brookfield and T. S. Ahlbrandt, pp. 223–245 (Amsterdam, the Netherlands: Elsevier, 1983).
73. H. A. McClure, The Rub' al Khali, in *Quaternary Period in Saudi Arabia*, Eds. S. Al-Sayari and J. G. Zötl, pp. 252–263 (Wien, Germany: Springer, 1978); see also Figure 12.15 in R. Cooke, A. Warren, and A. Goudie, *Desert Geomorphology* (London, U.K.: University College of London Press, 1993).
74. F. A. Street-Perrott, N. Roberts, and S. Metcalfe, Geomorphic implications of late Quaternary hydrological and climatic changes in the Northern Hemisphere tropics, in *Environmental Change and Tropical Geomorphology*, Eds. I. Douglas and T. Spencer, pp. 164–183. (London, U.K.: Allen & Unwin, 1985).
75. M. Sarnthein, Sediments and history of the postglacial transgression in the Persian Gulf and the northwest Gulf of Oman, *Marine Geology* 12 (1972): 245–266.
76. J. C. Doornkamp, D. Brunsden, and D. K. C. Jones, Eds., *Geology, Geomorphology and Pedology of Bahrain* (Norwich, U.K.: Geobooks, 1980).
77. N. Roberts, Age, palaeoenvironments, and climatic significance of late Pleistocene Konya Lake, Turkey, *Quaternary Research* 19 (1983): 154–171.
78. W. R. Farrand, Blank on the Pleistocene map, *Geographical Magazine* 51 (1979): 548–554.
79. C. Sung-Chiao, The sandy deserts and the Gobi of China, in *Deserts and Arid Lands*, Ed. F. El-Baz, pp. 95–113 (Den Haag, the Netherlands: Martinus Nijhoff, 1984).
80. E. Huntington, *The Pulse of Asia* (Boston, MA: Houghton Mifflin, 1907).
81. E. Derbyshire, On the morphology, sediments, and origin of the Loess Plateau of central China, in *Megageomorphology*, Eds. R. A. M. Gardner and H. Scoging, pp. 172–194 (Oxford, U.K.: Oxford University Press, 1983).
82. F. Heller and L. T. Sheng, Magneto-stratigraphical dating of loess deposits in China, *Nature* 300 (1982): 431–433.
83. K. Pye, Loess, *Progress in Physical Geography* 8 (1984): 176–217.
84. L. T. Sheng, G. X. Fei, A. Z. Sheng, and F. Y. Xiang, The dust fall in Beijing, China, on April 18, 1980, *Geological Society of America Special Paper* 186 (1982): 149–158.
85. A. S. Goudie, H. M. Rendell, and P. A. Bull, The loess of Tajik SSR, in *The International Karakoram Project* (2 vols.), Ed. K. J. Miller, pp. 399–412 (Cambridge, U.K.: Cambridge University Press, 1984).
86. A. A. Lazarenko, The loess of Central Asia, in *Late Quaternary Environments of the Soviet Union*, Ed. A. A. Velichko, pp. 125–131 (Harlow, U.K.: Longman, 1984).

87. R. S. Davis, V. A. Ranov, and A. E. Dodonov, Early man in Soviet Central Asia, *Scientific American* 243 (1980): 92–102; A. A. Lazarenko, N. S. Bolikhovskaya, and V. V. Semenov, An attempt at a detailed stratigraphic subdivision of the loess association of the Tashkent region, *International Geology Review* 23 (1981): 1335–1346.
88. A. L. Chepalyga, Inland sea basins, in *Late Quaternary Environments of the Soviet Union*, Ed. A. A. Velichko, pp. 229–247 (Harlow, U.K.: Longman, 1984).
89. M. C. Benbow, Tertiary coastal dunes of the Eucla Basin, Australia, *Geomorphology* 3 (1990): 9–29; The importance of such relict features has been succinctly summarized thus by T. Langford-Smith, The geomorphic history of the Australian deserts, *Striae* 17 (1982): 4–19.
90. J. A. Mabbutt, Landforms of arid Australia, in *Arid Lands of Australia*, Eds. R. O. Slayter and R. A. Perry et al. (Canberra, Australia: ANU Press, 1969); J. A. Mabbutt, Landforms of Australian deserts, in *Deserts and Arid Lands*, Ed. F. El-Baz, pp. 79–94 (Den Haag, the Netherlands: Martinus Nijhoff, 1984).
91. W. G. Woolnough, Presidential address, part I: The chemical criteria of peneplanation; part II: The duricrust of Australia, *Royal Society New South Wales, Journal and Proceedings* 61 (1927): 1–53.
92. T. Langford-Smith, A select review of silcrete research in Australia, in *Silcrete in Australia*, Ed. T. Langford-Smith, pp. 1–12 (Armidale, New South Wales, Australia: University of New South Wales Press, 1978).
93. W. J. E. Van der Graaf, R. W. A. Crowe, J. A. Bunting, and M. J. Jackson, Relict early Cenozoic drainages in arid Western Australia, *Zeitshcrift für Geomorphologie* 21 (1977): 379–400.
94. S. A. Schumm, River adjustment to altered hydrologic regimen: Murrumbidgee River and paleochannels, Australia (Reston, VA: U.S. Geological Survey Professional Paper 598, 1968); see Ref. 105, Chapter 11.8.
95. J. M. Bowler and R. J. Wasson, Glacial age environments of inland Australia, in *Late Cainozoic Palaeoenvironments of the Southern Hemisphere*, Ed. J. C. Vogel, pp. 183–208 (Rotterdam, the Netherlands: A. A. Balkema, 1984).
96. R. J. Wasson, Late Quaternary palaeo-environments in the desert dunefields of Australia, in *Late Cainozoic Palaeoenvironments of the Southern Hemisphere*, Ed. J. C. Vogel, pp. 419–432 (Rotterdam, the Netherlands: A. A. Balkema, 1984).
97. J. N. Jennings, Desert dunes and estuarine fill in the Fitzroy Estuary (NW Australia), *Catena* 2 (1975): 215–262.
98. J. M. Bowler, G. S. Hope, J. N. Jennings, G. Singh, and D. Walker, Late quaternary climates of Australia and New Guinea, *Quaternary Research* 6 (1976): 359–394.
99. M. A. J. Williams, Pleistocene aridity in tropical Africa, Australia and Asia, in *Environmental Change and Tropical Geomorphology*, Eds. I. Douglas and T. Spencer, pp. 219–233 (London, U.K.: Allen & Unwin, 1985).
100. R. J. Wasson, Holocene aeolian landforms of the Belarabon area SW of Cobar, NSW, *Royal Society of New South Wales, Journal and Proceedings* 109 (1976): 91–101.
101. J. H. Mercer, Glacial history of southernmost South America, *Quaternary Research* 6 (1976): 125–166.
102. J. A. Naranjo and P. Cornejo, Avalanchas multiples del Volcan Chaco en el norte de Chile: Un mecanismo de degradacion de volcanes miocenos, *Revista Geologica Chile* 16 (1989): 61–72.
103. J. A. Naranjo and R. Paskoff, Evolution Cenozoica del piedemonte Andino en Ia Pampa del Tamarugal, Norte de Chile (18°–21°S), *IV Congresso Geologico Chileno* 5 (1985): 149–165.
104. K. Segerstrom, Matureland of northern Chile and its relationship to ore deposits, *Geological Society of America Bulletin* 74 (1963): 513–518.
105. R. Cooke, A. Warren, and A. Goudie, *Desert Geomorphology*, Chapter 5.2 (London, U.K.: University College of London Press, 1993).
106. A. H. Clark, R. U. Cooke, C. Mortimer, and R. H. Sillitoe, Relationships between supergene mineral alteration and geomorphology, southern Atacama Desert, Chile: An interim report, *Institute of Mining and Metallurgy, Transactions* B76 (1967): 89–96. This view is confirmed by Naranjo and Paskoff.[103]

107. C. N. Alpers and G. H. Brimhall, Middle Miocene climatic change in the Atacama Desert, northern Chile: Evidence from supergene mineralisation at La Escondida, *Geological Society of America Bulletin* 100 (1988): 1640–1656.
108. S. E. Hollingworth and J. E. Guest, Pleistocene glaciation in the Atacama Desert, northern Chile, *Journal of Glaciology* 6 (1967): 749–751.
109. C. N. Caviedies and R. Paskoff, Quaternary glaciations in the Andes of north-central Chile, *Journal of Glaciology* 14 (1975): 155–170; W. Weischet, Zur Geomorphologie des Glatthang Reliefs in der ariden Subtropenzone des Kleinen Nordens von Chile, *Zeitshcrift für Geomorphologie* NF13 (1969): 1–21.
110. C. M. Clapperton, The glaciation of the Andes, *Quaternary Science Reviews* 2 (1983): 83–155; See Ref. 101.
111. S. L. Hastenrath and S. Kutzbach, Late Pleistocene climate and water budget of the South American Altiplano, *Quaternary Research* 24 (1985): 249–256.
112. D. Wirrman and L. F. de O. Almeida, Low Holocene level (7,700 to 3,650 years ago) of Lake Titicaca (Bolivia), *Palaeogeography, Palaeoclimatology, Palaeoecology* 59 (1987): 315–323.
113. R. Paskoff, Sobre la evolucion geomorfologia del gran acantilado costero del Norte Grande de Chile, *Norte Grande* (Inst. Geogr. Univ. Catolica de Chile) 6 (1978–1979): 1–12.
114. I. C. Russell, Sketch of the geological history of Lake Lahontan, in *Third Annual Report of the U.S. Geological Survey*, pp. 189–235 (Washington, DC: U.S. Geological Survey, 1883); G. K. Gilbert, *Lake Bonneville*, Monograph 1 (Washington, DC: U.S. Geological Survey, 1890).
115. G. I. Smith and F. A. Street-Perrott, Pluvial lakes of the western United States, in *Late-Quaternary Environments of the United States, Volume 1: The Late Pleistocene*, Ed. S. C. Porter, pp. 190–212 (Harlow, England: Longman, 1983).
116. G. R. Brakenridge, Evidence for a cold, dry full-glacial climate in the American southwest, *Quaternary Research* 9 (1978): 22–40.
117. W. A. Price, Greater American deserts, *Texas Academy of Sciences, Proceedings and Transactions* 27 (1944): 163–170.
118. G. L. Wells, Late-glacial circulation over central North America revealed by eolian features, in *Variations in the Global Water Budget*, Eds. A. Street-Perrot, M. Beran, and R. Ratcliffe, pp. 317–330 (Dordrecht, the Netherlands: Reidel, 1983).
119. A. Warren, Morphology and sediments of the Nebraska Sand Hills in relation to Pleistocene winds and the development of eolian bedforms, *Journal of Geology* 84 (1976): 685–700.
120. T. S. Ahlbrandt, J. B. Swinehart, and D. G. Maroney, The dynamic Holocene dune fields of the Great Plains and Rocky Mountain Basins, in *Eolian Sediments and Processes*, Eds. M. E. Brookfield and T. S. Ahlbrandt, pp. 379–406 (Amsterdam, the Netherlands: Elsevier, 1983).
121. V. T. Holliday, The Blackwater Draw Formation (Quaternary): A 1.5 ± Ma record of eolian sedimentation and soil formation on the southern High Plains, *Geological Society of America Bulletin* 101 (1989): 1598–1607.
122. D. Winstanley, Rainfall patterns and general atmospheric circulation, *Nature* 245 (1973): 190–194.
123. S. E. Nicholson and H. Flohn, African climate changes in late Pleistocene and Holocene and the general atmospheric circulation, *IASH Publication* 131 (1981): 295–301.
124. J. Imbrie and K. P. Imbrie, *Ice Ages: Solving the Mystery* (London, U.K.: Methuen, 1979).
125. Lu, Y., Pleistocene climatic cycles and variations of $CaCO_3$ contents in a loess profiles, *Chinese Journal of Geology*, 2 (1981): 122–131.
126. M. J. Selby, *Earth's Changing Surface: An Introduction to Geomorphology* (Oxford, U.K.: Oxford University Press, 1985).
127. A. S. Goudie, *Great Warm Deserts of the World. Landscapes and Evolution* (Oxford, U.K.: Oxford University Press, 2002); A. S. Goudie, Global deserts and their geomorphological diversity, in *Geomorphology of Desert Environments*, 2nd edn., Eds. A. J. Parsons and A. D. Abrahams, pp. 9–20 (Berlin, Germany: Springer, 2009).
128. J. Sun and D. R. Muhs, Mid-latitudes, in *Encyclopedia of Quaternary Sciences*," Ed. S. Elias, p. 609 (New York: Elsevier, 2007).

129. P. Meigs, The world distribution of arid and semiarid homoclimates, in *Reviews of Research on Arid Zone Hydrology*, pp. 203–209 (Paris, France: UNESCO, 1953).
130. J. Quade, J. A. Rech, J. L. Betancourt, C. Latorre, B. Quade, K. A. Rylander, and T. Fisher, Paleowetlands and regional climate change in the central Atacama Desert, northern Chile, *Quaternary Research* 69 (2008): 343–360.

# 2

## *Geology and Soils in Deserts of the Southwestern United States*

**William L. Stefanov and Douglas Green**

**CONTENTS**



## 2.1 Introduction

Deserts are present on every continent and represent the dominant biome in the southwestern portion of the United States. Four major deserts are present within the continental United States—from north to south, they are the Great Basin (Idaho, Nevada, Oregon, and Utah); Mojave (Arizona, California, and Nevada); Sonoran (Arizona and California); and Chihuahuan (Arizona, New Mexico, and Texas) Deserts.[1] This chapter presents a general introduction to the major geological processes, landforms, and soils typically found in desert environments of the southwestern United States. Our intent is to provide an overview of those basic geological and soil processes that architects and developers should consider when designing with the desert rather than an exhaustive technical work. This overview draws from a number of references and should therefore be considered as only the "tip of the iceberg" or perhaps "the crest of the dune" for these subjects. These references—which include both introductory texts and more specialized works—are called out at appropriate points in the text to guide the interested reader to sources of further information on desert geology, geomorphology, and soils.

## 2.2 Characteristics and Distribution of Deserts

Deserts typically have an annual water budget deficit; in other words, more water is withdrawn for use by plants and animals (including humans) in a given year than is replaced by natural recharge mechanisms (like precipitation) (Chapters 3 and 7). Annual precipitation in arid regions is <25.4 cm/year; the Phoenix, Arizona, metropolitan area averages about 18 cm/year.

Desert regions typically have lower vegetation cover than landscapes in more temperate climates (Figure 2.1) (see Chapter 7). As deserts are defined primarily on the basis of low precipitation and not on average temperatures, there are warm (e.g., the Sonoran and Chihuahuan Deserts), cool (e.g., the Mojave Desert), and cold (e.g., Antarctica) deserts. We will focus on the geological and soil characteristics of the warm deserts of central and southern Arizona as these regions have experienced significant development pressures over the past few decades.

Most of the world's deserts are located around latitude 30°N and 30°S, essentially bracketing the equatorial regions.[2] Atmospheric circulation cells cause cool, water-laden air from the equator to condense out water as they rise and move northward or southward. These now dry and warm air packets descend into the desert regions, causing even more evaporation of surface water and contributing to the dry character of the landscape (see Chapter 3).

**FIGURE 2.1**
The Sonoran Desert in the Eagle Tail Mountains west of Phoenix, Arizona. Note the characteristic desert features of exposed bedrock mountains on the skyline surrounded by alluvial fans of sediment derived from the hillslopes in the foreground; dry washes developed on the fan surfaces; and sparse vegetation cover. (Courtesy of W.L. Stefanov.)

In addition, mountain ranges (e.g., the Sierra Nevada of California) cause clouds to cool and lose water as rain as they head inland from the oceans and climb over peaks. The resulting "rain shadow" contributes to the persistence of the deserts in Arizona and Nevada by decreasing the water available for precipitation on the leeward side of the ranges (see Chapter 3). Deserts can also occur near coastlines due to cool ocean air heating up as it travels inland, leading to increased evaporation. A more detailed discussion of the character and geographic distribution of deserts appears elsewhere in this book (see Chapter 1).

## 2.3 Geological Processes in Deserts

### 2.3.1 Fundamental Geological Concepts

Lithified geological materials at the Earth's surface are products of the *rock cycle*.[3] *Igneous* rocks are crystallized directly from molten magma, either quickly as the products of eruptions at the surface or slowly following intrusion into preexisting rock deep below the surface. Uplifted and exposed igneous rock is constantly being reduced to smaller particles, or eroded, by continual mineralogical and chemical changes (Figure 2.1). *Sedimentary* rocks are formed of lithified layers of these eroded materials after deposition and accumulation in basins by wind and/or water. Sedimentary rocks can also form, under the right conditions, directly from chemical precipitation of elements like calcium and sodium. Subjecting igneous and sedimentary rocks to elevated temperature and pressure creates *metamorphic* rocks with different minerals and structure than the original rocks. While the term "rock cycle" implies a circular progression through the different rock types, it is more like a network of potential pathways, for example, metamorphic rocks can be eroded to form sedimentary rocks, sedimentary rocks can be melted and recrystallized to form igneous rocks, and metamorphic rocks can undergo additional heat and pressure to form new metamorphic rocks.

The theory of plate tectonics explains the large-scale geological processes that actively shape the Earth's surface today and provides the larger conceptual framework within which the rock cycle operates. Rather than having a continuous outer shell of rigid rock material, the surface of the Earth is formed of numerous interlocking plates that comprise the continental and oceanic outer crust of the planet.[4] These rigid plates are supported by viscous mantle material, which enables them to interact with each other in a variety of ways in response to movement of the underlying mantle.

The three major types of plate boundaries—divergent, convergent, and transform—are recognized. Divergent boundaries form where new magma erupts along rifts in the older crust (e.g., the Mid-Atlantic Ridge) where new oceanic seafloor crust is being formed and the Atlantic Ocean is widening. Convergent boundaries occur when crustal plates collide; if both plates are formed of continental crust, the collision can create high mountain ranges. A good example of a convergent boundary is the Himalayas, which are formed by the ongoing collision of the Indian and Eurasian plates. If an oceanic plate collides with a continental plate, the denser oceanic crust is driven downward into the mantle to eventually remelt and form new magma. This process is called *subduction* and typically forms a deep trench along the boundary together with volcanoes on the continental plate above the descending oceanic crust. The Andes in South America are created by volcanoes that erupted above the descending Nazca oceanic plate. Finally, transform boundaries occur when two plates move tangentially against each other, with neither plate overriding the other. The classic example of this type of plate boundary is the San Andreas Fault system in California, caused by northwest motion of the Pacific plate against the North American plate. Frequent and sometimes strong earthquake activity is associated with both convergent and transform boundaries.

While the oldest exposed bedrock in the southwestern United States formed billions of years ago, the desert landscapes we live in today are mainly the result of climatic changes that only took place between 10,000 and 8,000 years ago (see Chapter 6). The fundamental concept here is that of "deep time" or the realization that the landscape contains an extensive geological history of past climates and tectonic events recorded in the rocks that may be quite different from the current environmental conditions. In desert environments, water and wind are the major natural agents that move unconsolidated materials across the landscape, with episodic events like flash floods capable of altering the local landscape significantly over time periods of hours to days. In addition to these natural agents, humans have become major agents of landscape change in desert regions by converting arid lands to agricultural and urban/suburban land uses.

The basic concepts presented here will be expanded upon in the following sections, including some discussion of the role humans play in geological processes. An overview of the major geological events that formed the state of Arizona, which includes sections of the Sonoran, Chihuahuan, and Mojave Deserts, provides a sense of the region's geological history, in which the current desert landscapes are the most recent chapter. Other desert regions of the United States have similarly complex geologic histories.

## 2.4 Physiography and General Geological History of Arizona

Arizona can be broadly divided up into three physiographic provinces that have distinctive geological and geographic characters. These provinces are the Colorado Plateau Province, the Transition Zone, and the Basin and Range Province (Figure 2.2).[5] The Colorado Plateau Province forms the region at the adjoining corners of Arizona, Colorado, New Mexico, and Utah and is characterized by high plateaus, deeply incised canyons, and isolated buttes. In northern Arizona, the exposed rocks of the Plateau are mostly horizontal Paleozoic and Mesozoic sedimentary strata between 550 and 66 millions of years old that have not been greatly folded, faulted, or metamorphosed. Subsequent volcanic activity on the Plateau produced a large stratovolcano in Arizona—the San Francisco Peaks—and numerous basaltic lava flows, with the most recent eruptive activity taking place less than 1000 years ago.

The Transition Zone, also known as the Central Highlands Province, defines a west–northwest trending band across central Arizona bounded by the Colorado Plateau to the north and the Basin and Range Province to the south. It is comprised mainly of uplifted and eroded Proterozoic igneous and metamorphic bedrock ranges between approximately 1.8 and 1.4 billion years old and erosional remnants of younger Colorado Plateau rocks (Figure 2.2).* Unlike the "layer-cake" Colorado Plateau strata, rocks in the Transition Zone are more extensively faulted and form rugged mountain masses.

The Basin and Range Province is located primarily in southern and western Arizona, and the Phoenix and Tucson metropolitan areas are located within this province. Proterozoic and younger igneous, metamorphic, and sedimentary rocks are exposed in dominantly

* For detailed discussions of Proterozoic geology in Arizona, the interested reader is referred to Karlstrom.[6]

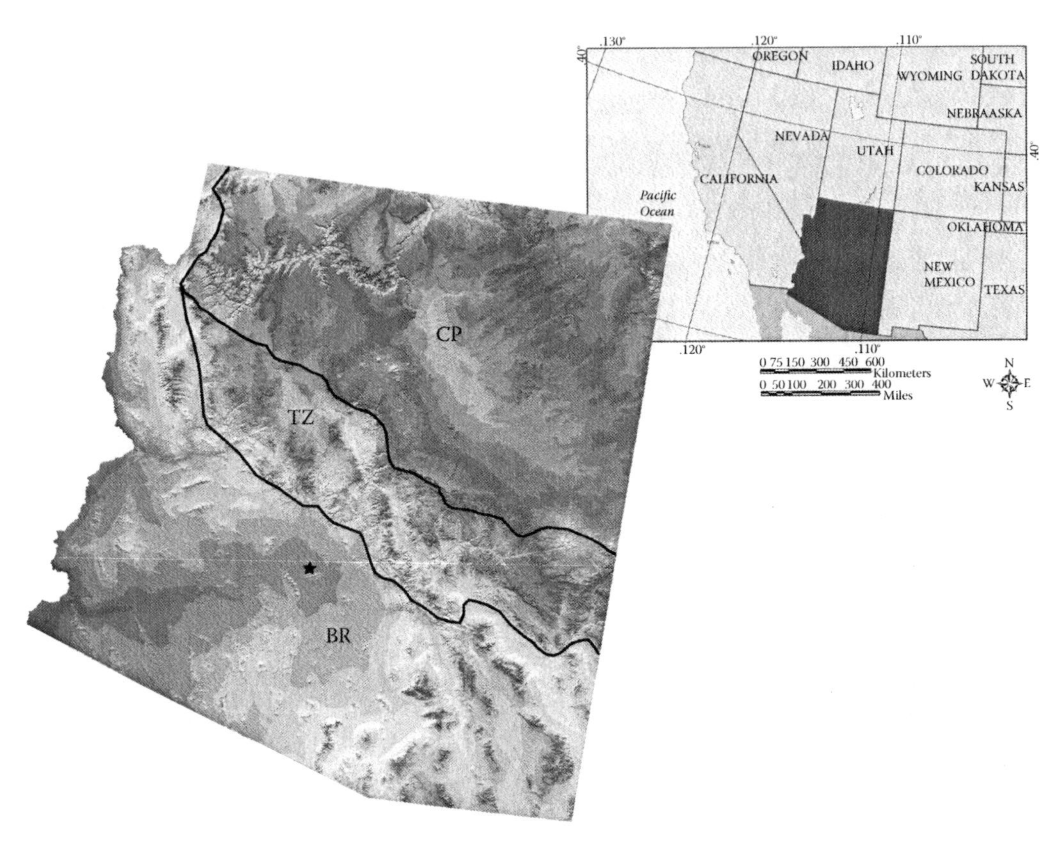


**FIGURE 2.2**
Arizona can be divided into three physiographic provinces, each with their own distinctive geological character. Approximate province boundaries are indicated by black lines: CP, Colorado Plateau Province; TZ, Transition Zone Province; BR, Basin and Range Province. The black star indicates the location of the state capitol of Phoenix. Shaded relief base map constructed from the U.S. Geological Survey National Elevation Dataset (1 arc second). Locator map of the southwestern United States indicates the location of Arizona (gray).

northwest–southeast trending elongated mountain ranges alternating with sediment-filled valleys. This area formed primarily due to episodes of extension of the Earth's crust in what is now Arizona from about 15 to 5 million years ago, which resulted in widespread faulting that caused uplift of mountain ranges and downdropping of adjacent valleys by fracturing and tilting of large blocks of crust.*

Formation of the oldest bedrock in Arizona took place approximately 1.8–1.7 billion years ago during a series of mountain-building events related to subduction along the southern boundary of the ancient core of the North American continent. The oldest of these crystalline intrusive and metamorphic basement rocks are visible in the bottom of the Grand Canyon. During the remainder of the Proterozoic Eon (>550 million years ago), the ancient basement was exposed, eroded, and eventually covered by layers of sedimentary and volcanic rocks between approximately 1.7 billion years and 550 million years ago. During that interval, the basement rocks and overlaying sedimentary rocks were again exposed, tilted, and eroded before being covered by the oldest of the horizontal Paleozoic sedimentary layers.

* For more detail on the geological history of Arizona, see Jenney and Reynolds.[7]

Arizona was located along the southwestern edge of the North American continental core between divergent plate boundaries by the beginning of the Paleozoic Era (550–248 million years ago), but by the end of the Paleozoic, these had become convergent boundaries during the assembly of the supercontinent of Pangaea. Sedimentary rocks deposited in Arizona during the Paleozoic record repeating cycles of inundation by shallow seas alternating with periods of land surface exposure; these cycles can be clearly recognized in the undeformed and nonmetamorphosed rocks of the Colorado Plateau.*

During the Mesozoic Era (248–65 million years ago), volcanism associated with a subduction zone to the west produced thick sequences of igneous rocks, both eruptive and intrusive, throughout much of the southern third of Arizona. Deposition of sediments derived from the volcanic rocks to the south, as well as sediments derived from source regions to the north, occurred in other areas of Arizona and indicated a complex variety of wind- and water-dominated depositional environments during the middle Mesozoic. The end of the Mesozoic and beginning of the Cenozoic Era (65 million years ago to the present) is marked by a significant tectonic event known as the Laramide Orogeny. This event is generally thought to have been caused by atypical shallow subduction of oceanic crust beneath the North American continent. The shallow subduction caused widespread volcanism and intrusion of igneous rock throughout much of the southern half of Arizona, and some areas of northern Arizona on the Colorado Plateau, between 80 and 55 million years ago. The crust was thickened and uplifted during the Laramide event such that Paleozoic and Mesozoic rocks deposited in central and western Arizona were subsequently removed by erosion and/or metamorphosed. The eroded material was transported to the northeast, as southern and western Arizona was a highland during Laramide time.

Another pulse of volcanism occurred between 37 and 15 million years ago, beginning in the southeast of Arizona and migrating westward through the southern half of the state. This volcanic activity is thought to be the result of renewed magma production as the subducted slab of oceanic crust beneath western North America increased its angle of descent. The compression of continental crust caused by the shallow slab subduction was now relaxed, and the crust began to extend and thin. The Colorado Plateau remained relatively unaffected by either the volcanism or crustal extension and became a highland during this time period reversing previous drainage patterns.

Basin and Range faulting in southern and western Arizona occurred 15–5 million years ago and may have been related to establishment of the San Andreas transform boundary along the continental margin, pulling apart of the crust by upwelling mantle material, or a combination of both mechanisms. Some volcanic fields were active at this time in southern and western Arizona. Three large and spatially separate volcanic fields located along the Transition Zone—Colorado Plateau boundary were also active during Basin and Range faulting; activity in the San Francisco field near Flagstaff has continued into historical times. Arizona is relatively quiescent today in terms of plate tectonics because it is not located on a convergent or divergent plate boundary. The extreme southwestern part of the state near Yuma, however, is at high risk from earthquakes associated with the nearby San Andreas transform boundary. The Colorado Plateau region north of Flagstaff has also experienced frequent moderate earthquakes and thus also has higher earthquake risk than other parts of the state.

River drainages across the state began to integrate after opening of the Gulf of California along the San Andreas transform boundary 6–5 million years ago. This is also when the

* For in-depth discussions of the geology of the Grand Canyon and Colorado Plateau, see Beus and Morales.[8]

Colorado River established its current course and began to build its delta in the Gulf of California with sediments derived from the Colorado Plateau. The Grand Canyon we see today is the result of downcutting and sediment removal by the Colorado River over the past several million years. The current arid climate became dominant in most areas of the southwestern United States between 10,000 and 8,000 years ago (see Chapter 6), after which time surficial geological processes typical of desert regions became important to further evolution of the landscape.

## 2.5 Surficial Weathering Processes

*Mechanical* and *chemical weathering* processes reduce masses of exposed rock to smaller and smaller particles, making sediments available for transport by wind and/or water (Figure 2.3). While the products of mechanical weathering are most evident in deserts, in the form of fragments that retain the mineralogy and chemistry of the source rocks, chemical weathering processes typically facilitate the physical reduction of rock masses by creating or exacerbating existing zones of weakness.* Common mechanical weathering processes active in deserts include heating and cooling over each day/night cycle, which leads to continual expansion and contraction of rock masses. This can weaken the rock mass over time by formation of microcracks and fractures. Fractured rock has greater permeability, providing infiltration pathways for water and salts. Salt weathering is caused by the formation and expansion of salt minerals in cracks, which can cause rock to flake apart or disintegrate over time. In deserts where freezing and thawing occurs, water in cracks

**FIGURE 2.3**
Mechanical weathering and mass wasting are the dominant processes that are gradually eroding the spires and mesas of Monument Valley in Utah and Arizona. Both the resistant cliff face and the less-resistant slope-forming rocks are undergoing mechanical erosion. The smaller rock fragments on the lower slope are derived both from the thinly layered siltstones forming the gently sloping base of the butte and the more massive, cliff-forming sandstone above. Mass wasting processes are moving the smaller rock fragments downslope onto the adjacent valley floor. (Courtesy of W.L. Stefanov.)

* For a more thorough presentation, see Cooke et al.[2]

can also freeze, expand, and fracture rock. Cycles of wetting and drying can also disaggregate rocks that are rich in clay minerals. Water can be taken into the structure of certain clay minerals, causing them to expand and exert stress on the surrounding rock fabric.

Another mechanism that may be involved in the spalling and reduction of large rock masses to smaller fragments is that of pressure release. Rocks formed at depth, for example, a mass of crystallized igneous rock such as granite are subject to confining pressures from the surrounding rock, but when the granite is uplifted and exposed at the surface as a hill or cliff face, the confining pressure is removed. The rock mass can now expand along preexisting fractures, leading to the formation of large sheets or slabs of easily detached material in a process known as *exfoliation*. Exfoliation can also occur at the scale of individual boulders, but this is thought to be due to thermal expansion/contraction and salt weathering processes.

*Chemical weathering*, or the chemical alteration of primary rock-forming minerals into new minerals and chemical substances, is considered less active in deserts due to the generally low availability of water but is still important in the breakdown of rock into sediments and soil following precipitation and dew formation.* Water passing over or through rock acts as a solvent, extracting various elements (Ca, Na, Mg, K, and to a lesser degree Si, Al, and Fe) from surfaces it contacts and transporting those elements away, resulting in changes to mineral structures and decreased rock strength. Hydration is a process where water enters certain mineral structures leading to volume changes as discussed in the previous section; water can also leave the mineral structure resulting in further volume changes. Hydrolysis includes numerous chemical reactions between water and rock-forming minerals, leading to the formation of secondary mineral species such as clays with weaker crystal structures. Water can also combine with carbon dioxide in soils (or the atmosphere) to form carbonic acid, which is an effective agent for dissolving calcium from carbonate rocks like limestone or dolomite under humid conditions. In semiarid to arid climates, this process is less effective, and limestone tends to be a resistant and cliff-forming rock type. Oxidation involves the exchange of electrons between oxygen and metallic elements in rock-forming minerals. This usually results in the formation of an oxide mineral; a common example is the mineral hematite (iron oxide), which can impart a reddish coloration to rocks, sediments, and soils.

Biological activity can also contribute to weathering of rock and formation of sediments and soil. Lichens and algae are common in deserts and can exist both on and beneath the rock surface. These organisms can create acids capable of etching minerals and can cause flaking of rock by expansion and contraction of tissues attached to the surface, or in pore spaces, or lodged in minute cracks. Large desert plants, such as saguaro cactus (*Carnegiea gigantea*), can disturb and mix the local surface sediments if they topple. Burrowing animals such as ants, lizards, and mice physically churn and mix the soil surface and enhance erosion, a process known as *bioturbation*. Human recreational activities such as off-road vehicle travel are another example of bioturbation.

Mechanical and chemical weathering, coupled with sediment-transport processes, has lowered the mountain ranges of the Basin and Range Province and is still transforming them into the thick sediments present in the adjacent valleys. The valleys that cradle the Phoenix and Tucson metropolitan regions are filled with sediments derived from the adjacent mountain ranges. The following section presents the major mechanisms involved in moving material from mountain slopes into adjacent valleys and river systems.

---

* A more detailed description of chemical reactions important to weathering can be found in Chapter 8 of Selby.[9]

## 2.6 Sediment-Transport Processes

*Mass wasting* is the general term used by geologists to describe the movement of material from bedrock hillslopes to valleys under the influence of gravity and facilitated by wind (*eolian*) and water (*fluvial*) processes. Gradual movement of material from highlands to lowlands in deserts is influenced by slope, the presence of surface water and groundwater, vegetation cover, disturbance by animals, and the characteristics of the sediments themselves. Large volumes of earth materials also can be moved catastrophically via landslides, debris flows, and rockfalls (Chapter 5). These events can be somewhat common in desert regions especially following high precipitation events, earthquakes, removal of vegetation cover, and/or failure along planes of weakness in the rock mass. The danger of catastrophic mass-wasting events can be mitigated using a variety of engineering controls such as subsurface water drainage, rock nets, rock bolts, and grouting.

Other sediment-transport processes are continually active in deserts at smaller scales than the processes mentioned earlier. Episodic rainfall in deserts dislodges sediments on hillslopes by rainsplash, essentially the raindrop impacts the soil surface and frees small particles. The sediment is then free to be transported by water flowing over the soil surface as overland flow or sheetwash. As more water flows over the soil surface, small channel networks (rills) began to form, which concentrates and speeds up the flow, allowing larger sedimentary particles to be transported.

As the rill networks combine and grow larger, they begin to form washes, which then deliver sediment from the hillslope into the mountain front region or *piedmont* described in the following section. At each stage, the ability of water to carry larger and larger sediment loads and particles increases. Sediments derived from bedrock are transported off the hillslopes and into valleys by these fluvial processes (Figure 2.4). In deserts, this form of sediment transport tends to occur in short amounts of time whenever high precipitation events occur, and as a result, most washes in deserts are dry most of the time yet represent the major transportation route of sediments (see Chapter 4). During large floods, sediments in the valleys also can be remobilized and deposited again further downstream from their original source areas.

(a)

(b)

**FIGURE 2.4**
A typical desert hillslope in the McDowell Mountains located in Scottsdale, Arizona, has relatively sparse vegetation cover and gravelly surficial soils (a). Infrequent but powerful precipitation events lead to the formation of rills and washes on desert hillslopes (light-colored linear features in (b), also in the McDowell Mountains); these are the major sediment transport pathways from hillslopes to the piedmont. (Courtesy of W.L. Stefanov.)

An additional force in movement of sediments is the wind. Dust storms can transport millions of tons of fine sediment over thousands of kilometers; Saharan dust can cross the Atlantic Ocean to deposit in North America. Wind can abrade, transport, and deposit small sedimentary particles through the process of saltation, and this process is thought to be the most important reason for erosion. In this process, particles are transported by the wind in short parabolic arcs that result in impacts with other particles on the ground. These impacts provide enough energy to dislodge other particles on the soil surface and make them available for transport by the wind. Saltation of particles can continue as long as the wind is blowing and there are sediments available for transport. Constant winds bearing small particles can also sculpt individual rocks, essentially sandblasting them into fluted and polished forms called *ventifacts*. Small oriented grooves in ventifacts left by the wind-borne particles can often be used to map past wind patterns in desert environments. Sedimentary particles transported and deposited by eolian processes are also thought to be a major source of material for new soil development in deserts by providing a constant source of new sediments to the soil surface.[10]

## 2.7 Desert Landforms

As with desert geological processes, landforms found in deserts are a subject of much previous and continuing study. The major landforms found in desert piedmonts are presented here; these have been further classified into many distinct subtypes, which are not discussed here.* The mountain front region, or *piedmont*, is defined as the interface between the hillslope and fluvial (river) environments (Figure 2.5). Landforms in the piedmont reflect the balance between the availability of source material from hillslopes versus the capacity of eolian and fluvial processes to transport the material away. If more

**FIGURE 2.5**
This piedmont region along the northeast side of the McDowell Mountains, Scottsdale, Arizona, is the interface between hillslope and fluvial (river channel) transport processes. (Courtesy of W.L. Stefanov.)

* For a more thorough presentation, see Cooke et al.[2]

material is available than can be readily evacuated, depositional features like alluvial fans form at the base of the mountain. Bedrock surfaces can be exhumed, and channels incised, into the piedmont if more material is transported away than is delivered from the upslope source regions. This balance between degree and pattern of erosion of the hillslopes, and amount and form of sedimentation in the piedmont, is controlled by several factors including tectonic uplift, as illustrated by the formation of Basin and Range mountains; change in base level of adjacent river systems, as shown by the integration of river systems following the opening of the Gulf of California; lithology and structure of the eroding mountains; morphology of the mountain front, particularly the presence of embayments; and changes in climate, such as from cool and wet to warm and dry conditions.

Major landforms in the piedmont are *alluvial fans* and *pediments*. Alluvial fans—so-called because of their characteristic fan shape in plan view—are the depositional sites for sediments carried in mountain streams and washes and frequently originate from a specific point along the mountain front. In general, coarser sediments are deposited near the mountain base, with finer sediments deposited further out on the fan surface. Excellent examples of alluvial fans can be seen in Death Valley National Park of California's Mojave Desert (Figure 2.6). Several alluvial fans can coalesce along the mountain front to form a continuous zone of sediments called a *bajada*.

*Desert pavements* are flat surfaces formed of interlocked gravels, typically found on alluvial fans, which can be tens to hundreds of thousands of years old. They are thought to be formed by eolian deposition on the surface and a combination of vertical motion of sedimentary particles and removal of small particles by wind and water.[11–13] This landform is important as it forms a protective armor for the underlying sediments and soils and reduces erosion. Frequently the gravel forming desert pavements is a deep brown to black color regardless of the actual colors of the rocks comprising the gravel (Figure 2.7). A coating of *desert varnish* on the gravel causes this dark coloration. Desert varnish is composed of iron- and manganese-bearing clay minerals and is thought to form by the deposition of iron-bearing clay minerals in dust onto individual gravel surfaces. Microbial activity also builds varnish coatings by fixing manganese into clay mineral structures. Desert varnish is not limited to desert pavements—it can develop on any

**FIGURE 2.6**
Alluvial fans are found at the outlets of large washes on mountain ranges, such as this example from Death Valley, California. (Courtesy of W.L. Stefanov.)

**FIGURE 2.7**
Well-developed desert pavements typically also have well-developed desert varnish coatings on rock fragments. This pavement is located in the White Tank Mountains west of Phoenix, Arizona. The scale is 6 in. long. (Courtesy of W.L. Stefanov.)

exposed rock surface if appropriate environmental conditions and iron and manganese-bearing dust are available.

Pediments are characterized by an erosional surface that slopes away from a mountain front, usually but not always covered with a thin veneer of sediments or soil. The pediment surface is typically cut into the same bedrock as the adjacent mountain but can also truncate both bedrock and alluvium. Recent work based on modeling of sediment production and transport in the piedmont junction suggest that pediment form is governed by feedback mechanisms that balance the amount of bedrock erosion with sediment deposition and transport.[14] While pediments are typically found in semiarid to arid environments, it is probable that they originally formed under different climatic and tectonic regimes than exist today; for example, pediment surfaces in granitic rock in the Mojave Desert have been interpreted as representing a "weathering front" or exhumed boundary between fresh bedrock and deeply weathered material formed during wetter climatic conditions.[15]

*Dunes* are another landform associated with deserts, caused by the transport and accumulation of predominantly sand-sized particles due to wind (Figure 2.8). Individual dunes range in size from tens of centimeters to hundreds of meters high and can be from 1 m to 1 km in width. There are many different forms of dunes, but they all are adjusted to the prevailing wind patterns.* The majority of dunes are located in sand seas or *ergs*; these are essentially continuous regions of windblown sand that are tens of thousands of square kilometers in area. Ergs are located in Africa, Arabia, Australia, and Eurasia. Dunefields contain at least 10 sand dunes but are smaller in size than ergs. Some active dune fields in the southwestern United States include the Algodones Dune Field in southeastern California; the White Sands National Monument, New Mexico; and the Kelso Dunes in the Mojave National Preserve of eastern California. Common sand sources in deserts include eroding sandstone and granitic bedrock, river sediments, and sediments from dry lake

* For a more thorough presentation, see Cooke et al.[2]

**FIGURE 2.8**
Dunes and dune fields, such as the Kelso Dunes of California, pictured here, can develop wherever there is constant wind and a source of sediments movable by the wind. Vegetation is in the process of stabilizing this dune, but sand ripples in the foreground indicate that motion of particles by saltation is still occurring. (Courtesy of W.L. Stefanov.)

beds (playas). If enough sediment is available and steady winds are present, mobile dunes can migrate (in some cases quite rapidly) over the landscape. Dunes can lose mobility or become "anchored" if vegetation or soil forms following climate change, the dune becomes cemented by carbonate or clay-rich sediments, or if sediment is no longer available to feed continued dune formation.

## 2.8 Desert Soils

The unconsolidated products of weathering can be considered simply as sediment on its way to eventual consolidation as rock, but this would be a narrow view that diminishes the role of biological processes in the formation of a material that serves as the interface between rocks, minerals, and living things—soil. Soil is a necessary component of agriculture and, by extension, urban civilization as we know it today. Soil is defined in various ways by different user groups, but we will define it as the collection of natural bodies occupying parts of the earth's surface that support plants and that have properties due to the integrated effect of climate and living matter action upon parent material, as conditioned by relief, over periods of time. This statement incorporates elements common to many definitions of soil, such as unconsolidated mineral matter at the surface of the earth, and the natural medium for plant growth. Soils are considered to be a product of the five soil-forming factors of climate, parent material, organisms, relief, and time.[16] Young soils in deserts tend to be "immature" in that they contain many primary rock minerals that have not chemically weathered into mineral types that are more stable at surface conditions. These soils also tend to have low content of organic carbon due primarily to the low above- and below-ground biomass of plants. Eolian (or wind-driven) processes are important in desert soil formation (see Chapter 1), but in-place formation of soil from weathering of bedrock is also important.

### 2.8.1 Climate

Many people can appreciate that soils vary among climatic regions. Climate influences soil formation by its control on factors such as precipitation amounts and intensity as well as temperature regime. Both moisture and temperature are important drivers of abiotic activity within soils; examples include (but are not limited to) leaching of materials down through soil profiles, fracturing of rocks and minerals by heating and cooling processes, and the dissolution of minerals by water as discussed in the surficial weathering processes section earlier. Climate also drives biological activity in the soil profile and at the soil surface.

### 2.8.2 Parent Material

Parent material is the sediment or bedrock that weathers to form soils. It can range from river alluvium to wind-deposited materials, to rock materials such as granites or basalts, just to name a few. The mineral composition and hardness of a parent material influence soil fertility and rate of soil formation. Soils developed from parent materials high in iron and magnesium, such as basalts, tend to be rich in secondary clay minerals, fine-textured, and highly fertile. In contrast, soils developed from silica-rich parent materials, such as granites and sandstones in arid environments, typically retain more primary minerals such as quartz and feldspar, have coarse textures, and have relatively low fertility.

### 2.8.3 Organisms

Organisms, one of the most important drivers of soil formation, range from macrofauna such as burrowing mammals, earthworms, and ants to microfauna, which include fungi and bacteria. Macrofauna are essential as this group disturbs soil materials by burrowing and bringing materials at depth to the surface (*bioturbation*), by incorporation of organic matter into the soil profile, and by their primary role in physically reducing the particle size of organic matter. Microfauna play a pivotal role in cycling of nutrients, particularly nitrogen, phosphorus, and potassium, in soil systems. Bacteria and fungi convert nutrients from their organic form to inorganic forms. This is important, as the inorganic form of nutrients is required for uptake by organisms. Plants are also important to the formation of soils, in that they provide organic matter to the soil surface from litter and to the soil profile by root inputs. In desert ecosystems, patchy shrub cover leads to the formation of "islands of fertility" that concentrate nutrients below the individual shrubs. This is due to a variety of interacting processes, including litter fall, uptake of certain nutrients by the plant, interception of air- and water-borne dust, soil formation, soil erosion, animal decomposition, and mineralization by soil biota.*

### 2.8.4 Relief

*Relief* includes the slope and aspect (facing direction) of a particular site. Relief influences climatic affects, for example, soils on south-facing slopes, are often shallower than those on north-facing slopes where temperatures are cooler and soils are moister for a greater period of time. This promotes greater biological activity and formation of thicker soils. Slope plays a role in soil formation by its influence on soil erosion. Typically, soils on moderate to steep slopes are shallower due to constant erosion carrying particles down slope, which relates to the previously described sediment-transport processes.

* For a more detailed discussion of spatial patterns of soil nutrient distribution, see Whitford.[1]

### 2.8.5 Time

Soil formation in deserts is a slow process often requiring thousands of years. All of the above factors are integrated over time. The soil-forming factors working together produce the tremendous variety of soils we see on our landscapes. Ancient soil horizons, known to geologists as *paleosols*, can sometimes be preserved in stratigraphic sequences. Paleosols are typically recognized by the presence of root traces and burrows and can be very useful in recognizing environmental change in the geological record.[17]

Soils play a vital role in the natural and built environments. Some of the functions of soils include (a) water retention, which provides water on site for plants and other organisms and reduces downstream flood peaks; (b) nutrient processing, in which all nutrients are processed through the soil system by the macrofauna and microfauna; and (c) habitat that soil provides to many species of burrowing organisms and also forms a major portion of plant habitat.[18] From the human perspective, soils are important as (a) the agricultural medium, because almost all crops are grown in soils and (b) the foundation material that soils provide for houses, buildings, and roads.[19]

### 2.8.6 Soil Composition

A soil can be thought of as being composed of the components of mineral matter, organic matter, air, and water. A typical loam soil consists of about 45% mineral matter and 5% organic matter. The remaining pore space is occupied by about 25% water and 25% atmosphere. Soil solids consist of gravels, sands, silts, and clays, which typically are referred to as the *mineral matter*. The percentage of sand, silt, and clay is the *soil texture*. The clay fraction provides a soil with its ability to retain nutrients for later use by plants or other organisms; sands and silts do not have a significant role in nutrient retention. Clays also influence soil structure and strength, which control how difficult it is to deform the soil. Soils high in clays are difficult to dig or work and form hard clods when dry. Organic matter rarely exceeds 8% in mineral soils and consists of root, leaves, dead microorganisms, and other detritus. Organic matter represents a very important nutrient source in soil systems. In addition, organic matter contributes to soil structure and makes the soil easier to work. Decomposition of organic matter yields nutrients and humus. Humus, like clay, is important for the retention of nutrients in soils.

Pore space in soils is either filled with water or air. Pores play two critical roles in soils: first, pores form the major pathway by which water and air enter and exchange with the soil system; second, pores are where soil water is retained for future use by organisms, providing room for root elongation and pathways for nonburrowing soil organisms.[20] Pore space is easily reduced by management activities resulting in a compacted soil. Compaction of a soil is indicated by bulk density. Bulk density of soils ranges from about 1.0 to 1.5 Mg $m^{-3}$ in natural soils to 1.65–2.0 Mg $m^{-3}$ or higher if the soil is compacted. For comparison, most rocks have bulk densities on the order of 2.65 Mg $m^{-3}$. Organic matter, texture, and degree of compaction influence bulk density.

### 2.8.7 Soil Chemical Properties

Of the many chemical properties that a soil may possess, three of the most common and important are pH, cation exchange capacity, and salt content. The measure of the acidity of a soil is known as its *pH*. Although pH by definition ranges from 1.0 to 14.0, soil pH typically ranges from 4.0 to 8.0 in most environments and 7.0 to 8.4 in desert environments. Soils with the pH < 7.0 are termed acid soils, and soils with the pH > 7.0 are termed basic soils. Soil pH is a very important chemical parameter of soils and is often termed the "master variable." It influences soil-weathering processes, the type of plants that can survive in a site, and the availability of nutrients and the activity of soil microflora, just to name a few factors. Under certain chemical conditions, such as high concentrations of sodium, soil pH can be as high as 9.5. Many typical eastern landscaping plants require arid soils with pH less than those commonly found in desert soils. Successful establishment and growth of these plants in desert soils typically require adjustment of the pH.

Cation exchange capacity is an indication of a soil's ability to retain nutrients. Cation exchange capacity is influenced by clay content, humus content, and pH.* Salt content of a soil is often measured and reported, particularly in deserts and even more commonly in agricultural lands subjected to long-term irrigation. Due to low precipitation and high evaporation, salts are not removed from soils as they are in other regions and may actually accumulate on the surface. This is particularly common in enclosed dry lakebeds in deserts known as *playas*, and salt accumulation can be related to irrigation of agricultural lands, particularly when the water used is high in sodium or other major cations. Salts in semiarid and arid regions are typically carbonates, sulfates, and chlorides of calcium, magnesium, and sodium. A number of techniques and methods are used to measure salt content.[21]

### 2.8.8 Internal Soil Structure

Soils are composed of distinct layers termed horizons. Five primary or "master" horizons are recognized, in order for increasing depth; they are given as follows: O—a layer composed of organic materials—typically rare in desert soils; A—the first mineral horizon darkened by decomposition residues of organic materials; E—horizon where materials are leached from, usually light in color and rich in sands, and this horizon generally does not occur in desert soils; B—this horizon is a region of accumulation in the soil profile and can be enriched with clay, carbonates, or other materials; C—mineral horizon of little alteration of parent materials; and R—consolidated bedrock.

Unlike other regions of the United States, desert soils exhibit fewer horizons, and these horizons are often not distinct. This is due to low precipitation, little biological activity, and low-organic matter production when compared to other regions. Many desert soils have an A/C profile or an A/B/C with the B-horizon weakly developed because of little movement of materials such as clays in the soil profile. More characteristic of desert soils however is the accumulation of carbonates in the soil profile. Carbonates are water soluble and are translocated in the profile upward by evaporation and downward by the movement of water. Evidence of carbonate accumulation in the profile can be seen as white lenticular masses, coatings on rocks contained within the soil, or as discrete layers of indurated (hardened) carbonate (Figure 2.9). Soil horizons that have significant accumulation of carbonates are designated with the lower case letter k (e.g., Bk). In semiarid to arid regions, however, enough soil carbonate can accumulate in subsurface horizons that it dominates the morphology of the horizon. These carbonate-rich horizons are given the designation K.[10] In the southwestern United States, K horizons are

* For an introduction to the measurement of soil cation exchange capacity, readers are referred to Brady and Weil.[21]

**FIGURE 2.9**
A well-indurated K, or *caliche*, horizon (black arrows) forms the resistant ledge at the top of this stream wash in the White Tank Mountains west of Phoenix, Arizona. (Courtesy of W.L. Stefanov.)

referred to as calcretes or *caliche* and can be so well indurated that heavy equipment or dynamite is necessary to dig through them.

## 2.9 Landscape Organization of Soils

Soils vary across the landscape as a result in variation of the soil-forming factors from one landscape position to another. In the hierarchical organization of soils, what is informally known as a soil type is in fact a taxonomic group or "order." In the metropolitan area of Phoenix, the most common soil types are in the order Aridisols, which are soils of hot arid regions, or Entisols, which are young soils that show little soil profile development (e.g., little development of soil horizons).

If soils were examined along a transect in the Sonoran Basin and Range from mountains to the valley axis, several broad generalizations based on landscape position can be made (Figure 2.10). In the mountainous regions, soils are typically shallow and dominated by gravels or cobbles. The shallow nature of these soils is due in part to the slow rates of soil formation in desert areas, a function primarily of the lack of water. Interestingly, the shallow nature of these soils is also driven by water erosion associated with monsoon rains falling on surfaces of naturally low vegetative cover. These rains are responsible for the removal and deposition of soil particles in lower landscape positions, especially alluvial fans and pediments. Soils found on older alluvial surfaces (fans) are moderately coarse textured, frequently with B horizons and clay dominated layers reflecting their great age. It is the soils of these areas that are most likely to have well-developed K horizons or caliche layers. In younger alluvial deposits, the clay layers thin out and soil horizons are less developed. Soils of river floodplains are thick and range from coarse to fine-textured. Little horizon development is apparent due to reworking of the soil materials by periodic river flooding, leading to general classification of these soils as Entisols.

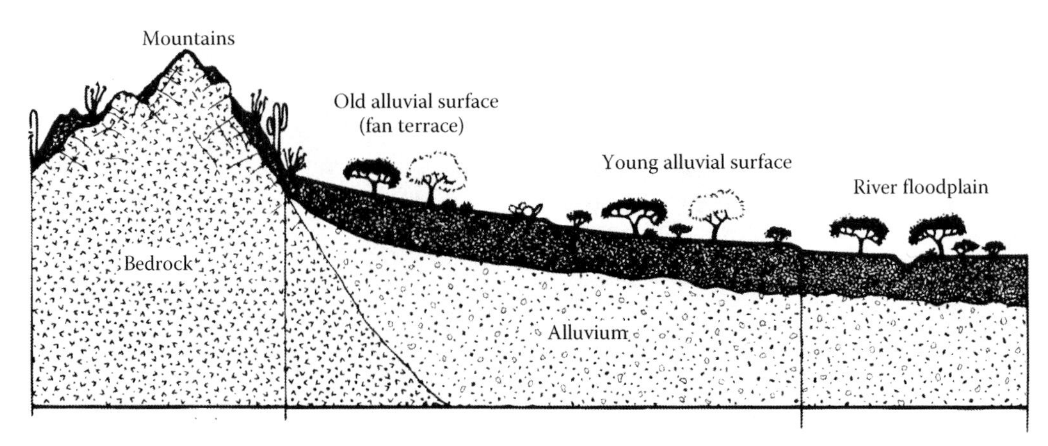


**FIGURE 2.10**
Schematic representation of typical soil-landscape position correspondence in the Basin and Range geological province. (Modified from Hendricks, D.M., *Arizona Soils*, University of Arizona College of Agriculture, Tucson, AZ, 1986.)

## 2.10 Soils of the Built Environment

Most soils literature and classification schemes treat soil as a naturally occurring body at the surface of the earth. Soils in the built environment are very different from those of natural systems, and these soils are often termed "urban soils." Definitions of what constitutes urban soils usually include the following: (a) presence of a nonagricultural man-made layer >50 cm thick; (b) significant mixing, filling, and removal of natural soil materials; (c) possible contamination either by chemical or other wastes, such as bricks and cement; and (d) occurrence in urban or suburban areas. There are several features that make urban soils unique.

Urban soils are very heterogeneous in space due to human activities, which include filling over existing soils, removing existing soils, and imposing a multitude of management regimes in relatively small areas. Vertical heterogeneity reflects contrasting layers from fill and deposition of waste products in the soil system. Soil structure in urban soils is often weakened due to reduced inputs of organic matter. Organic matter contributes to soil structure formation by influencing soil acidity, biogeochemical weathering of minerals, and presence or absence of soil biota. Soil structure in urban soils is also subjected to repeated compaction in places such as parks where human activity can be very intense.

Deterioration of soil structure and compaction leads to reduced aeration and drainage because of the reduction in porosity. Soil-moisture movement and retention can also be impeded by the presence of compacted layers within the profile. Proper infiltration and aeration are critical for soil fauna and plants to thrive in urban environments. Organic matter inputs from plants represent a very important component of a healthy soil ecosystem because it is both an energy and nutrient source for soil organisms. Plants benefit from the nutrient content of organic matter after microbial processing. Organic-matter cycles are truncated when surfaces are paved and when organic matter is removed by homeowners or landscapers.

Soil temperature controls the activity of soil organisms and also the respiration rates of plant roots. Soils in urban systems are typically warmer than those of nonurbanized systems because of decreased surface shading, decreased albedo (or reflectance), and increased compaction. Higher temperatures may lead to faster organic matter decomposition and to stressed plants.

Urban soils often contain considerable amounts of manmade materials such as construction debris (e.g., bricks, pipe cement, metals) that, while not immediately hazardous, may reduce water-holding capacity and interfere with root penetration and water movement within the soil profile. Contaminants that may be present include pesticide residues, hydrocarbons, and other materials (see Chapter 5).

These differences from naturally developed soils make urban soils a unique management challenge for urban planners and soil scientists alike. Most soil data in the United States are based on soils of natural systems or agricultural soils, and the applicability of much of the data to all soils in urban areas may be questionable.*

## 2.11 Accessing Geological and Soils Information

The primary source of information for geological surface and subsurface maps and profiles, hydrologic information, geological hazard assessments, and mineral resource information for the United States—including the desert regions—is the U.S. Geological Survey (USGS).† State agencies are another important source of geological information for the desert regions of the continental United States.‡ The USDA Natural Resources Conservation Service (NRCS) is the primary source of information for soils data in the United States. The NRCS publishes soil surveys for counties or parts of counties. These soil surveys contain a great deal of useful soils information including location maps of different soil types, soil classifications, and interpretation for different uses. The most current soil survey information is available through the Web Soil Survey online mapping and data query tool, which provides interactive access to the U.S. General Soil Map (STATSGO2) and Soil Survey Geographic (SSURGO) databases. Soil data can also be downloaded in digital formats suitable for use in Geographic Information System (GIS) software environments through the Web Soil Survey interface. The NRCS also provides information on soil-related hazards such as shrink-swell potential, which can cause severe damage to house foundations.§

* As a starting point for information about urban soils, the reader is referred to Craul.[22]

† The majority of data available from the USGS can be accessed at http://www.usgs.gov (accessed August 21, 2012).

‡ All of these state geological surveys provide some data access through their websites: Arizona Geological Survey—http://www.azgs.az.gov/ (accessed August 21, 2012); California Geological Survey—http://www.conservation.ca.gov/cgs/Pages/Index.aspx (accessed August 21, 2012); Idaho Geological Survey—http://www.idahogeology.org/ (accessed August 21, 2012); New Mexico Bureau of Geology and Mineral Resources—http://geoinfo.nmt.edu/ (accessed August 21, 2012); Nevada Bureau of Mines and Geology—http://www.nbmg.unr.edu/ (accessed August 21, 2012); Oregon Department of Geology and Mineral Industries— http://oregongeology.com/sub/default (accessed August 21, 2012); Texas Bureau of Economic Geology—http://www.beg.utexas.edu/ (accessed August 21, 2012); Utah Geological Survey—http://geology.utah.gov/ (accessed August 21, 2012).

§ NRCS data is available online at http://websoilsurvey.nrcs.usda.gov/app/HomePage.htm (accessed August 21, 2012).

## 2.12 Human Impacts on Desert Geology and Soils

Although deserts have existed for thousands of years, they are quite easily disrupted and damaged by human activities.[19] Large-scale conversion of natural desert landscapes to urban and agricultural landscapes leads to loss of native species, enhancement of urban heat islands, creation of fugitive (airborne) dust, and increased exposure to geological hazards.[23,24] Development on alluvial fans, pediments and hillslopes, emplacement of dams, and grazing have altered existing drainage patterns, soil patterns and development, vegetation patterns, sediment-transport processes, and groundwater recharge.[25] Some of the resulting effects from these changes include increased flooding potential, boulder fall hazards, slope failure, stream bank erosion, and soil loss. Groundwater withdrawal for agricultural and residential use has caused earth fissures, subsidence, and structural failures in buildings (see Chapter 5). Increased human presence in deserts for recreational use has caused degradation of soil surfaces and soil/groundwater contamination.

As we continue to develop our understanding of deserts and the physical, chemical, and biological processes that are active within them, we gain a better understanding of how fragile these geosystems and ecosystems are. Increased collaboration between scientists in many disciplines, architects, city and landscape managers, urban planners, and a concerned and educated public is necessary to ensure that deserts remain functioning biomes for future generations to experience.[26] It is hoped that the results of such collaborations will help us to decrease our impact on deserts, foster preservation and conservation efforts, and perhaps allow us to truly "design with the desert."

## References

1. W.G. Whitford. *Ecology of Desert Systems* (New York: Academic Press, 2002).
2. R. Cooke, A. Warren, and A. Goudie. *Desert Geomorphology* (London, U.K.: University College London Press, 1993).
3. E.J. Tarbuck, F.K. Lutgens, and D. Tasa. *Earth: An Introduction to Physical Geology*, 9th edn. (Upper Saddle River, NJ: Prentice-Hall, 2007).
4. P. Kearey, K.A. Klepeis, and F.J. Vine. *Global Tectonics*, 3rd edn. (Cambridge, MA: Wiley-Blackwell, 2009).
5. D. Nations and E. Stump. *Geology of Arizona*, 2nd edn. (Dubuque, IA: Kendall/Hunt Publishing Co., 1996).
6. K.E. Karlstrom, ed. *Proterozoic Geology and Ore Deposits of Arizona*, Digest 19 (Tucson, AZ: Arizona Geological Society, 1991).
7. J.P. Jenney and S.J. Reynolds, eds. *Geologic Evolution of Arizona*, Digest 17 (Tucson, AZ: Arizona Geological Society, 1989).
8. S.S. Beus and M. Morales, eds. *Grand Canyon Geology*, 2nd edn. (New York: Oxford University Press, 2003).
9. M.J. Selby. *Hillslope Materials and Processes*, 2nd edn. (New York: Oxford University Press, 1993).
10. P.W. Birkeland. *Soils and Geomorphology*, 3rd edn. (New York: Oxford University Press, 1999).
11. L.D. McFadden, J.B. Ritter, and S.G. Wells. Use of multiparameter relative-age methods for age estimation and correlation of alluvial fan surfaces on a desert piedmont, eastern Mojave Desert, California, *Quaternary Research* 32: 276–290, 1987.

12. L.D. McFadden, E.V. McDonald, S.G. Wells, K. Anderson, J. Quade, and S.L. Forman. The vesicular layer and carbonate collars of desert soils and pavements: Formation, age and relation to climate change, *Geomorphology* 24: 101–145, 1998.
13. E.V. McDonald, L.D. McFadden, and S.G. Wells. The relative influence of climatic change, desert dust, and lithologic control on soil-geomorphic processes on alluvial fans, Mojave Desert, California: Summary of results, *San Bernardino County Museum Association Quarterly* 42: 35–72, 1995.
14. M.W. Strudley, A.B. Murray, and P.K. Haff. Emergence of pediments, tors, and piedmont junctions from a bedrock weathering-regolith thickness feedback, *Geology* 34: 805–808, 2006.
15. T.M. Oberlander. Landscape inheritance and the pediment problem in the Mojave Desert of southern California, *American Journal of Science* 274: 849–875, 1974.
16. H. Jenny. *Factors of Soil Formation* (New York: McGraw-Hill, 1941).
17. J. Reinhardt and W.R. Sigleo, eds. Paleosols and weathering through geologic time: Principles and application, Special Paper 216 (Boulder, CO: Geological Society of America, 1988).
18. E.A. Paul, ed. *Soil Microbiology, Ecology, and Biochemistry*, 3rd edn. (New York: Academic Press, 2007).
19. E.A. Keller. *Environmental Geology*, 6th edn. (New York: Macmillan Publishing, 1992).
20. D.C. Coleman, D.A. Crossley, Jr., and P.F. Hendrix. *Fundamentals of Soil Ecology*, 2nd edn. (New York: Academic Press, 2004).
21. N.C. Brady and R.R. Weil. *The Nature and Properties of Soils*, 13th edn. (Upper Saddle River, NJ: Prentice-Hall, 2002).
22. P.J. Craul. *Urban Soil in Landscape Design* (New York: John Wiley & Sons, 1992).
23. H. Chamley. *Geosciences, Environment, and Man* (Amsterdam, the Netherlands: Elsevier Science, 2003).
24. G. Heiken, R. Fakundiny, and J. Sutter, eds. *Earth Science in the City: A Reader* (Washington, DC: American Geophysical Union, 2003).
25. R.U. Cooke and J.C. Doornkamp. *Geomorphology in Environmental Management*, 2nd edn. (New York: Oxford University Press, 1990).
26. M. Gray. *Geodiversity: Valuing and Conserving Abiotic Nature* (Chichester, U.K.: John Wiley & Sons, 2004).
27. D.M. Hendricks, *Arizona Soils* (Tucson, AZ: University of Arizona College of Agriculture, 1986).

# 3

## *Scales of Climate in Designing with the Desert*

**Anthony J. Brazel**

**CONTENTS**



### 3.1 Introduction

Climate is considered an ensemble of weather processes and varies in its characteristics depending on the time scale chosen and spatial area considered.[1] Table 3.1 provides an example of the types of meteorological motion systems and space/time scales. If one were to study wind gusts as extreme events that might affect buildings, observational methods or calculations would have to address processes that happen over seconds to minutes and resolve effects over areas less than a square mile (e.g., a microburst from a cloud). On the other hand, forecasting midlatitude cyclones requires only a coarse time and space domain to assure awareness of impending storms across a region (e.g., a spatial resolution of hundreds of miles).

Climatology consists of concepts such as how frequent these variable processes occur; their magnitudes and dimensions—including frequencies and magnitudes over time; the mean state; and whether there are significant cycles, step jumps, or subtle long-term trends that are evident in the climate system. The expression "climate system" is typically used to indicate a series of complex physical and dynamic processes between earth's surface and the atmosphere that interact to characterize a climate. The theoretical climatologists must get at the fundamental causes for these frequencies/magnitudes and their changes over time using physical and mathematical principles.

The applied climatologist and planner, who is asked to advise on issues of desert living, is faced with what might be perceived as an equally daunting set of issues. He or she asks what difference a climate makes to citizens, cities, towns, companies, agencies, and governments, and what is it about the climate system that must be understood so that we can provide meaningful strategies to mitigate or adapt to the negative consequences of climate and take advantage of the positive consequences. Contemporary applied problems (design-related or otherwise), more often than not, start out with an appreciation of the

**TABLE 3.1**

Spatial Systems of Climate

| System | Horizontal Scale (Miles) | Vertical Scale (Miles) | Time Scale |
|---|---|---|---|
| Global wind belts | >1200 | 2–6 | 1–6 months |
| Regional macro | 300–650 | 0.5–6 | 1–6 months |
| Local climate | 0.5–6 | 1/60–1/6 | 1–24 h |
| Microclimate | <1/6 | <1/60 | <24 h |

*Source:* Barry, R.G., *Trans. Inst. Brit. Geogr.*, 49, 61, 1970.

space scale of any issue and the time scale required for consideration. The time scale in this chapter is basically contemporary, that is to say twentieth century, and the near future. Space limits the fascinating realm of long-term change of the climate system back before the major impact of humans.[2]

## 3.2 Global Scale

Figure 3.1 shows the global and central Arizona annual temperature anomalies (annual differences from long-term mean) since 1950. The climate of the Southwest is one of past and current variability, especially of precipitation, and an excellent review has been provided in the literature of the southwestern United States.[3] In the realm of future planning, whether these past trends persist at the same rate, abate, or amplify are important to several design

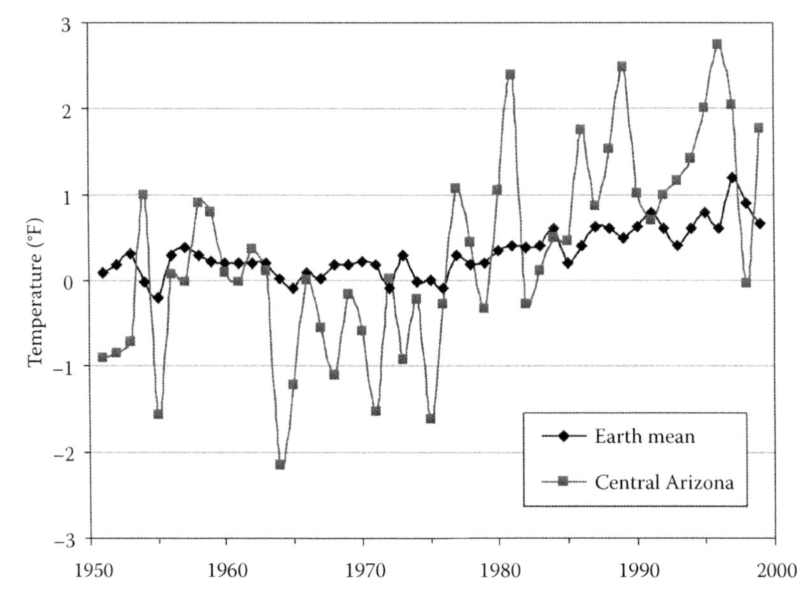


**FIGURE 3.1**
Earth mean and central Arizona temperature anomaly trends (base period 1950–2000; year minus the mean) showing warming at two scales, especially since the mid-1970s. Central Arizona is a 328,000 mile$^2$.

and environmental applications. These applications include water resources, energy policy, environmental planning, transportation and emissions, and ecosystem sustainability.

Projections of future climate at the global scale, the verification of climate models that produce these projections, and the realm of policy-makers' responses to these important findings can be gleaned from browsing many websites* and certainly by keeping abreast of mainline scientific journals and magazines, if not simply the more popular literature (e.g., *Science, Nature, Scientific American, US News,* and *World Report*). Designers for the future should access scenario data of global climate change from the IPCC4,[4] learn what is forecast for our region, always understand the uncertainties, assume some level of user risk, and plan accordingly.

Climate records on the global scale are sparse in many regions, and existing climate models only resolve part of the complexity of the climate system (and at gross resolutions of spatial scale; e.g., grid cells the size of Arizona). There is now a consensus that anthropogenic factors play a major part of these trends, and that there are strong possibilities for major changes over the next 50 years.[5] We must keep a watch for possible future effects, even when planning at the local scale. News surveys are indicating that the public is increasingly aware of these global changes. The ultimate challenge is how to find the middle ground between under-using or over-using what we know imperfectly.[6] Climate users who are not climatologists vary greatly in their understanding of how accurate climate records may be, and how they may be used to help solve problems. Many users of information initially feel there is a weather site for every purpose and for every location. Most clients do not know how to interpret or extrapolate data for different locales and situations. Part of the issue is what data are available. This problem is lessening. For example, over the last decade in Arizona, there have been many special automated systems put in place for specific purposes (Table 3.2). These data are useful for local/regional scale analyses. To assess the long-time scales related to global change, we still have fewer sites that are accurate and reliable from say 100 years ago than we do now. These are mostly sites run by the federal government as part of their national system of weather data through the auspices of the National Oceanic and Atmospheric Administration (NOAA). Currently, at the National Climatic Data Center, there are efforts at reassessing the entire observational

**TABLE 3.2**

Examples of Central Arizona Special Automated Networks

| Network | Purpose | Elements | Record Longevity |
|---|---|---|---|
| AZMET | Agricultural | Hourly temperature, humidity, wind, solar, soil temperature, moisture, precipitation | >10 years |
| PRISMS | Electricity | 5 min temperature, humidity, storms, wind, solar, pressure, precipitation | >10 years |
| MCFCD | Floods | Hourly temperature, humidity, precipitation, wind | Several years |
| USGS | Floods | Precipitation, wind, temperature | Many years |
| State | Air quality | Wind, temperature, air quality elements | Many years |

AZMET, Arizona Meteorological network, run by the University of Arizona; PRISMS, Phoenix Real-time Instrumentation for Surface Meteorological Studies, maintained by Salt River Project, in cooperation with ASU and National Weather Service; MCFCD, Maricopa County Flood Control District; USGS United Stated Geological Survey.

* Examples of websites include http://www.epa.gov and http://www.noaa.gov (accessed July 15, 2009).

**TABLE 3.3**

Global Climate Model Estimates for the Colorado Basin at Various Years in the Future

| | Temperature Change | Precipitation (%) | Runoff (%) | Reservoir Storage (%) | Hydropower Production (%) |
|---|---|---|---|---|---|
| CTRL | +0.9 | −1 | −10 | −7 | −16 |
| Period 1 | +1.8 | −3 | −14 | −36 | −56 |
| Period 2 | +3.1 | −6 | −18 | −32 | −45 |
| Period 3 | +4.9 | −3 | −17 | −40 | −53 |

CTRL is a control run based on 1995 greenhouse emissions. Temperature is in °F.
Periods 1, 2, 3 are so-called Business as Usual increases in emission scenarios projected for years 2010–2039, 2040–2069, and 2070–2098. Business as Usual assumes no addressing of increased greenhouse gas emissions. For downscaling procedures and model constructs, see Christensen.[8]

climate network in the country because of the stimulus of global change and the need to envision scenarios of the future.

Recent assessment of the world's future climate is contained in the U.S. Global Climate Program reports by region of the country,[7] in many publications,* and in the more recent Intergovernmental Panel on Climate Change Assessment (IPCC4).[5] As an example, Table 3.3 illustrates general scenarios for the Colorado River Basin due to global climate change in the future.[8] The results point to a future climate of warming temperatures, less precipitation and decreased runoff, and storage and hydropower in the Colorado Basin over time in the twenty-first century. In a more local study[9] for the Salt-Verde Watershed of central Arizona, a 6.7°F warming is projected and a decrease of 0.1 in. per month precipitation, yielding a scenario of over 15% less runoff from the watershed by mid-century. It should be emphasized that these are scenarios and not precise predictions, and there is much uncertainty involved, especially for smaller-scale systems such as the Southwest monsoon and summer precipitation. Nevertheless, these scenarios may sway decision makers to become aware that planning strategies used during the last 100 years (based on historical variations) may not sustain us for the next 100 years because a major shift in the climate system has a better than random chance of occurring.[9]

## 3.3 Regional Scale

The regional scale in climatology spans a considerable range of space, such as continental, suboceanic regions (e.g., areas of contiguous ocean temperatures), plains, plateaus, mountains, broad basins, watersheds, earth biome regions, and often what is classified on world climate maps as earth's climate regions (tropical rain forest, savanna, desert, steppe, humid subtropical, marine west coast, Mediterranean, humid continental, subarctic, tundra, ice cap). The dynamic processes that create these differences are many. It should be emphasized that the majority of the explanations of regional climate relate first of all to

* See, for example, Stott et al.[5] and Christensen et al.[8]

the earth–sun relationships, the resultant earth-atmosphere's general circulation pattern (trade winds and Hadley cells, jet streams, the polar front activity, air masses that develop and collide), and the two-thirds-oceans, one-third-land global configuration, with extreme differences in the northern and southern hemispheres.

The key to desert design, and coping with desert climates on earth, relates to regions of the subtropics as part of the general atmospheric circulation due to the Hadley Cell motion system between the equator and the Tropics of Cancer and Capricorn (23.5° north and south of the equator). In these subtropical latitudes, the air is hot and dry due to subsidence (downward moving air to the surface) and compressional warming. In the context of the deserts in the American Southwest, surface high pressure and clear skies exist most of the time, interrupted only by dynamics of a summer monsoon from the south, fall tropical storms and hurricanes, and upper-level low-pressure systems occasionally, or higher-latitude intrusions of frontal storms in winter. Thus, we are used to expecting persistent diurnal, seasonal, and annual climate rhythms in desert regions, as well as in other climates. When there is a change in the timing, intensity, and persistence of these rhythms due to global reverberations and changes as well as local changes, regional changes may be induced that must be understood in order to anticipate and cope with any anomalies of normally expected climate variability.

One of the best examples is our increased appreciation in only the last decade or so of the impacts on the desert southwest of the Southern Oscillation and Pacific Decadal variations in the Pacific Ocean basin.[10] The Southern Oscillation is a surface atmospheric pressure difference across the southern Pacific that reverses atmospheric circulation and, as a result, affects surface ocean currents.* The El Niño (EN) and La Niña (LN) phenomena represent extensive areas of warm and cold water, respectively, off the west coast of South America, notably Peru, and induce regional changes in the supply of moisture to our region on a quasi-periodic and variable time scale, often less than a decade in length. The U.S. government now monitors and forecasts these conditions on a seasonal basis since their impacts are sizeable. In fact, several scientists consider these oscillations good analogs for study on possible future societal impacts due to expected future regional climate change. The Pacific Decadal Oscillation is longer than the El Niño–La Niña cycle and, in concert with the Southern Oscillation, may cause longer-term changes in drought or flooding on a time scale over decades.

In Arizona, when EN events are observed, winds reverse and come toward South America. There is a low pressure, and we tend to receive above-normal precipitation in the winter months. When LN occurs, drought may be more extensive (see Figure 3.2 as an example). The teleconnections between EN/LN and Southwest precipitation are not strong correlations but are statistically significant (e.g., the $r^2$ in Figure 3.2 is only 0.272, thus a large unexplained variance still exists). Coping with moisture extremes of the summer monsoon and variable winter supplies of moisture is a challenge to designing livable spots in the desert and to avoid dangerous zones that may flood periodically. Community developments must not only protect against these conditions, but should develop further schemes to store water during the wet times to combat the dry times. This surely would be important in the rapid growth scenario of our future, with possible limited reservoir and ground water supplies.

* When El Niño occurs, the Southern Oscillation Index (SOI) is negative—Tahiti minus Darwin pressure normalized—meaning that in Darwin, Australia air pressure is higher than at Tahiti in mid-Pacific. When La Niña occurs, the SOI index positive.

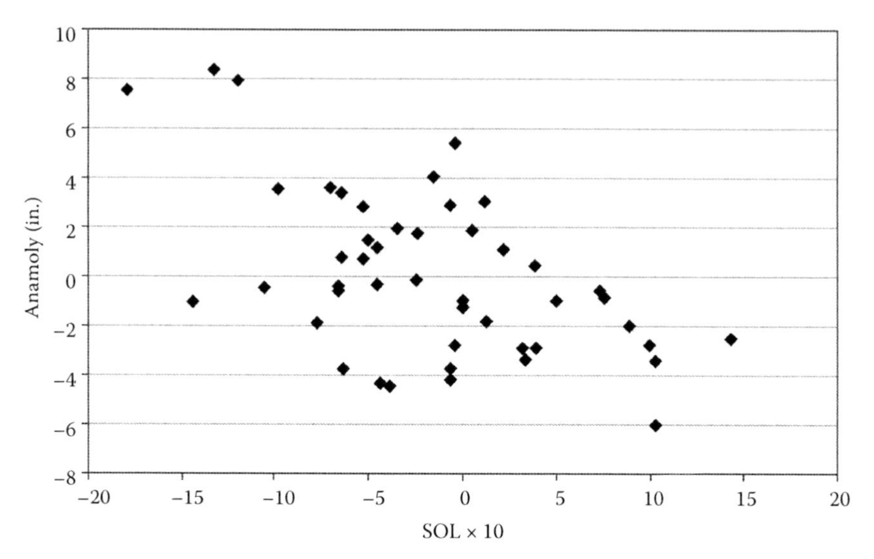


**FIGURE 3.2**
Precipitation anomaly for the central Arizona region (in inches for the water year—October for 1 year to September of the next year) from mean values (1950–1995) versus the Southern Oscillation Index (normalized pressure at Tahiti, central South Pacific Ocean minus Darwin, Australia × 10). Note inverse relation (negative SOI means an El Nino and tendency toward more precipitation; positive SOI means La Nina and less precipitation). R value is 0.272 and is significant at the 0.05 level of significance.

## 3.4 Regional to Local Scale

If we consider the whole contiguous desert Southwest as some uniform region and dismiss the notion of variability and factors of climate that operate at the local scale for designing and living with the desert, we would ignore sustainability-threatening variability within the region. Much of this with-region variability is due to several factors: (1) elevation, slope, and aspect of terrain; (2) watershed variation, orientation of river channels, and wind drainage paths; (3) kind of surfaces, soil, vegetation, heating and cooling rates over these surfaces, and their impermeability to water; and (4) climate within and over the built or human-impacted lands (e.g., urban, desert, agriculture) of the region as examples of local factors. Each of these factors requires careful thought and analysis as to significance for living in the desert.

There are four North American deserts to consider: (1) The Great Basin, (2) Sonoran, (3) Mojave, and (4) Chihuahuan.* Different temperatures, elevations, and quantities of precipitation occur among these deserts (Figures 3.3 through 3.5). Generally, the Great Basin, centered over the state of Nevada, is considered a cold desert (winter temperatures well below freezing), with snowfall totals ranging from 4 to 11 in. Austin (elevation 6661 ft, latitude 39°30′N, longitude 117°5′W) and Las Vegas (elevation 2170 ft, latitude 36°5′N, longitude 115°10′W) represent ranges of climate conditions in this desert region, with even more variable conditions at higher elevations in mountain areas. Total precipitation at Austin is 12.6 in. and Las Vegas is 4.2 in. Annual average maximum and minimum temperatures are 60.9°F and 34.3°F at Austin and 80.0°F and 53.6°F at Las Vegas.

* These deserts are also discussed from different perspectives in Chapters 1, 2, and 7.

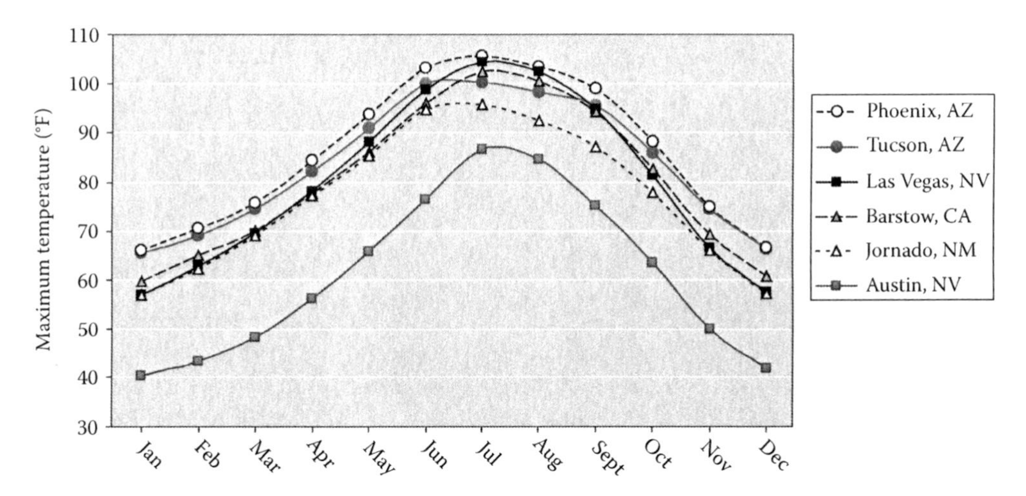


**FIGURE 3.3**
Mean monthly maximum temperature for selected stations in the four main southwestern U.S. desert regions: Austin and Las Vegas, Nevada (Great Basin); Barstow, California (Mojave); Phoenix and Tucson, Arizona (Sonoran); and the Jornada Experimental Range, New Mexico (Chihuahuan).

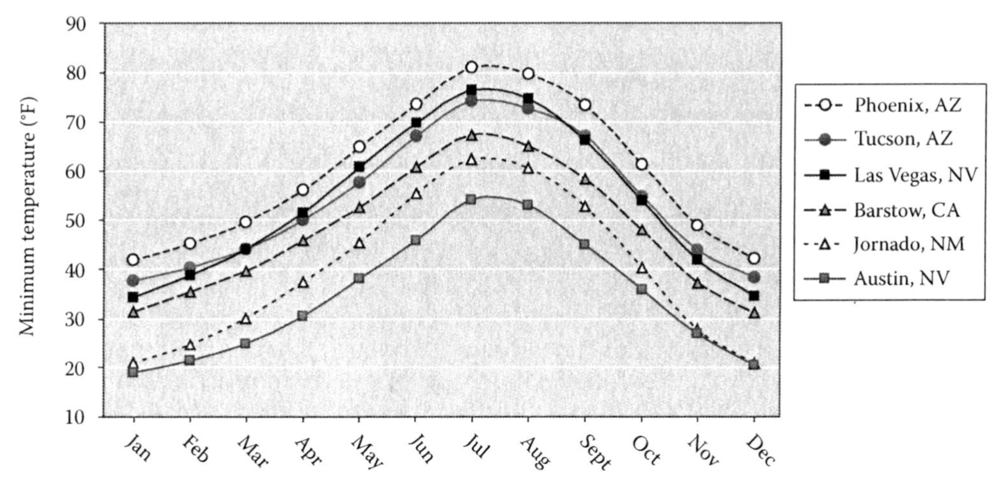


**FIGURE 3.4**
Mean monthly minimum temperature for selected stations in the four main southwestern U.S. desert regions: Austin and Las Vegas, Nevada (Great Basin); Barstow, California (Mojave); Phoenix and Tucson, Arizona (Sonoran); and the Jornada Experimental Range, New Mexico (Chihuahuan).

The Great Basin is sagebrush country, with pinyon and juniper woodlands higher up and mountaintops getting double the lowland precipitation.

The Sonoran area, mostly in Arizona and extending into Mexico along both sides of the Gulf of California, is biologically diverse and experiences the summer monsoon and also receives substantial winter moisture, especially at higher elevations. A distinctive vegetation form is, of course, the saguaro cactus (*Carnegiea gigantea*). Intervening mountain regions host sky islands of rich biological diversity. Phoenix (elevation 1110 ft, latitude 33°26′N, longitude 112°2′W) and Tucson (elevation 2420 ft, latitude 32°14′N,

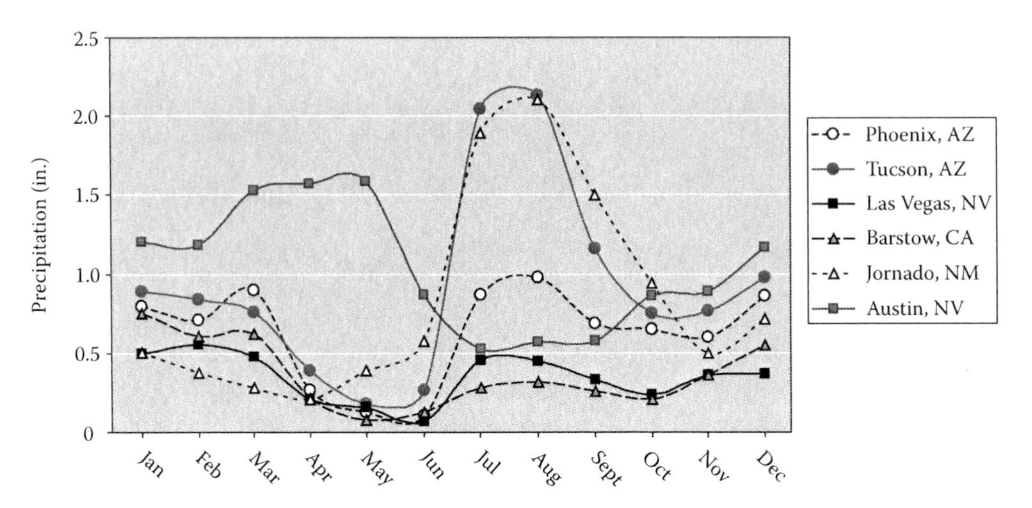


**FIGURE 3.5**
Mean monthly precipitation (inches) for selected stations in the four main southwestern U.S. desert regions: Austin and Las Vegas, Nevada (Great Basin); Barstow, California (Mojave); Phoenix and Tucson, Arizona (Sonoran); and the Jornada Experimental Range, New Mexico (Chihuahuan).

longitude 110°57′W) climate data illustrate conditions of the more lowland sites in the region. Total annual precipitation at Phoenix and Tucson averages 7.57 and 11.16 in., respectively. Average annual maximum and minimum temperatures at Phoenix are 85.8°F and 59.6°F, respectively; for Tucson, maximum and minimum temperatures are 83.4°F and 53.7°F. Note the pronounced onset of the summer monsoon rainfall pattern, an abrupt change to a moist summer regime (Figure 3.5).

The Mojave Desert is a transition between the Sonoran and Great Basin Deserts (Brady, this volume), but the Mojave Desert has a unique and characteristic bioform in the Joshua tree (*Yucca brevifolia*). Because it is a transitional desert, it receives both less winter and summer monsoon rains as illustrated by climate conditions at Barstow, California (elevation. 2140 ft, latitude 34°54′N, longitude 117°1′W). Its precipitation is only 4.4 in./year, with maximum and minimum temperature averages 80.2°F and 47.5°F, respectively.

The Chihuahuan Desert occurs primarily in Mexico in the intermountain basin in the Rocky Mountains and between the Sierra Madre Oriental and Sierra Madre Occidental. It is at a higher elevation and receives more summer precipitation than the Mojave and Sonoran Deserts. A characteristic form is the maguey or century plant (*Agave*). Data from the Jornada Experimental Range of New Mexico (elevation 4270 ft, latitude 32°37′N, longitude 106°44′W) is used to approximate climate conditions in this desert area. Precipitation averages 10.0 in./year, with average maximum and minimum temperatures of 76.7°F and 39.7°F, respectively. The three plots of the climate station data for these deserts illustrate considerable variations of temperatures (over 20°F for maximums, 30°F for the minimums) and moisture among the deserts (4–13 in. with much more variability in higher elevations). The timing of precipitation is quite variable as a function of rain shadow effects, impacts of Pacific storms, and exposure to the Southwest summer monsoon. Major gradients of these conditions occur with elevation and location in relation to the terrain of the areas within the region.

## 3.5 Human-Dominated Effects within the Region

Embedded among the diverse desert climates and large ranges in temperature and precipitation in the Southwest terrain are human settlements with their own distinctive emerging local climatic regimes. One of the more recent interdisciplinary foci in climatology today is a good example of the importance of appreciating processes at the local scale and, at the same time, appreciating the time-varying state at this scale: that of studying the urban environment and how cities are creating new local climates through time.[11] Scholars who publish in this rapidly emerging area, and others, appreciate that well over half the population now reside in cities. The contributors to this field are a diverse group that includes climatologists foresters, weather forecasters, air quality and fluid dynamics specialists, architects and planners, transportation and material science specialists, construction industry specialists, urban ecologists, and policy-makers.

Cities in desert areas are growing rapidly, are within climate regimes dominated by local-scale processes (stable air, less storms, terrain influences), and experience heat island and precipitation effects a large portion of the time during a seasonal cycle, more so than storm-dominated, moist climates on earth. As shown in Figure 3.6, minimum temperatures in downtown Phoenix and the Phoenix Sky Harbor airport have increased since 1960, in comparison to a rural location at Sacaton, Arizona, some 20 miles to the southeast of the Phoenix area.

Factors that contribute to urban heat excesses include: (1) increased surface area absorbing the sun's energy due to vertical buildings and decreased albedo (surface reflectivity), (2) the absorption rate of materials and storage of heat during a day, (3) the lack of wetness of the surface and amount of vegetation and lakes, (4) the geometry of building arrangements and the canyon-like heat trapping effect, (5) the emitted heat from buildings and roofs, (6) transportation emissions and air-quality effects on heating and cooling within the city,

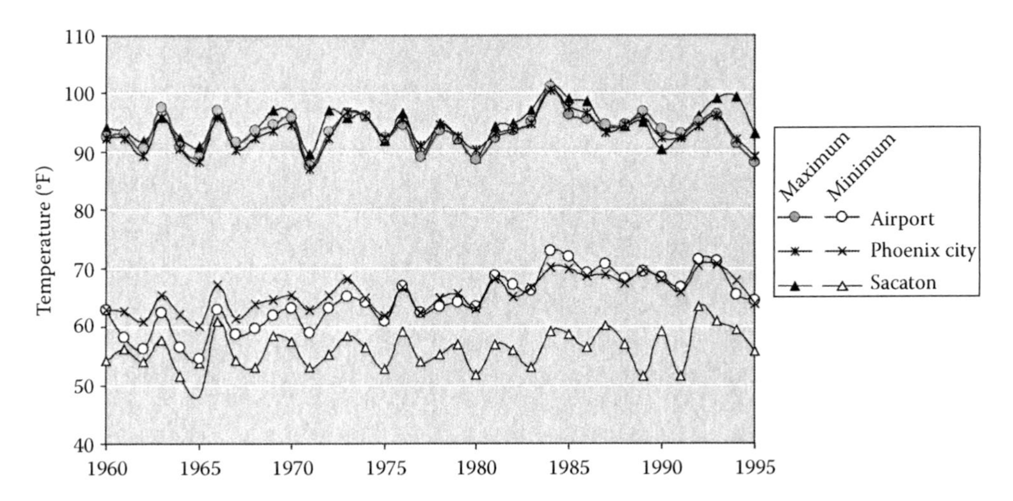


**FIGURE 3.6**
Two urban-dominated surfaces (Phoenix downtown and Sky Harbor International Airport) versus the more rural site of Sacaton, Arizona, about 20 miles southeast of Phoenix in the desert. Data are mean monthly maximum and minimum temperatures for May. Note the pronounced difference in the minimum temperature time trend between urban sites and rural site.

**TABLE 3.4**

Radiative Surface Temperatures (10:00 a.m.) and Areal Coverage in Phoenix, on June 24, 1992

| Surface Type | Surface Temp. (°F) | Albedo (%) | NDVI Index* | Area (Miles$^2$) |
|---|---|---|---|---|
| Water | 86.1 | 12.3 | 0.04 | 2 |
| Irrigated agric. | 88.8 | 21.1 | 0.56 | 80 |
| Irrigated res. | 100.2 | 18.4 | 0.30 | 126 |
| Dry res. | 110.6 | 19.4 | 0.13 | 178 |
| Commercial | 111.9 | 18.7 | 0.02 | 355 |
| Desert | 116.9 | 19.3 | 0.07 | 226 |
| Barren | 113.7 | 25.1 | 0.03 | 53 |
| Unclassified | — | — | — | 669 |

*Source:* Derived from Lougeay, R. et al., *Geocarto Int.*, 11, 79, 1996.

*NDVI, normalized difference vegetation index = Landsat TM (band 4 − band 3)/(band 4 + band 3).

Albedo, reflected incoming solar radiation from the surface.

and (7) the land-use geography of the city. All of these aspects require shared knowledge from specialists across an array of disciplines.

Table 3.4 provides results from a remote-sensing project, in which three characteristics—surface temperature, albedo, and a Normalized Difference Vegetation Index (NDVI)—were calculated from Landsat 1992 radiance imagery on June 24, 1992, at 10:00 a.m. over Phoenix.[12] The imagery allows us to estimate an integrated surface ground temperature for a 394 ft resolution scale. NDVI is an estimate of surface wetness and biomass that can be derived from specific wavelengths of energy received by the satellite system (except values for a water surface are inappropriate and should be ignored). Note considerable range in wetness across the region. Generally, the hotter surfaces are the drier surface types at this time of the day (10:00 a.m.). Typically, the city commercial and industrial zones retain heat later in the day and at night due to thermal heat storage properties and canyon-like trapping of heat.[13]

Let us simulate the local effects at this scale by (1) increasing the amount of shade in the city; (2) changing areas of parks, watered areas, and open areas; (3) changing albedo by increasing the amount of pavement on the desert; (4) spreading out buildings and affecting the wind-affected roughness at ground level over differing surfaces; and (5) creating more and more deep urban "canyons" of the city.[14] Tables 3.5 and 3.6 demonstrates these typical urban effects on the local climate. The bottom line in response is: (1) increasing the area of pavement increases energy absorption and raises the temperature later in the day, (2) less shade can cause significant heating, (3) reducing greenspace heats the city, (4) actually increasing the overall roughness could aid in increasing wind flow and pushing away bad air (but this is a complex question), and (5) creating deep urban canyons may shade the surface during the day, but trap heat to higher levels at night than would normally occur. Wind ventilation is also a key to ameliorating the street-level urban canyon heat excesses and developing more daytime coolness in the hot desert. Table 3.7 shows the importance of wind processes and street-level ventilation in relation to above roof flows.[15] Whether heat excess is ventilated depends on the speed and type of flow across the variable urban landscape.

**TABLE 3.5**

Changing Local Climate (on a Clear Mid-July, Mid-Morning Period in Phoenix—Same Time as Landsat Overfly Times)[a]

| Parameter Changed | Surface Temperature (°F) | | |
|---|---|---|---|
| % Shade on surface area | | | |
| 25 | 113.5 | | |
| 50 | 102.9 | | |
| 75 | 90.1 | | |
| 100 | 74.3 | | |
| % Wetting of surface | | | |
| 30 | 107.9 | | |
| 50 | 102.2 | | |
| 70 | 98.1 | | |
| % Albedo | | | |
| 10 | 118.8 | | |
| 15 | 116.2 | | |
| 20 | 113.7 | | |
| 30 | 108.7 | | |
| **Surface type and aerodynamic roughness around the site**[b] | | | |
| Irrigated smooth | 95.0 | Irrigated rough | 77.5 |
| Desert smooth | 107.4 | Desert rough | 98.6 |
| Asphalt smooth | 124.5 | Asphalt rough | 104.9 |
| Water smooth | 93.9 | Water rough | 77.0 |

[a] Values from Brazel and Howard[14] using equilibrium surface temperature energy budget model.
[b] Roughness is aerodynamic roughness ranging from smooth at close to 0.0 in.; rough = 3.9 in.

**TABLE 3.6**

Urban Canyon Effects: Height/Width (H/W) of Buildings, and Street Temperature Excess/Reductions (°F)

| | Nighttime | | Daytime |
|---|---|---|---|
| H/W | Observed Temp. Excess[a] | Model Temp. Excess[b] | Temp. Reduction (obs) |
| 0.25 | | +0.9 | |
| 0.5 | 1–1.8 | +1.8 | −4.5 (−10.0 at well ventilatedsite) (in Tempe, Arizona) |
| 1.0 | 1–1.8 | +3.6 | −9.0 (Negev Desert Study, Israel[c]) |
| 2.0 | 1–1.8 | +5.4 | −7.9 (Tempe, Arizona) |
| 3.0 | | +6.3 | |
| 5.0 | | +8.1 | |
| 10.0 | | +9.9 | |

[a] Modeled nighttime data after Oke.[13]
[b] Studies in Tempe (difference is street canyon minus ground site outside the canyon).
[c] After Pearlmutter[16] (difference is street canyon minus roof top).

**TABLE 3.7**

Modeled Urban Canyon Geometries and Anticipated Flow Regimes[a]

| L/H Ratios[c] | W (ft) | H/W Ratios | Type of Flow |
|---|---|---|---|
| Cubic canyon (66/66 ft) | 66 | 1.0 | Skimming flow |
| | 131 | 0.5 | Wake interference |
| | 262 | 0.25 | Isolated roughness |
| Short canyon (197/66 ft) | 66 | 1.0 | Skimming flow |
| | 131 | 0.5 | Wake interference |
| | 262 | 0.25 | Isolated roughness |
| Medium-length canyon (328/66 ft) | 66 | 1.0 | Skimming flow |
| | 131 | 0.5 | Wake interference |
| | 262 | 0.25 | Isolated roughness |
| Long canyon (459/66 ft) | 66 | 1.0 | Skimming |
| | 131 | 0.5 | Wake interference |
| | 262 | 0.25 | Transition to isolated roughness |

*Source:* Hunter, I. et al., *Energy Build.*, 15, 315, 1990/1991.

L, canyon length (first number in parenthesis); H, height of canyon (second number in parenthesis); W, canyon width.

[a] Wind direction assumed perpendicular to the canyon. Well-spaced flow elements create isolated flow of turbulent air moving around buildings; moderate compactness leads to what is called "wake interference" as disturbed air has insufficient distance to readjust before encountering next building; very compact building arrays cause mesoscale skimming over the top of the canyon and air is decoupled from within the canyon and above roof height. Data given earlier illustrates the geometries and flow types.

For each of these seemingly simple scenarios, an integrated set of views of benefits and liabilities for citizens and their quality of life in the city should be evaluated. This is often done by considering mitigation schemes (e.g., increase albedo and amount of vegetation) to cool down city temperatures and reduce pollution problems as well.* The acceptance of strategies for mitigation depends on the will of cities to design, plan, and evaluate the significance of local-induced climate change in the context of ongoing global changes.[16] This will is more and more emerging among cooling-community advocates for many cities.[17]

## 3.6 Conclusions

Climate is often assumed as a background environmental factor against which we can plan for today and tomorrow. I hope that the aforementioned gives some appreciations for the fact that climate can be considered to be arranged according to scale effects which cause interactions we should be studying in depth. Climate is not static, contrary to its apparent definition as simply an ensemble of weather. Climate may change due to anthropogenic as well as natural forces. Climate records for the past 100 years vary in their quality and quantity, both in space and time. Employing records to solve real-world problems can be one of the most challenging applied endeavors, requiring a close association between

* http://www.heatislandmitigationtool.com/ (accessed July 15, 2009).

climate data consumers and the community of researchers and scholars who are the disseminators and facilitators of information flow. The climate system is:

> ... highly complex ... interacting with an enormous number of at least equally complicated ecological systems, each with innumerable degrees of freedom, and a large number of potential modes of activity...there are ample opportunities for the small, seemingly insignificant, factors to combine in unexpected ways, and produce surprise ... (Akbari et al.[17]).

## References


1. R.G. Barry, A framework for climatological research with particular reference to scale concepts, *Transactions and Papers of the Institute of British Geographers* 49 (1970): 61–70.
2. M.E. Mann, Climate during the past millennium, *Weather* 56 (2001): 91–102.
3. P.R. Sheppard, A.C. Comrie, G.D. Packin, K. Angersbach, and M.K. Hughes, The climate of the US Southwest, *Climate Research* 21 (2002): 219–238.
4. S. Solomon, D. Qin, M. Manning, Z. Chen, M. Marquis, K.B. Averyt, M. Tignor, and H.L. Miller, Eds., Climate change 2007: The physical science basis, in *Contribution of Working Group I to the Fourth Assessment Report of the Intergovernmental Panel on Climate Change* (Cambridge, U.K.: Cambridge University Press, 2007).
5. P.A. Stott, S.F.B. Tett, G.S. Jones, M.R. Allen, J.F.B. Mitchell, and G.J. Jenkins, External control of 20th century temperature by natural and anthropogenic forcings, *Science* 290 (2000): 2133–2137; T.R. Karl and K.E. Trenberth, Modern global climate change, *Science* 302 (2003): 1719–1723.
6. K. Redmond, Climate change issues in the mountainous and intermontane west, Regional Climate Change Workshop, Rocky Mountains/Great Basin, Salt Lake City, Utah, unpublished manuscript, 1998.
7. W.A. Sprigg and T. Hinkley, Preparing for a changing climate—The potential consequences of climate variability and change (Tucson, AZ: University of Arizona, Institute for the Study of Planet Earth, National Oceanic and Atmospheric Administration, and United State Geological Survey, Report of the Southwest Regional Assessment Group, 2000).
8. N.S. Christensen, A.W. Wood, N. Voisin, D.P. Lettenmaier, and R.N. Palmer, The effects of climate change on the hydrology and water resources of the Colorado river basin, *Climatic Change* 62 (2004): 337–363.
9. A.W. Ellis, T.W. Hawkins, R.C. Balling, Jr., and P. Gober, Estimating future runoff levels for a semi-arid fluvial system in central Arizona, USA, *Climate Research* 35 (2008): 227–239.
10. K.T. Redmond and R.W. Koch, Surface climate and streamflow variability in the Western United States and their relationship to large-scale circulation indices, *Water Resources Research* 27 (2001): 2381–2399; A.J. Brazel and A.W. Ellis. The climate of the Central Arizona and Phoenix long-term ecological research site (CAP LTER) and links to ENSO, in *Climate Variability and Ecosystem Response at Long-Term Ecological Research Sites*, D. Greenland, D.G. Goodin, and R.C. Smith, Eds., pp. 117–140 (Oxford, U.K.: Oxford University Press, 2003).
11. A.J. Brazel, Future climate in central Arizona: Heat and the role of urbanization (Tempe, AZ: Arizona State University, Consortium for the Study of Rapidly Urbanizing Regions, Research Vignette No. 2, 2003).
12. R. Lougeay, A.J. Brazel, and M. Hubble, Monitoring intraurban temperature patterns and associated land cover in Phoenix, Arizona using Landsat thermal data, *Geocarto International* 11 (1996): 79–91.
13. T.R. Oke, Canyon geometry and the nocturnal urban heat island: Comparison of scale model and field observations, *Journal of Climatology* 1 (1981): 237–254.

14. A.J. Brazel and D. Howard, Surface climate and remote sensing of Scottsdale, Arizona (Tempe, AZ: Arizona State University, School of Planning & Environmental Design, Report to City of Scottsdale, 1998).
15. I. Hunter, I.D. Watson, and G.T. Johnson, Modeling air flow regimes in urban canyons, *Energy and Buildings* 15–16 (1990/1991): 315–324.
16. D. Pearlmutter, Street canyon geometry and microclimate: Designing for urban comfort under arid conditions, in *Environmentally Friendly Cities, Proceedings of PLEA* '98, E. Maldonado and S. Yannas, Eds., pp. 163–166 (Lisbon, Portugal: James & James Publishers Ltd, 1998).
17. H. Akbari, S. Davis, S. Dorsano, J. Huang, and S. Winnett, Eds., *Cooling Our Communities: A Guidebook on Tree Planting and Light-Colored Surfacing* (Washington, DC: U.S. Protection Agency, 1992), 217 pp.

# 4

# *Water Resources in the Desert Southwest*

**Robert H. Webb and Stanley A. Leake**

**CONTENTS**



## 4.1 Introduction

As the old saying goes, there is nothing more precious than water in the desert. The Ancestral Puebloans, Hohokam, and other pre-Columbian cultures knew this and built their civilizations near guaranteed water supplies. When the Spaniards arrived in present-day Arizona, they found that the Tohono O'odham and Piman cultures had settled in prime riverine sites, turning perennial flow through lush riparian ecosystems into irrigation water for productive agriculture. The Spaniards followed suit, building their missions along perennial reaches of the Santa Cruz River, including at one place aptly named "Punta de Agua" (Point of Water) south of Tucson. When the Mormons spread southward from Utah in the 1870s, their destinations were riverside settings on the Little Colorado, Salt, and San Pedro Rivers (Figure 4.1).[1]

At the beginning of the twenty-first century, water is even more precious in the desert Southwest. The single greatest source of water in the region, the Colorado River, is overallocated and drought depleted,[2] leaving an elaborate agricultural economy and growing metropolitan areas at the mercy of sustained drought. From the late 1970s through the mid-1990s, severe flooding in the Gila River system had reservoirs brimming while unplanned releases forced draconian flood protection and zoning downstream; now, some of those reservoirs store more sediment than water as the first decade of the early twenty-first century

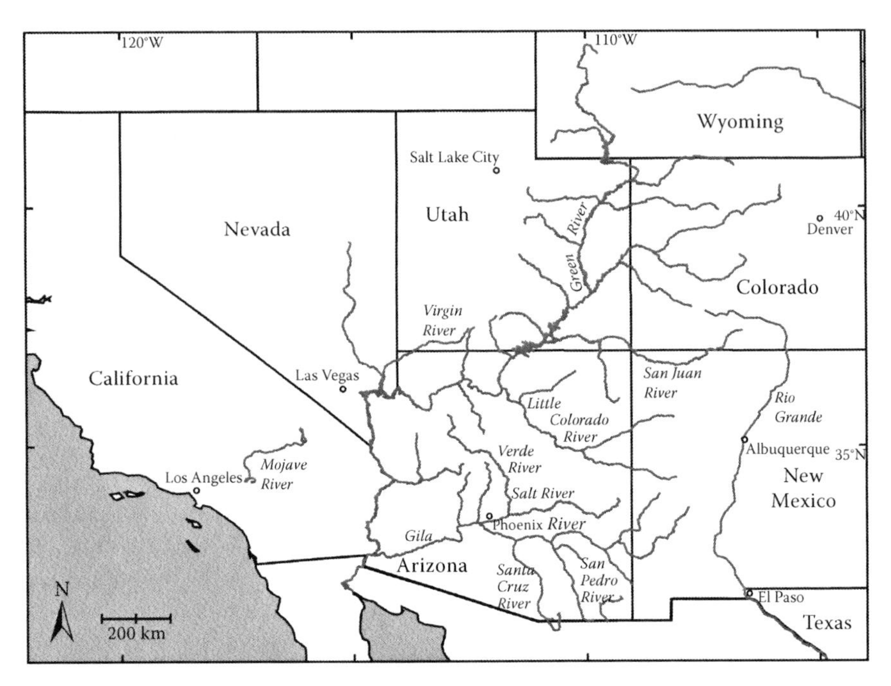


**FIGURE 4.1**
Map of the southwestern United States showing the locations of principal rivers.

is dominated by drought.* What once was thought to be an endless dependable water supply beneath the ground has turned into declining water tables and land subsidence. The desert cycle of floods and droughts leaves water managers and residents alike scratching their heads and wondering just what in the hydrologic world they can depend on.

Understanding of the desert hydrologic cycle begins with a recognition of climate and the influence of climate variability on surface water and groundwater. It extends to rainfall, runoff, and recharge, the three "r's" of hydrology. Rainfall is spatially and temporally variable in arid and semiarid regions, and its deficit defines the deserts. Runoff can turn into floods, which can both damage and nurture. Recharge is the deposit in our hydrologic savings account, hopefully not to be withdrawn at a higher rate than the additions. Interactions between surface water and groundwater, often neglected under legal protections, both help and hurt riparian ecosystems. Taken together, these processes can tell us what to expect in the hydrologic future in the desert Southwest, and how we might help thwart the persistent attempts of the harsh desert to keep our water-loving civilization at bay.

## 4.2 Hydrologic Settings in the Desert Southwest

Especially given its complex systems of water development, the desert region in Arizona and parts of adjacent states (Figure 4.1) have the most complex hydrologic systems in North

* Considerable uncertainty exists in future climate predictions; see Chapter 3. The early twenty-first century drought may or may not be over, depending upon one's definition of drought as a specific period of concurrent time with below-average rainfall, or whether a more general drought period, punctuated by wet years but overall with below-average precipitation is used. In this sense, we are applying the latter concept.

America[3] due, in no small part, to the complexity of the regional geologic framework (see Chapter 2). Here, we restrict our discussion to the Mojave and Sonoran Deserts and the Colorado Plateau, although our information generally applies to the other deserts as well. The Mojave and Sonoran Deserts are part of the Basin and Range Province, and fault-block mountains bound deep alluvial basins. The Colorado Plateau is a physiographic province characterized by low tectonic activity and few active faults, and the landscape either is bare bedrock or consists of shallow sediments over bedrock.

Most of the drainage systems on the Colorado Plateau and in the Sonoran Desert drain to the Colorado River and ultimately to the Gulf of California, while most in the Mojave and Great Basin Deserts drain to closed basins called playas (see Chapter 1). Precipitation in closed basins does not runoff directly to the oceans; instead, water pools in intermittent lakes and either recharges to groundwater or evaporates. The principal through-flowing rivers include the Green, Colorado, and San Juan Rivers in the north; the Gila River system in the south, the principal drainage of most of Arizona, consists of the Verde, Salt, and Gila Rivers with the lesser San Pedro and Santa Cruz River draining southeastern Arizona (Figure 4.1). In the Mojave Desert, the Owens and Mojave Rivers are the only major drainages, and each terminates in a major playa system.

All of the principal drainage basins feeding into the Colorado River originate in highland areas that receive far more annual precipitation than the desert lowlands. For the entire Colorado River drainage, the area of highest precipitation is in the Rocky Mountains of Colorado, and the area of lowest precipitation is in the delta area south and west of Yuma, Arizona (Figure 4.1). In the adjacent Great Basin, again the highest precipitation is in the higher elevation mountains and the lowest precipitation is in low-elevation closed basins such as Death Valley. Given this disparity in higher precipitation in headwaters areas, combined with more agriculture and urbanization in the desert lowlands, it is not surprising that extensive surface-water regulation networks exist to capture, store, and transfer water from where it is generated to where it is used.

In terms of areal extent, groundwater systems generally do not correspond to surface-water drainage basins.* Instead, groundwater basins correspond to subsurface geologic structure and rock permeability. For example, major springs in Death Valley are an outflow from an extremely large groundwater basin that extends into central Nevada.[6,7] On the Colorado Plateau, sandstone units form the principal aquifers, and these can be isolated from other aquifers by intervening shale or limestone units.[8] Through most of the Mojave and Sonoran Desert, however, groundwater basins are smaller and more closely related to the Basin and Range structural framework with basin-fill sediments adjacent to bedrock of the mountains (Figure 4.2).

## 4.3 Hydroclimatology

Climate, as previously discussed in this book (see Chapter 3), strongly affects the hydrology of the Southwest. Hydroclimatology is the study of interactions between climate and hydrologic processes. An understanding of the origins and interannual and seasonal fluctuations of surface water, as well as longer-term fluctuations in groundwater levels,

* For a comparison of surface-water drainage basins with groundwater basin areas in Arizona, see Seaber et al.[4] and Arizona Department of Water Resources.[5]

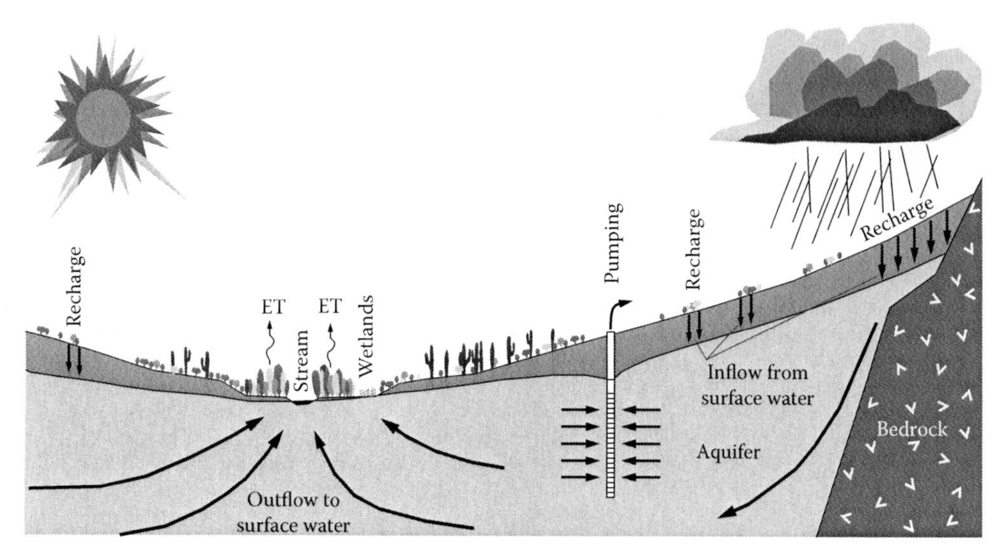


**FIGURE 4.2**
Generalized hydrologic cycle in the southwest depicting surface water-groundwater interactions that are typical of many alluvial aquifer systems in the Basin and Range. ET, evapotranspiration.

requires recognition of global- and hemispheric-scale climatic processes that affect delivery of moisture to the region. Flood frequency, in particular, is affected by hydroclimatic processes (see Chapter 5).[9]

In terms of land area, most of the region receives less than 10 in. of precipitation, most of which occurs as rainfall.[10] However, headwaters of many rivers in the region are above 8000 ft and receive more than 20 in. of precipitation, most of which arrives as snowfall in late fall, winter, and spring. Summer storms are common, with the exception of the western Mojave Desert and the headwaters of the Mojave River. Reliable runoff, generated during the winter and spring months, has high sediment concentrations; summer-storm runoff is more episodic, generally is in lower volume, and contains very high sediment concentrations. As a general rule, the climatic and physiographic setting dictates that most of the runoff is generated at higher elevations while most of the sediment comes from lower elevations.

Summer rainfall generally results from intense thunderstorms of local extent. This seasonal precipitation tends to be reliable and strongly affects regional vegetation patterns, essentially defining the spatial extent of the Sonoran Desert (see Chapter 7).[11,12] In addition to the normal summer thunderstorms, tropical cyclones, a generic term that includes tropical depressions, tropical storms, and hurricanes, contribute a significant, although unreliable, amount of precipitation at any time from June through October.[13] These storms form in either the eastern North Pacific Ocean, the Gulf of Mexico, or the Atlantic Ocean. Few tropical storms or hurricanes have made landfall in the southwestern United States; most dissipate over the ocean, and their leftover moisture is advected into the region, either in weak monsoonal flow or strong systems from the North Pacific.

The El Niño-Southern Oscillation (ENSO) phenomenon of the Pacific Ocean strongly affects interannual variation in precipitation in the Southwest (see Chapter 3 of this volume).[9,12,14–17] ENSO effects can be separated into three general categories: warm ENSO events (commonly known as El Niño conditions), cool ENSO events (commonly known as La Niña conditions), and other conditions (also see the discussion in Chapter 3). Precipitation during El Niño conditions generally is high, although notable dry periods have occurred as well. Winter precipitation is the most affected; the effects of El Niño on summer precipitation are weak at best. The largest floods in the region tend to be associated with El Niño conditions.[9] La Niña conditions are predominantly dry, and most significant regional droughts are associated with this climate state. Anything except extreme wet or dry periods can occur during other conditions.

Despite geographic variation in hydroclimatology, decadal-scale climatic fluctuations appear to affect the region in a relatively uniform fashion (see Chapter 3). While all of the region may simultaneously experience dry or wet conditions, the magnitude and persistence of unusual climatic conditions varies. While wet conditions generally are uniform from the Mojave River through southern Arizona, droughts seldom are uniform in severity or length. Orographic effects and seasonality of precipitation add to the complexity, making general statements about climatic variability difficult. Finally, average temperatures are related to precipitation; temperatures tend to be annually or seasonally high during droughts and may be relatively low or high during wet periods.

Analyses of climatic trends provide a general framework of decadal climatic fluctuations affecting hydrology of the Southwest.[2,12,17,18] The period of 1880 through 1891 was generally wet, with numerous regional-scale storms that caused channel downcutting and generally led to the observation that "rainfall follows the plow." The most severe drought, and the one that affected the largest amount of the region, occurred between 1891 and 1904. The combination of overstocking of the range and the drought led to the death of half of the cattle in the region between 1891 and 1896. El Niño conditions in 1904 and 1905 ended the drought, and the wettest period in the region's history began in 1909 and extended through about 1920. This period continues to cause water problems in the southwestern United States because above-average flows in the Colorado River were divided among seven western states according to the Colorado River Compact, resulting in the current problem of overallocation of water supplies in this critical river.[2]

Climate was regionally variable between 1920 and the early 1940s, ending with the strong El Niño conditions of 1941 through 1942. In southern Arizona, conditions were relatively dry with few significant winter storms. From the Mojave River through southern Utah, conditions were generally wet, punctuated with a mild drought during the Dust Bowl years of the early 1930s. Between the mid-1940s and the early 1960s, drought conditions prevailed with considerable regional variation in intensity. The mid-century drought, centered on the La Niña conditions of 1954 through 1956, was most severe in the Mojave Desert, in southern Utah, and to the east in New Mexico. Normal and above-average summer precipitation mitigated this drought in central and southern Arizona.

Beginning in the early 1960s, and fueled by several significant El Niño periods, the climate of the region became significantly wetter and warmer. Numerous strong storms in fall and winter occurred between 1970 and 1995, leading to significant floods in central and southern Arizona and above-average precipitation in the Mojave Desert and the Colorado Plateau. Notable periods of El Niño conditions occurred from 1978 through 1980, 1982 and 1983, and 1993 through 1995. Brief droughts interrupted this wet period in 1986 and 1989 through 1991, with the latter event having severe effects in the Mojave Desert.

Despite El Niño conditions in 1997 through 1998 and 2002 through 2003, winter drought generally prevailed at the end of the twentieth century through 2004. The drought centered on 2002 created several record extremes, including the lowest flow in the Colorado River in southern Utah and record low annual rainfall at many stations. The reasons for switching of interdecadal periods between generally wet and generally dry conditions remains speculative, but research centers on hemispheric-scale, low-frequency oceanic processes in the North Pacific and North Atlantic Oceans.[19] In 2005, despite predictions of a 30 year drought beginning in 1996, record winter rainfall occurred in the desert Southwest, further complicating our abilities to understand and predict interannual variation in precipitation.

## 4.4 Geomorphology

Several types of channels can be defined in the Southwest, depending on the substrate under the channel and flow regime. Bedrock channels occur in canyons that span a complex array of geomorphic configurations. Some canyons, notably those carved by Kanab Creek and the Virgin River, consist of a thin veneer of alluvium over bedrock. Most "bedrock canyons" have a relatively thick alluvial fill that has both coarse-grained and fine-grained terraces. For example, Grand Canyon, although usually considered to be a bedrock canyon, has considerable alluvial fill beneath the channel.[20] Alluvial channels form in deep alluvial fills with little or no bedrock constraints on lateral channel migration or vertical downcutting. This general class of channels has more variation than bedrock canyons owing to the complex interactions among surface water, groundwater, and subsurface geology. Most alluvial channels in the Southwest are arroyos, or channels that deeply incised into alluvium at the end of the nineteenth century (Figure 4.3). The depth of these channels is dependent on regional base-level control, which regionally is sea level in the Gulf of California but locally can be a variety of geologic structures or units and usually those that affect the major rivers.

Because of their geomorphic history, alluvial channels develop terraces at varying heights and distances away from the channel thalwegs, or deepest points. The heights of terraces reflect both their depositional age and stability. Whether streamflow is perennial or not in alluvial channels is dependent on bedrock structure, the age of alluvial terraces, variation in the particle size of the fill sediment, and regional groundwater flow. Faults create discontinuities in bedrock and alluvial aquifers and can force groundwater to the surface. Cementation of alluvial terraces, a common occurrence with increasing age, or where groundwater has high concentrations of calcium carbonate, can restrict downward water movement. Fine-grained alluvial fills can also restrict downward movement of groundwater, locally raising water level above the regional water table.

Over most of the Southwest, and particularly in the Basin and Range, bedrock canyons become alluvial channels at the mountain front (Figure 4.2). On the Colorado Plateau, the complex structural geology creates a variety of transitions between alluvial and bedrock reaches. On the Escalante River, an alluvial channel transitions into a bedrock canyon downstream from Escalante, Utah, creating an abrupt change in the type and stability of riparian ecosystems.[21] This type of transition is less common in the Basin and Range, but smaller-scale examples occur in the middle reach of the San Pedro River and along several reaches of the Gila River upstream from the confluence with the San Pedro River and downstream from the confluence with the Salt River (Figure 4.1). The lower Colorado River represents a large-scale example of a river with alternating bedrock canyons and alluvial fills.

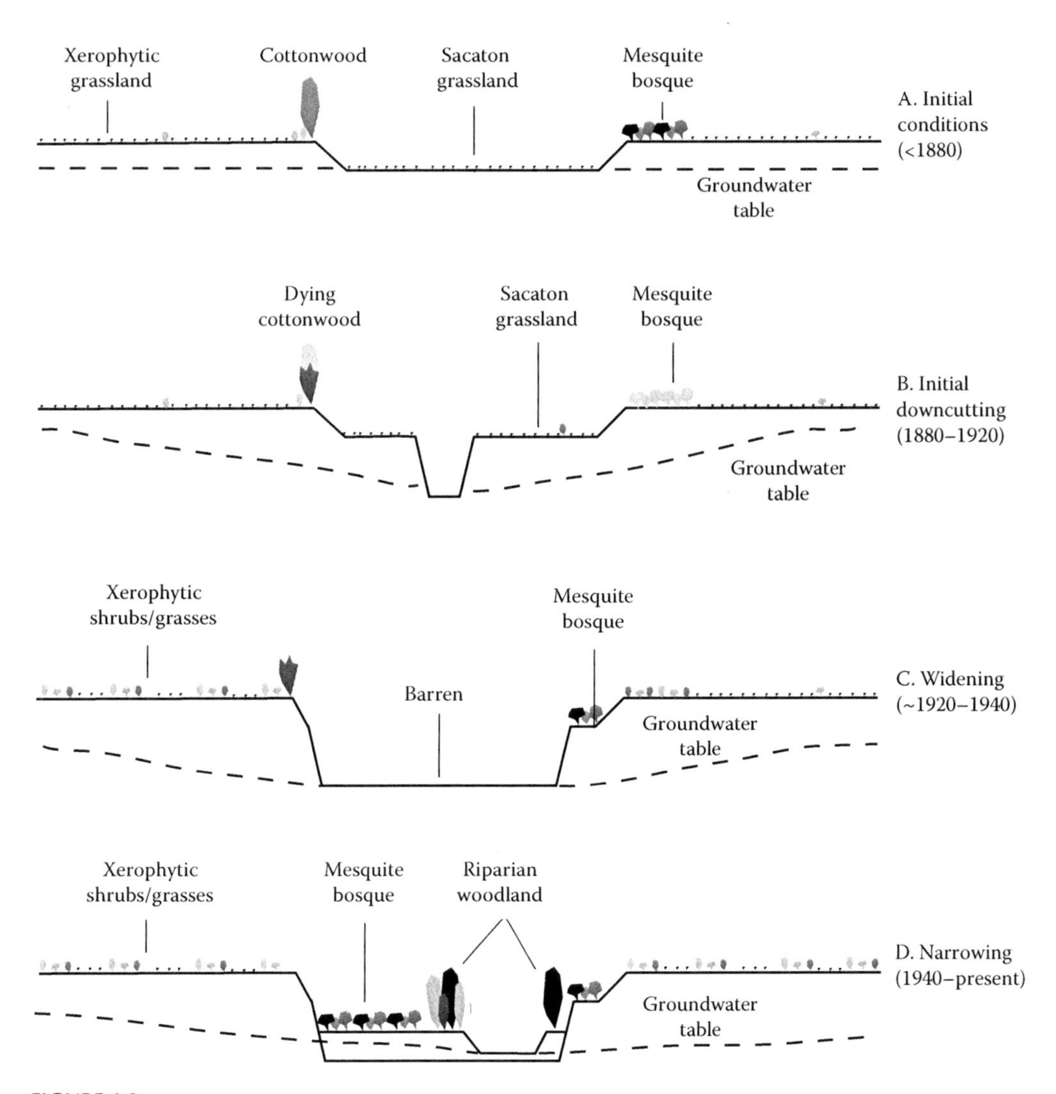


**FIGURE 4.3**
Schematic diagram showing the interaction between arroyo downcutting and riparian vegetation in southern Arizona.

## 4.5 Surface Water

Several metrics describe flow in rivers. Flow is segregated by water years, normally defined from October 1 through September 30; because the traditional definition splits the critical early fall tropical-cyclones season, we redefine the water year in the Southwest as November 1 through October 31.[9] Discharges are calculated using stage-discharge relations, developed from a combination of streamflow measurements and indirect-discharge calculations following floods.[22] Stage, the height of water above an arbitrary datum, is the only continuously measured variable at gauging stations, usually recorded at 15 min intervals. Over a 24 h period, unit discharges are averaged to create daily discharge, the basic measurement of surface-water resources.

Watercourses in the Southwest have three general categories of surface-water flow. Only the largest rivers have perennial flow, and these include the Green, Colorado, San Juan, Escalante, upper Virgin, Verde, upper Salt, and upper Gila Rivers (Figure 4.1). During some extreme droughts, some perennial rivers may have interrupted flow in reaches that cross deep alluvial basins. Intermittent streams may have periods of sustained flow, particularly during wet periods, but are dry at other times. This type of watercourse generally flows over an alluvial basin with high groundwater levels. Ephemeral streams only flow during storm runoff. Most mid- to low-elevation drainage basins are ephemeral, although intersection with local or regional groundwater systems may create short reaches of perennial flow. Owing to the combination of human influences and climatic fluctuations, intermittent and ephemeral streams have become drier historically, and certain reaches that once may have been perennial, such as the lower Gila River, are now mostly dry except during storm runoff, urban wastewater discharge, or irrigation returns.

In intermediate-sized drainages, and particularly those with relatively low-elevation headwaters, have alternating perennial and ephemeral reaches. Excellent examples include the San Pedro River, where perennial flow extends northward from south of the U.S.–Mexico border to downstream of Charleston and ephemeral flow occurs northward, except in the reaches near Cascabel, north of Redington, and near the confluence with the Gila River. Similarly, the Mojave River has perennial reaches in and north of Victorville, historically near Camp Cady, and within Afton Canyon. As one moves downstream, influent reaches increase in discharge owing to groundwater additions, transitioning to effluent reaches where flow infiltrates into the aquifer. Alternatively, gaining streams are influent while losing streams are effluent. Influent and effluent sections are associated with geologic structures, including faults and shallowly buried bedrock. Rivers that once had this natural configuration, such as the Santa Cruz and lower Gila Rivers, now have artificial perennial-ephemeral reaches owing to excessive groundwater use, irrigation returns, and wastewater effluent discharge.

Most towns and cities in the desert Southwest do not use significant surface water for domestic or municipal supplies. Exceptions include municipalities served by Central Arizona Project (CAP) water from the Colorado River and Salt River Project (SRP) water from the Salt and Verde Rivers. The Colorado River Compact, signed in 1922 and modified several times afterward, allocates 2.8 million acre-feet of water to Arizona in addition to the 1 million acre-feet supplied by the Gila River (including the Salt and Verde Rivers). The CAP reached the Phoenix metropolitan area in 1983 and was completed to its current length in 1993.* The SRP delivers about a million acre-feet of water to the Phoenix metropolitan area via a network of canals patterned, at least in part, after Hohokam canals built a millennium before settlement. Roosevelt Dam, the largest of a series of dams on the Salt and Verde Rivers, is one of the first large dams built in the United States and was completed in 1911.[23]

Water development has far greater effects on streamflow than does climatic variation. The lower Colorado River is a reach that has extremely high flow regulation and diversion below Hoover Dam.[21] Comparison of the annual flow volume at gauging stations downstream from this dam shows the magnitude of water diversion (Figure 4.4). At Yuma (Figure 4.4D), the record before 1935 represents the pre-dam water flow of the lower Colorado River. As dams and diversion structures are added, more water is removed until flow at Yuma is less than 10% of the pre-dam volume. Despite intensive flow regulation and flood control, the large regional floods in the late 1970s and early 1980s caused a small but notable increase in annual flow at Yuma.

* http://www.cap-az.com/static/index.cfm?contentID=20 (accessed February 9, 2009).

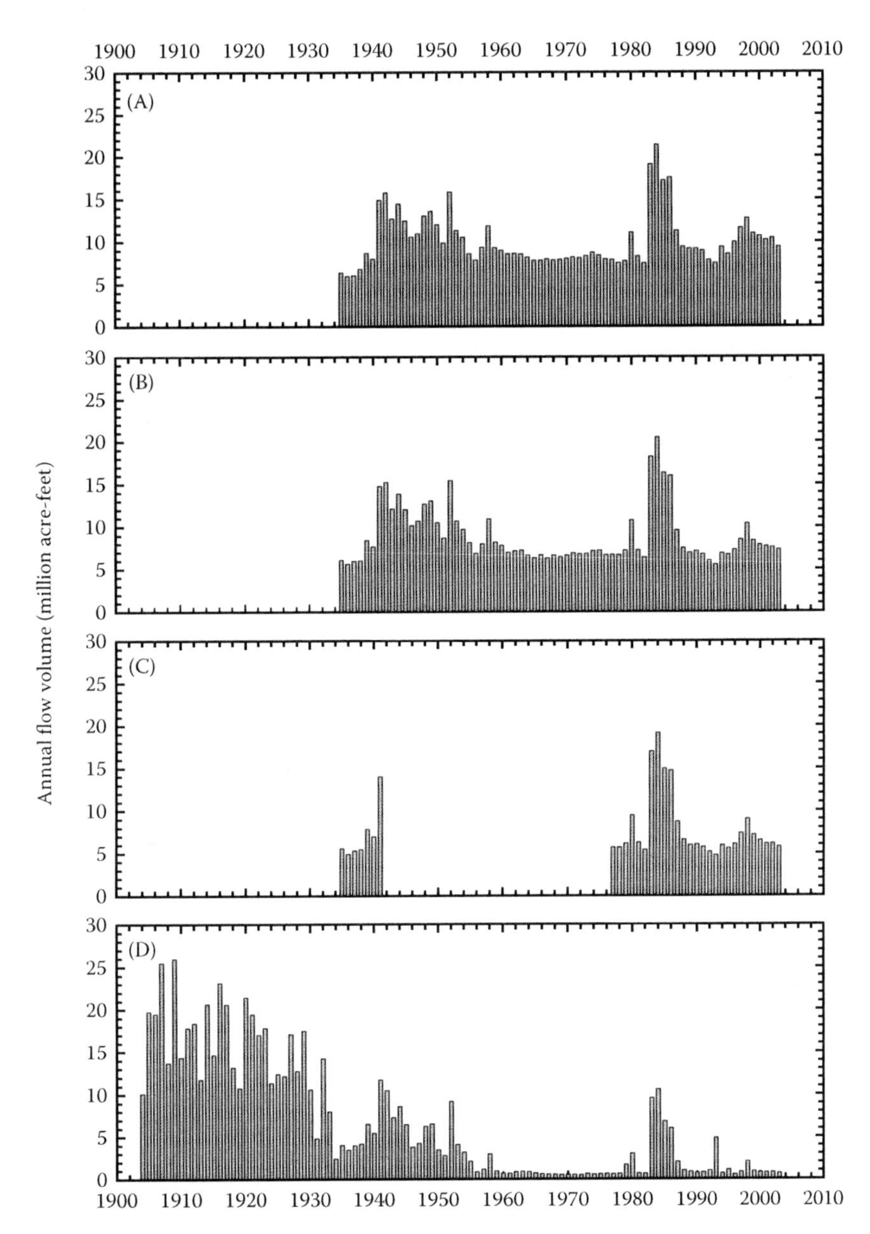


**FIGURE 4.4**
Annual flow volumes for the Colorado River. (A) Colorado River below Hoover Dam, Arizona—Nevada (USGS gauging station 09421500). (B) Colorado River below Parker Dam, Arizona—California (09427520). (C) Colorado River above Imperial Dam, Arizona—California (09429490). (D) Colorado River below Yuma Main Canal Wasteway at Yuma, Arizona (09521100).

Increasingly, surface-water supplies in arid and semiarid regions are overallocated owing to the demands of agricultural production competing with the extreme growth and water demands of urban areas. Water availability has become an increasing scientific challenge because of the interconnectedness of surface-water and groundwater systems and the legal framework of water allocations.[24] Surface-water availability is likely to become less dependable in the future as a result of the influence of sustained drought, increasing water needs for rapidly growing urban areas, and growing demands for sustainable water supplies for riparian ecosystems.

## 4.6 Groundwater

Groundwater consists of the two fundamental states of unsaturated and saturated zones delineated vertically by the water table. The unsaturated zone, as the name implies, is a three-phase system of gases, water, and sediment, with both gases and water in motion. In the unsaturated zone, water generally moves downward, but water may move upward either by capillary action, vapor transport, or hydraulic lift by plants. A contiguous saturated zone in permeable rocks, regardless of extent, is called an aquifer. Like surface water, groundwater moves according to elevation and pressure gradients, only the rates of movement are orders of magnitude lower than surface water.*

Major aquifers in the Basin and Range part of the Southwest deserts are characterized by having large volumes of water stored in pore spaces in the gravel, sand, silt, and clay that fill the basin. Most natural recharge occurs along edges of the basin fill where runoff crosses the interface between low-permeability rocks of the mountains and runs down coarse-grained alluvial channels. In areas with permeable rocks in the mountains, water can enter the aquifers as mountain-block recharge.† Influent stream reaches can also be considered a source of recharge, although a mass balance would likely indicate that the net change water available to an alluvial aquifer traversed by a stream is negative when both influent and effluent reaches and use by riparian vegetation is considered. Regardless of the mechanism, annual recharge to aquifers is usually small in relation to the volume of fresh water stored in the aquifers. Natural discharge from these aquifers occurs by flow to streams, springs, and wetlands; uptake by plants; and groundwater underflow to adjacent basin aquifers.

Major aquifers in the Colorado Plateau are characterized as having large volumes of water in areally extensive sandstone and other consolidated rock units. Recharge occurs through direct infiltration of rainfall and snowmelt, as well as through infiltration of runoff into narrow alluvial channels incised into the consolidated rocks. Like the basin aquifers, annual recharge to aquifers beneath the Colorado Plateau is small in relation to the volume of water in storage. From recharge areas, groundwater moves toward regional drains that include streams and springs below the Mogollon Rim (along the south edge of the Colorado Plateau), the Verde River, the Little Colorado River, and springs and streams in the Grand Canyon and tributary canyons. In addition to discharging to these features, some water also is used by phreatophytic vegetation where the water table is near land surface. The water table is relatively close to the land surface adjacent to much of the Verde and Little Colorado Rivers, but it can be at great depths exceeding several thousand feet below land surface under parts of the plateau adjacent to deep canyons and in structural basins. For individual aquifers underlying the Colorado Plateau, natural discharge also can occur as vertical movement to underlying or overlying aquifers.

Base flow in most perennial streams and rivers in the region ultimately is generated from groundwater discharge, either locally or, in the case of the Colorado River, from a long distance away. For rivers solely benefiting from groundwater discharge, base flow has little interannual variation but may have long-term trends owing to a variety of conditions,

* For a recent review of groundwater, see Alley et al.[25]

† An excellent source of information on groundwater recharge is contained in Stonestrom et al.[26]

including decadal-scale climatic variation, groundwater pumping, and changing water use for agriculture and riparian vegetation.

Groundwater levels fluctuate naturally and in response to water development. Riparian vegetation is known to cause diurnal fluctuations of as much as 1 ft during the growing season. Natural fluctuation in water levels may exceed 5–10 ft over a period of years, and low-frequency positive trends may be present owing to recharge beneath rivers or at mountain fronts, or in response to decreases in pumping. Negative trends usually signal effects of human withdrawals of groundwater, but persistent drought, channel downcutting, or increasing use by riparian plants may also cause water-level declines.

In the arid and semiarid southwestern United States, groundwater has been an important source of supply for agriculture, industry, and public use. Groundwater withdrawals in Arizona escalated rapidly in the middle part of the twentieth century when large-capacity turbine pumps became available[27] and electricity was brought to rural areas suitable for agriculture.[28–30] Approximately 80% of groundwater withdrawal in Arizona is used for agriculture,[28] although municipal use is increasing greatly owing to the rapid increase in the state's population.* The volume of pumped groundwater statewide peaked around 1984 for several reasons. First, the landmark Arizona Groundwater Management Act, passed in 1980, regulates groundwater in Active Management Areas (AMAs) around major metropolitan areas, such as Tucson and Phoenix.* Pumping intensified just prior to this Act, which created an artificial high spike as water users attempted to justify their allocations. Introduction of CAP water from the Colorado River, beginning in 1983, delivered significant amounts of water for irrigation, municipal uses, and groundwater recharge.[29] Finally, agricultural priorities shift with commodity markets, which can lead to less water-intensive crops or suspension of farming.[30]

Much of the large-scale development of groundwater in Arizona that began in the mid-twentieth century was in basin-fill aquifers The conventional wisdom at that time was that the "safe yield" of an aquifer was the rate of annual recharge and that an aquifer with pumping exceeding safe yield was in a state of "overdraft." Nonetheless, pumping in many of the desert-basin aquifers greatly exceeded the rate of annual recharge. Furthermore, undesired consequences came about in many cases, even at pumping at rates less than the safe yield of an aquifer. Various methods have been developed to assess change in groundwater conditions and address the question of safe yield,[31] but the most important characteristic is the rate of decline in groundwater levels. Major problems associated with large-scale pumping included falling water tables and increased pumping lift over large areas; loss of water available to connected streams, springs, and wetlands; land subsidence and surface fissuring[29]; and degraded water quality.

Groundwater development on the Colorado Plateau has lagged that of development in the basins in both timing and magnitude of withdrawals, but the needs for agriculture, public supply, and industry have increased in recent decades. Major consequences of groundwater development on the plateau have included loss of base flow in streams and deepening water tables.

As previously discussed, reduced water availability to streams, springs, wetlands, and riparian plants is a consequence of groundwater pumping both in the alluvial basins and in the Colorado Plateau parts of the southwestern deserts. Concerns for preserving surface-water flows and riparian systems stem from the need to protect surface-water rights, as well as aquatic and terrestrial ecosystems associated with desert streams. As illustrated in Figure 4.5, the effects of pumping on a groundwater-dependent riparian system can be

* http://www.azwater.gov/dwr/WaterManagement/Content/AMAs/default.htm (accessed February 9, 2009).[30]

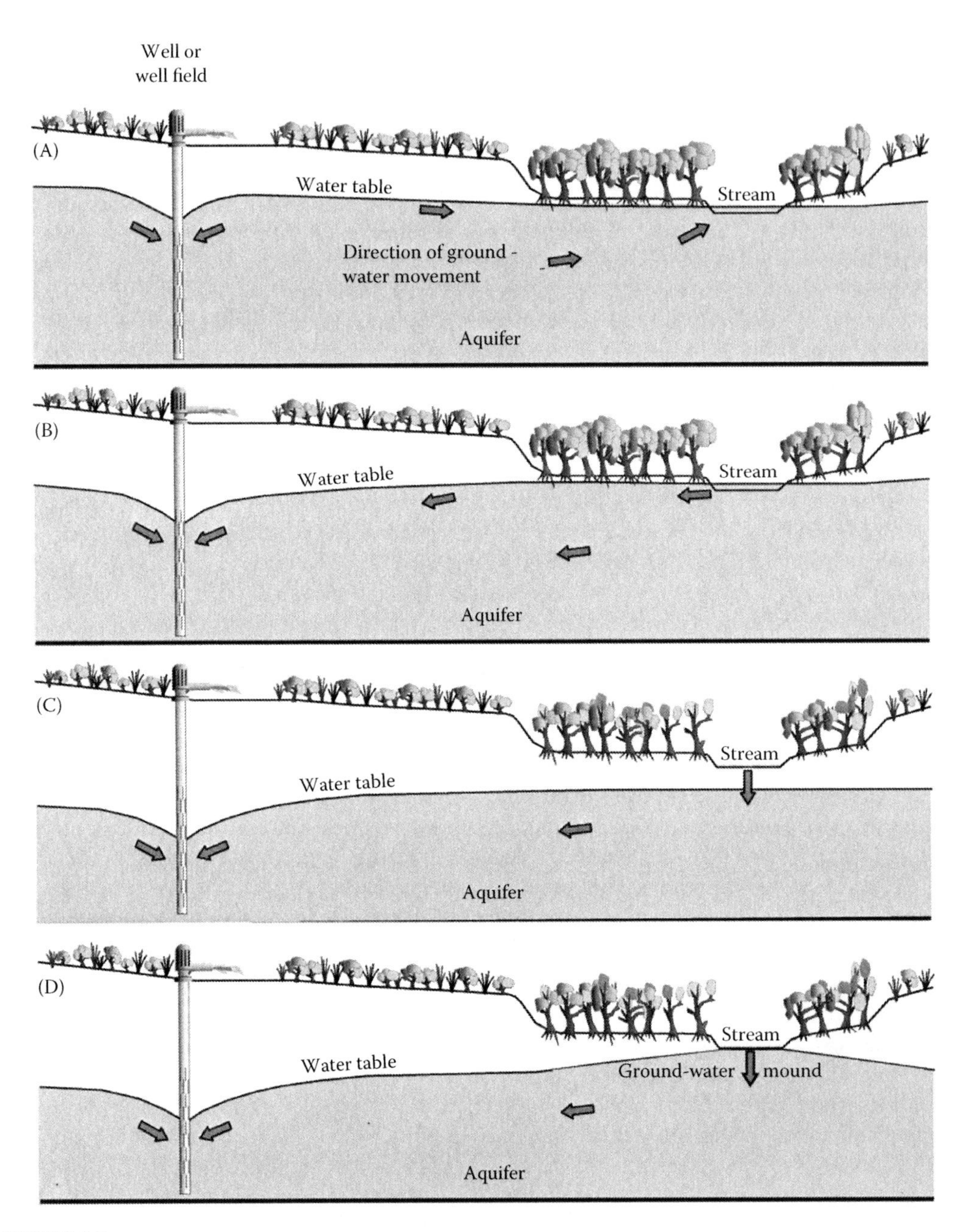


**FIGURE 4.5**
Schematic diagram showing the expected stages of capture of groundwater outflow to a perennial stream.

categorized into three stages. Stage I begins with the cone of depression that develops at the onset of pumping, locally affecting water stored around the well and not significantly impacting the reach-scale stream/aquifer system. Stage II begins after a substantial withdrawal draws down the aquifer sufficiently to create a water-level gradient away from the stream and floodplain. Finally, after a substantial period of pumping in excess of rate of groundwater flow from up-gradient areas, surface-water and groundwater systems may become disconnected during Stage III if streamflow cannot provide enough recharge to maintain water levels in the alluvial aquifer.

Groundwater in the region is important in sustaining unique desert ecosystems and in allowing humans to live and thrive in arid and semiarid areas with insufficient or no surface water. During the latter part of the twentieth century, the region experienced the most rapid population growth within the nation. Continued population growth and reliance on groundwater for public supply will require careful planning and management of resources to minimize undesired consequences of groundwater pumping.

## 4.7 Groundwater–Surface Water Connection

In the desert Southwest, most water rights for surface and groundwater are set by state law. In states such as Arizona and California, surface water and groundwater are considered legally to be distinct entities despite the hydrologic fact that they are intimately linked.[32,33] As discussed previously, surface water interacts in a complex way with geomorphology and groundwater in the Southwest. Discharge from bedrock or alluvial aquifers creates perennial or intermittent streams, and subsurface geologic structures or bedrock can force groundwater levels to the surface. Abrupt termination of such features allows surface water to infiltrate back into the aquifer. Finally, channel downcutting can reduce bed levels below the water table of the alluvial aquifer, releasing stored groundwater; conversely, aggradation can increase the distance between channel bed and the water table.

Continued groundwater pumping in alluvial basins can create a water table that slopes away from connected streams. Instead of effluent flow conditions, where groundwater discharges to a channel, influent conditions creates what is known as a groundwater mound immediately beneath the channel (Figure 4.5). This type of water table is common beneath channels with artificial flow, such as wastewater effluent released into natural channels. New riparian ecosystems may be able to take advantage of this limited-extent alluvial aquifer depending upon surface-water discharge and the extent of groundwater level declines from pumping in the vicinity.

Groundwater recharge beneath alluvial channels can be substantial during floods. For example, following the floods of December 1978, water levels rose up to 82 ft in wells in alluvial aquifers in southeastern Arizona.[34] Similar rises occurred following runoff in 1979, 1980, and 1983. Streamflow measurements during a February 1978 flood on the Gila River indicated that 112,000 ac ft (17%) of the inflow recharged the alluvial aquifer; an earlier measurement in January 1966 indicated that 175,000 ac ft (29%) of inflow was recharged.[35] Groundwater rises attributed to flood discharges can also be a problem. In 1979, rises in groundwater levels owing to flood-related dam releases to the lower Gila River caused waterlogging in agricultural lands, temporarily removing them from production.[34]

## 4.8 Water Quality

Water quality is extremely important in determining the suitability of water for domestic supplies, irrigation, or industrial use. Dissolved and suspended constituents vary considerably across the desert Southwest and depend in large part on geologic and soil characteristics in the headwaters of watersheds, where water is recharged into aquifers, and/or the characteristics and age of groundwater.[36] Although many inorganic constituents and organic compounds are regulated in drinking water,* here we will focus on four of the most commonly cited indicators of water quality: pH, hardness, salinity, and total dissolved solids (TDS).

pH of water determines much of its chemistry and suitability as a domestic supply. Typical waters in the western United States range in pH from slightly acid (around $pH = 6$) to highly basic ($pH > 10$) or neutral ($pH \approx 7.0$). Rainwater tends to be slightly acidic to very acidic, depending upon the amount of certain pollutants in the atmosphere, but its chemistry changes immediately upon contact with soils and vegetation. Throughout much of this region, pH is buffered by calcium carbonate and various salts that are dissolved as rainwater passes over rocks, through soils, and into aquifers, which raises the pH to 7–8 or higher. Hardness is a measure of the carbonate and bicarbonate content of water, whereas salinity is a measure of salt content.

Hardness is particularly important to the buildup of calcium carbonate ("scale") in pipes and plumbing fixtures in buildings. The carbonates that comprise hardness generally are highest in rivers and aquifers emanating from limestone and dolomite bedrock, sandstones with calcareous cement, or old soils bearing significant caliche (pedogenic carbonate). Salinity, which includes all salts but generally is dominated by sodium chloride, has many sources but is highest in sedimentary units—particularly shales and evaporites—that either accumulated salts under marine conditions or evaporation of freshwater lakes. Water high in salinity affects human health, can corrode pipes and water-delivery systems, and eventually will degrade agricultural lands subject to irrigation.

Perhaps the most important measure of water quality is total dissolved solids, or the amount of inorganic and organic compounds carried in solution. Generally, half to nearly two-thirds of the aquifers in the southwestern United States contain less than 500 parts per million (ppm) of TDS, and 70% of aquifers had less than 1,000 ppm, but some brackish aquifers can reach 10,000 ppm.[37] TDS in surface water varies considerably with the amount of runoff and the size of rivers; dissolved solids tend to be higher during drought periods and in larger rivers, such as the Colorado.

A good example of the impact of water quality on the urban environment is the effect of CAP water deliveries on Tucson in the 1990s. Colorado River water delivered in CAP canals and mixed with some surface water in central Arizona has significantly higher TDS than groundwater in the Tucson basin, as well as different chemical constituents.† Old pipes in some houses, which were conditioned to the lower TDS groundwater, burst after short exposure to CAP water, in part because the different water chemistry caused dissolution of compounds lining those old pipes. The problem appears to have been solved by mixing groundwater and CAP water (blended water),‡ thereby making a compromise water chemistry that doesn't impact domestic users.

## 4.9 Riparian Vegetation

Riparian vegetation is a substantial environmental resource in the southwestern United States that both provides a major esthetic element, provides ecosystem services for aquatic ecosystems, decreases flow velocities and flood peak discharges in headwater areas, but also can increase potential flood hazard in urban areas. Riparian vegetation generally increased

* http://www.epa.gov/waterscience/standards/ (accessed June 23, 2009).
† http://www.cap-az.com/operations/water-quality (accessed June 11, 2009).
‡ http://www.uswaternews.com/archives/arcquality/ttucwat11.html (accessed June 23, 2009).

in the late twentieth century, although major declines occurred in areas with major groundwater development.[21] In many reaches of perennial and ephemeral flow, riparian vegetation increased in the twentieth century, although much of that increase was in nonnative tamarisk or other invasive species. Riparian vegetation, while promoted by above-average winter precipitation and flooding in the late twentieth century, is easily removed by groundwater overdraft, construction associated with channelization and installation of bank protection for flood mitigation, diversion of most or all surface-water flow, construction of reservoirs and the resultant downstream alteration of seasonal flow, or a combination of these factors.

Following its removal, and at least initially, riparian vegetation reestablishment is promoted by wastewater effluent and bank protection. Xerophytic riparian vegetation—trees and shrubs that thrive with extra water but do not need it—colonized bank-protected channels through the Phoenix metropolitan area, and riparian vegetation has thrived in the Santa Cruz and San Pedro Rivers. Bank protection thwarts lateral channel change and focuses flow and groundwater discharge on a fixed channel location, creating a semblance of stability of water availability. As a result, dense stands of vegetation increases flow roughness and decreases channel conveyance, thereby increasing flood hazard. Because many people would rather see green riparian vegetation than bare channels, a tension has developed between environmental concerns and flood-hazard mitigation; soil-cemented reaches as originally designed can only effectively pass floods if vegetation is reduced or eliminated. One of the central design issues for urban areas is to create channels that can serve the dual purpose of environmental esthetics and flood-hazard mitigation.

## 4.10 Conclusions

River channels, surface water, and groundwater are integral to the southwest desert landscape, on the one hand representing a natural wildness and on the other a resource to be tamed and used. High variability of precipitation and surface runoff is one strong characteristic of arid regions; some hydrologic records display nonstationarity with less predictable water-supply characteristics. As a result, channels through urban areas are modified to account for this variability and uncertainty, leaving the seeming paradox of wide channels that are dry most of the time. The largest rivers in the region are regulated, either solely for flood control, for both flood control and water diversion or storage, or just for water use downstream. Groundwater systems, if unaffected by withdrawal, are less variable, but development has caused substantial lowering of water levels in the large alluvial basins of the region. Water quality in the region is highly variable but may be decreasing owing to declining surface-water supplies, high evaporation rates in water-delivery systems, and depletion of groundwater aquifers.

## References

1. McClintock, J.H., *Mormon Settlement in Arizona* (1985 reprint) (Tucson, AZ: University of Arizona Press, 1921).
2. Webb, R.H., R. Hereford, and G.J. McCabe, Climatic fluctuations, drought, and flow in the Colorado River, in S.P. Gloss, J.E. Lovich, and T.S. Melis, eds., *The State of the Colorado River Ecosystem in Grand Canyon* (Flagstaff, AZ: U.S. Geological Survey Circular 1282, 2005), pp. 59–69.

3. Hirsch, R.M., J.F. Walker, J.C. Day, and R. Kallio, The influence of man on hydrologic systems, in M.G. Wolman and H.C. Riggs, eds., *Surface Water Hydrology* (Boulder, CO: Geological Society of America, Decade of North America volume O-1, 1990), pp. 329–359.
4. Seaber, P.R., F.P. Kapinos, and G.L. Knapp, Hydrologic unit maps (Washington, DC: U.S. Geological Survey Water-Supply Paper 2294, 1987).
5. Arizona Department of Water Resources, Map showing Arizona groundwater basins with index of cities, towns, settlements and sites (Phoenix, AZ: Arizona Department of Water Resources, Open File Report 7, 1990).
6. Harrill, J.R. and D.E. Prudic, Aquifer systems in the great basin region of Nevada, Utah, and adjacent states (Washington, DC: U.S. Geological Survey Professional Paper 1409-A, 1998).
7. Belcher, W.R., ed., Death valley regional ground-water flow system, Nevada and California—Hydrogeologic framework and transient ground-water flow model (Las Vegas, NV: U.S. Geological Survey Scientific Investigations Report 2004–5205, 2004).
8. Leake, S.A., J.P. Hoffmann, and J.E. Dickinson, Numerical ground-water change model of the C aquifer and effects of ground-water withdrawals on stream depletion in selected reaches of Clear Creek, Chevelon Creek, and the Little Colorado River, Northeastern Arizona (Sacramento, CA: U.S. Geological Survey Scientific Investigations Report 2005–5277, 2005).
9. Webb, R.H. and J.L. Betancourt, Climatic variability and flood frequency of the Santa Cruz River, Pima County, Arizona (Washington, DC: U.S. Geological Survey Water-Supply Paper 2379, 1992).
10. Sellers, W.D., R.H. Hill, and M. Sanderson-Rae, *Arizona Climate* (Tucson, AZ: University of Arizona Press, 1985).
11. Turner, R.M., J.E. Bowers, and T.L. Burgess, *Sonoran Desert Plants* (Tucson, AZ: University of Arizona Press, 1995).
12. Turner, R.M., R.H. Webb, J.E. Bowers, and J.R. Hastings, *The Changing Mile Revisited* (Tucson, AZ: University of Arizona Press, 2003).
13. Smith, W., The effects of eastern North Pacific tropical cyclones on the southwestern United States (Salt Lake City, UT: National Oceanic and Atmospheric Administration Technical Memorandum NWS WS-197, 1986).
14. Andrade, E.R. and W.D. Sellers, El Niño and its effect on precipitation in Arizona, *Journal of Climatology* 8: 403–410, 1988.
15. Cayan, D.R. and R.H. Webb, El Niño/southern oscillation and streamflow in the western United States, in H.F. Diaz and V. Markgraf, eds., *El Niño, Historical and Paleoclimatic Aspects of the Southern Oscillation* (Cambridge, U.K.: Cambridge University Press, 1992), pp. 29–68.
16. Hereford, R., R.H. Webb, and S. Graham, Precipitation history of the Colorado plateau region, 1900–2000 (Flagstaff, AZ: U.S. Geological Survey Fact Sheet 119–02, 2002).
17. Hereford, R., R.H. Webb, and C. Longpré, Precipitation history and ecosystem response to multidecadal precipitation variability in the Mojave Desert and vicinity, 1893–2001, *Journal of Arid Environments* 67: 13–34, 2006.
18. Hereford, R. and R.H. Webb, Historic variation in warm-season rainfall on the Colorado Plateau, U.S.A., *Climatic Change* 22: 239–256, 1992.
19. McCabe, G.J., M.A. Palecki, and J.L Betancourt, Pacific and Atlantic ocean influences on multidecadal drought frequency in the United States, *Proceedings of the National Academy of Science* 101: 4136–4141, 2004.
20. Hanks, T.C. and R.H. Webb, Effects of tributary debris on the longitudinal profile of the Colorado River in Grand Canyon, *Journal of Geophysical Research, Earth Surface* 111: F02020, doi:10.1029/2004JF000257, 2006.
21. Webb, R.H., S.A. Leake, and R.M. Turner, *The Ribbon of Green: Change in Riparian Vegetation in the Southwestern United States* (Tucson, AZ: University of Arizona Press, 2007).
22. Rantz, S.E. et al., Measurement and computation of streamflow (Reston, VA: U.S. Geological Survey Water-Supply Paper 2175, 1982).
23. Rogge, A.E., D.L. McWatters, M. Keane, and R.P. Emanuel, *Raising Arizona's Dams* (Tucson, AZ: University of Arizona Press, 1995).

24. Anderson, M.T. and L.H. Woosley, Jr., Water availability for the western United States—Key scientific challenges (Reston, VA: U.S. Geological Survey Circular 1261, 2005).
25. Alley, W.M., R.W. Healy, J.W. LaBaugh, and T.E. Reilly, Flow and storage in groundwater systems, *Science* 296: 1985–1990, 2002.
26. Stonestrom, D.A., J. Constantz, T.P.A. Ferré, and S.A. Leake, Ground-water recharge in the arid and semiarid southwestern United States (Menlo Park, CA: U.S. Geological Survey Professional Paper 1703, 2007).
27. Green, D.E., *Land of Underground Rain: Irrigation on the Texas High Plains, 1910–1970* (Austin, TX: University of Texas Press, 1973).
28. Anning, D.W. and N.R. Duet, Summary of ground-water conditions in Arizona, 1987–1990 (Phoenix, AZ: U.S. Geological Survey Open-file Report 94–476, 1994).
29. Carpenter, M.C., South-central Arizona, in D. Galloway, D.R. Jones, and S.E. Ingebritsen, eds., *Land Subsidence in the United States* (Reston, VA: U.S. Geological Survey Circular 1182, 1999), pp. 65–78.
30. Tadayon, S., Water withdrawals for irrigation, municipal, mining, thermoelectric-power, and drainage uses in Arizona outside of active management areas, 1991–2000 (Reston, VA: U.S. Geological Survey Scientific Investigations Report 2004–5293, 2005).
31. Tillman, F.D., S.A. Leake, M.E. Flynn, J.T. Cordova, K.T. Schonauer, and J.E. Dickinson, Methods and indicators for assessment of regional ground-water conditions in the southwestern United States (Tucson, AZ: U.S. Geological Survey, Scientific Investigations Report 2008–5209, 2008).
32. Glennon, R., *Water Follies: Ground-Water Pumping and the Fate of America's Fresh Waters* (Washington, DC: Island Press, 2002).
33. Winter, T.C., J.W. Harvey, O.L. Franke, and W.M. Alley, *Ground Water and Surface Water: A Single Resource* (Reston, VA: U.S. Geological Survey Circular 1139, 1998), p. 79.
34. Aldridge, B.N. and T.A. Hales, Floods of November 1978 to March 1979 in Arizona and West-Central New Mexico (Reston, VA: U.S. Geological Survey Water-Supply Paper 2241, 1984), 36.
35. Aldridge, B.N. and J.H. Eychaner, Floods of October 1977 in Southern Arizona and March 1978 in Central Arizona (Reston, VA: U.S. Geological Survey Water-Supply Paper 2223, 1984).
36. Cordy, G.E., D.J. Gellenbeck, J.B. Gebler, D.W. Anning, A.L. Coes, R.J. Edmonds, J.A.H. Rees, and H.W. Sanger, Water quality in the Central Arizona Basins, Arizona, 1995–98 (Reston, VA: U.S. Geological Survey Circular 1213, 2000).
37. Anning, D.W., N.J. Bauch, S.J. Gerner, M.E. Flynn, S.N. Hamlin, S.J. Moore, D.H. Schaefer, S.K. Anderholm, and L.W. Spangler, Dissolved solids in basin-fill aquifers and streams in the Southwestern United States (Reston, VA: U.S. Geological Survey Scientific Investigations Report 2006–5315, 2007).

# 5

# *Geologic, Hydrologic, and Urban Hazards for Design in Desert Environments*

**Robert H. Webb, Stanley A. Leake, and Richard A. Malloy**

**CONTENTS**



## 5.1 Introduction

Settlement of the arid parts of the western United States required access to reliable water supplies, and the location of towns and the design of structures paid little heed to hazards inherent in this environment. Beyond the rattlesnakes and scorpions, a variety of geologic hazards pose substantial threats to urban and rural infrastructure. Settlers who chose to build along channels that offered dependable water quickly learned that flash floods are a significant problem when their irrigation dams and houses washed away. Others had their homes, often made of mud bricks, fall down during earthquakes. Waste was dumped into the ground or in waterways, or burned, fouling the air. Much of our knowledge of geologic and hydrologic hazards in the desert Southwest comes from the cumulative experience of settlers and their descendents with hazards, either natural or human-caused, and the net effect is that considerable regulatory authority is in place to manage and/or guide urban design in this region.

All hazard design ultimately is based on the concept of risk, or, more explicitly, acceptable risk. Clearly, we do not know the size of the absolute maximum flood that may impact a river basin or can we predict the largest possible earthquake that might occur in a given place; furthermore, elimination of risk from a given hazard would be prohibitively expensive in design and construction. Specific hazards are regulated under the concept of acceptable risk; for example, the acceptable risk for flood hazard is greater than the 100-year flood for floodplain construction or infrastructure, or, for critical infrastructure, greater than the 500- or 1000-year flood.[1] Understanding the level of risk in regard to geologic and hydrologic hazards is critical to design and sustainability of urban landscapes.

This chapter uses examples of hazards from the southwestern United States with a focus on the authors' experiences in Arizona to illustrate the array of hazards that may be present and affect urban and landscape design. Here, we discuss the regional risk for a variety of geologic, hydrologic, and urban hazards, focusing on the ones with greatest regional impact, and we refer readers to other publications by urban designers for unique ways to mitigate against these hazards.* We emphasize that while there may be commonality in hazards that affect floodplain management, urban design, and hazard mitigation, in most cases, local approaches based on scientific or engineering analyses of the hazard threat may be more important regionally than the typical "one-size-fits-all" approach used in the past.

## 5.2 Flood Hazards

Flooding on major rivers is the most common hydrologic hazard in arid and semiarid regions of the southwestern United States, and societal measures to protect against flood damage are readily apparent in the banks of most rivers. Nationwide, flood damage increased through the twentieth century (Figure 5.1)† and is statistically related to increases in multiday precipitation.[3] The reason for increasing amounts of flood damage involves a complex interaction among societal pressures to develop river floodplains, climatic variation, or change that influences storm intensities and our ability to estimate flood hazards and regulate flood-plain development.

All of the rivers in the southwestern United States have annual floods with discharges significantly higher than typical flows, which is by definition zero for most of the year in ephemeral channels (see Chapter 4). Although floods generally are considered as single-phase flow (water), desert rivers are decidedly two-phase flow with significant entrained sediment. Rivers with high-elevation headwaters, such as those crossing the Colorado Plateau, the Humboldt and Truckee Rivers in Nevada, the Rio Grande in New Mexico, and the Salt and Gila Rivers in Arizona, have annual flood peaks in response to snow melt or rarely as rain falling on existing snow packs. Lower-elevation rivers, such as the Santa

* Numerous publications discuss current and future designs for hazard mitigation, including Burby[2]; additional planning documents are available from the National Association of Floodplain Managers, http://www.floods.org/inex.asp?menuID=298&firstlevelmenuID=188&siteID=1 (accessed June 17, 2010).

† http://www.nws.noaa.gov/oh/hic/flood_stats/Flood_loss_time_series.shtml (accessed June 24, 2009).

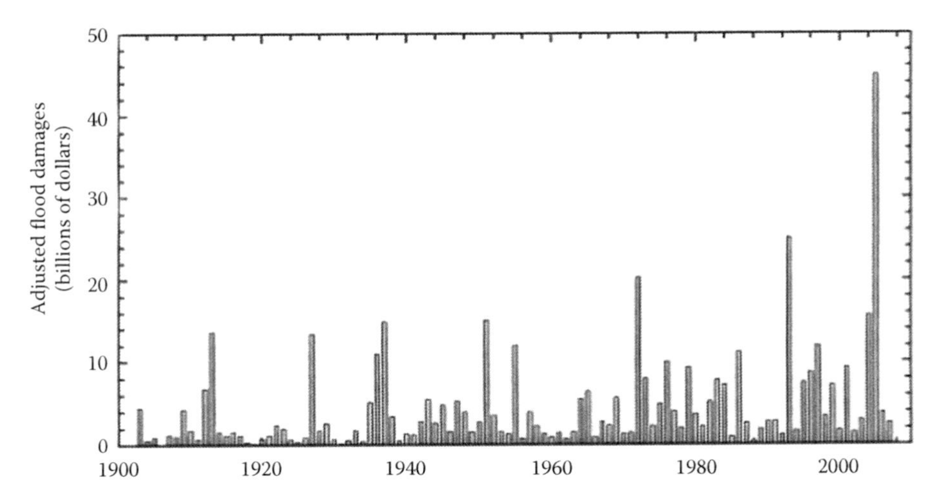


**FIGURE 5.1**
Flood damages in the United States, 1903–2007. (From http://www.nws.noaa.gov/oh/hic/flood_stats/Flood_loss_time_series.shtml, accessed June 24, 2009.) The peak in 2005 primarily reflects damages caused by Hurricanes Katrina and Rita.

Cruz River in southern Arizona, respond strongly to seasonal rainfall, and as a result flow and flood records may be related to decadal-scale climatic processes.[4] In some watersheds, such as the San Juan River, the largest floods are storm-induced while most annual peaks are snowmelt-related.

Large rivers, such as the Colorado and Rio Grande Rivers, span a number of climatic regimes and respond in a complex way to interannual or decadal-scale climatic processes (see Chapter 3). For example, many of the largest floods on the Colorado River occurred during El Niño (warm ENSO) conditions,* but some notable floods, including the largest in the gaging record (1922), occurred during years not associated with ENSO. The Rio Grande in New Mexico has a similar flood response to ENSO conditions, as do smaller rivers such as the Santa Cruz River at Tucson, Arizona.[4,5] At present, our knowledge of regional hydroclimatology is insufficient to forecast large floods well in advance, and flood-hazard analyses are all retrospective and dependent upon gaging records and historical observations.

For intermediate to large channels, flood damage in the arid region results from three interrelated effects. Overbank flooding occurs when flow overtops channel banks and spreads laterally over the top of floodplains, inundating buildings and infrastructure. This type of damage is the most common in most of the United States, and the National Flood Insurance Program[6] and its flood-hazard maps[7] were developed to protect from this type of flood impact by regulating floodplain development. While overbank flooding is common in arid and semiarid regions of the United States, the most damaging effects of floods typically result from lateral channel change, which can destroy bridges, buildings, and other infrastructure by undercutting their abutments or foundations. This type of damage was particularly severe during the 1983 flood on the Santa Cruz River in Tucson, Arizona[8] and prompted continuous installation of bank protection on channels through most of the metropolitan areas. Bank protection is now prominent along channels in the western United States, particularly in the vicinity of bridges and other infrastructure.

* El Niño–Southern Oscillation (ENSO) years during which large floods occurred on the Colorado and other rivers in the southwestern United States include 1862, 1884, 1891, 1905, 1916, 1941, 1957, 1984, 1993, and 1998.

Distributary flow systems, which are common on alluvial fans or in shallow ephemeral channels, offer a third type of flood hazard.* This hazard is severe because channel avulsion, or abrupt shifting of channel location,† on relatively flat, unchannelized surfaces of alluvial fans frequently occurs during floods.[11,12] Avulsions combine the damaging effects of overbank flooding and lateral channel change because channels shift abruptly into positions either previously unoccupied or abandoned, and they have the added damaging effect of sediment deposition.[13] Online resources are available to determine hazard mitigation for this type of hazard.‡

Flood frequency, which is the regulatory standard for the National Flood Insurance Program, can be estimated in two ways. The most common method uses the annual flood series, which is a subset of gaging records.§ The annual flood series represents the largest instantaneous discharge in the water years of the gaging record (Figure 5.2); the water

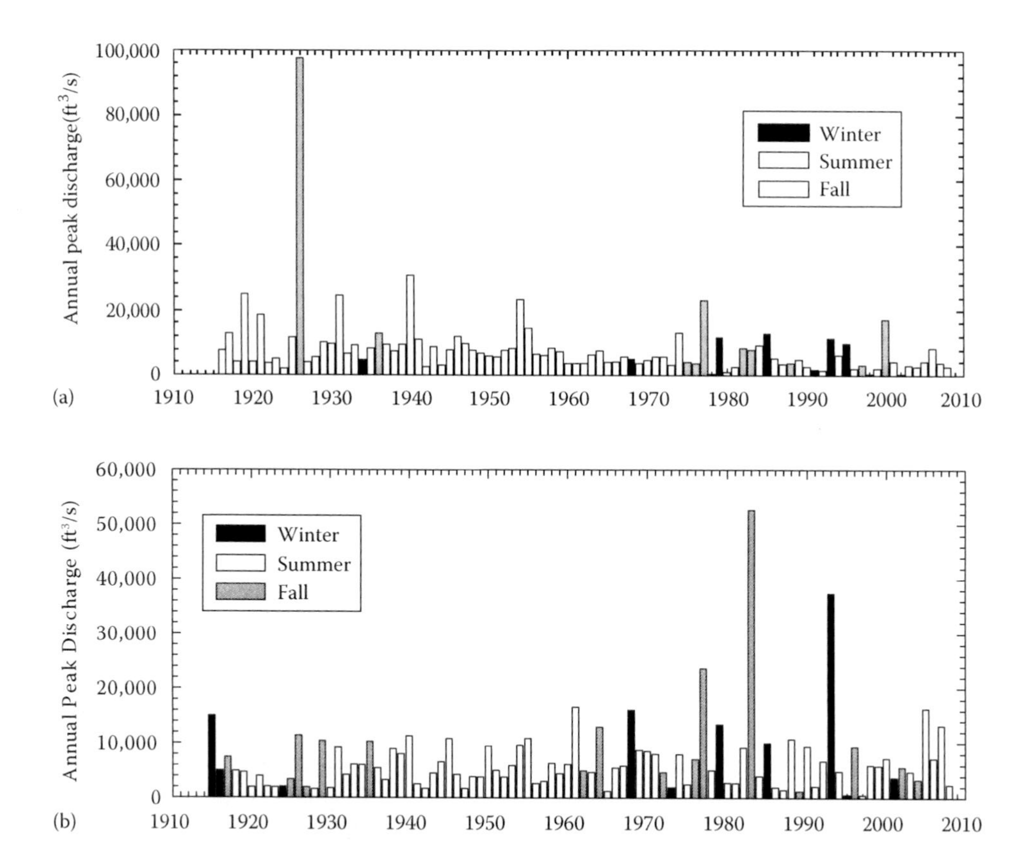


**FIGURE 5.2**
Annual flood series for two rivers in southern Arizona. (a) The San Pedro River at Charleston, Arizona. (b) The Santa Cruz River at Tucson, Arizona.

* For Arizona, some preliminary analyses of distributary flood hazard are in Hjalmarson and Kemna.[9]
† For general definitions of hydrologic or geologic terms, the reader should see Osterkamp.[10]
‡ http://www.fema.gov/plan/prevent/floodplain/nfipkeywords/alluvial_fan_flooding.shtm (accessed June 24, 2010).
§ Most gaging records in the United States created and maintained by the U.S. Geological Survey can be accessed on-line using http://waterdata.usgs.gov/nwis/sw (accessed June 24, 2009).

year is defined, purely for convenience, as the federal government fiscal year (October 1 through September 30). Another useful subset of flood data is the partial duration series, which contains all floods above an arbitrary and fixed base discharge; some years have many floods while some years have only one or no floods that exceed the base discharge. Flood frequency can be estimated from the annual flood series using standardized techniques that are applied throughout the United States.*

Not all streams and rivers that can cause flood damage have streamflow gaging records; in fact, the number of streamflow gaging stations in the United States has declined significantly since its peak in the 1970s and 1980s.[15] Flood frequency at ungaged sites commonly is estimated using regional-regression techniques that estimate flood frequency as a function of drainage area, climate, elevation, and geographic area.[16] Online geographic information system tools are in development to allow a user to click on a map showing rivers and streams in the United States and access flood frequency information for that ungaged site.†

Flood-frequency analysis, whether performed directly on gaging data or indirectly using regional-regression techniques, requires some assumptions that have been severely criticized as unrealistic, particularly given the specter of future climate changes. As currently applied worldwide, flood-frequency analysis requires some data assumptions that may not be reasonable, given past and potential future trends in climate and flood-producing storms (see Chapter 3). The critical assumption is that the annual flood series is stationary, a term defined as a series with time-invariant mean and variance over the period of record. Some have argued that historical gaging records are nonstationary[4]; others claim that stationarity is no longer a viable assumption in hydrologic time-series analyses,[18,19] particularly in flood-frequency analysis. Proponents of nonstationarity in surface-water hydrology either offer different techniques for analyzing flood records or advocate models that rely on future climate predictions.

The San Pedro and Santa Cruz Rivers well illustrate the response of middle-elevation watercourses to climate variation (Figure 5.2). Both records illustrate the concept of nonstationarity in flood frequency,[4,5] although for different reasons. The statistical concept of stationarity holds that the moments of a time series are temporally invariant; hydrologic nonstationarity occurs when the mean and (or) the variance of flow or flood magnitude changes with time (e.g., has a trend or fluctuation). The San Pedro River had a large flood in 1926 that was considerably larger than the second largest flood; but, otherwise, the annual flood series does not show significant trend in magnitude through the twentieth century (Figure 5.2a) and therefore is considered to be a stationary time series. However, the season during which the annual flood occurred has changed, with a shift from predominantly summer floods in the middle of the twentieth century to a mixture of mostly fall and winter floods since the mid-1960s. In contrast, the annual flood series of the Santa Cruz River (Figure 5.2b) shows changes in seasonality of flooding as well as flood magnitude.

The issue of stationarity, as applied to flood-frequency analysis, is as complicated as the surface-water hydrology of the southwestern United States. Large rivers, such as the Colorado and Rio Grande, are subject to decreasing snowpack[20–22] and its impact on spring flood flows; these rivers therefore would be expected to have a nonstationary series of

* Techniques for estimating flood frequency in the United States are contained in U.S. Water Resources Council.[14] Software for estimating flood frequency is available from the U.S. Geological Survey at http://water.usgs.gov/software/PeakFQ/ (accessed June 24, 2009).

† An example is the State of Utah, one of the first states that has implemented the StreamStats program within U.S. Geological Survey; see http://water.usgs.gov/osw/streamstats/ (accessed March 16, 2009).[17]

decreasing annual floods. Other rivers, such as the Santa Cruz, respond to different storm types, and some of these types may well generate floods from a stationarity distribution while others change with time. For example, using the partial-duration series, floods caused by summer thunderstorms in the region demonstrably have time-invariant mean and variance over the period of record, while floods caused by regional storms—either in winter or dissipating tropical cyclones—are related to global-scale climate variation and are nonstationary.[4] Because regional storms generally create the largest floods, the annual flood series therefore is nonstationary as well.

The Santa Cruz River at Tucson is perhaps the best example to illustrate the impact of nonstationarity in an annual flood series on the estimation of flood hazard. Starting in 1970, the 100 year flood is 20,870 $ft^3/s$ based on the station statistics from 55 years of record (Figure 5.3). Adding each new annual flood peak with each passing year and calculating the new 100 year flood result in an extremely interesting time series of flood hazard. Because of large annual peak floods in 1977, 1983, and 1993, the 100 year flood rises from 20,870 $ft^3/s$ to a maximum of 38,180 $ft^3/s$ in 1993 (Figure 5.3). What is most interesting is that in 2007, the 100 year flood is 31,610 $ft^3/s$ and higher than the +95% confidence limit estimated in 1970. In other words, the statistics of flood frequency calculated in 1970 and 2007 is significantly different, which would only occur if the time series is nonstationary. In addition, the period of increased flood frequency is only for part of the record—from about 1977 through 1995 (Figure 5.2B)—with what appears to be time-invariant flood frequency for the gaging record before 1977 and after 1995.

In general, changes in surface-water flow illustrated by the gaging records of the San Pedro and Santa Cruz Rivers represent some trends in southern Arizona.[23] Streamflow in the San Pedro River at Charleston has declined by 50% in the period of record,[24] but the annual flood series does not have an apparent trend (Figure 5.2A), illustrating the decoupling of flood frequency with typical flows. Trends of increasing flood frequency decrease northward in the region, and trends in streamflow decline are highly variable in the region, although some general patterns remain. Large, regional floods occurred in 1862, 1884, 1891, 1963, 1977 through 1979, 1983, and 1993; similarly, streamflow was lower during the mid-century drought than streamflow between 1960 and 1995. Differences are also significant; although floods that scoured channels were common in southern Arizona

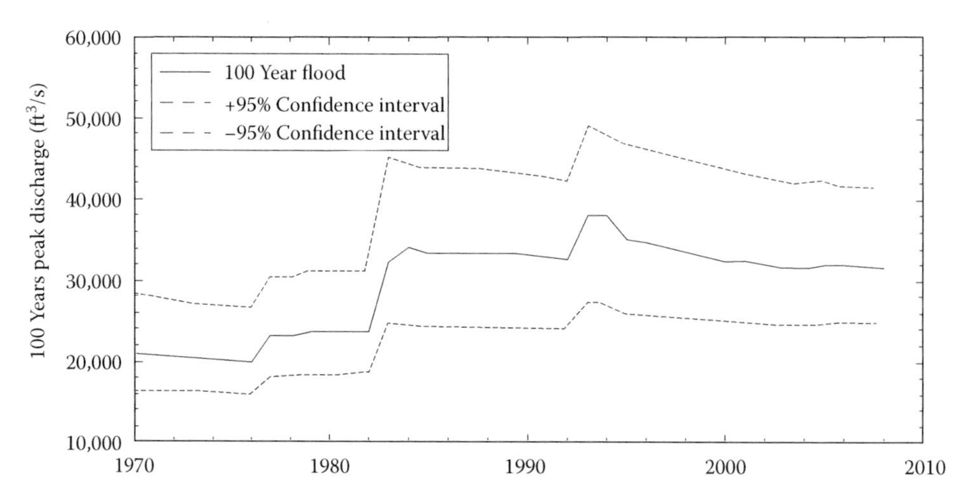


**FIGURE 5.3**
Estimated 100 year flood as a function of time, 1970–2007, Santa Cruz River at Tucson, Arizona.

in the late 1970s through the mid-1990s, these floods did not occur in southern Utah, southern and central Nevada, or the Mojave Desert.

The primary lesson from the regional flooding of the1970s and 1980s in central and southern Arizona is that lateral channel change is the dominant flood hazard,[25] not overbank inundation as is regulated under the National Flood Insurance Program. To minimize lateral channel change, several types of bank protection were installed through metropolitan or other areas where lateral channel shifts would cause significant damage to infrastructure (such as at bridge crossings). For example, in 1982, soil cement—a weak form of concrete created by mixing a small amount of cement with ambient channel sediments—was installed on both channel banks along the Santa Cruz River through Tucson. Combined with grade-control structures to minimize the potential for channel downcutting, these flood-control measures effectively minimized damages in treated reaches during large floods in 1983 and 1993 except at the beginning and end points of bank protection. Similar types of continuous bank protection are used throughout the Phoenix metropolitan area and other urban areas in the region.

Nonstationarity in flood frequency can have severe economic impacts. In the case of the Santa Cruz River, bank protection was found to be necessary through much of the Tucson metropolitan area in response to the large floods from 1977 through 1993. If design of that bank protection was completed in 1970, when the 100 year flood was 20,870 $ft^3/s$, the floods of 1983 and 1993 (peak discharges of 52,700 and 37,400 $ft^3/s$, respectively) likely would not have been contained within the engineered channel. Conversely, if the high flood peaks of 1977–1993 do not continue (Figure 5.2B), the flood protection along the channel may have been too expensive for the level of protection required under the National Flood Insurance Program. The central question of flood-hazard mitigation is whether it is a risk-based system (i.e., floodplain regulation and design based on a 100 year flood or other return period) or whether it is a hazard-avoidance system with minimal risk (i.e., overdesigned floodplain structures).

Installation of bank protection along an engineered channel is by no means the end of flood-hazard mitigation, as illustrated by the Santa Cruz River at Tucson. On August 23, 2005, thunderstorms in the early morning hours caused a substantial rise in the normally dry channel. Based on water height within the engineered and soil-cemented channel, observers believed that the flood had a discharge of greater than 40,000 $ft^3/s$, which would make it the second largest flood in recorded history, and emergency responders began to close bridges and issue alerts. Careful measurements established the peak discharge at 16,300 $ft^3/s$, which is high but not extraordinary for this river. The combination of sedimentation and establishment of riparian vegetation in the engineered channel had reduced flood conveyance, thereby increasing the stage for a given discharge and increasing flood hazard.

One of the key challenges to design of channels is the trade-off of flood-hazard mitigation in a watercourse that is mostly dry with the public expectation and environmental value of riparian corridors through urban environments. If the objective solely is hazard mitigation, then engineered channels with stabilized banks and periodic maintenance to remove vegetation and sediment buildup is appropriate, but these channels are viewed as ugly and maintenance of them is considered to be environmental degradation. However, if expectations of environmental quality in river channels are desirable, then engineering design must account for the changing amounts of riparian vegetation and its influence on sediment aggradation between the stabilized banks. One way that this is accomplished in Arizona is creating an abnormally wide channel with low-flow controls on lateral channel change and high-flow areas that sustain riparian vegetation.

## 5.3 Mass Wasting: Debris Flows and Rockfalls

Mass wasting is a general term used by geologists and hydrologists to describe a continuum of geologic processes ranging from avalanches to soil creep (see Chapter 2). These processes involve the three-phase media of solids (e.g., rocks and soil), air, and water moving at velocities ranging over several orders of magnitude.[26] Although the full range of mass wasting represents general hazards, the principal hazards of this group in arid and semiarid regions are rockfalls, avalanches, and debris flow. Landslides, perhaps the most widely recognized mass-wasting hazard, generally do not occur in this region under present-day climate except in certain localities after extreme rainfalls.

Rockfalls are relatively small failures of bedrock or slopes that generate relatively fast-moving two-phase mixtures of solids and air. The resulting coarse-grained, poorly sorted deposits are called colluvium and are stored on steep hillsides known as colluvial or talus slopes. Rockfalls are an isolated hazard for certain buildings or infrastructure below cliffs or steep slopes throughout arid regions (Figure 5.4), but this hazard becomes nonexistent away from mountain fronts. Avalanches are larger-scale slope failures. The Blackhawk Landslide, in Lucerne Valley of the Mojave Desert of California, is an air-cushion avalanche that traveled about 5 miles from its source area to its depositional area more than 17,000 years ago.* Air-cushion avalanches are large enough that they trap air beneath the mass of moving debris, much like a hovercraft, and move over the landscape with less friction and higher speeds than if the avalanche moved in contact with the ground surface.* Although this type of extremely large avalanche occurred in a different climatic regime during the Pleistocene, smaller avalanches remain a common occurrence in arid regions, particularly where road cuts or other construction activities undermine slopes. Snow avalanches are

**FIGURE 5.4**
Debris-flow deposition in Soldier Canyon, Tucson basin, Arizona. This debris flow, which occurred in July 2006, filled an existing channel with coarse sediment. (From Webb, R.H. et al., Debris flows and floods in southeastern Arizona from extreme precipitation in late July 2006: Magnitude, frequency, and sediment delivery, U.S. Geological Survey Open-File Report 2008-1274, Denver, CO, 2008.)

* http://3dparks.wr.usgs.gov/landslide/big/43.htm (accessed June 26, 2009).[27]

a different problem where desert cities abutt steep mountain ranges that accumulate significant snowpack in winter.

Debris flow is a second type of flood hazard recognized by the National Flood Insurance Program[6] that occurs in arid and semiarid areas in the southwestern United States (Figure 5.4). This type of hazard typically is restricted to mountainous areas, bedrock canyons, and the heads of alluvial fans and typically does not occur on master streams that pass through urban centers. Numerous historical debris flows have occurred in the arid Southwest, but perhaps the most famous is the flood of September 14, 1974, in El Dorado Canyon, an ephemeral stream draining an arid part of the Mojave Desert leading into Lake Mohave on the Arizona–California border. This debris flow killed 9 people and destroyed at least 5 homes, 38 vehicles, and most of a marina on Lake Mohave.[28]

Debris flows are two-phase slurries of water and sediment with the consistency of wet concrete—albeit concrete that contains extremely large boulders—that initiate during intense rainfall on steep terrain. Grand Canyon, the epitome of steep terrain subjected to intense rainfall, has one of the highest frequencies of debris flows in the region with five occurring in an average year throughout the canyon.[29] On the Colorado Plateau, debris-flow frequency is highest where fine-grained rock known as shales occurs in steep cliffs.[30] Although debris flows are known from a variety of geologic terranes in the Sonoran Desert, they are most common from granitic mountains, such as the Santa Catalina Mountains north of Tucson.[31] Debris flows can destroy roads, bridges, and houses, but their influence on filling channels with coarse sediment, and thereby decreasing flow capacity for future floods, is perhaps their most important hazard.

Growth of housing developments on the alluvial fans skirting the mountains of the southwestern United States portend increased future risk from debris flows and underscore the need for new tools for floodplain management. Along the northern edge of Tucson (Figure 5.1), debris-flow deposition on the apices of alluvial fans primarily is of Pleistocene age, but areas with significant recent or Holocene debris-flow deposits have been identified.[32] Modeling of debris-flows deposition, which would be needed to quantitatively map hazard areas, is challenging and inaccurate over complex topography. One of the better ways of predicting mobility and inundation potential of debris flows is to use stochastic modeling. One such model, called LAHARZ,[33,34] requires the assumption of a simplified debris-flow event consisting of initiation, transport, and deposition zones for each event as well as a high-resolution digital-elevation model of the deposition zone.

An example of debris-flow deposition mapping as estimated using LAHARZ appears in Figure 5.5. Using a range of volumes of sediment mobilized during intense rainfall, a range of depositional areas is estimated at the apex of an alluvial fan. While this technique creates a map of debris-flow deposition, and therefore could possibly be used to regulate floodplain development, it does not provide information on risk because no return-period information is used. Debris-flow frequency little known and is not measured at gaging stations, unlike flood frequency, and estimation of risk from debris flows requires more information than is available at present. The central question for debris-flow hazard is whether hazard avoidance consisting of restrictions on construction along channels with a past history of debris-flow occurrence is preferable to a risk-based system where the amount of risk is unknown or poorly known.

Debris flows pose another, perhaps greater, risk downslope from where the slurry stops and streamflow drains the remainder of the runoff. As with flood hazards, channel conveyance is assumed to be unchanging from its original design characteristics, and debris-flow deposition can greatly decrease conveyance and increase overbank flood

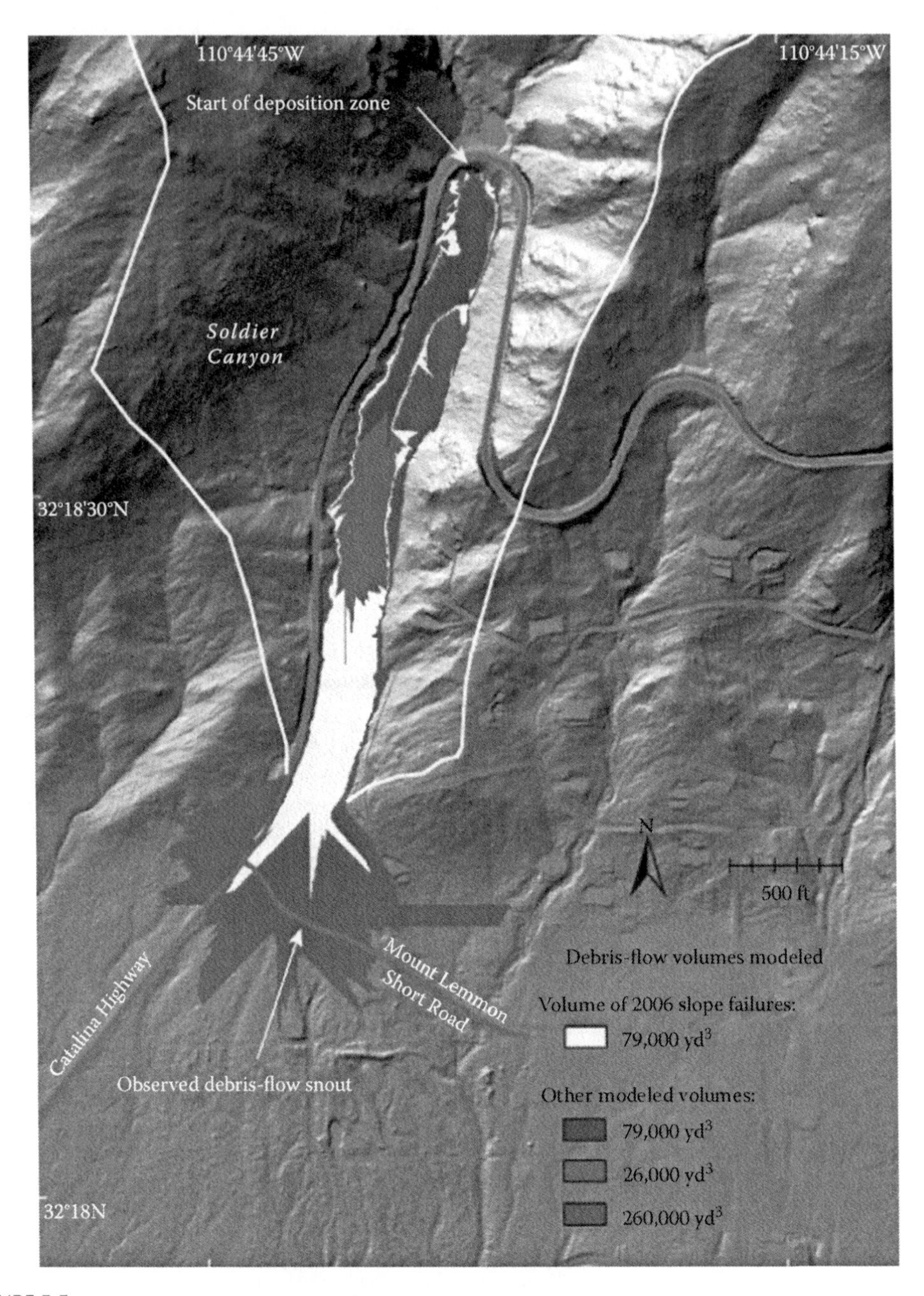


**FIGURE 5.5**
Modeling of debris-flow hazard at Soldier Canyon, southern Santa Catalina Mountains, using LAHARZ. (From Magirl, C.S. et al., *Geomorphology*, 119, 111, 2010.)

potential. Careful consideration should be given to the potential for debris flows upslope of subdivision development when hydrologic design is used for constructed drainage channels to mitigate the potential for channel aggradation, either from debris flows or the typical processes that deliver sediment from headwaters to terminus in ephemeral streams.

## 5.4 Land Subsidence

Land subsidence is the lowering of the land surface from changes that occur underground. One of the most common human causes of land subsidence is compaction of fine-grained sediments induced by extraction of water from the subsurface. It also can occur in certain areas, notably in Los Angeles, California, and Houston, Texas, where oil extraction causes similar compaction of underground strata.[35] Where significant amounts of these sensitive subsurface strata are present, substantial ground-surface lowering can occur as water levels decline. In some cases, land subsidence can continue for decades after a cessation or reduction in pumping that stabilizes groundwater levels.[35] In Arizona, surface-elevation subsidence of as much as 18 ft has been recorded west of Phoenix.[36] In Tucson, as much as 4 in. of subsidence was associated with a water-level decline of 45 ft between 1989 and 2005.[37] Subsidence fissures (Figure 5.6), which typically occur near the margins of alluvial basins between zones of differential subsidence, can rupture roads, canals, buildings, and other infrastructure.[38]

Extraction of groundwater is the major cause of subsidence in the deserts of the southwestern United States, and as pumping increases, land subsidence also increases. In alluvial aquifers, groundwater is pumped from the pore space between grains of sand and gravel. If an aquifer has beds of clay or silt within or next to it, the lowered water pressure in the sand and gravel causes slow drainage from the clay and silt beds. The reduced water pressure is a loss of support for the clay and silt strata; because these beds are compressible,

**FIGURE 5.6**
Subsidence crack in Benson, Arizona. (Courtesy of S.R. Anderson.)

(a)

(b)

**FIGURE 5.7**
Public-supply well in Las Vegas, Nevada damaged by land subsidence. (Modified from Baum, R.L. et al., Landslide and land subsidence hazards to pipelines, U.S. Geological Survey Open-File Report 2008-1164, Reston, VA, 2008, http://geochange.er.usgs.gov/sw/changes/anthropogenic/subside, accessed June 26, 2009.) (a) By 1964, pumping had caused the land surface to sink from below the well head and concrete slab, which originally were at the land surface. (b) By 1997, distance from the concrete slab to the land surface was even greater.

they compact by decreasing the pore space. This compaction results in a land surface that is lower in elevation, usually permanently. In other words, land subsidence caused by lowered groundwater levels cannot be reversed by recharging the aquifer.

Land subsidence causes many problems to urban environments and infrastructure including (1) changes in elevation and slope of streams, canals, and drains; (2) damage to bridges, roads, railroads, storm drains, sanitary sewers, canals, and levees; (3) damage to private and public buildings; and (4) damage to wells and casings (Figure 5.7), from forces generated by compaction of fine-grained materials in aquifer systems.[38] In some coastal areas, subsidence has resulted in tides moving into low-lying areas that were previously above high tide. Earth fissures can be more than 100 ft deep and several thousand feet in length; one extraordinary fissure in central Arizona is 10 miles long.[36] These features start out as narrow cracks, an inch or less in width, and they intercept surface drainage and can erode to widths of tens of feet at the surface.

In Arizona, as in other states, areas subject to land subsidence are known and mapped.* In areas subject to subsidence from groundwater pumping, mitigation has involved (1) reduction in groundwater pumping and switching to surface water where such supplies are available, (2) moving groundwater pumping to areas less susceptible to land subsidence, and (3) moving groundwater pumping to areas where subsidence is likely to cause less damage. The State of Arizona has programs to map land subsidence† and to map subsidence-related earth fissures.‡ The relevance of these programs cannot be better illustrated than with the Central Arizona Project aqueduct and its interaction with subsidence zones in central Arizona.[39]

* http://www.azgs.az.gov/EFC.shtml (accessed June 23, 2010).

† http://www.azwater.gov/AzDWR/Hydrology/Geophysics/LandSubsidenceInArizona.htm (accessed July 7, 2009).

‡ http://www.azgs.state.az.us/EFC.shtml (accessed July 7, 2009).

## 5.5 Earthquakes

Most of the arid and semiarid regions of the southwestern United States are in the Basin and Range physiographic province (see Chapter 2), which formed as a result of extensional tectonic activity beginning about 20–30 million years ago.[40] Crustal extension caused numerous faults in this region, and many of them are still active and are the source of considerable seismic hazard to urban areas. Although ground motion is the largest hazard, surface rupture and movement can also cause considerable damage. The southern extension of the most famous fault in western United States, the San Andreas, crosses the western Sonoran Desert, extending from Palm Springs, California, through the Salton Sea and beneath the Gulf of California. Other well-known faults that pose high seismic hazards include the Owens Valley fault, which is parallel to and east of the Sierra Nevada Mountains in California; the complex system of faults along the Wasatch Front in Utah, which extends southward via the Hurricane fault into Arizona; and the central Nevada seismic zone. These faults are prominently displayed in maps of recent seismic activity in the western United States.*

Numerous historical earthquakes have occurred in the desert Southwest. In Arizona, the most famous of these was a magnitude 7.2 event that occurred on May 3, 1887, with an epicenter just south of the U.S.–Mexico border in the San Bernardino Valley of Sonora.[41] This earthquake essentially leveled Tucson, Arizona, then a town composed of mud brick buildings, killed 51 people in Sonora, and caused numerous hydrologic changes in the region. The March 26, 1872, earthquake on the Owens Valley fault at Lone Pine, California, (magnitude 7.4) killed 27 people and leveled most of the houses in town.† Other large earthquakes in the region include the February 24, 1892, event in the Imperial Valley (magnitude 7.9), the December 21, 1932, earthquake near Cedar Mountain, Nevada (magnitude 7.2), and the March 12, 1934, earthquake near Kosmo, Utah (northwest of Salt Lake City), the largest in that state's history (magnitude 6.6). The recent magnitude 7.2 earthquake centered south of Mexicali, Baja California,‡ caused substantial damages in that city and in Calexico, California.§

Earthquake prediction remains a goal of seismologists, and although promising approaches have been identified in recent years, these scientists do not have a reliable method for forecasting earthquake occurrence. Instead, seismologists rely on earthquake hazard assessments developed using geophysical models that incorporate the best scientific data and models to predict hazards.[42] This type of approach is used to predict the probability of occurrence of ground motion that occurs during earthquakes and are useful for design purposes. Seismic hazard maps¶ depict earthquake probabilistic data at a variety of geographic scales of use to urban designers.**

Several metrics of ground motion are used to describe earthquake hazards.†† Peak horizontal acceleration is ground motion expressed as percent of gravity; vertical acceleration is generally lower than horizontal acceleration and is not used. For moderate to large earthquakes, the pattern of peak horizontal acceleration is complicated and extremely variable over short distances because small-scale geologic differences significantly change accelerations. Peak horizontal velocity reflects the pattern of the earthquake faulting geometry, with largest

* http://earthquake.usgs.gov/regional/states/seismicity/us_west_seismicity.php (accessed June 26, 2009).
† http://earthquake.usgs.gov/regional/states/events/1872_03_26.php (accessed June 27, 2009).
‡ http://earthquake.usgs.gov/earthquakes/recenteqsww/Quakes/ci14607652.php (accessed June 23, 2010).
§ http://en.wikipedia.org/wiki/2010_Baja_California_earthquake (accessed June 23, 2010).
¶ http://pubs.usgs.gov/fs/1996/fs183–96/fs183–96.pdf (accessed June 26, 2009).[43]
** http://earthquake.usgs.gov/research/hazmaps (accessed June 27, 2009).
†† http://earthquake.usgs.gov/eqcenter/shakemap/background.php (accessed June 27, 2009).

velocities nearest and parallel to the fault.* Variability in soils and bedrock again causes complexities but the overall pattern of horizontal velocity normally is simpler than peak accelerations. Severe damage and damage to flexible structures are best related to ground velocity.

Following large earthquakes, spectral response maps are used to portray the response of a damped, single degree-of-freedom oscillator to the recorded ground motions. This data is useful for engineers to determine how structures respond to a range ground motions caused by periodic motion; specific reference periods are defined within the Uniform Building Code (UBC) to define the design spectra for infrastructure. ShakeMap spectral response maps† are made for the response at the three UBC reference periods of 0.3, 1.0, and 3.0 s.

Numerous tools are available to assess earthquake hazards, and almost all of these are internet-based with frequent updates. For most states, the state geologic surveys have prepared fault maps,‡ amplified ground shaking maps, and liquefaction potential maps,§ as well as earthquake-induced inundation, or tsunami, maps. Maps of peak horizontal acceleration (Figure 5.8) are the most common means for describing regional seismic

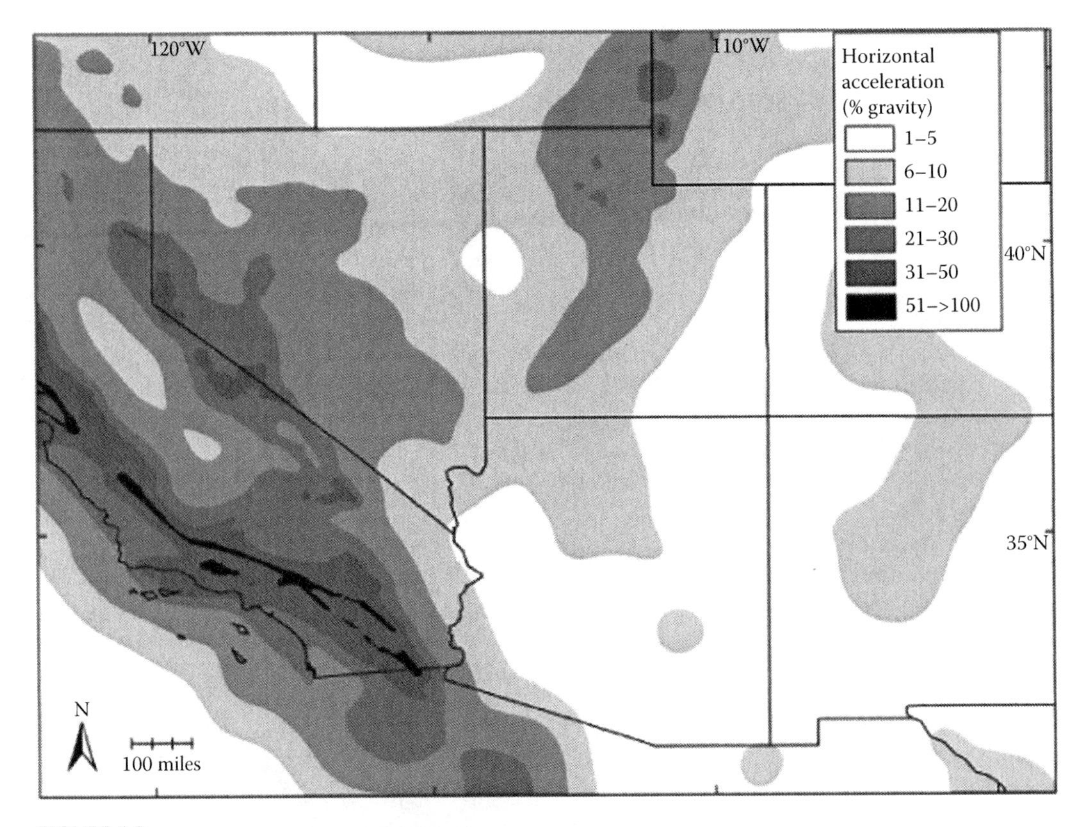


**FIGURE 5.8**
Seismic hazard map for the southwestern United States. (From http://gldims.cr.usgs.gov/nshmp2008/viewer.htm, accessed June 29, 2009.) The units are peak horizontal acceleration, as a percentage of gravity, presented as a 10% probability of occurrence in the next 50 years. Seismic hazard, therefore, increases with increasing shades of gray.

* http://earthquake.usgs.gov/earthquakes/shakemap/background.php#velmaps (accessed June 23, 2010).
† http://earthquake.usgs.gov/eqcenter/shakemap/background.php#velmaps (accessed June 27, 2009).
‡ http://www.data.scec.org/faults/faultmap.html (accessed June 23, 2010).
§ http://geology.utah.gov/utahgeo/hazards/liquefy.htm (accessed June 23, 2010).

hazard. These maps clearly show that seismic hazard in the arid and semiarid areas of the western United States is highly localized, and some areas (i.e., west of Phoenix, Figure 5.8) have essentially no seismic hazard, while others (e.g., Yuma, Arizona) have a high seismic hazard. As previously noted, the details of soils, geology, and substrate determine local responses to seismic shaking, and these factors generally are available to designers through soil engineering reports that may be required in the zoning process.

## 5.6 Other Geologic Hazards

### 5.6.1 Radon

Radon is an urban hazards issue because this noble gas, which can cause lung cancer, can accumulate in poorly ventilated houses and buildings, particularly in basements.* Radon is one of the daughter products of the radioactive decay of uranium, and its distribution in Arizona, as one example, is related to certain specific lithologies that underly urban areas.† This noble gas enters buildings through cracks in floors and foundations or through release from water, and its presence can be mitigated by proper ventilation or prevented by eliminating cracks in foundations or other points where gas may seep from soils or rock into occupied areas. In Arizona, as in other desert regions, the problem of radon may be relatively low[44] owing to a climate that encourages higher ventilation of structures.

### 5.6.2 Expansive Soils

Expansive soils are a problem throughout the United States, and some assessments assert that damages from expansive soils exceed all other geologic hazards combined.‡ Although these soils are common in arid regions, the problem is not as extensive as in other regions. Expansive soils typically result from soil-forming processes that create expansive clays,§ although some of the larger problems in Arizona result from shales or marine origin that contain multilayer clays, most commonly smectites. Shrinking and swelling of soils in response to rainfall, irrigation, or water leaks cause buckling of roadways, sidewalks, and other infrastructure and are the most common cause of cracked floors, foundations, and walls in arid regions not subject to subsidence or frequent earthquakes.

## 5.7 Urban and Man-Made Hazards

Until this point, we have discussed the hazards occurring from forces of nature to the environment as a result of an imbalance in a natural system or state. In many cases, human activities can cause hazards of nature by disrupting the natural hydrologic, geologic, or climatic processes. For example, the removal of upland vegetation by fire, construction, or other human activities can lead to downstream flooding because runoff, unimpeded by

* http://www.epa.gov/radon/ (accessed June 23, 2010).
† http://www.azgs.az.gov/HomeOwners-OCR/radongasinarizona.pdf (accessed June 23, 2010).[44]
‡ http://geology.com/articles/expansive-soil.shtml (accessed June 23, 2010).
§ http://www.azgs.az.gov/hazards_problemsoils.shtml (accessed June 23, 2010).

plants, increases instead of infiltrating into the soil. In this case, the human activity *caused* an alteration in the flow of water in the watershed, and the increased flooding was the *result*.

How can one differentiate between a natural hazard or a human caused hazard? Take, for instance, a scenario where a high level of arsenic found in a municipal water well. Differentiating this as the result of anthropogenic activities or a natural concentration can be difficult and will require a formal investigation and the involvement of experts to determine the source of this unwanted constituent, and the necessary studies could take years and thousands of dollars to complete. In the remainder of this chapter, we discuss the impacts of anthropogenic activities in land use and development that can have significant impact in Southwestern deserts.

Hazards caused by human activities in the built environment generate concerns that can lead to adverse impacts on human health, contamination of air, water or land, or cause the loss of property use or value from this activity. Numerous hazards result from the construction and maintenance of urban environments, and we discuss some of those that are related to pollution of land and water and address some hazards that are relatively well-defined spatially and can either be mitigated or avoided by effective management or planning.*

Pollution-related impacts generally involve two- or three-phase media as combinations of air, water, and soils. Every operation that produces any kind of waste in desert environments has the potential for releasing pollutants that can contaminate these resources. Although mining is perhaps the largest source of pollution in the southwestern United States, other sources of pollution with lasting impacts include manufacturing, military bases (particularly Air Force facilities), retail, auto shops, drycleaners, laboratories, and fueling stations. Most of the information in this section is derived from websites that are regularly updated, providing users with up-to-date information on urban hazard issues, and several excellent summary volumes also are available.[46–48]

### 5.7.1 Regulatory Environment

The United States has a multitiered level of oversight and enforcement of environmental regulations related to hazards that can or may impact the public. On the federal level, the U.S. government has designated the Environmental Protection Agency (EPA)† as the principal agency to oversee concerns about environmental contamination or hazards to human health. The EPA has broad powers to enforce federal laws on environmental protection and bring regulated activities in compliance with national standards. Each state has their own department of environmental quality or compliance that sets statewide guidelines for compliance with EPA for health or contamination standards. Detailed information can be obtained on environmental regulations from each state's department of environmental quality websites.‡ The county government is given authority to regulate some aspects of environmental quality, such as air quality monitoring and dust control or other responsibilities delegated by the State or created by environmental policy. On the local level, municipal governments have the responsibility to set land-use guidelines through zoning laws and regulate local land uses and manage municipal solid waste and wastewater operations. Together, these public agencies comprise the police power of enforcement of

* For a comprehensive overview of environmental land management see Raldolph.[45]

† http://www.epa.gov/ (accessed June 23, 2011).

‡ For Arizona, http://www.adeq.gov; for Nevada, http://www.ndep.nv.gov; for California, http://www.calepa.ca.gov/, for New Mexico, http://www.nmenv.state.nm.us/; for Texas, http://www.tceq.state.tx.us/ (all accessed June 25, 2010).

environmental laws on development activities by private sector. In practice, the fragmentation of government oversight of environmental regulations is not resulted in a seamless operation, and an uneven application and enforcement of environmental standards occurs in certain districts. In addition, conflicts *within* the government agencies over enforcement activities may lead to inaction in some instances.

Contaminants of water, air, and land are divided into two major categories based on how they are released into the environment and thus how they are regulated. The first category is called *point source* pollutants, which are those pollutants that have a fixed location and known type and quantity of release into the environment. Point source locations can be power plants, factories, or industrial sites; construction sites with high off-site production of dust or sediment can also be considered as point sources. They are often required to apply for a permit to the EPA or other regulatory agencies and have stringent reporting requirements to quantify the amount of pollutants released into the environment.

The second category of release is called *non-point source* pollution, where releases are either from a mobile source or are otherwise widely dispersed into the air, land, or water. These releases are harder to regulate because the location or operator that released the pollutant cannot be directly identified from the point of contamination. An example is the dust production from dirt roads maintained by several jurisdictions or private parties in a county. Cases of nonpoint source pollution have resulted in lengthy legal challenges to hold private parties accountable for diffuse releases of pollutants that have proven to be difficult to litigate and require establishing a burden of proof on an operator to prosecute this type of violation.

The federal government plays a large role in the regulation of environmental laws in the United States through agencies such as the EPA, the Department of Transportation, the Department of Homeland Security, and others. Of the numerous federal laws that deal with definitions, descriptions, and cleanup of pollution and contamination, we believe that the most important are the Resource Conservation Recovery Act (RCRA)*; the Comprehensive Environmental Response, Compensation, and Liability Act (CERCLA); the Emergency Planning and Community Right-to-Know Act; the Clean Water Act (CWA); the Clean Air Act; Toxic Substance Control Act; and the Hazardous Material Transportation Act.† In addition, state, county, and local laws and ordinances that attempt to address and manage environmental contamination and pollution issues are too numerous to list in a brief overview. In the following section, each environmental concern is discussed in general terms, and we stress that whole volumes have been written on each of these subjects. Our goal is to provide basic information that some planners or designers may find useful for discussion or consideration.

### 5.7.2 Mining Pollution

In the arid and semiarid parts of the United States, particularly in Arizona, parts of California, Nevada, and New Mexico, mining has been a vital and essential part of the economy since settlement. Mining can be subdivided by substrate into hard-rock and soft-rock sources and by technique: strip mining, surface mining, and subsurface mining. The impacts vary according to the quality of the material that is extracted, whether it is processed onsite or transported to a processing facility, and the type of processing that is

* http://www.epa.gov/waste/inforesources/online/index.htm (accessed June 28, 2010).
† http://ncseonline.org/nle/crsreports/briefingbooks/laws/e.cfm(accessed June 25, 2010). Another excellent source of information is *Environmental Law Handbook*.[49]

required. Minerals mined in the region include metals and metalloids, such as gold, silver, and copper; fossil fuels, particularly coal, natural gas, and oil; materials used in industrial processes, such as clays, diatomite, and zeolites; gemstones; and materials used to develop building materials, such as building stone, lime, limestone, gypsum, and sand and gravel.

For nonfuel resource extraction, California is ranked third behind Arizona and Nevada in the southwestern United States. The economic impact to these three States is enormous: in Arizona, mining contributes $6.71 billion to the state's economy, which is the largest contribution of the 50 states and about 10% of U.S. total mineral production.* Copper is the largest mineral industry, followed by molybdenum products, sand and gravel, and limestone products. In Nevada and California, mining contributes $5.24 billion (second) and $4.5 billion (third), respectively.[50] Online map servers provide location information for most of these mines.† Although the economic benefits of mining are large, the environmental impacts particularly waste management and site rehabilitation associated with the mining industry are also substantial.

Both surface and subsurface mining have substantial impacts on landscapes that range from subsidence or collapse features associated with subsurface mining and diversion or blockages of watercourses with mines and their attendant tailings piles. Viewscape impacts are substantial and may have some of the largest impacts in terms of regional planning and design. The environmental impact of mining includes acid mine drainage, increased sedimentation in channels, contamination of surface and ground water by metals, release of toxins such as cyanide, increased dust emission and deposition, and local impacts, such as contamination from solvents, petroleum, and chemicals used in processing operations.‡ Water and air pollution are the largest offsite impacts, requiring special mitigation and (or) warning systems. For example, the town of Green Valley, Arizona, has special wind forecasts to warn of dust emissions from nearby tailings piles accumulated from strip mining for copper.§

Mining operations for common building materials typically create large depressions along the streambeds on the outskirts of urban areas and create a significant amount of fine particulates, which often contributes to an increased $PM_{10}$ levels in the areas surrounding these operations. In-stream mining for sand and gravel is largely banned in Arizona because of its impact on channel geometry, particularly bed degradation during floods, both upstream and downstream from the mining site. Channel erosion has numerous negative environmental impacts that range from effects on shallow groundwater to destruction of riparian vegetation.

Pollution from mining operations has lasting impacts to lands, groundwater, surface water, and air quality, and lengthy and costly rehabilitation is necessary in many cases to rehabilitate the affected lands to mitigate impacts, particularly those that transport contaminants offsite.¶ Mining wastes from both active and inactive mines cause negative environmental impacts on the land, air, and water.** These include waste generated during the extraction, beneficiation, and processing of minerals. Most extraction and beneficiation

* http://minerals.usgs.gov/minerals/pubs/mcs/2007/mcs2007.pdf (accessed June 25, 2010)[50]; http://www.admmr.state.az.us/ (accessed June 28, 2010).

† For example, mines in Nevada can be mapped using http://www.nbmg.unr.edu/dox/mm/mm97/mm97.htm (accessed June 25, 2010).

‡ http://www.techtransfer.osmre.gov/NTTMainSite/Library/hbmanual/epa530c/chapter3.pdf (accessed June 28, 2010).

§ http://www.azdeq.gov/environ/air/ozone/gvwind.pdf (accessed June 23, 2010).

¶ http://www.epa.gov/osw/nonhaz/industrial/special/mining/index.htm (accessed June 25, 2010).

** http://www.epa.gov/osw/nonhaz/industrial/special/mining/index.htm (accessed June 28, 2010).

wastes from hardrock mining—especially the mining of metallic ores and phosphates—and 20 specific mineral processing wastes are categorized by EPA as "special wastes" and have been exempted by the Mining Waste Exclusion from federal hazardous waste regulations under Subtitle C of the RCRA.* This distinction is important to understand when redevelopment of sites affected by these types of mining.

### 5.7.3 Hazardous Wastes

*Hazardous wastes* are defined by the RCRA as materials that exhibit characteristics such as ignitability, corrosivity, reactivity, and toxicity; are included on the EPA list; or are spent materials or other inherently waste like material that affects human health and environment when disposed improperly.† Hazardous wastes are regulated by RCRA and it follows the concept of cradle-to-grave management. The act covers not only the generator of hazardous wastes but also the transporter and the disposal facility. The generators are classified as large quantity generator (LQG), small quantity generator (SQG), and conditionally exempt small quantity generator (CESQG), depending on the amount of hazardous wastes generated and or stored onsite prior to offsite disposal.

CESQGs generate no more than 220 lb of hazardous waste in any month. CESQGS are exempt from hazardous-waste management regulations, provided that certain basic requirements are met. SQGs generate between 220 and 2200 lb of hazardous waste in any month, while LQGs generate more than 2200 lb of hazardous waste per month. LQGs must comply with more extensive hazardous waste rules, particularly with some wastes that are considered to be so dangerous that they are called acutely hazardous wastes. If a business generates or accumulates more than 2.2 lb of acutely hazardous waste in a calendar month, all of the acutely hazardous waste must be managed according to the regulations applicable to LQGs.‡ In 2007 alone, Arizona generated 54,091 tons of RCRA wastes from 175 generators, California generated 608,654 tons of RCRA wastes from 2,312 generators, and Nevada generated 10,041 tons of RCRA wastes from 73 generators.§

Both SQGs and LQGs of hazardous waste are required to have an EPA ID number for their waste-generation facilities. LQGs are required to have waste minimization and pollution prevention plans to reduce the amount of hazardous waste generated. Annual reports and audits are also required under RCRA. Using an RCRAInfo query,¶ Arizona, California, Arizona, and Nevada have more than 11,000, 8,455, and 4,148 facilities, respectively, in 2010. This database includes small and large quantity generators of hazardous wastes, transporters of hazardous wastes, and treatment storage and disposal facilities. Large-scale agricultural and meat processing operations add to the contamination of the water and soil because discharges and runoff from these operations contain pesticides, insecticides, fertilizers, and animal wastes. In spite of CWA regulations and discharge restrictions, off-site contamination and pollution that affect natural and urban environments continue and are factors affecting urban design locally.

Department of Defense (DOD) facilities are a major source of pollution, and closure of bases offers the opportunity for redevelopment and the problem of mitigation of wastes and pollution. A 1992 hazardous waste survey by DOD[51] revealed that 59 of the nation's

* http://www.epa.gov/compliance/civil/rcra/rcraenfstatreq.html (accessed June 25, 2010).
† 40 CFR Part 261, Hazardous waste identification and listing, Subparts C and D; http://www.epa.gov/wastes/inforesources/pubs/hotline/training/hwid05.pdf (accessed June 28, 2010).
‡ http://www.epa.gov/waste/inforesources/online/index.htm (accessed June 25, 2010).
§ http://www.epa.gov/waste/inforesources/data/br07/state07.pdf (accessed June 25, 2010).
¶ http://www.epa.gov/enviro/html/rcris/ (accessed June 23, 2010).

**FIGURE 5.9**
Soil and water remediation equipment on the Williams Air Force base superfund site in Mesa, Arizona. (Courtesy Richard A. Malloy.)

military bases had signification contamination, typically from fuel spills and dumping of trichloroethylene and other solvents.* In the Phoenix metropolitan area, the former Williams Air Force Base (AFB) (Figure 5.9) and Luke AFB have environmental contamination, while in the desert areas of California, Edwards AFB, George AFB, Norton AFB, and Miramar NAS are on the list of contaminated military bases. Several of these bases have either been closed or are proposed to be closed and made available for redevelopment or turned into public lands. The most typical types of contamination include soils and groundwater containing solvents and fuel oils.

### 5.7.4 Superfund Sites

CERCLA, also known as the *Superfund*, provides for the cleanup of highly contaminated sites.† There have been many sites in the desert Southwest that have made the list including active and abandoned mines, manufacturing sites, abandoned sites, and military bases. CERCLA allows for recouping the cleanup costs from the "potentially responsible parties." A contaminated site is first scored using the Hazard Ranking System (HRS) to be placed on the National Priorities List (NPL), and this makes the site eligible for cleanup using Superfund monies. The NPL is a list of national priorities among the known releases or threatened releases of hazardous substances, pollutants, or contaminants and guides determination of which sites warrant further investigation for cleanup.

The HRS uses a numerically based screening system that uses information from initial limited investigations—the preliminary assessment and the site inspection‡—to score the relative potential of sites to pose a threat to human health or the environment. The structured-analysis approach assigns numerical values to factors that relate to risk based on conditions at the site. The factors are grouped into three categories: (1) the likelihood that a site has released or has the potential to release hazardous substances into the environment,

* http://www.gmasw.com/ao_bases.htm (accessed June 25, 2010).
† http://www.epa.gov/superfund/about.htm (accessed June 25, 2010).
‡ http://www.epa.gov/superfund/cleanup/pasi.htm (accessed June 25, 2010).

(2) characteristics of the waste (e.g., toxicity and waste quantity), and (3) people or sensitive environments (targets) affected by the release. Four pathways can be scored under the HRS: (1) groundwater migration (drinking water), (2) surface water migration (drinking water, human food chain, sensitive environments), (3) soil exposure (resident population, nearby population, sensitive environments), and (4) air migration (population, sensitive environments). After scores are calculated for one or more pathways, they are combined using a root-mean-square equation to determine the overall site score.

The electronic scoring tool Quickscore* can be used to do the scoring calculations. If all the pathway scores are low, the site HRS score is low, although the site score can be relatively high even if only one pathway score is high. This is an important requirement for HRS scoring, because some extremely dangerous sites pose threats through only one pathway.† This distinction and weakness with HRS is important to understand when these sites are chosen for new development or redevelopment, because a low score may mean a low national priority but a high risk to environmental and(or) human health could still be present.

### 5.7.5 Brownfields

The term *brownfield site* refers to lands that are contaminated, and for which the expansion, redevelopment, or reuse is complicated by the presence of hazardous substances, pollutants, or contaminants.‡ Projects are leveraged $18.68 per dollar expended by the EPA, and nationwide, a total of 61,023 jobs are attributed to the Brownfields Program. In addition, stormwater runoff from brownfields redevelopment is 47%–62% lower than alternative greenfields scenarios, and residential property values are expected to increase by 2%–3% when nearby brownfields are rehabilitated. Section 101 of CERCLA provides several exclusions and amendments. In general, the term brownfield site does not include facilities that are the subject of planned or ongoing removal action under CERCLA; facilities on the NPL list or is proposed for listing; facilities that are subject to unilateral administrative order, a court order, or judicial consent decree; or facilities for which a permit has been issued by the United States or an authorized state under Solid Waste Disposal Act, Federal Water Pollution Act, Toxic Substances Control Act, or the Safe Drinking Water Act. There are additional provisions in the Act that allow site-by-site determination of what may be defined as a brownfield site and qualify for redevelopment assistance.

Sites that meet the definition of a brownfield site are contaminated by a controlled substance§; are contaminated by petroleum or a petroleum product excluded from the definition of "hazardous substance" under section 101; are sites determined by the Administrator or the State as appropriate to be relatively low risk as compared to other petroleum-only sites in the State; are sites for which there is no viable responsible party and which will be assessed, investigated, or cleaned up by a person or corporation that is not potentially liable for cleaning up the site; are not subject to any order issued under section 9003 (h) of the Solid Waste Disposal Act¶; or is mine-scarred land. In general, the cleanup costs for brownfield sites are much lower than a typical NPL sites, the cleanup times are faster, and numerous grants are available from the EPA to assist with the additional costs compared

* http://www.epa.gov/superfund/programs/npl_hrs/quickscore.htm (accessed June 25, 2010).
† http://www.epa.gov/superfund/programs/npl_hrs/quickscore.htm (accessed June 28, 2010).
‡ http://www.epa.gov/brownfields/ (accessed June 28, 2010).
§ A controlled substance is defined in section 102 of the Controlled Substances Act (21 U.S.C. 802); http://www.justice.gov/dea/pubs/csa/802.htm (accessed June 28, 2010).
¶ 42 U.S.C. 6991 b(h); http://www.epa.gov/brownfields/overview/glossary.htm (accessed June 28, 2010).

**TABLE 5.1**

Designated Nonattainment Areas by the U.S. Environmental Protection Agency

| State | Location | 8 h Ozone | 2006 PM 2.5 | 1997 PM 2.5 | PM 10 |
|---|---|---|---|---|---|
| AZ | Ajo | | | | Moderate |
| AZ | Douglas | | | | Moderate |
| AZ | Hayden/Miami | | | | Moderate |
| AZ | Nogales | | Nonattainment | | Moderate |
| AZ | Phoenix | Sub par | | | Serious |
| AZ | Rillito (Pima Co.) | | | | Moderate |
| AZ | Yuma | | | | Moderate |
| CA | Los Angeles South coast Air basin | Extreme | Nonattainment | Nonattainment | Serious |
| CA | Southeast desert Modified AQMA | Moderate | | | |
| NV | Las Vegas | Sub par | | | Serious |
| NV | Reno | | | | Serious |
| NM | Anthony | | | | Moderate |
| TX | El Paso | | | | Moderate |

*Source:* http://www.epa.gov/oaqps001/greenbk/ancl3.html, accessed July 13, 2011.

with redevelopment of other sites. Since the inception of the program in 2002, several sites nationwide have been rehabilitated and are back in use as productive real estate.

### 5.7.6 Air Quality

The Southwestern deserts were once promoted as a destination for health and rehabilitation for those people suffering from tuberculosis or respiratory disorders. Development in the Southwest over the past 50 years has significantly altered the quality of the air to a point of noncompliance with standards set forth by the Clean Air Act and administered by the EPA. The rapidly urbanizing cites of the Southwest have a mixture of tailpipe emissions from automobiles that release carbon monoxide (CO) and nitrous oxide ($NO_3$), industrial pollution from power plants and factories, and particulate matter ($PM_{2.5}$ and $PM_{10}$) from disturbing the ground or driving on unpaved roads and high levels of ozone ($O_3$). The minimal rainfall in deserts allows contaminants released into the environment to remain and accumulate in the atmosphere above urban areas, where under certain atmospheric conditions (such as inversions), they may remain for days or weeks. Southwest cities are constantly faced with attempting to create rational policies to manage the airshed below EPA threshold for these criteria pollutants. Table 5.1 outlines the areas of nonattainment as listed in the EPA's *Greenbook*.*

In some states, the EPA has threatened the state to take direct enforcement actions because the state failed to meet targeted pollution levels during a noncompliance period. Future development activities in the Southwest will continue to test the line between meeting EPA air-quality standards and allowing activities that will support the current and future growth of the region.

* http://www.epa.gov/oaqps001/greenbk/ancl3.html (accessed July 7, 2011).

### 5.7.7 Stormwater Pollution

The built environment differs distinctly from the natural setting by the type of surface. Urban lands are covered with large expanses of impervious materials such as roads, concrete building foundations, driveways, parking lots, and rooftops. Water cannot penetrate these hardened surfaces and will runoff unless it is captured. As rainwater falls on urban lands, it picks up contaminants such as sediment, oil, and grease; toxic chemicals and pesticides residues; viruses, bacteria, and nutrients from animal waste and failing septic systems; road salts; and heavy metals from roof shingles, motor vehicles, and other sources.* This potent mix of contaminants is transported offsite in the stormwater runoff and may significantly affect the downstream water quality in the watershed.

The EPA has established the National Pollutant Discharge Elimination System (NPDES) Stormwater Program to regulate stormwater outflow from three areas: municipal separate storm sewer systems, construction activities, and industrial activities. Municipal waste-water treatment plants from all urbanized areas in the United States are required to operate under an NPDES permit. All construction sites greater than 1 ac in size are required to have a general construction permit that regulates the clearing and grading of the land, movement of soil, and stockpiling of materials. Industrial sites must have a Multi-Sector General Permit for industrial stormwater discharges. These operations represent the main sources of discharge of contaminants into the water system, and the EPA delegates the oversight of the permitting of these operations to the states.†

Stormwater discharges are generally considered point sources of pollution, and developers or operators of discharge sites may be required to receive an NPDES permit before they can release stormwater runoff. The intent of these regulations is to prevent or minimize stormwater runoff carrying harmful pollutants to discharge into streams, rivers, reservoirs, or coastal waters. The oversight of stormwater runoff by the EPA presents the developer or operator of a discharge site a real challenge to manage the onsite activities to avoid violating the conditions of the permit of operation. The EPA has created several guidance documents to help manage stormwater runoff, such as the National Menu of Best Management Practices.‡

### 5.7.8 Vegetation Hazard

Fire is a natural process in the southwestern United States that cleanses and rejuvenates the land of dead and dying plant material. Some types of ecosystems in this region are dependent upon fire to reverse senescence of plants and rejuvenate vegetation assemblages, while other ecosystems appear to have seldom burned and are potentially irreversibly harmed by fire. Moreover, this natural process has been suppressed through the management of wildfires by the U.S. Forest Service and other federal agencies (see Chapters 11 and 12). This practice has resulted in a buildup of dry plant material on the forest floor on thousands of acres of Southwestern forests and increases in woody vegetation in areas that once were grasslands.

The once sparsely populated forested areas of the Southwest are increasingly being subdivided to allow the construction of retirement or vacation homes for urban dwellers. Similarly, retirement subdivisions have created a checkerboard of dwellings in formerly continuous grassland areas in southeastern Arizona, and the once common fires must be

* http://water.epa.gov/polwaste/nps/urban.cfm (accessed June 23, 2011).
† http://cfpub.epa.gov/npdes/stormwater/swbasicinfo.cfm (accessed June 23, 2011).
‡ http://cfpub.epa.gov/npdes/stormwater/menuofbmps/index.cfm (accessed June 23, 2011).

suppressed to minimize destruction of property. Over time, the increase of development along the wildland–urban interface has increased the hazard for fires and the potential loss of lives and property. In 2011, Arizona and New Mexico experienced the largest and most severe wildfires in their long histories with wildlands, resulting in the loss of many homes, destruction of property, and increase in flood hazards, both in forested areas and former desert grassland.* Effective interventions are urgently needed around many homes and communities to avoid catastrophic losses in populated areas of the Southwest; otherwise, fires of equal or greater magnitude can be expected in the future. Design on the urban–wildland interface must address questions of fire safety for new developments and retrofitting of existing ones to minimize this hazard.

## 5.8 Summary

Urban design and planning require knowledge of the myriad hydrologic and geologic hazards posed by the desert environment, how these hazards are posed spatially, and the concept of acceptable risk. The desert hydrologic environment provides an excellent example. River channels, surface water, and groundwater are integral to the desert landscape, on the one hand representing a natural wildness and on the other a resource to be tamed and used. High variability of precipitation and surface runoff is one strong characteristic of arid regions; some flood records display nonstationarity with less predictable flood characteristics and uncertain levels of risk. Both mass wasting and more common storm runoff can lead to long-term sedimentation problems that can exasperate flood hazard by defeating engineered channels. Riparian vegetation, desirable in the urban environment, uses considerable water for growth and can exacerbate sedimentation in channels, leading to tradeoffs between hazard mitigation and aesthetics.

Other hazards may be more predictable or understandable, particularly within their context of causality. Overdraft of groundwater systems has caused substantial subsidence in the large alluvial basins of the region that affects buildings and infrastructure. This hazard is not expected in areas without substantial groundwater extraction or outside of large alluvial basins. Seismic hazards are localized in the southwestern United States along existing fault zones in predictable areas. While earthquake prediction is in its infancy, knowledge of the hazard is well developed, and this hazard can be mitigated using existing building codes with acceptable risk.

Few areas of the arid Southwest are subject to all these natural hazards simultaneously, but Palm Springs, California, provides one example of an urban environment that faces all of them. The San Andreas fault passes north and east of town, making this urban center subject to extreme seismic hazards. Drainages from Mount San Jacinto to the west periodically produce debris flows, and the White Water River, which flows through town, can produce large floods from rainfall and(or) snowmelt from both the San Jacinto and San Gorgonio Mountains. Finally, excessive groundwater development in the Coachella Valley, particularly south and east of town, has caused significant subsidence. Urban design in an area such as Palm Springs would benefit from multihazard risk assessment

* http://www.azcentral.com/news/articles/2011/07/09/20110709arizona-fire-wallow-fire-100-percent-contained.html (accessed July 13, 2011).

and mitigation, which is one of the cornerstones of the Federal Emergency Management Agency's hazard reduction program.[1]

Hazards inherent in the built environment have long been identified with various regulations and ordinances designed to mitigate their impacts. A plethora of environmental regulations apply to the issue of waste management, cleanup, and mitigation, particularly where a contaminated area is to be rehabilitated and changed into a different land use. Off-site pollution, whether considered to be point source or nonpoint source, is one of the most significant hazards designers need to contend with as dust generation, in particular, can affect quality of life in urban settings. Wildland fire is gaining in importance as a hazard, particularly in an era when climates are predicted to change and increase fire danger.

Designers should consider for "all hazards" vulnerability assessments to make sure the most vulnerable land uses and populations avoid the highest hazard locations. Urban settlements in the Southwest will require places to store wastes and industries that will generate hazards from their industrial operations. Therefore, it will require of the political, industry, and community leaders a firm commitment to effectively handle, store, or transport toxic materials generated by development and industrial operations without adverse harm to people who live in this region.

## References

1. Federal Emergency Management Agency (FEMA), Multi hazard identification and risk assessment: A cornerstone of the national mitigation strategy (Washington, DC: Federal Emergency Management Agency, 1997).
2. Burby, R.J., ed., *Cooperating with Nature: Confronting Natural Hazards with Land-Use Planning for Sustainable Communities* (Washington, DC: Joseph Henry/National Academy Press, 1998).
3. Pielke, R.A. Jr. and M.W. Downton, Precipitation and damaging floods: Trends in the United States, 1932–97, *Journal of Climate* 13: 3625–3637, 2000.
4. Webb, R.H. and J.L. Betancourt, Climatic variability and flood frequency of the Santa Cruz River, Pima County, Arizona (Washington, DC: U.S. Geological Survey Water-Supply Paper 2379, 1992).
5. Cayan, D.R. and R.H. Webb, El Niño/southern oscillation and streamflow in the western United States, in H.F. Diaz and V. Markgraf, eds., *El Niño, Historical and Paleoclimatic Aspects of the Southern Oscillation* (Cambridge, U.K.: Cambridge University Press, 1992), pp. 29–68.
6. FEMA, National flood insurance program: Summary of coverage (Washington, DC: Federal Emergency Management Agency, Report F-679, 2007).
7. FEMA, Guide to flood maps (Washington, DC: Federal Emergency Management Agency, Report FEMA 258, 2006).
8. Saarinen, T.F., V.R. Baker, R. Durrenberger, and T. Maddock, Jr., *The Tucson, Arizona, Flood of October 1983* (Washington, DC: National Academy Press, 1984).
9. Hjalmarson, H.W. and S.P. Kemna, Flood hazards of distributary-flow areas in Southwestern Arizona (Tucson, AZ: U.S. Geological Survey Water-Resources Investigations, Report 91-4171, 1991).
10. Osterkamp, W.R., Annotated definitions of selected geomorphic terms and related terms of hydrology, sedimentology, soil science and ecology (Reston, VA: U.S. Geological Survey Open-File Report 2008-1217, 2008).
11. National Research Council, *Alluvial Fan Flooding* (Washington, DC: National Academy Press, 1996).
12. FEMA, Alluvial fans, hazards and management (Washington, DC: Federal Emergency Management Agency, Report FEMA 165, 1989).

13. Committee on Alluvial Fan Flooding, *Alluvial Fan Flooding* (Washington, DC: Water Science and Technology Board, Commission on Geosciences, Environment, and Resources, National Research Council, National Academy Press, 1996).
14. U.S. Water Resources Council, *Guidelines for Determining Flood Flow Frequency* (Washington, DC: U.S. Interagency Advisory Committee on Water Data, Hydrology Subcommittee, Bulletin 17B, 1981).
15 Wahl, K.L., W.O. Thomas, and R.M. Hirsch, The stream-gaging program of the U.S. Geological Survey (Reston, VA: U.S. Geological Survey Circular 1123, 1995).
16. Kenney, T.A., C.D. Wilkowske, and S.J. Wright, Methods for estimating magnitude and frequency of peak flows for natural streams in Utah (Reston, VA: U.S. Geological Survey Scientific Investigations, Report 2007-5158, 2007).
17. Ries, K.G. III, J.D. Guthrie, A.H. Rea, P.A. Steeves, and D.W. Stewart, StreamStats: A water resources web application (Reston, VA: U.S. Geological Survey Fact Sheet FS 2008-3067).
18. Milly, P.C.D., J. Betancourt, and M. Falkenmark, Climate change: Stationarity is dead: Wither water management?, *Science* 319: 573–574, 2008.
19. Pielke, R. Jr., Collateral damage from the death of stationarity, *GEWEX Newsletter* 5–7, May 2009.
20. Mote, P.W., A.F. Hamlet, M.P. Clark, and D.P. Lettenmaier, Declining mountain snowpack in western North America, *Bulletin of the American Meteorological Society* 86: 39–49, 2005.
21. Stewart, I.T., D.R. Cayan, and M.D. Dettinger, Changes toward earlier streamflow timing across western North America, *Journal of Climate* 18: 1136–1155, 2005.
22. Knowles, N., M.D. Dettinger, and D.R. Cayan, Trends in snowfall versus rainfall in the western United States, *Journal of Climatology* 19: 4545–4559, 2006.
23. Webb, R.H. and S.A. Leake, Ground-water surface-water interactions and long-term change in riverine riparian vegetation in the southwestern United States, *Journal of Hydrology* 320: 302–323, 2006.
24. Thomas, B.E. and D.R. Pool, Trends in streamflow of the San Pedro River, Southeastern Arizona, and regional trends in precipitation and streamflow in Southeastern Arizona and Southwestern New Mexico (Tucson, AZ: U.S. Geological Survey Professional Paper 1712, 2006).
25. Baker, V.R., Questions raised by the Tucson flood of 1983, *Journal of the Arizona-Nevada Academy of Sciences* 14: 211–219, 1984.
26. Varnes, D.J., Slope movement types and processes, in R.L. Schuster and R.J. Krizek, eds., *Landslides: Analysis and Control* (Washington, DC: National Academy of Sciences, National Research Council, NRC Publication Number 2904, 1978), pp. 11–33.
27. Shreve, R.L., The Blackhawk landslide (Boulder, CO: Geological Society of America, Special Paper 108, 1968).
28. Glancy, P.A. and L. Harmsen, A hydrologic assessment of the September 14, 1974, flood in El Dorado Canyon, Nevada (Washington, DC: U.S. Geological Survey Professional Paper 930, 1975).
29. Webb, R.H., P.G. Griffiths, C.S. Magirl, and T.C. Hanks, Debris flows in Grand Canyon and the rapids of the Colorado River, in S.P. Gloss, J.E. Lovich, and T.S. Melis, eds., *The State of the Colorado River Ecosystem in Grand Canyon* (Reston, VA: U.S. Geological Survey Circular 1282, 2005), pp. 119–132.
30. Webb, R.H., P.G. Griffiths, and L.P. Rudd, Holocene debris flows on the Colorado Plateau: The influence of clay mineralogy and chemistry, *Geological Society of America Bulletin* 120: 1010–1020, 2008.
31. Webb, R.H., C.S. Magirl, P.G. Griffiths, and D.E. Boyer, Debris flows and floods in southeastern Arizona from extreme precipitation in late July 2006: Magnitude, frequency, and sediment delivery (Denver, CO: U.S. Geological Survey Open-File Report 2008-1274, 2008).
32. Youberg, A., M.L. Cline, J.P. Cook, P.A. Pearthree, and R.H. Webb, Geologic mapping of debris flow deposits in the Santa Catalina Mountains, Pima County, Arizona (Tucson, AZ: Arizona Geological Survey Open-File Report 08-06, 2008).
33. Iverson, R.M., S.P. Schilling, and J.W. Vallance, Objective delineation of lahar-inundation hazard zones, *Geological Society of America Bulletin* 110: 972–984, 1998.

34. Schilling, S.P., *LAHARZ: GIS Programs for Automated Mapping of Lahar-Inundation Hazard Zones* (Vancouver, WA: U.S. Geological Survey Open-File Report 98-638, 1998).
35. Galloway, D., D.R. Jones, and S.E. Ingebritsen, *Land Subsidence in the United States* (Reston, VA: U.S. Geological Survey Circular 1182, 1999).
36. Carpenter, M.C., South-central Arizona, in D. Galloway, D.R. Jones, and S.E. Ingebritsen, eds., *Land Subsidence in the United States* (Reston, VA: U.S. Geological Survey Circular 1182, 1999), pp. 65–78.
37. Carruth, R.L., D.R. Pool, and C.E. Anderson, Land subsidence and aquifer-system compaction in the Tucson Active Management Area, south-central Arizona, *1987–2005* (Tucson, AZ: U.S. Geological Survey Scientific Investigations, Report 2007-5190, 2007).
38. Baum, R.L., D.L. Galloway, and E.L. Harp, Landslide and land subsidence hazards to pipelines (Reston, VA: U.S. Geological Survey Open-File Report 2008-1164, 2008), http://geochange.er.usgs.gov/sw/changes/anthropogenic/subside (accessed 26 June 2009).
39. Schumann, H.H., B.M. Wrege, and W.D. Meehan, Hydrogeology, land-subsidence projections, and earth-fissure hazards along the Tucson Aqueduct alignment of the Central Arizona Project in Pinal and Pima Counties, Arizona (Tucson, AZ: U.S. Geological Survey Water-Resources Investigations Report 02-4028, 2002).
40. Nations, D. and E. Stump, *Geology of Arizona* (Dubuque, IA: Kendall/Hunt Publishing Company, 1981).
41. DuBois, S.M. and A.W. Smith, The 1887 earthquake in San Bernardino Valley, Sonora: Historic accounts and intensity patterns in Arizona (Tucson, AZ: Bureau of Geology and Mineral Technology, Special Paper No. 3, 1980).
42. Petersen, M.D., T. Cao, K.W. Campbell, and A.D. Frankel, Time-independent and time-dependent seismic hazard assessment for the state of California: Uniform California Earthquake Rupture Forecast Model 1.0, *Seismological Research Letters* 78: 99–109, 2007.
43. Brown, W.M. III, D.M. Perkins, E.V. Leyendecker, A.D. Frankel, J.W. Hendley II, and P.H. Stauffer, Hazard maps help save lives and property (Reston, VA: U.S. Geological Survey Fact Sheet 183-96, 2001).
44. Spencer, J.E., *Radon Gas: A Geologic Hazard in Arizona* (Tucson, AZ: Arizona Geological Survey, Down-to-Earth Series 2, 1992).
45. Raldolph, J., *Environmental Land Use Planning and Management* (Washington, DC: Island Press, 2004).
46. Blackman, W.C. Jr., *Basic Hazardous Waste Management*, 3rd edn. (Boca Raton, FL: CRC Press, 2001).
47. Cox, D.B. (editor-in-chief) and A.P. Borgias (technical editor), *Hazardous Materials Management Desk Reference* (New York: McGraw-Hill, 2000).
48. Healy, M.J., D. Watts, L.L. Battista, T. Grosvenor, D.A. Kretkowski, J. Lewis, and. J.D. Pico, *Pollution Prevention Opportunity Assessments: A Practical Guide* (New York: John Wiley & Sons, 1998).
49. Sullivan, T.F.P., *Environmental Law Handbook*, 19th edn. (Blue Ridge Summit, PA: Government Institutes, 2009).
50. U.S. Geological Survey, Mineral commodity summaries 2007 (Washington, DC: U.S. Geological Survey, U.S. Government Printing Office, 2007).
51. Committee on Remedial Action Priorities for Hazardous Waste Sites, *Ranking Hazardous-Waste Sites for Remedial Action* (Washington, DC: Commission on Geosciences, Environment, and Resources, Board on Environmental Studies and Toxicology, National Research Council, National Academy Press, 1994).
52. Magirl, C.S., P.G. Griffiths, and R.H. Webb, Modeling debris flows in southeastern Arizona using the stochastic transport model LAHARZ, *Geomorphology* 119: 111–124, 2010.

# Part II

# The Living Desert

John H. Brock

The "living desert" is the place where the physical environment creates habitats for organisms and communities to survive. These living communities may be ephemeral or sustained by the ecosystem they reside, largely due to the fragile balance of climate, hydrology, and the inherent ability of the land to sustain life. To explore the dynamics of these relationships, one can gain a deeper understanding of this theme through the field of ecology, which is the study of the interaction between the biotic and physical worlds, and can be defined as "the structure and function of nature."

In ecological terms, a niche is the organism's function, and the morphological variation among the organisms provides community structure. The assemblages of plants and animals living in a relatively stable environment are termed communities, and several communities may comprise ecosystems. For example, a warm-desert ecosystem will have shrub/grass/succulent plants dominating areas, with variation related to microsites and drainages, such as riparian communities along perennial streams and washes. Desert ecosystems are much more diverse and resilient than people give them credit. However, people must manage desert ecosystems carefully, because of their unique characteristics, and need to learn to design and live in human-modified areas based on ecological principles rather than a self-focused human environment.

The unifying characteristics of the warm deserts, discussed in this book, are long periods of dryness during the year and high temperatures, with soils low in nitrogen but often high in salts. The dominant character of individual desert ecosystems within this bioregion is that rainfall is seasonal. In the Sonoran Desert, rainfall is found to fit a bimodal distribution with some precipitation during the winter season and also in the summer from convectional thunderstorms driven by monsoon winds from south of the region. In the Mojave Desert, rainfall comes primarily in the winter, with little summer thunderstorm activity. The Chihuahuan Desert receives the majority of its rainfall during the warm summer months during the growing season and little during the winter. The paleohistory of the southwestern desert region discussed in Chapter 6 by Thomas provides us with a window into the past to gain a better understanding of the ancient communities that once inhabited the region. The difference in desert seasonality of rainfall explains, in part, the structure of plant communities as outlined in Chapter 7 by Ward Brady, adaptations of desert plants as outlined in Chapter 8 by Mark Dimmitt, and animals as discussed in Chapter 9 by Mark Sullivan et al.

Unique plant communities can be found in the desert bioregion. Vegetation zonation is found with increase of elevation on isolated mountain ranges and has been termed "the sky islands" discussed in Chapters 7 and 10. These habitats are biological islands, because of the geographic separation of small mountain ranges in the warm part of the basin and range geological province. With increasing altitude, the general weather associated with the blocks of mountains becomes cooler, with increased rainfall compared to the surrounding warm deserts. The sky islands are the areas where Merriam's biotic zones were developed and can be easily seen. In the sky islands, desert shrublands or desert grasslands give way to "evergreen" oak and juniper woodland, and at the highest elevations to coniferous forest dominated primarily by pine trees.

This part on the living desert establishes the biological and ecological foundation for humans living in arid landscapes of the southwestern part of the North American continent. Through the knowledge of desert ecology, persons may design a lifestyle in a fashion that is more sustainable in a region with limits in natural resources, like water. Knowing about the vegetative and animal history of the region provides an appreciation for current ecosystems making up this part of North America. Historical knowledge also helps people adapt to changes that are inevitable in the ecological context. Current vegetation zones or types are reflections of the past and point to the future. In addition, animals of all kinds inhabit the vegetated ecological communities, and the chapter dealing with animals begins to show the human interaction with desert ecosystems. The wildlife chapter points out anthropogenic changes including some public policies established to help provide for the continuation of animal species on the landscapes.

Humans have effected major changes in desert ecosystems and accordingly have disturbed or "wounded" the habitats available to plants and animals. These wounds are discussed in Chapter 10 by Dave Foreman, as are objectives to help recovery of native species. Humans have changed the southwestern landscapes by acts of commission and omission. Acts of commission include modifying the hydrologic cycle by reservoir construction, placing transportation corridors across the bioregion, mining mineral and energy resources, poorly controlling grazing by livestock, and allowing urban areas to sprawl across the land. Acts of human omission that have negatively impacted southwestern landscapes include the suppression of wildfires and introduction of some invasive species. Two chapters deal with this subject: Chapter 11 by Stephen Pyne entitled "Built to Burn" begins the discussion of fire and managing its effect on the landscape, and Chapter 12 on ecological restoration by William Wallace Covington deals, in part, with the lack of natural fires, discusses forest restoration, and shows how humans can change ecological trajectories by the use of land management tools. This part provides readers a window into the ecological concepts that can lead to sustainable habitation of the semiarid and arid regions of southwestern North America.

# 6

# *Deep History and Biogeography of La Frontera**

**Thomas R. Van Devender**

**CONTENTS**



## 6.1 Introduction

The border between the United States and Mexico traverses some of the most spectacular and interesting landscapes of the American Southwest, from the Gulf of Mexico dunes and Tamaulipan thornscrub of the Río Grande Valley, through the Chihuahuan Desert in the Big Bend Country of Texas and Coahuila, across the desert grasslands of the Continental Divide of New Mexico and Chihuahua through the "sky island" country of isolated Madrean mountains and the saguaro-studded Sonoran Desert of Arizona and Sonora, and the Mediterranean chaparral of California and Baja California to reach the Pacific Ocean (see Chapter 7). The dramatic vegetation gradients along the border are summarized in *Biotic Communities of the American Southwest—United States and Mexico*,[1] the accompanying vegetation map,[2] Webster,[3] and various regional floras. Historic vegetation changes related to human activities along the border have been the subject of considerable discussion.[4–6] In this chapter, I will discuss various aspects of the fossil record, paleoenvironmental

* Adapted with permission from Van Devender, T., in G. L. Webster and C. J. Bahre, eds., *Vegetation and Flora of La Frontera: Vegetation Change along the United States–Mexican Boundary* (Albuquerque, NM: University of New Mexico Press, 2001), pp. 56–83.

reconstructions, and speculations that help understand the dynamic nature of the biotic communities of *La Frontera* in their deep historical and biogeographical context.

## 6.2 Out of the Tropics

Long before the deserts of North America evolved, the climate and vegetation of *La Frontera* were tropical. In the Paleocene, geological time period that began with the extinction of the dinosaurs 65 million years ago (mya), temperate evergreen and tropical rainforests were widespread across the continent with little regional differentiation.[7] The climates of North America were warm and humid with forests with strong Asian affinities; primitive ferns (*Anemia*), cycads (*Dion, Zamia*), and palms occurred as far north as Alaska and 70°N Lat. in Greenland. Eocene fossils of several palms (*Phoenicites, Sabalites*) were found in Alaska.[7,8] An alligator, a soft-shell turtle, a primitive tortoise, a primitive monitor lizard, and a ground boa were reported from Eocene sediments at 78°N Lat. on Ellesmere Island, Canada.[9,10]

Through the Eocene (54–35 mya), deciduous trees became increasingly more common, providing the first evidence of a dry season and the presence of tropical deciduous forests.[11–13] There were evolutionary radiations in many plants and animals as they adapted to new heat and moisture regimes as more sunlight penetrated the forest canopy to the ground. Many of the adaptations to aridity in modern desert plants and animals evolved in the dry seasons of these early Tertiary tropical forests.

## 6.3 Miocene Revolution

### 6.3.1 Mountain Building

From the late Oligocene to the middle Miocene (about 30–15 mya), a series of enormous volcanic eruptions changed the climates and established the modern biogeographic provinces of North America.[14] The Rocky Mountains were uplifted at least 5003 ft near Florissant, Colorado, while more than 7004 ft of volcanic rocks were deposited in the Jackson Hole area of west-central Wyoming.[15] During the same period, a kilometer-thick layer of rhyolitic ash settled in the Sierra Madre Occidental of western Mexico—overlaying a kilometer of early Tertiary andesites![16]

With the uplift of the mountains, there were profound climatic and biotic consequences. The upper flow of the atmosphere was blocked for the first time, preventing tropical moisture from both the Pacific Ocean and the Gulf of Mexico from reaching the mid-continent, thereby drying out the modern Great Plains and Mexican Plateau. Harsher climates initiated evolutionary radiations in the modern successful plant and animal groups as well as segregated drought- and cold-tolerant species into new environmentally limited biotic communities, or biomes, including tundra, conifer forests, and grasslands that were distributed along elevational and latitudinal environmental gradients.[13,17] Tropical forests were restricted to ribbons along the lowlands of Mexico and central America. Thornscrub, the dry vegetation found today at the lower, drier edges of tropical deciduous forest, was likely the regional vegetation covering the drier areas in the present Chihuahuan and

Sonoran deserts as far north as *La Frontera*. Unfortunately, the vegetation terms "thornscrub," "thorn forest," and "tropical deciduous forest" have often been confused in the literature; for example, throughout the excellent historical biogeographical discussion in Morafka et al.,[18] "thornscrub" was mistakenly used instead of "tropical deciduous forest."

A new method of estimating paleoelevations has challenged the aforementioned scenario. Wolfe[19] used correlations of leaf morphology in fossil floras in modern floras to modern climates to estimate the history of temperature in the Tertiary. Recently, his analyses were expanded to include regressions between foliar morphology and various aspects of climate, and were developed into a method of estimating the elevations of fossil floras at the time of deposition.[20] In this approach, the physiological tolerances and limits of the living populations or closest relatives of the fossil taxa are not considered because they are extinct species that might have been different in the past. The leaf morphology–climate relationship based on worldwide floras is thought to be a better indicator of climate than the physiological tolerances of the living relatives of the fossil taxa. The first studies using this methodology reached dramatically different paleoelevation estimates than previous studies based on floristic affinities. For example, MacGinitie[21] inferred a paleoelevation of 3001 ft (now at 8202 ft) for the latest Eocene (35 mya, formerly called Oligocene) Florissant Beds in Colorado based on a paleoflora closely allied with the highlands of northeastern Mexico. The flora was a mixture of plants now found in tropical and montane areas and a sequoia (*Sequoia affinis*). In contrast, Gregory[22] using Wolfe's multivariate climate analysis techniques[20] estimated that the Florissant Beds were at 7,545–10,826 ft elevation. The climatic implications of the additional 4543–7824 ft elevation in the late Eocene of the Rocky Mountains are profound, especially the inferences about cold temperatures. All of the ecological, evolutionary, and biogeographic changes in the biota discussed for the Oligocene–middle Miocene in the Axelrod model should have occurred earlier if Gregory were correct. It is difficult to accept this in light of the persistence of tropical plants such as cedar (*Cedrela*), palms, and piocha (*Trichilia*) into the early Oligocene, or the gradual modernization of the Rocky Mountain flora from the Oligocene to the early Miocene.[15]

A reexamination of Gregory's study[22] of the Florissant flora is enlightening. The living relatives of at least 34% of the 29 taxa used in her paleoclimatic analysis are today restricted to elevations and latitudes lower than Florissant. Cedar, hopbush (*Dodonaea*), mesquite (*Prosopis*), and soapberry (*Sapindus*) are genera with tropical affinities whose extant species live in areas with higher mean annual temperatures than the upper elevations of the Rockies. *Trichilia* in particular is a tropical grass genus in the Meliaceae that reaches its northern limit in southern Sonora at 27°N Lat., where *Tilia americana* and *Tilia hirta* live in tropical deciduous forest below about 3280 ft elevation. The inescapable conclusions of a paleoelevation of 7,545–10,826 ft for the Florissant Beds are that a third of the flora had greater cold tolerances than their living relatives, and they were more vulnerable to extinction than their tropical descendants.

Wolfe[23] envisioned a "Big Chill," a cold period of 1–2 million years in length in the earliest Oligocene that had a profound impact on the flora. Other climatic reconstructions for the Eocene\Oligocene boundary, primarily based on the climatic relationships of surviving taxa were very different. For example, analyses of fossil reptiles and amphibians[24] and of pollen and leaf floras[13] indicated increasing aridity and seasonality with little cooling. If Wolfe's multivariate climate analyses of leaf floras systematically underestimate mean annual temperatures, then paleoelevations are overestimated, resulting in questionable landscape and paleoclimatic reconstructions. I feel that important paleoecological signals from plant and animal taxa in these fossil deposits must be considered. For the present, the Axelrod model of landscape evolution is the most useful, although it is susceptible to revisions in the timing of the uplift of the Sierra Madre Occidental.

### 6.3.2 Evolution of the Deserts

Although the modern North American climatic regimes and biotic provinces were established in the Miocene Revolution, the deserts were not yet in existence. Based on a series of fossil floras in California, Axelrod[14] inferred that the Sonoran Desert formed as the result of a drying trend in the middle Miocene (15–8 mya), displacing thornscrub to more southerly latitudes. Some of the species in the new desertscrub communities such as brea (*Cercidium praecox*), guayacán (*Guaiacum coulteri*), tree ocotillo (*Fouquieria macdougalii*), organpipe cactus (*Stenocereus thurberi*), and senita (*Lophocereus schottii*) were probably segregated out of thornscrub. Bradley[25] demonstrated that the behavioral and physiological adaptations of reptiles that allow them to thrive in hot, dry environments are not unique to deserts and evolved in more mesic habitats. Other plants such as foothills paloverde (*Cercidium microphyllum*), ironwood (*Olneya tesota*), and saguaro may have evolved with the Sonoran Desert. However, saguaro is not closely related to the columnar cacti of tropical deciduous forest (etcho, *Pachycereus pecten-aboriginum* or saguira, *Stenocereus montanus*) or thornscrub (organpipe or senita). It was derived from *Neobuxbaumea*, a genus of columnar cacti restricted to central Mexico south of the Sierra Madre Occidental.

Another important chapter in the history of the Sonoran Desert pertains to Baja California, which was attached to the Mexican mainland. As the Gulf of California formed, a strip of land stocked with tropical plants and animals drifted in splendid isolation northwestward to meet California and form the Baja California Peninsula.[14,26,27] Natural selection shaped them into many unique endemics including boojum tree or cirio (*Fouquieria columnaris*). As with the mainland Sonoran Desert species, the biogeographical affinities of many Baja California plant and animals are with central Mexico south of the Sierra Madre Occidental. The nearest relatives of many Baja California reptiles are found today on the Pacific coast of south-central Mexico in the Balsas Basin or Sierra Madre Sur.[27] The geologic history of Baja California was reconstructed differently in each of the three papers cited earlier. The initial rifting was completed by 10–12 mya with the formation of the proto-Gulf of California. However, Grismer[27] argues that Baja California formed later as the modern Gulf of California formed in the latest Miocene and that most of the evolution occurred after 5.5 mya. At least some of the marine sediments in the Imperial Formation of southern California appear to be older than this, suggesting greater age of the Gulf of California.[28]

Thus, the Sonoran Desert was in existence by the late Miocene (5–8 mya). Despite the absence of fossils, the regional geologic events that reshaped western North America likely affected the Chihuahuan Desert as well. Aridity intensified on the Mexican Plateau due to increasing rainshadow effects of the Sierra Madre Occidental to the west, the Sierra Madre Oriental to the east, and the Rocky Mountains to the north. Cold had a greater role in the evolution of the Chihuahuan Desert biota than in the Sonoran Desert as the Continental Divide was uplifted, and there were no barriers to the icy "blue northers "(incursions of Arctic air). Apparently, the deserts are perhaps the youngest biotic communities of North America. The geologic events that shaped the landscape and altered regional climates were mostly in the early-middle Miocene, suggesting that the speciation of many of the prominent desertscrub plants occurred earlier in tropical deciduous forest or thornscrub. Subsequently, changing climates and immigration rather than evolution were the major impacts on the desert biota—dramatically shifting species ranges and community compositions.

### 6.3.3 Pliocene Climates

During the late Miocene and early Pliocene, sea level rose enough that the Gulf of California expanded north into the Salton Trough of southeastern California to deposit the marine sediments of the Imperial Formation.[27,28] For part of this time, a marine embayment in the Los Angeles Basin almost connected with the Gulf of California, again effectively isolating Baja California. The extensive sediments of the roughly contemporaneous Bouse Formation in the lower Colorado River basin have been interpreted as either estuarine based on invertebrate fossils, including barnacles, or freshwater lakes on strontium isotopic analyses.

In Anza Borrego Desert State Park in southern California, the Pliocene-early Pleistocene terrestrial sediments of the Palm Springs Formation overlie the marine Imperial Formation. A fossil lizard skull from the Pliocene (ca. 4.3–2.5 mya) in the Vallecito Creek local fauna 25 miles north of the international boundary was described, although it could have been easily placed in the extant genus *Iguana*.[29] It was associated with the extant desert iguana. The green iguana is a tropical lizard that today occurs no farther north than southern Sinaloa, ca. 932 miles to the southeast. Tropical species in desert at high latitudes and much higher sea levels indicate warmer global temperatures and oceans, enhanced monsoonal summer rainfall, and the northward expansions of tropical thornscrub and deciduous forests in Sonora and the Sonoran Desert in *La Frontera*.

### 6.3.4 Historical Biogeography

The distributions of a number of plant and animal species or closely related species pairs suggest past connections between the Chihuahuan and Mohave\Sonoran deserts. This region was called Mojavia[30] and modified for herpetofauna.[31] Different distribution patterns in the vicariant species pairs likely reflect different separation times and evolutionary mechanisms. One type of east–west species pairs reflects the evolution of similar species from common subtropical ancestors. For example, the Big Bend gecko in the Chihuahuan Desert and the barefoot gecko in the Sonoran Desert were both derived from a tropical ancestor (very close to *Coleonyx mitratus*[32]; living in tropical deciduous forest or thornscrub). The arborescent yuccas, *Yucca filifera*, in the southern Chihuahuan Desert and *Y. valida* in Baja California are a similar example in plants. Presumably the early-middle Miocene orogeny restricted the ancestor's ranges into a northerly "u"-shape straddling the Sierra Madre Occidental; the living descendants occur at the northern tips of the "u." In these particular cases, the Sonoran Desert species were further isolated as Baja California separated from the mainland.

Other species pairs likely reflect simple range splits that evolved into eastern and western species with the uplift of the Continental Divide and the initiation of glacial climates about 2 mya. Examples in reptiles of east–west species pairs are banded geckos, horned lizards,[33] and rat snakes; justifications for not using Herndon Dowling's generic name *Bogertophis* for these two species or *Senticollis* for *Elaphe triaspis* are in Van Devender and Bradley.[34] In plants, the relationship between the Chihuahuan and Mohave deserts is particularly strong, in part because of the physical connection across central Arizona along the Mogollon Rim. Closely related (vicariant) species in Texas–Coahuila–Chihuahua and Arizona–California include Torrey and Mohave yuccas (*Yucca torreyi/Y. schidigera*), Joshua tree/Whipple and Thompson yuccas (*Yucca brevifolia, Yucca whipplei/Yucca thompsoniana* and relatives), heath and Burro Creek cliff roses (*Cowania ericaefolia/Cowania subintegra*), canotias (*Canotia wendtii/Canotia holacantha*), crucifixion thorns (*Castela emoryi/Castela holacantha*), and Texas and blue paloverdes (*Cercidium texanum/Cercidium floridum*).

Bullsnake, western diamondback rattlesnake, creosotebush (*Larrea tridentata*), and many other species are widespread in the warm deserts and desert grasslands of North America. During Pleistocene glacial periods, their ranges were separated by nondesert habitat and were only established during the present interglacial (the Holocene) in the last 6000 years.[17] In many species, the distributions of infraspecific taxa are often closely tied to the modern biotic provinces, and likely expanded and contracted as those biomes responded to glacial-interglacial fluctuations. An interesting possibility is that the evolution of many, but not all, infraspecific taxa was related to the formation of the modern biomes that they inhabit, reflecting natural selection that occurred millions of years ago. For example, the desert grassland kingsnake is widespread in desert grasslands from Texas to Zacatecas and Arizona but is abruptly replaced by the desert kingsnake in transition to Sonoran Desert near Tucson.

## 6.4 Desert Ice Ages

The warmth of the Pliocene ended suddenly at the beginning of the Pleistocene when the earth entered a new climatic era of cool, continental conditions. Traditionally, four ice ages or glacial periods, based on terrestrial sedimentary deposits, were recognized in North America and widely correlated between Europe and South America. However, recent studies of isotopic climatic indicators in continuous sediment cores from the ocean floors record 15–20 glacial periods in the last 2.4 million years with the ice ages about 10 times as long as interglacials, which lasted 10,000–20,000 years.[35]

In the last glacial period (the Wisconsin), the massive Laurentide ice sheet covered most of Canada and extended as far south as New York and Ohio. The mixed deciduous forest in much of the eastern United States was displaced by boreal forest with spruce (*Picea* spp.) and jack pine (*Pinus banksiana*).[36] Glaciers covered the highest elevations of the Rocky Mountains and the Sierra Nevada as well as the peaks of the Sierra Madre del Sur in south-central Mexico. Now-dry playas in the Great Basin and on the Mexican Plateau were large lakes, and enough water was tied up in ice on land to lower sea level about 328 ft.

### 6.4.1 Packrat Curators

Although *La Frontera* is distant from glaciated areas, past glacial climates have resulted in profound changes in climate and vegetation. Paleoenvironmental insights of remarkable power have been gleaned from an unlikely source—"packrat middens." Packrats or wood rats are medium-sized rodents that carry plant materials and other objects back to their houses or dens. Some of this material may become cemented into the hard, dark organic masses called middens that can be preserved indefinitely in dry rockshelters.[37] Detailed reconstructions of local communities on rocky slopes for the last 45,000 years have been completed for many desert areas using the abundant, well-preserved plant remains in radiocarbon-dated middens.[38] Midden assemblages are excellent for documenting past floras and reconstructing the vegetation on rocky slopes within about 30 m of the rock shelters.

The midden record extends back for several tens of thousands of years of the Wisconsin glacial and the Holocene. Plant remains in middens document widespread expansions of woodland trees and shrubs down to elevations that now support deserts from

45,000 to 11,000 years ago during the Wisconsin.[38] Under glacial climates, with cooler summers and greater winter rainfall, warm desertscrub communities dominated by creosotebush were restricted to the Bolson de Mapimi area in the southern Chihuahuan Desert[39]; using the climatic criteria of Schmidt[40]; and below 984 ft elevation in the Lower Colorado River Valley in the Sonoran Desert.[41]

Packrat middens in rockshelters in the Hueco Mountains 35 km northeast of the international boundary at 1,340–1,430 m elevation just west of El Paso, Texas, in the northern Chihuahuan Desert yielded a 42,000 year series of packrat middens.[38,39] Samples dated from 42,000 to 11,000 year B.P. (radiocarbon years before 1950) were dominated by woodland species including pinyon pines (*Pinus edulis, Pinus remota*), juniper (*Juniperus* sp.), and shrub oak (*Quercus pungens*).[39] The middle Holocene (ca. 8000–4500 year B.P.) vegetation was desert grassland with Chihuahuan desertscrub developing about 4500 years ago with the arrival of creosotebush, lechuguilla (*Agave lechuguilla*), and ocotillo (*Fouquieria splendens*).

Midden series from Maravillas Canyon and Río Grande Village on the Texas–Coahuila boundary at 600–835 m elevation in the Big Bend of Texas provide vegetation records for the last 45,600 years.[39] As in the Hueco Mountains, the late Wisconsin (22,000–11,000 year B.P.) samples documented pinyon–juniper–oak woodlands in areas that now support Chihuahuan desertscrub. However, the middle Wisconsin (45,600–22,000 year B.P.) was more xeric with little or no papershell pinyon (*P. remota*) and increased desert grassland elements. The Holocene vegetational history differed from the northern Chihuahuan Desert as desertscrub developed by 10,500 year B.P. without the middle Holocene desert grassland. The oak in the Wisconsin samples was Hinckley oak (*Quercus hinckleyi*), now a rare endemic shrub only found in the Solitario and near Shafter north of Big Bend National Park.

In Arizona, woodlands with single leaf pinyon (*Pinus monophylla*), junipers, shrub live oak (*Quercus turbinella*), and Joshua tree (*Y. brevifolia*) were widespread in the present Arizona Upland subdivision of the Sonoran Desert.[41] Ice age climates with greater winter rainfall from the Pacific and reduced summer monsoonal rainfall from the tropical oceans likely favored woody cool-season shrubs with northern affinities[42] rather than the summer-rainfall trees, shrubs, and cacti of tropical forests and subtropical deserts.

In the Puerto Blanco Mountains of Organ Pipe Cactus National Monument (9 miles north of the Sonora border), saguaro and brittlebush returned to Arizona soon after the beginning of the Holocene about 11,000 years ago but were living in xeric woodlands. Although Sonoran desertscrub vegetation formed about 9000 years ago when displaced woodland species finally retreated upslope, the modern climatic regime and community composition were not established until foothills paloverde, ironwood, and organpipe cactus arrived about 4500 years ago. However, the plant communities never achieved an equilibrium state because the climate continued to fluctuate, although on lesser scales. Notable late Holocene climatic events were the wet period about 1000 years ago coincident with the development of the sophisticated Anasazi and Hohokam Indian cultures, and the relative aridity of this century.

Similar successional stages likely occurred during each interglacial. Although the late Holocene desertscrub communities likely resembled the original late Miocene Sonoran Desert, relatively modern desertscrub communities were developed for about 5%–10% of the 2.4 million years of the Pleistocene. Ice age climates with woodlands in the desert lowlands typical of about 12,000 year B.P. persisted for about 80%–90% of this period.[43,44]

## 6.5 Tropical Interglacials

Although the general environmental history of *La Frontera* involves the transition from tropical deciduous forests in the early Tertiary to more temperate ice age woodlands and interglacial deserts in the Pleistocene, the vertebrate fossil record suggests that there were more tropical interglacials than the Holocene.

### 6.5.1 El Golfo, Sonora

The first fossil record of the giant anteater in North America was in early Pleistocene sediments from El Golfo de Santa Clara in northwestern Sonora (47 miles southwest of the international boundary[45,46]). The nearest populations of this large tropical mammal are 1864 miles to the southeast in the humid, tropical lowlands of Central America. As for many large mammals, the modern distribution may not accurately reflect their physiological range limits because of human predation in the last 11,000 years.[47] Other large extinct animals in the El Golfo fauna include antelope, a bear, camels, cats, horses, proboscidians, and a tapir. Fossils of extant species in the fauna included Sonoran Desert toad, slider, boa constrictor, California beaver, and jaguar. The Sonoran Desert toad is a regional endemic while the slider and boa constrictor occur today in Sonora in wetter, more tropical areas to the southeast. The California beaver was a larger species than the extant beaver but much smaller than the bear-sized giant beaver.

El Golfo is at the head of the Gulf of California in the Lower Colorado River Valley subdivision of the Sonoran Desert.[48] Today, hyperarid desert at El Golfo is too dry to support any of these animals, although historically the delta of the Colorado River was a very wet area; extensive cottonwood (*Populus fremontii*) gallery forests supported abundant beaver. There is an account of a large spotted cat (likely jaguar) that entered James Ohio Pattie's camp on the Colorado south of Yuma in December of 1827 to feed on drying beaver skins.[49] Although the Sonoran Desert is the most "tropical" of the North American deserts, periodic catastrophic freezes are the most important climatic factor setting the northern limits of tropical species and communities.[50,51] Hastings and Turner[52] stated that there is probably no part of the Sonoran Desert that is free from freezing temperatures.

The El Golfo fossil fauna was deposited during the Irvingtonian Land Mammal Age that began 1.8 mya. This fauna reflects an early Pleistocene interglacial when the climate was much more tropical than it is today, i.e., frost free, much greater rainfall in the warm season, and higher humidity. Greater summer rainfall would have been coupled with intensified summer monsoonal circulation patterns and warmer sea surface temperatures, unlike most of the Pleistocene when colder water in the northern Pacific intensified the Aleutian Low, augmenting winter rainfall.[53] The giant anteater, capybara, and ground sloths in the fauna were members of ten families of mammals that immigrated into North America in the late Pliocene or early Pleistocene after the opening of the Panamanian land bridge during the Great American Interchange.[54] In contrast, the imperial mammoth and a hyena in the fauna were Eurasian immigrants.

### 6.5.2 Rancho La Brisca, Sonora

The Rancho La Brisca fossil locality is located in a canyon north of Cucurpe, Sonora (56 miles south of the international boundary).[55] The fauna was dominated by Sonoran mud-turtle and fish, reflecting a wet *ciénega* paleoenvironment. The presence of the sabinal frog

149 miles north of the northernmost extant population on the Río Yaqui indicates that the climates were once more tropical than at the site today. Teeth of a large extinct bison, which only immigrated to North America from Siberia in the late Pleistocene between 170,000 and 150,000 years ago,[56] were also found. The coexistence of bison and tropical species indicates that sediments near Rancho La Brisca were deposited about 80,000 years ago in the last interglacial (the Sangamon), which was considerably more tropical than the Holocene.

## 6.6 Ice Age Mammals and Grassland Dynamics

Sediments in playas, springs, and caves provide insights into the ice age environments of the highlands along *La Frontera* between the Chihuahuan and Sonoran deserts. Rich bone beds in sites from the Great Plains to southeastern Arizona reveal a late Pleistocene fauna rivaled today only by African savannas, like the Serengeti Plains of Tanzania.[57] However, Wisconsin-aged sediments scattered in sites from the Great Plains to Arizona were dominated by pollen forest trees instead of grasses, illustrating the peril in inferring the existence of grassland biomes from the presence of large vertebrates. Although conifers are such prodigious pollen producers that ice age communities may have been relatively open pine parklands rather than closed forests, typical grassland assemblages were not encountered until the Holocene.

Sediments in U-Bar Cave, a desert grassland area in southwestern-most New Mexico (ca. 6 miles north of the international boundary) yielded bones and teeth of a very diverse fauna.[58] Extinct large mammals included dire wolf giant short-faced bear, horses, mountain deer, pronghorns, and shrub ox. The Willcox Playa and Murray Springs (50 and 8 miles north of the international boundary, respectively) are desert grassland sites in southeastern Arizona where the ice age portions of the pollen profiles were dominated by pine pollen. Murray Springs is best known because elegant flint spear points were imbedded in mammoth bones provided glimpses of Clovis hunters. Other large mammals in this rich Pleistocene fauna included American lion, bison, camel, horses, llama, and tapir.[59]

### 6.6.1 Pleistocene Overkill

About 11,000 years ago, approximately two thirds of the large mammal species of North America became extinct.[47] Common, widespread grazers including horses and mammoths seem to have disappeared at the very time spruce and pine retreated and grasslands expanded from Canada to Arizona. Martin[47] forcefully presents the case that big game hunters, rather than changing climate, caused widespread extinctions within a few hundred years after their entry into North America from Siberia via the Bering Strait. Whether the "overkill" model is accepted or not, there is no paleobotanical evidence of climatic changes severe enough to cause extinction in biotic regimes ranging from the boreal forests and conifer parklands to the deserts of the Southwest, the Mediterranean chaparral of southern California, and throughout the New World tropics.

Changes in climate and vegetation at the time of extinction in the southwestern United States were not greater than similar fluctuations in other interglacials.[35] The tropical nature of the Rancho La Brisca fossil fauna suggests that the magnitude of climate change in last interglacial was greater than in the Holocene.[55] Moreover, plant remains in ancient packrat middens record the survival of woodland plants in desert lowlands for several thousand years in the early Holocene after the megafaunal extinction and before the formation of desert grassland in the northern Chihuahuan Desert.[39] It is clear that the large herbivore faunas of *La Frontera* wereare dramatically reduced compared to previous interglacials.[57] The ecological impacts of these missing herbivores on Holocene biotic communities were probably substantial.

A 12,000 year old stratified vertebrate fauna from Howell's Ridge Cave in the Little Hatchet Mountains of southwestern New Mexico (12 miles north of the international boundary)[60] helps place the historical increase in shrubs in desert grasslands in New Mexico in perspective. Bones of an extinct horse were found in the lower Wisconsin levels close to a California condor bone radiocarbon dated to 13,460 year B.P.[61] A pinyon–juniper–oak woodland was probably on the ridge in the late Wisconsin.

Thousands of bones and teeth of small vertebrates were carried to the ridge by owls and hawks, and deposited in the cave in regurgitated pellets, providing faunal samples from all nearby habitats. Percentages of the minimum numbers of individuals of species with specific habitat requirements allowed environmental changes through time to be inferred. Tiger salamander and Colorado chub bones, and teeth of voles in the deposit suggest that the playa contained perennial water more often from 12,000 to 4,000 years ago, and again at about 3,000 and 1,000 years ago.

Declines in typical grassland animals and increases of typical desert species such as Couch's spadefoot toad and round-tailed horned lizard about 3900, 2500, and 990 years ago reflect shift to habitats with more exposed soil—increases of shrubs in the desert grassland similar to that recorded in the last century.

## References


1. Brown, D. E., Biotic communities of the American Southwest—United States and Mexico, *Desert Plants* 4: 1–342, 1982.
2. Brown, D. E. and C. H. Lowe, *Biotic Communities of the Southwest* (Fort Collins, CO: U. S. Department of Agriculture, Forest Service, General Technical Report RM-78 (map), Rocky Mountain Forest and Range Experimental Station, 1980).
3. Webster, G. L., A reconnaissance of the vegetation and flora of La Frontera, in G. L. Webster and C. J. Bahre, eds., *Vegetation and Flora of La Frontera: Historic Vegetation Change along the United States/Mexico Boundary* (Albuquerque, NM, University of New Mexico Press, 2001), pp. 6–38.
4. Bahre, C. J., *A Legacy of Change: Historic Human Impacts on the Vegetation of the Arizona Borderlands* (Tucson, AZ: University of Arizona Press, 1991).
5. Bahre, C. J., Human impacts on the grasslands of southeastern Arizona, in M. P. McClaran and T. R. Van Devender, eds., *The Desert Grassland* (Tucson, AZ: University of Arizona Press, 1995).
6. Bahre, C. J. and C. F. Hutchinson, Historic vegetation change along the United States–Mexico boundary west of the Río Grande, in G. L. Webster and C. J. Bahre, eds., *Vegetation and Flora of La Frontera: Historic Vegetation Change along the United States/Mexico Boundary* (Albuquerque, NM, University of New Mexico Press, 2001), pp. 67–83.
7. Wolfe, J. A., Paleogene floras from the Gulf of Alaska Region (U.S. Geological Survey Professional Paper No. 997, 1977).
8. Wolfe, J. A., An interpretation of Alaskan Tertiary floras, in A. Graham, ed., *Floristics and Paleofloristics of Asia and Eastern North America* (Amsterdam, the Netherlands: Elsevier, 1972), pp. 201–233.

9. Dawson, M. R., R. M. West, W. Langston, Jr., and J. H. Hutchinson, Paleogene terrestrial vertebrates: Northernmost occurrence, Ellesmere Island, Canada, *Science* 192: 781–782, 1976.
10. Estes, R. and J. H. Hutchison, Eocene lower vertebrates from Ellesmere Island, Canadian Archipelago, *Palaeogeography, Palaeoclimatology, Palaeoecology* 30: 325–347, 1980.
11. Wolfe, J. A. and D. Hopkins, Climatic changes recorded by tertiary land floras in northwestern North America, in K. Hatai, ed., *Tertiary Correlations and Climatic Changes in the Pacific* (Tokyo, Japan: *Symposium, 11th Pacific Scientific Congress*, 1967), pp. 67–76.
12. Axelrod, D. I. and H. P. Bailey, Paleotemperature analysis of tertiary floras, *Palaeogeography, Palaeoclimatology, Palaeoecology* 6: 163–193, 1969.
13. Leopold, E. B., G. Liu, and S. Clay-Poole, Low-biomass vegetation in the Oligocene, in D. R. Prothero and W. A. Berggren, eds., *Eocene-Oligocene Climatic and Biotic Evolution* (Princeton, CT: Princeton University Press, 1992), pp. 399–420.
14. Axelrod, D. I., Age and origin of the Sonoran Desert, *California Academy of Sciences Occasional Paper* 132: 1–74, 1979.
15. Leopold, E. B. and H. D. MacGinitie, Development and affinities of tertiary floras in the Rocky Mountains, in A. Graham, ed., *Floristics and Paleofloristics of Asia and Eastern North America* (Amsterdam, the Netherlands: Elsevier, 1972).
16. Yetman, D. A., T. R. Van Devender, P. Jenkins, and M. Fishbein, The Río Mayo: A history of studies, *Journal of the Southwest* 37: 294–345, 1995.
17. Van Devender, T. R., Desert grassland history: Changing climates, evolution, biogeography, and community dynamics, in M. P. McClaran and T. R. Van Devender, eds., *The Desert Grassland* (Tucson, AZ: University of Arizona Press, 1995), pp. 68–99.
18. Morafka, D. J., G. A. Adest, L. M. Reyes, G. Aguirre, and S. S. Liberman, Differentiation of North American desert: A phylogenetic evaluation of a vicariance model, *Tulane Studies in Zoology and Botany (New Orleans, LA)*, 1(Suppl.): 195–226, 1992.
19. Wolfe, J. A., Tertiary climatic fluctuations and methods of analysis of tertiary floras, *Palaeogeography, Palaeontology, and Palaeoecology* 9: 27–57, 1995.
20. Wolfe, J. A., A method of obtaining climatic parameters from leaf assemblages (U. S. Geological Survey Professional Paper No. 1964, 1993).
21. MacGinitie, H. D., *Fossil Plants of the Florissant Beds, Colorado* (Washington, DC: Carnegie Institution of Washington Publication No. 599, 1953).
22. Gregory, K. M., Palaeoclimate and palaeoelevation of the 35 ma Florissant flora, Front Range, Colorado, *Palaeoclimates* 1: 23–57, 1994.
23. Wolfe, J. A., Climatic, floristic, and vegetational changes near the Eocene/Oligocene boundary in North America, in D. R. Prothero and W. A. Berggren, eds., *Eocene-Oligocene Climatic and Biotic Evolution* (Princeton, CT: Princeton University Press, 1992), pp. 421–436.
24. Hutchison, J. H., Western North American reptile and amphibian record across the Eocene/Oligocene boundary and its climatic implications, in D. R. Prothero and W. A. Berggren, eds., *Eocene-Oligocene Climatic and Biotic Evolution* (Princeton, CT: Princeton University Press, 1992), pp. 451–463.
25. Bradley, S. D., Desert reptiles: A case of adaptation or pre-adaptation? *Journal of Arid Environments* 14: 155–174, 1988.
26. Murphy, R. W., Paleobiogeography and genetic differentiation of the Baja California herpetofauna, *California Academy of Sciences Occasional Paper* 137: 1–137, 1983.
27. Grismer, L. L., The origin and evolution of the peninsular herpetofauna of Baja California, Mexico, *Herpetological Review* 2: 51–106, 1994.
28. Spencer, J. E. and P. J. Patchett, Sr isotope evidence for a lacustrine origin for the upper Miocene to Pliocene Bouse Formation, lower Colorado River trough, and implications uplift, *Geological Society of America Bulletin* 109: 767–778, 1997.
29. Norrell, M. A., Late tertiary fossil lizards of the Anza Borrego Desert, California, U.S.A., *Contributions in Science, Natural History Museum Los Angeles County* 414: 1–31, 1989.
30. Axelrod, D. I., Evolution of the Madro-Tertiary geoflora, *Botanical Review* 24: 433–509, 1958.

31. Morafka, D. J., *A Biogeographical Analysis of the Chihuahuan Desert through Its Herpetofauna* (The Hague, the Netherlands: W. Junk, 1977).
32. Grismer, L. L., A reevaluation of the North American gekkonid genus *Anarbylus Murphy* and its cladistic relationships to Coleonyx Gray, *Herpetologica* 39: 394–399, 1983.
33. Montanucci, R. R., A phylogenetic study of the horned lizards, genus *Phrynosoma*, based on skeletal and external morphology, *Contributions in Science, Natural History Museum Los Angeles County* 390: 1–36, 1987.
34. Van Devender, T. R. and G. L. Bradley, Late quaternary amphibians and reptiles from Maravillas Canyon Cave, Texas, with discussion of the biogeography and evolution of the Chihuahuan Desert herpetofauna, in P. R. Brown and J. W. Wright, eds., *Herpetology of the North American Deserts* (Van Nuys, CA: Southwestern Herpetologists Society, Special Publication No. 5, 1994), pp. 23–53.
35. Imbrie, J. and K. P. Imbrie, *Ice Ages. Solving the Mystery* (Hillside, NJ: Enslow, 1979).
36. Delcourt, P. A. and H. R. Delcourt, Paleoclimates, paleovegetation, and paleofloras during the late quaternary, in Flora of North America Editorial Committee, ed., *Flora of North America North of Mexico* (New York: Oxford Press, Vol. 1, 1993).
37. Betancourt, J. L., T. R. Van Devender, and P. S. Martin, eds., *Packrat Middens: The Last 40,000 Years of Biotic Change* (Tucson, AZ: University of Arizona Press, 1990).
38. Van Devender, T. R., R. S. Thompson, and J. L. Betancourt, Vegetation history of the deserts of southwestern North America; the nature and timing of the late Wisconsin-Holocene transition, in W. F. Ruddiman and H. E. Wright, Jr., eds., *North America and Adjacent Oceans during the Last Deglaciation* (Boulder, CO: Geological Society of America, 1987).
39. Van Devender, T. R., Late quaternary vegetation and climate of the Chihuahuan Desert, United States and Mexico, in J. L. Betancourt, T. R. Van Devender, and P. S. Martin, eds., *Packrat Middens: The Last 40,000 Years of Biotic Change* (Tucson, AZ: University of Arizona Press, 1990), pp. 104–133.
40. Schmidt, R. H. Jr., A climatic delineation of the 'real' Chihuahuan Desert, *Journal of Arid Environments* 2: 243–250, 1979.
41. Van Devender, T. R., Late quaternary vegetation and climate of the Sonoran Desert, United States and Mexico, in J. L. Betancourt, T. R. Van Devender, and P. S. Martin, eds., *Packrat Middens: The Last 40,000 Years of Biotic Change* (Tucson, AZ: University of Arizona Press, 1990), pp. 134–165.
42. Nielson, R. P., High-resolution climatic analysis and Southwest biogeography, *Science* 232: 27–34, 1986.
43. Porter, S. C., Some geological implications of average Quaternary glacial conditions, *Quaternary Research* 32: 245–261, 1989.
44. Winograd, I. J., J. M. Landwehr, K. R. Ludwig, T. B. Coplen, and A. C. Riggs, Duration and structure of the past four interglaciations, *Quaternary Research* 48: 141–154, 1997.
45. Lindsay, E. H., Late Cenozoic mammals from northwestern Mexico, *Journal of Vertebrate Paleontology* 4: 208–215, 1984.
46. Shaw, C. A. and H. G. McDonald, First record of giant anteater (Xenartha, Myrmecophagidae) in North America, *Science* 26: 186–188, 1987.
47. Martin, P. S., Pleistocene overkill: The global model, in P. S. Martin and R. G. Klein, eds., *Quaternary Extinctions* (Tucson, AZ: University of Arizona Press, 1984), pp. 354–403.
48. Turner, R. M. and D. E. Brown, Sonoran desertscrub, *Desert Plants* 4: 121–181, 1982.
49. Davis, G. P. Jr., *Man and Wildlife in Arizona: The American Exploration Period 1824–1865* (Phoenix, AZ: Arizona Game and Fish Department, 1982).
50. Shreve, F., Vegetation of the northwestern coast of Mexico, *Bulletin of the Torrey Botanical Club* 61: 373–380, 1934.
51. Turnage, W. V. and A. L. Hinckley, Freezing temperatures in relation to plant distributions in the Sonoran Desert, *Ecological Monographs* 8: 529–550, 1938.
52. Hastings, J. R. and R. M. Turner, *The Changing Mile* (Tucson, AZ: University of Arizona Press, 1965).

53. Van Devender, T. R., T. L. Burgess, J. C. Piper, and R. M. Turner, Paleoclimatic implications of Holocene plant remains from the Sierra Bacha, Sonora, Mexico, *Quaternary Research* 41: 99–108, 1994.
54. Marshall, L. G., S. D. Webb, J. J. Sepkoski, and D. M. Raup, Mammalian evolution and the Great American interchange, *Science* 215: 1351–1357, 1982.
55. Van Devender, T. R., A. M. Rea, and M. L. Smith, The Sangamon interglacial vertebrate fauna from Rancho la Brisca, Sonora, *Transactions of the San Diego Society of Natural History* 21: 23–55, 1985.
56. Repenning, C. A., Personal communication, 1984.
57. Martin, P. S., Vanishings, and future, of the prairie, *Geoscience and Man* 10: 39–49, 1975.
58. Harris, A. H., Preliminary report on the vertebrate fauna of U-Bar Cave, Hidalgo County, New Mexico, *New Mexico Geology* 84: 74–77, November 1985.
59. Lindsay, E. H., Late Cenozoic vertebrate faunas, southeastern Arizona, in J. F. Callender, J. C. Wilt, and R. E. Clemons, eds., *Land of Cochise-Southeastern Arizona* (Tucson, AZ: *New Mexico Geological Society Guidebook, 29th Field Conference, Santa Fe, NM*, 1978), pp. 269–275.
60. Van Devender, T. R. and R. D. Worthington, The herpetofauna of Howell's Ridge Cave and the paleoecology of the northwestern Chihuahuan Desert, in R.H. Wauer and D. H. Riskind, eds., *Transactions of the Symposium on the Biological Resources of the Chihuahuan Desert, United States and Mexico* (Washington, DC: United States National Park Service, 1977), pp. 85–106.
61. Emslie, S. D., Extinction of condors in Grand Canyon, Arizona, *Science* 237: 768–770, 1987.

# 7

# *Vegetation Zones of the Southwest*

**Ward W. Brady**

CONTENTS



## 7.1 Introduction

Part of the fascination and beauty of the southwestern region of the United States and northern Mexico arises from the complex mosaic of ecological communities that are spread across the landscape. While it is apparent to any traveler that the dominant theme of this mosaic is desert, patches of grassland, chaparral, forest, and other vegetation types intertwine as elevation changes and watercourses dissect the region. Previous chapters in this book (see Chapter 1) have given a general overview of the deserts of the world and have provided brief discussions of geology, soils, climate, and hydrology specific to the southwestern arid region (see Chapters 2 through 4). The objective of this chapter is to discuss how these environmental factors interact with plant strategies to produce the observed mosaic of communities and to proceed to a broad discussion of the communities themselves. This then will provide a context for discussion, in subsequent chapters, of native wildlife populations (see Chapter 9) and anthropogenic changes that are rapidly changing the fabric of this landscape.

From a landscape ecology perspective, desert communities can be considered the matrix of the landscape. That is, desert communities are the dominant and background ecological system across the area. Because this matrix is itself quite complex and variable across the southwestern region, it is important to define its common features. First, while the term desert itself is difficult to rigorously define, it nevertheless generally includes (1) the idea of dryness due to the combination of low precipitation and high temperatures and (2) the idea of a resultant sparseness of vegetation. UNESCO and UNEP[1,2] formalized the concept of aridity when they defined *arid* regions based on the relationship between water input to and potential water loss from the ecosystem. The aridity of an ecosystem increases as water gain from precipitation falls below potential water loss from evaporation and transpiration. The desert regions of the southwest span the *hyperarid* to *semiarid* range under the UNESCO classification and generally have precipitation/evapotranspiration (P/ET) ratios of less than 0.50.

Gleason and Cronquist[3] focus on the amount of precipitation relative to demand as well as the seasonal distribution of precipitation when looking at major patterns of correlation between vegetation and climate. Their *Desert* type occurs in climates where water is a limiting factor during both the summer and winter seasons (summer and winter are here used in the context of the northern hemisphere temperate zone). As will be discussed in more detail subsequently, the distribution of precipitation does vary by season within deserts, but in all cases low year-round availability of water severely limits vegetation productivity.

While the foundational signature of deserts is aridity as defined by the P/ET ratios, the precipitation that does occur in them typically arrives in scattered and unpredictable episodes. While this is true to varying extents in all ecosystems, the aridity of deserts amplifies its importance. Thus, Noy Mier[4] notes that, in addition to being characterized by aridity, deserts typically (1) have precipitation that is highly variable throughout the year and which often occurs in spatially and temporally intermittent events and (2) have rainfall totals that from year to year are highly unpredictable. Thus, to the foundational aridity component, Noy Mier adds an uncertainty component. Deserts are then described as "water controlled ecosystems with infrequent, discrete, and largely unpredictable water inputs." Like other ecosystems, but very noticeably in desert ecosystems, production is triggered by a rainfall event and the size of the subsequent production pulse is dependent on the magnitude and seasonal timing of the pulse. The biomass produced by this pulse then is either lost to mortality and consumption or put into a reserve such as seeds or energy stores in roots and stems.[5]

## 7.2 Desert Plants and Life Histories

The species that populate desert ecosystems were described by Noy Mier[4] as *arido-passive*. These are annual and perennial plant species that are dependent on rainfall events to trigger growth and reproduction and which then pass into a resistant or dormant stage during dry periods. Noy Mier refers to this life history pattern as the *pulse-reserve* paradigm. Grime, from a slightly different perspective, describes such plants as exhibiting either *stress-tolerant* or *ruderal* primary strategies. Primary strategies here refer to similarities in genetic characteristics that recur widely among species and that cause them to exhibit similarities in ecology.[6]

The concept of stress tolerant and ruderal strategies can best be understood relative to the *competitive* strategy. Competition between plant species occurs in all ecosystems; however, in ecosystems with abundant resources, the primary strategic response to resource limitation is morphological (see Chapter 8). For instance, if plants with a competitive strategy are shaded, they initiate shoot growth in an attempt to maximize the capture of photons. Likewise, the competitive response to water stress is to produce new roots.[6] In desert ecosystems, a morphological response to water shortage is not always appropriate because soil moisture is often simply not available in quantities that would reward growth. Survival in desert ecosystems, therefore, requires different responses to resource limitations.

Plants having *stress tolerance* as their primary strategy show an amazing variety of adaptations that allow them to endure unfavorable conditions. Walter[7–9] reviewed the diversity of adaptations to aridity that have evolved and these are further discussed in Chapter 8. For our purposes, it is enough to emphasize that the common factor among plants having the stress-tolerant primary strategy is that they cease active growth and reproduction when environmental conditions become unfavorable and then endure until conditions again become suitable. It is their ability to endure during times of resource limitation and then quickly respond when resources become available that characterizes this strategy. Grime[6] lists common attributes of stress tolerators found in arid regions as long lived, slow growing, evergreen, and having mechanisms allowing rapid uptake of resources when they are available. Thus, unlike plants having the competitive strategy, the response of stress tolerators to environmental variation is physiological rather than morphological.

Plants that have the *ruderal* strategy accommodate resource limitations in a fundamentally different manner. In general, the ruderal strategy is considered to be an adaptation to disturbances that limit plant biomass by partial or total destruction.[6] Numerous ephemeral desert plants have developed life histories that tolerate the "destruction" of vegetative biomass through desiccation and which then endure between suitable rainfall events as seeds. Ruderals differ from both competitors and stress tolerators in that they tend to be short lived, fast growing, and, in most cases, seed production is followed immediately by the death of the parent plant.

Across the southwestern deserts, species composition varies but plant strategies of the component species are remarkably consistent. Desert ecosystems are dominated by stress tolerators and ruderals. As the P/ET ratio increases in the less arid communities (such as those found at higher elevations), one begins to observe plant life histories that increasingly incorporate aspects of the competitive strategy (resulting in life histories that are compromises between the dominant stress-tolerant and ruderal strategies and the competitive strategy—secondary strategies in Grime's system[6]). Thinking about plants from the perspective of what strategies best allow them to cope with the environmental constraints they experience allows one to see fundamental patterns that knit together the diverse mosaic of ecosystems covering the landscape.

## 7.3 Desert Ecosystems of the Southwestern Region

A vegetation formation is a large unit of vegetation that has similar form and structure (*physiognomy*) of the most conspicuous plants.[10] Brown et al.[11] describe four major subdivisions within the *Desertscrub Formation* of the western United States and northern Mexico. Here, I shorten Brown's classification of "desertscrub" to the more common "desert." It is important to note that this common physiognomy, which characterizes all four North American deserts, reflects the common stress-tolerant life history shared by dominant species in all deserts. The Desertscrub Formation is subdivided into four major subdivisions based on climate and dominant species.[5,11] These are the *Great Basin Desert* (roughly centered in Nevada, Utah, and northeastern Arizona), the *Mojave Desert* (centered in southeastern California and southern Nevada), the *Chihuahuan* Desert (centered in north-central Mexico but extending into southern Arizona and southwestern Texas), and the *Sonoran Desert* (centered in Arizona and Sonora, Mexico). Because of its location at higher latitude, the Great Basin Desert is often referred to as a *cold desert*, while the remaining deserts in our region are regarded as *hot deserts*.

### 7.3.1 The Sonoran Desert

Because of its biotic diversity, the Sonoran Desert is typically further subdivided into at least five subdivisions: the Lower Colorado River, Arizona Upland, Plains of Sonora, Gulf Coast, and Vizcaino subdivisions.[11–13] The richness of the flora in the Sonoran Desert arises in part because of climatic conditions (see Chapter 3) but also because its location allows a mixing of species from both temperate zone and tropical zone floras. The Sonoran Desert is quite geographically complex with numerous mountain ranges, often called *sky islands*, occurring throughout the area. Although each of these subdivisions is important, our primary focus will be on the Lower Colorado River and Arizona Upland subdivisions. The Lower Colorado River surrounds the lower Colorado and lower Gila Rivers and is the hottest and driest subdivision. The Arizona Upland, on the other hand, is located on the eastern and northern portions of the Sonoran desert and is the highest and coldest subdivision.

Across all of the Sonoran Desert, the biological diversity is a function of its unique climate. The climate ranges from semiarid in the higher elevations to hyperarid along the lower Colorado River and has two distinct seasons of rainfall. Winds originate from the west during the winter, bringing moisture from the Pacific Ocean across the desert, resulting in storms that are often widespread and of moderate intensity (see Chapter 3). During the summer months, the direction of the wind shifts to a prevailing southern and eastern flow, bringing moisture in from the Gulf of California, the tropical eastern Pacific, and the Gulf of Mexico. When this increased humidity is combined with intense summer temperatures, the results are convective storms that are often localized and intense. Another distinctive feature of the Sonoran Desert is its mild winters with frost seldom occurring except in the Arizona Upland.

The result of the bimodal pattern of precipitation is a desert that is both rich in species diversity and complex structurally compared to many, if not most, other deserts. This diversity is expressed through two principal themes. First, because of the abundance of large shrubs or small trees, in contrast to the low shrubs characterizing the other deserts, the Sonoran Desert is often considered an arboreal desert (Figure 7.1), or even a thornscrub at its upper-elevation limits. These arboreal species serve as the signature species for the desert and, for instance, in the Arizona Upland, include such species as saguaro cactus (*Carnegiea gigantea*; Figure 7.2) commonly growing to a height of 40–50 ft, and the palo verde (*Parkinsonia* spp.), which grow to 12–16 ft. Second, two distinct ephemeral floras exist corresponding to the bi-seasonality of precipitation. These ephemeral species comprise approximately half the species in the region and include a large number of ruderal, invasive species of Mediterranean origin that have recently entered the region representing a vegetation change[14] equal to geologic time-frame changes.[15] These invasive species depress native wildflowers and promote wildfires that may lead to the loss of some fire-sensitive species such as cacti. The topic of invasive species and development is discussed in a later chapter.

Across the entire Sonoran Desert, creosote bush (*Larrea tridentata*) is the dominant shrub (Figure 7.3). The creosote bush range also extends into the Mojave Desert to the west and the Chihuahuan Desert to the east. Creosote bush is a small (3–10 ft in height) evergreen shrub that provides an excellent example of the stress-tolerance strategy. It combines attributes of resisting water loss (e.g., thick resinous leaves with high stomatal resistance and

**FIGURE 7.1**
Arboreal character of the Upper Sonoran Desert is illustrated in this photo with saguaro (*C. gigantea*) and ocotillo (*F. splendens*) in the background and littleleaf palo verde (*Parkinsonia microphyllum*) in the foreground.

the ability to lose leaves and/or stems to reduce water needs during drought), the ability to function at very low water potentials, and the ability to quickly take up available water using an extensive, shallow, fibrous root system that may extend 4 m from the plant. Individual creosote bush plants are long lived (estimates are 100–200 years) and because the plants are clonal, rings of genetically identical creosote bush plants have been estimated to be much older.

The Arizona Upland includes such species as mesquite (*Prosopis* spp.), jojoba (*Simmondsia chinensis*), ocotillo (*Fouquieria splendens*), and bear grass (*Nolina microcarpa*). Yuccas (*Yucca* spp.) occur as the Arizona Upland transitions into more mesic chaparral and grassland communities (Figure 7.4). Nitrogen-fixing, leguminous trees and shrubs (family Fabaceae) are especially common and reflect the tropical origin of many common Sonoran Desert species. Perennial grasses also become more important in the flora in these regions as well. The Arizona Upland is also noted for its rich cactus flora from several genera including the prickly pears (*Opuntia* spp.), the cholla (*Cylindropuntia* spp.) (Figure 7.5), and the barrel cacti (*Ferocactus* and *Echinocactus* spp.).

**FIGURE 7.2**
Saguaro (*C. gigantea*) is one of the signature species of the Sonoran Desert.

While the Arizona Upland is an arboreal desert, the drier and warmer Lower Colorado subdivision is dominated by sparse stands of creosote bush, white bursage (*Ambrosia dumosa*), and brittle bush (*Encelia farinosa*). Frost-sensitive species that are able to survive in this subdivision, particularly around the numerous ephemeral washes, include ironwood (*Olneya tesota*), brittle bush, and smoke tree (*Psorothamnus spinosa*).

The Sonora Plains, located entirely within the state of Sonora, Mexico, has a greater percentage of summer precipitation with the attendant denser vegetation. The arboreal species that have increased abundance include legumes, especially mesquite, while species common to much of the Sonoran Desert, such as creosote bush, have limited distribution. The Central Gulf Coast subdivision, located around the Gulf of California, is hyperarid with undependable summer and winter rains. The flora is dominated by large stem succulents including the massive cardón (*Pachycereus pringlei*) and trees that often remain leafless throughout the year including palo verde, ocotillo (*Fouquieria macdougallii*), and elephant tree (*Bursera* spp.). The Vizcaino subdivision, on the Pacific Ocean, or western side of the Baja California peninsula, has undependable rains that fall primarily in the winter. The flora has many unique species; the most notorious being the eccentric-looking

**FIGURE 7.3**
Creosote bush (*L. tridentata*) is the most common species across the hot deserts of North America. Note the abundance of ruderal species around the shrub including the invasive grass red brome (*Bromus rubens*). (Courtesy of WW Brady.)

cirio (*Fouquieria columnaris*) whose English name, the boojum tree, was taken from Lewis Carroll's poem "The Hunting of the Snark."

Across all of its subdivisions, the Sonoran Desert is drained by a series of streams and rivers that ultimately empty into the Gulf of California. Along these watercourses, arboreal species dominate, particularly along perennial streams where well-developed riparian forests occur that are dominated by such species as the Fremont cottonwood (*Populus fremontii*), Goodding willow (*Salix gooddingii*), and velvet ash (*Fraxinus velutina*). Salt cedar (*Tamarix ramosissima*), an invasive species from southern Eurasia,[16] is now a dominant species in many riparian areas. The riparian corridors that cross the Sonoran Desert form critical elements of the wildlife habitat of the region. A distinctive habitat type found primarily in the Sonoran Desert but also in adjacent portions of the Mojave Desert is a *fan palm oasis* (Figure 7.6) dominated by native California fan palm (*Washingtonia filifera*). Like riparian areas, the fan palm oases are critical habitats for wildlife.

The Sonoran Desert is among the most studied deserts of the world. The Carnegie Institute established the Desert Laboratory in Tucson in 1903 for the study of desert ecosystems.[17] The world's oldest and most regularly monitored vegetation plots are located on this facility and have been instrumental in shaping our understanding of the structure and function of desert species and ecosystems.[18]

### 7.3.2 The Chihuahuan Desert

The Chihuahuan Desert lies to the east and south of the Sonoran Desert. This large desert, which lies primarily in the Mexican state of Chihuahua, extends further to the south than any other North American desert. It lies on the large intermountain plateau of northern Mexico and thus much of the area is at a fairly high elevation compared to the Sonoran Desert. Because of its location and elevation, the climate is one of hot summers with sporadic rains and cold, dry winters, often with hard frosts. While the Sonoran Desert is

**FIGURE 7.4**
*Yucca elata* is a common species in desert grassland and the Upper Sonoran and western Chihuahuan Deserts.

**FIGURE 7.5**
Chainfruit or jumping cholla (*Cylindropuntia fulgida*) is a common cactus in the Sonoran Desert. Saguaro (*C. gigantea*) and littleleaf palo verde (*P. microphyllum*) are in the background.

**FIGURE 7.6**
Mojave Desert on the Arizona–California border. The dominant shrub on the landscape is creosote bush (*L. tridentata*). Small gray shrubs are white bursage (*A. dumosa*), a valued forage plant for the feral burros seen in the photo.

drained by streams and rivers that ultimately empty to the sea, the Mojave Desert contains numerous closed-basin playas (or bolsons) with no outlet to the sea. The northeastern part of the Mojave Desert, however, does drain to the Rio Grande River and, ultimately, to the Gulf of Mexico.

Creosote bush occurs throughout much of the Chihuahuan Desert, and other species shared with the Sonoran Desert include mesquite and ocotillo. The northern portions of the desert are sparsely populated by stress-tolerant perennials such as creosote bush, mesquite, and yucca, and these species are complemented by a population of ephemerals that grow following the summer rains.[19,20] To the east and south, small shrubs such as tarbush (*Flourensia cernua*), semi-succulent plants including bear grass (*Nolina* spp.), sotol (*Dasylirion wheeleri*), agave (*Agave* spp.), and yucca become more common and conspicuous. Some perennial grasses such as burrograss (*Scleropogon brevifolius*) and tobosa (*Hilaria mutica*) are also found. As one progresses to the southern and eastern reaches of the Chihuahuan Desert, succulents from genera including *Opuntia*, *Mammillaria*, *Echinocereus*, and *Echinocactus* are increasingly important in the flora.

The Chihuahuan Desert is largely lacking in arboreal species except in wash communities where acacia (*Acacia* spp.), desert willow (*Chilopsis linearis*), mesquite, and other small trees occur. Overall, the flora of the Chihuahuan Desert is surprisingly rich given the harsh environment, yet it remains visually monotonous in comparison to the Sonoran Desert due to the low structural diversity of life forms.

### 7.3.3 The Mojave Desert

To the northwest of the Sonoran Desert lies the Mojave Desert. Like the Chihuahuan Desert, the Mojave Desert is visually rather monotonous compared to the Sonoran Desert (Figure 7.7). It is also, for the most part, an upland desert. The notable exception is the Death Valley region, which has a low point of −86 m and is the lowest elevation point in North America. The extreme dryness of Death Valley results largely from its being in the rain shadow of the Panamint Mountains and Mount Whitney (the highest peak in the contiguous United States at 14,504 ft), which lie directly to the west. Over the

**FIGURE 7.7**
Joshua tree (*Y. brevifolia*) is the signature species of the Mojave Desert. Shrubs in the background are creosote bush (*L. tridentata*).

western portions of the Mojave region, rainfall occurs primarily during the cool season, while precipitation becomes bi-seasonal as one moves to the east.[21] Hard freezes may occur during the winter and temperatures become very high during the typically dry summers. The Mojave Desert, like the Chihuahuan Desert, is characterized by playas surrounded by desert mountains. Within this larger pattern, patterns of soil fertility exist where "fertile islands" form as a result of essential soil nutrients concentrating under individual perennial shrubs.[22]

Creosote bush again is an important species across the desert. Other species that commonly occur include white bursage and the Mojave Desert's most distinctive plant, the Joshua tree (*Yucca brevifolia*; Figure 7.8). Other small stress-tolerant shrubs occur along with an ephemeral flora, which becomes evident if and when winter rains occur. Neither arboreal species nor succulents make up a significant portion of the flora. Limited riparian habitats support species such as Fremont cottonwood, Goodding willow (black willow), velvet ash, mesquite, and the exotic salt cedar.[19,20]

**FIGURE 7.8**
Desert palm (*W. filifera*) occurs in discrete palm oases in the Sonoran and Mojave Deserts, in this case in the Kofa Mountains of southwestern Arizona.

### 7.3.4 The Great Basin Desert

To the north of the Mojave and Sonoran Deserts lies the Great Basin Desert.[11] This vast desert extends through Nevada and Utah to southern Wyoming, Idaho, and Oregon. Most of the Great Basin Desert has no drainage to the sea, resulting in a landscape with basins existing between mountain ranges. Some basins are playas like those found in the Chihuahuan and Mojave Deserts that contain water following spring runoff but which are dry during the summer. Other basins contain shallow permanent salt lakes, the most famous of which is the Great Salt Lake in Utah. Gradients in soil conditions from the center of these basins outward significantly influence vegetation patterns across the landscape.

Compared to the other deserts, the Great Basin Desert is very cold in the winter, and growth of plants is limited to the summer months. Rainfall occurs primarily during the winter, although summer thunderstorms occur in the mountains. Dominant plants in this desert include sagebrush (*Artemisia* spp.), shadscale (*Atriplex* spp.), and greasewood (*Sarcobatus vermiculatus*). Big sagebrush (*Artemisia tridentata*) is sensitive to salt accumulation in the soil

**FIGURE 7.9**
Big sagebrush (*A. tridentata*) dominates large areas of the Great Basin Desert.

**FIGURE 7.10**
Shadscale saltbush (*Atriplex confertifolia*) is one of the signature species of the Great Basin Desert.

and thus is found in the higher portions of the basins (Figure 7.9). In the lower portions of the basins where salts have accumulated, chenopods (*Atriplex*, *Grayia*, and *Ceratoides* spp.) become dominant (Figure 7.10). Limited amounts of perennial grass species such as bluebunch wheatgrass (*Pseudoroegneria spicata*) and sheep fescue (*Festuca ovina*) occur in sagebrush stands. Ephemeral species also occur. Two annual invasive species of importance are cheatgrass (*Bromus tectorum*) and halogeton (*Halogeton glomeratus*). Because of the dominance of the low-shrub life form and the uniform gray-green color of these species, the landscape across the Great Basin Desert is also rather visually monotonous. Close inspection, however, reveals a complex mosaic of ecosystems with a surprising richness of plant and animal species.

### 7.3.5 Correlation of Vegetation and Climate above the Desert Background

There is, of course, more to the relationship between vegetation and climate than a simple measure of water availability. Gleason and Cronquist[3] discuss correlation of climate and vegetation as a function of both the availability of water and its seasonal distribution. These two variables are used as a simple rubric for delineating four broad vegetation types within the temperate zone, all of which occur within the southwestern region. This rubric gives only a general guide to the correlation between climate and vegetation as numerous ecologically interesting exceptions occur. The broad pattern, however, is valid and useful for understanding the patterns of vegetation across the landscape. As this pattern is considered, one must keep in mind that other environmental factors, particularly geology, soils, and disturbance history, will add additional levels of complexity to the landscape mosaic.

We will focus on topographically defined patches that rise above the background matrix of the hyperarid to semiarid desert. In these topographically defined patches, the general pattern observed is that precipitation increases and temperature decreases as elevation increases. The result is that these topographic patches have increasing P/ET ratios, and the vegetation communities that develop are more diverse and have higher vegetative abundance (measured as either primary productivity or standing biomass). Some of these patches, which often abruptly rise out of the surrounding desert background, have engendered considerable scientific interest and their study has been important in the development of ecological theory. For instance, Merriam[23] chose to study changes in plant and animal communities on Humphrey's Peak (12,635 ft) in northern Arizona "because of its southern position, isolation, great altitude, and proximity to an arid desert." His studies led to the "life zone" concept, which is still a common approach to describing altitudinal and latitudinal vegetation changes as a series of discrete biotic communities.

Whittaker and Niering[24] took a different approach on the Santa Catalina Mountains, a sky island dominated by Mount Lemmon (9156 ft) in southern Arizona. Here they described vegetation patterns as a continuum along which the distribution of individual plant species independently varied in response to their specific adaptations to the elevation/climatic gradient. Their work was fundamental in the development of the gradient concept of vegetation distribution. When discussing sky islands, in the context of development, it is also important to note that many of these mountain top ecosystems have been relatively isolated at least since the late Pleistocene (Chapter 6) and contain many unique and typically endangered species. The presence of these species presents special challenges to landscape development apart from other ecological considerations.

Merriam[23] and Whittaker and Niering both describe the changes in vegetative communities that occur over elevation gradients. The specific changes, however, also depend on seasonal distribution of precipitation. In the western region of the North American deserts, increases in elevation initially result in increases in winter precipitation relative to summer precipitation due to the influence of an oceanic climate. Gleason and Cronquist[3] refer to the vegetation type developing under this climatic regime as the *Sclerophyllous Forest* type (commonly referred to as chaparral communities). Dominant plants of this type share the common theme of being able to withstand severe summer drought but also being able to quickly begin photosynthesis as temperatures rise in the spring using moisture accumulated during the wetter winter season. Common plant species are large shrubs with thick, leathery, evergreen leaves that persist through dry periods but are available for immediate photosynthesis when conditions are suitable. Because of the summer drought characterizing these communities, fire is a common disturbance feature and species are well adapted to regenerate following burns either through resprouting from the plant's base or from fire-adapted seeds. While species in this type are well adapted to periodic burning, these same fires become unacceptable as development occurs (see Chapter 11). As in other fire-prone vegetation types, suppression of fire often results in accumulation of fuel and the potential for higher intensity fires.

Chaparral communities in the coastal regions (California) are very diverse and include numerous habitats and common plant species include chamise (*Adenostoma fasciculatum*), several species of manzanita (*Arctostaphylos* spp.), and several oaks (*Quercus* spp.).[3] Species of pine including Coulter pine (*Pinus coulteri*) and digger pine (*Pinus sabiniana*) often occur at the upper altitudinal limits of the California chaparral type.

Chaparral communities resembling those of California also occur in parts of Arizona and New Mexico. The physiognomy of these communities is very similar; however, the general affinities of the flora are with the Sonoran Province as opposed to the Californian Province.[3] Dominant species include turbinella oak (*Quercus turbinella*), pointleaf manzanita (*Arctostaphylos pungens*), and desert ceanothus (*Ceanothus greggii*).

The seasonal distribution of precipitation on elevated patches is reversed in the eastern regions of the North American deserts. Here, increases in elevation result in increases in summer precipitation relative to winter precipitation due to the presence of a continental climate. This creates conditions under which the *Grassland* type develops. During the warmer months, water is available for growth and development of the dominant grasses, and during the drier, cooler months, the aboveground portion of most grass plants die. Regrowth occurs the next season from perennating buds that are located near or below the ground surface and that are well protected against water loss by the dead litter of previous years' leaves. Raunkiaer[25] referred to plants of this life form as hemicryptophytes (literally *half-hidden* plants) since the perennating buds are well protected from the elements leading to either stress (water loss) or disturbance (fire and grazing in this case). Fire has historically been a common disturbance feature in many grasslands and its suppression along with the associated overgrazing by livestock has contributed to invasion of shrubby species (see Chapters 11 and 12).[26] While the extent of grasslands has diminished over the last century, spectacular grasslands, nevertheless, remain around many sky islands in southern Arizona and northern Mexico. Common species in these grasslands include several gramas (e.g., *Bouteloua gracilis*, *B. hirsuta*, and *B. eriopoda*), three-awns (*Aristida* spp.), curly mesquite (*Hilaria belangeri*), and tobosa.

Across the entire region, at elevations above sclerophyllous forest or grassland types, water becomes much less limiting during either the summer or winter season. This allows for the development of a variety of *Forest* types. Trees, which dominate the forest type, have their perennating buds elevated above the ground and exposed to the elements year round. Raunkiaer[25] referred to this life form as phanerophytes (literally *exposed* plants). While the upper limit for the distribution of trees (elevation or latitude) is typically set by the length and warmth of the summer season, the lower limit is set by the ability of the trees to avoid desiccation during any season. The piñon–juniper woodland occurs at lower elevations within the forest type and is a xeric forest in which several species of juniper (*Juniperus* spp.) and piñon pine (*Pinus edulis*) occur with a highly variable understory. Above the piñon–juniper woodlands, are large areas of forest dominated by ponderosa pine (*Pinus ponderosa*). Associated species are highly viable and include junipers and oaks in drier, warmer, lower-elevation areas and aspen (*Populus tremuloides*) and Douglas fir (*Pseudotsuga menzeisii*) in wetter, cooler, higher-elevation areas. The understory again is highly variable but often includes grasses such as blue grama or Arizona fescue (*Festuca arizonica*). Both the piñon–juniper woodland and ponderosa pine forests are well adapted to frequent fires (see Chapter 11).

At increasing elevations with higher P/ET ratios, the forest types correspond to those also found at higher latitudes. A spectacular example of this is the spruce-fir forest atop Mount Graham, a sky island located in southern Arizona (10,718 ft). Merriam[23] classified this forest type as a Hudsonian boreal forest equivalent to ecosystems occurring in

Canada. Common species in the these mixed-conifer forests include Douglas fir, white fir (*Abies concolor*), aspen, and at the higher elevation portions also include other species of fir (*Abies*) and spruce (*Picea*) including blue spruce (*Picea pungens*). Above the forest type *Alpine Tundra* occurs to a very limited extent. The climate of this type is defined not by precipitation but by temperature and will not be considered in detail here due to the limited extent of its occurrence in the southwestern region.

## 7.4 Conclusion

The landscape diversity that exists within the borders of the North American deserts is immense as is the number of plant and animal species that inhabit these communities. These communities are at once both very robust and very fragile. They are robust in that the component species have evolved attributes (life histories) that allow them to survive under very harsh conditions of limited and undependable rainfall. On the other hand, these same life histories have not made them either resistant or resilient in the face of new anthropogenic environmental challenges. The challenges—including old challenges such as grazing and the newer challenges of urbanization and invasive species—will be discussed in detail in later chapters. The challenge of development is to proceed and meet the needs of human populations while at the same time protecting and preserving the unique species and communities of the deserts. While this is not a small challenge, it is critical if the life of the desert is to continue.

## References

1. UNESCO, *World Map of Arid Regions* (Paris, France: United Nations Educational, Scientific, and Cultural Organization, 1977).
2. UNEP, *World Atlas of Desertification* (New York: John Wiley & Sons, 1992).
3. Gleason, H.A. and A. Cronquist, *The Natural Geography of Plants* (New York: Columbia University Press, 1964).
4. Noy Mier, I., Structure and function of desert ecosystems, *Israeli Journal of Botany* 28: 1–19, 1979/1980.
5. Whitford, W., *Ecology of Desert Systems* (New York: Academic Press, 2002).
6. Grime, J.P., *Plant Strategies, Vegetation Processes, and Ecosystem Properties*, 2nd edn. (Chichester, U.K.: John Wiley & Sons, 2001).
7. Walter, W., *Vegetation of the Earth in Relation to Climate and the Ecophysiological Conditions* (London, U.K.: English Universities Press, 1973).
8. Slatyer, R.O., *Plant-Water Relationships* (London, U.K.: Academic Press, 1967).
9. Levitt, J., *Responses of Plants to Environment Stresses* (New York: Academic Press, 1975).
10. Burrows, C.J., *Processes of Vegetation Change* (London, U.K.: Unwin Hyman, 1990).
11. Brown, D.E., C.H. Lowe, and C.P. Pase, *Biotic Communities of the Southwest*, General Technical Report RM-41, map, 1:1,000,000 scale (Fort Collins, CO: U.S. Forest Service, Rocky Mountain Forest and Range Experiment Station, 1977).
12. Jaeger, E.C., *The North American Deserts* (Stanford, CA: Stanford University Press, 1957).
13. Phillips, S.J. and P.W. Comus, *A Natural History of the Sonoran Desert* (Tucson, AZ: Arizona-Sonora Desert Museum Press, 2000).

14. Burgess, T.L., J.E. Bowers, and R.M. Turner, Exotic plants at the desert laboratory, Tumamoc Hill, Tucson, Arizona, *Madroño* 38: 96–114, 1991.
15. McClaran, M.P. and W.W. Brady, Arizona's diverse vegetation and contributions to plant ecology, *Rangelands* 16: 208–217, 1994.
16. Horton, S., The development and perpetuation of the permanent tamarisk type in the phreatophyte zone of the southwest, in *Importance, Preservation and Management of Riparian Habitat: A Symposium*, Tucson, AZ (Washington, DC: U.S. Forest Service, 1977), General Technical Report RM-43, pp. 124–127.
17. Bowers, J.E., A debt to the future: Scientific achievements of the desert laboratory, Tumamoc Hill, Tucson, Arizona, *Desert Plants* 10: 9–12, 25–47, 1990.
18. Goldberg, D.E. and R.M. Turner, Vegetation change and plant demography in permanent plots in the Sonoran Desert, *Ecology* 67: 695–712, 1986.
19. Brown, D.E. ed., Biotic communities of the American Southwest—United States and Mexico, *Desert Plants* 4: 3–341, 1982.
20. Shelford, V.E., *The Ecology of North America* (Urbana, IL: University of Illinois Press, 1963).
21. Hereford, R., R.H. Webb, and C. Longpré, Precipitation history and ecosystem response to multidecadal precipitation variability in the Mojave Desert and vicinity, 1893–2001, *Journal of Arid Environments* 67: 13–34, 2006.
22. Bolling, D. and L.R. Walker, Fertile island development around perennial shrubs across a Mojave Desert chronosequence, *Western North American Naturalist* 62: 88–100, 2002.
23. Merriam, C.H., *Life Zones and Crop Zones of the United States*, Survey Bulletin 10 (Washington, DC: U.S. Department of Agriculture, Division of Biology, 1898).
24. Whittaker, R.H. and W.A. Niering, Vegetation of the Santa Catalina Mountains, Arizona. I. Gradient analysis of the south slope, *Ecology* 46: 429–452, 1965.
25. Raunkiaer, C., *The Life Forms of Plants and Statistical Plant Geography: Being the Collected Papers of C. Raunkiaer* (Oxford, U.K.: Clarendon Press, 1934).
26. Drewa, P.B., D.P.C. Peters, and K.M. Havstad, Fire, grazing, and honey mesquite invasion in black grama-dominated grasslands of the Chihuahuan Desert: A synthesis, in K.E.M. Galley and T.P. Wilson, eds., *Proceedings of the Invasive Species Workshop: The Role of Fire in the Control and Spread of Invasive Species. Fire Conference 2000: The First National Congress on Fire Ecology, Prevention, and Management*, San Diego, CA (Tallahassee, FL: Tall Timbers Research Station, 2001) Miscellaneous Publication No. 11, pp. 31–39.

# 8

# *Plant Ecology of the Sonoran Desert Region**

**Mark A. Dimmitt**

**CONTENTS**



## 8.1 Introduction

You could easily recognize a desert even if you were blindfolded. You would discover that you can walk fairly long distances without bumping into plants, and when you do the encounter is likely to be painful. Even standing still there are unmistakable clues about your location. You can feel the arid atmosphere pulling moisture out of your body. The intense sunlight actually creates a sensation of pressure on your skin. On really hot, dry days you can smell the parched vegetation that is literally toasting and filling the air with pungent terpenes and aromatic oils.

Most desert plants also look different from those in other habitats; they are often spiny, almost always tiny-leafed, and rarely "leaf green." Many have bold, sculptural growth forms characterized by swollen stems or starkly exposed stems unconcealed by foliage. At the other extreme is a unique desert growth form that landscape architect Iain

* Adapted with permission from Plant ecology of the Sonoran Desert Region, in Phillips, S.J. and Comus, P.W. eds., *A Natural History of the Sonoran Desert*, Arizona-Sonora Desert Museum Press/University of California Press, pp. 128–151, 2001.

Robertson calls diaphanous plants. Their stems and foliage are so fine-textured and sparse that the eye has difficulty focusing on them instead of looking right through them.

These tactile, olfactory, and visual experiences offer clues to desert plants' adaptations to their rigorous environment. Before exploring these adaptations, it is necessary to understand something about plant structures, functions, and classification.

## 8.2 Basic Plant Anatomy and Classification

Many people mistakenly identify ocotillo, agaves, African euphorbias, and numerous other plants as types of cacti because of their succulent or spiny stems. In fact these plants, despite their almost identical in outward appearance, are unrelated to each other and to cacti. Similarities of outward appearance are often examples of convergent evolution and are not reliable indicators of relationship. Convergent evolution results when unrelated organisms develop similar adaptations to similar environmental conditions. The sexual parts of plants (flowers and fruits) are used almost exclusively to determine their interrelationships. The parts of flowers and fruits are easier to identify and describe than the vegetative organs (leaves, stems, and roots), and these complex floral patterns remain more readily traceable as they evolve. On the other hand, leaves and other vegetative parts can also be measured, but it's difficult to determine relationships among plants from such measurements. Moreover, qualitative vegetative characters are difficult to describe precisely even when the overall appearance (gestalt) is distinctive. For example, nearly every hiker knows poison-ivy on sight. But try to describe the foliage so precisely that someone who has never seen one can distinguish it from skunk bush (*Rhus aromatica*). It's quite difficult to describe the leaves' different shades of green, degrees of hairiness, and the scalloping of the margins, especially if you lack the minutely detailed vocabulary of the botanist. For example, pubescent, puberulent, lanate, villous, hirsute, hirsutulous, ciliate, tomentose, strigose, pilose, and hispid are just some of the terms describing different degrees of hairiness. Vegetative parts are also more plastic, that is, they vary greatly under environmental influences. The leaves of brittlebush grow much larger and greener in shade or during rainy periods than in sun or drier conditions.

The complexity of flowers and fruits creates distinctive patterns that can be characterized exactly. Petals, stamens, and other parts can be counted, their lengths and widths measured (and these are usually less variable than the dimensions of a leaf). Where the stamens are attached to the petals (or other part) can be described unambiguously. For example, a flower that has many (more than 10) petals and sepals that intergrade into one another, many stamens (usually hundreds), a multilobed stigma, and an inferior ovary unequivocally identifies a member of the cactus family. All 2000 species of cacti possess some variation of this basic pattern, and no other plant group does.

To recognize floral patterns you must be able to identify the parts of a flower Figure 8.1. The following drawing identifies the anatomy of a generalized flower.

### 8.2.1 Classification and Plant Identification

In the game "Twenty Questions" players attempt to identify an unknown by asking the person who knows the answer a series of yes-or-no questions. If done well, 20 questions

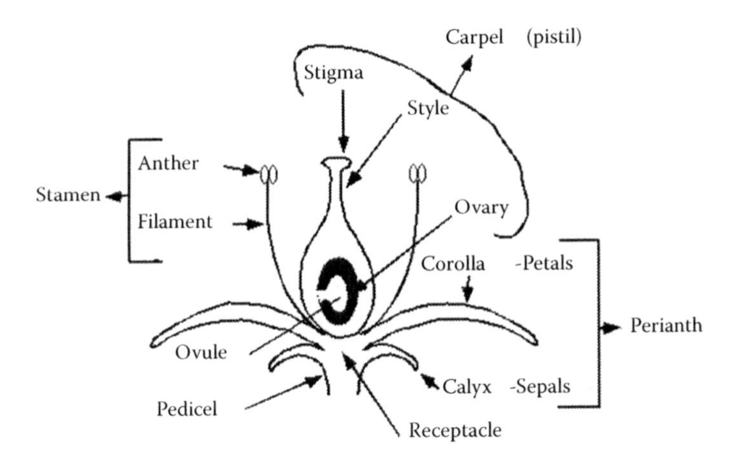


**FIGURE 8.1**
Generalized diagram showing the major parts of an angiosperm flower.

are sufficient to eliminate every other possibility in the world and leave the correct answer standing. Assume, for example, that the unknown thing is a dog. First question: "Is it a concept?" (No—therefore it's an object.) Second question: "Is it alive?" (Yes.) Third question: "Is it a plant?" (No.) Fourth question: "Is it a vertebrate? (Yes.) Fifth question: "Is it an herbivore?" (No.) The enormous inventory of the universe has been narrowed to a very short list in only five questions.

Botanists identify plants (and zoologists animals) unknown to them, with a Twenty Questions-like procedure called a *dichotomous key* (or simply key). A key is a nested series of dual choices that quickly narrows the possibilities to a single species. For example, the first pair of choices might ask you whether the flower has three petals versus four or five. Each of the two possible answers leads to another pair of choices, and so on until you have identified your quarry out of 300,000 species of flowering plants. But before you can use such a key effectively, or before you can describe your unknown to someone who will identify it for you, you must know the parts of the flower and plant you are examining.

The sepals collectively make up the *calyx*. They enclose all other flower parts in the bud, usually completely concealing the rest of the flower until it opens.

The petals collectively make up the *corolla*. Petals are frequently the visual advertising banner that attracts pollinators. Petals and sepals look similar in many flowers, such as in lilies and agaves. By definition the sepals are the parts on the outside; petals are typically concealed in the bud.

The corolla and calyx make up the *perianth*. The perianth parts may be separate or fused together for part or all of their length. Often there is only one series of perianth parts. Of necessity these must be on the outside and therefore they are sepals, even if they are large and colorful.

The female part of a flower is the *pistil*, composed of stigma, style, and ovary. The ovary contains ovules, which develop into seeds when fertilized by the sperm in pollen. Seeds are plant embryos encased in a protective membrane and usually contain stored energy to fuel germination. If the ovary is visible beneath the calyx, it is said to be inferior. It is superior if you must look inside the flower to see it (i.e., it is above the calyx).

The male part of a flower is the *stamen*, composed of the anther and the filament. Anthers produce pollen grains, which contain sperm cells.

## 8.3 Photosynthesis

Chlorophyll is one of the most consequential chemicals in the biosphere. Nearly all life on the planet depends on it. Living organisms seem to defy the law of entropy, which describes a universal tendency toward increasing disorder. By using energy acquired from outside they prevent themselves—temporarily—from dying and disintegrating into simple, dissociated molecules (becoming disordered). A small number of species derive their energy from metabolizing sulfur compounds. All others, including all the organisms that we encounter in everyday life, depend on solar energy (light) to maintain their orderly existence. Light, however, is unmanageable; it can't be concentrated and stored for later use. (Outside of science fiction there is no such thing as a photon battery.) Enter photosynthesis. Green plants use light energy to combine low-energy molecules (carbon dioxide and water) into high-energy molecules (carbohydrates), which they accumulate and store as energy reserves. Chlorophyll—the green pigment in plants—is the only known substance in the universe that can capture volatile light energy and convert it into a stable form usable for biological processes (chemical energy).

> See it with your eyes: Earth reenergized by the sun's rays every day.—Moody Blues

Almost without exception, living organisms—plants, animals, and the other three kingdoms—obtain energy for sustaining life from carbohydrates (sugars and starches) by the metabolic process of respiration. (Respiration is colloquially and medically used to mean breathing. The mechanical act of breathing, however, is only the first step in the physiological process of respiration—the intake of oxygen.) Respiration is the chemical pathway through which carbohydrate is broken down (oxidized) into carbon dioxide and water, releasing the energy stored in the carbohydrate molecules. This is represented by the formula: carbohydrate + $O_2$ → $H_2O$ + $CO_2$ + energy. (The multiple arrows indicate many sequential chemical reactions.) Green plants manufacture carbohydrates by photosynthesis. Animals acquire their carbohydrates by eating plants or other animals.

Photosynthesis is the opposite of respiration: Carbon dioxide and water are combined to form larger molecules of carbohydrate, with the addition of energy from sunlight: $H_2O$ + $CO_2$ + energy → carbohydrate + $O_2$. Water is absorbed through the roots, and $CO_2$ diffuses into the leaves through the *stomates* (valved pores in leaf and stem surfaces). The plant joins several carbon dioxide molecules and adds hydrogen atoms split from water molecules to form molecules of sugar (simple carbohydrate). Surplus oxygen atoms from the water molecules are released through the stomates as oxygen gas ($O_2$).

When you see the word "carbohydrate," think "stored energy" and "calories." Plants store energy for long-term use in the form of starch, which is a complex carbohydrate consisting of long chains of sugar molecules. When a plant needs energy to grow new leaves or flowers, it does exactly what animals do—it respires carbohydrate to release the stored energy. The complex respiratory pathway of scores of individual chemical reactions is nearly identical in all life forms: bacteria, mushrooms, higher plants, up to the highest life forms such as toads.

In contrast to plants, animals use fat as their main energy store; it has twice the calories per gram as carbohydrate and protein. When animals in need of energy run low on the small amount of carbohydrate stored in the liver or circulating in the blood, they convert fat (or protein if they run out of fat) into carbohydrate and then respire it.

The most common form of photosynthesis creates a three-carbon sugar as its first stable product, so it's called C3 photosynthesis. Other sugars with more carbon atoms are later synthesized from this first one. About 96% of all plant species use C3, but there are two specialized variations.

One variant is called C4 photosynthesis because the first product is a four-carbon sugar. C4 plants actively transport carbon dioxide to localized bundles of photosynthetic tissue. This process offers improved efficiencies under hot, sunny conditions. C4 plants use carbon dioxide more efficiently (by bypassing photorespiration) and lose less water through *transpiration* (water evaporated from inside plants) per unit of carbohydrate made. The overall result is that C4 plants can grow much faster under high temperatures than most C3 plants. The majority of summer-growing grasses in warm climates are C4. So are many of the summer-growing plants, especially weeds (invasive pioneer plants) that seem to spring up overnight such as pigweed (*Amaranthus* spp.), summer spurges (*Euphorbia hyssopifolia* and others), devil's claw (*Proboscidea* spp.), and many saltbushes (*Atriplex* spp.). Only 0.4% of all the Earth's plant species are C4, but a number of them are vital crops, such as corn (*Zea mays*), sorghum (*Sorghum* spp.), sugar beets (*Beta vulgaris*), and sugar cane (*Saccharum officinarum*). Another variant of photosynthesis, crassulacean acid metabolism (CAM), is discussed in Section 8.4.1.

## 8.4 Coping with Desert Climate

The impression that the desert environment is hostile is strictly an outsider's viewpoint. Adaptation enables indigenous organisms not merely to survive here, but to thrive most of the time. Furthermore, specialized adaptations often result in a requirement for the seasonal drought and heat. For example, the saguaro (*Carnegiea gigantea*), well adapted to its subtropical desert habitat, cannot survive in a rain forest or any other biome, not even a cold desert. In these other places it would drown, freeze, or be shaded out by faster growing plants.

Aridity is the major—and almost the only—environmental factor that creates a desert, and it is this functional water deficit that serves as the primary limitation to which desert organisms must adapt. Desert plants survive the long rainless periods with three main adaptive strategies: succulence, drought tolerance, and drought evasion. Each of these is a different but effective suite of adaptations for prospering under conditions that would kill plants from other regions.

### 8.4.1 Succulence

As a group, succulents are the most picturesque desert plants (Figure 8.2). They capture our attention because they look nothing like the familiar plants of the temperate zone where most people live. Their vernacular names suggest how they command our attention: elephant tree (*Bursera microphylla*), boojum (*Idria columnaris*), jumping cholla (*Opuntia fulgida*), creeping devil (*Stenocereus eruca*), and shindagger (*Agave schottii*). Spanish names translate into such as dragon's blood, child-killer, and *old man's head. Even some scientific names are inspired by the plants' characteristics: Ferocactus* (as in ferocious), *Opuntia molesta* (the molesting-spined cactus), *Opuntia invicta* (the inflicting one), and *Agave jaiboli* (as in highball, because liquor is made from it) (Figure 8.3).

Succulent plants store water in fleshy leaves, stems, or roots in compounds or cells where it is not easily lost. All cacti are succulents, as are such non-cactus desert dwellers as agaves,

**FIGURE 8.2**
Succulence as a plant adaptation to life in arid environments. The large columnar cactus is saguaro (*C. gigantea*).

**FIGURE 8.3**
Succulence of an *Agave* that is a member of the Liliaceae family.

aloes, elephant trees, and many euphorbias. Several other adaptations are essential for the water storing habit to be effective.

### 8.4.2 Getting Water

Succulents must be able to absorb large quantities of water in short periods, and they must do so under unfavorable conditions. Because roots take up water by passive diffusion, succulents can absorb water only from soil that is wetter than their own moist interiors. Desert soils seldom get this saturated and don't retain surplus moisture for long. Desert rains are often light and brief, barely wetting the soil surface that may dry out after just a day or two of summer heat. To cope with these conditions, nearly all succulents have extensive, shallow root systems. A giant saguaro's root system is just beneath the soil surface and radiates as far as the plant is tall. The roots of a two foot tall cholla in an extremely arid site may be 30 ft (9 m) long. Most succulents in fact rarely have roots more than 4 in. (10 cm) below the surface and the water-absorbing feeder roots are mostly within the upper half inch (1.3 cm). Agaves are an exception in lacking extensive root systems; they rarely extend much beyond the spread of the leaf rosette. Instead, the leaves of these plants channel rain to the plants' bases.

### 8.4.3 Conserving Water

Succulents must be able to guard their water hoards in a desiccating environment and use it as efficiently as possible. The stems and leaves of most species have waxy cuticles that render them nearly waterproof when the stomates are closed. Water is further conserved by reduced surface areas; most succulents have few leaves (agaves), no leaves (most cacti), or leaves that are deciduous in dry seasons (elephant trees, boojums). The water is also bound in extracellular mucilages and inulins that hold tightly onto the water.

Many succulents possess a water-efficient variant of photosynthesis called CAM. The first word refers to the stonecrop family (Crassulaceae) in which the phenomenon was first discovered. *Dudleya* is in this family, as are hen-and-chickens (*Sempervivum tectorum*) and jade plant (*Crassula ovata*). CAM plants open their stomates for gas exchange at night and store carbon dioxide in the form of an organic acid. During the day the stomates are closed and the plants are nearly completely sealed against water loss; photosynthesis is conducted using the stored carbon dioxide. At night the temperatures are lower and humidity higher than during the day, so less water is lost through transpiration. Plants using CAM lose about one-tenth as much water per unit of carbohydrate synthesized as those using standard C3 photosynthesis. But there is a trade-off: The overall rate of photosynthesis is slower, so CAM plants grow more slowly than most C3 plants. (An additional limitation is the reduced photosynthetic surface area of most succulents compared with "ordinary" plants.)

The equilibrium between gaseous carbon dioxide and the organic acid is dependent on temperature. Acid formation (carbon dioxide storage) is favored at cool temperatures; higher temperatures stimulate release of carbon dioxide from the acid. Thus CAM works most efficiently in climates that have a large daily temperature range (i.e., arid lands). Cool nights allow much carbon dioxide to be stored as acid, and the warm days cause most of the carbon dioxide to be released for photosynthesis. (A note of interest: A plant in CAM mode will store enough acid to impart a sour taste in early morning; the flavor becomes bland by afternoon when the acid is used up. But don't taste indiscriminately—many succulents are poisonous!)

Many succulents possess CAM, as do semisucculents such as some yuccas, *epiphytic* (growing on trees or rocks) orchids and *xerophytic* (arid-adapted) bromeliads. Exceptions are stem succulents with deciduous, non-succulent leaves, such as elephant trees, limberbushes (*Jatropha*), and desert rose (*Adenium*). Succulents from hot, humid climates that lack substantial daily temperature fluctuations also are usually not CAM. Some succulents such as *Agave deserti* can switch from CAM to C3 photosynthesis when water is abundant, allowing faster growth. Some 3.5% of all plant species spread among about 25 plant families use CAM.

Another crucial attribute of CAM plants is their idling metabolism during droughts. When CAM plants become water-stressed, the stomates remain closed both day and night and the fine (water-permeable) roots are sloughed off. The plant's stored water is essentially sealed inside and gas exchange greatly decreases. However, a low level of respiration (oxidation of carbohydrate into water, carbon dioxide, and energy) is carried out within the still-moist tissues. The carbon dioxide released by respiration is recycled into the photosynthetic pathway to make more carbohydrate, and the oxygen released by photosynthesis is recycled for respiration. Thus the plant never goes completely dormant but is metabolizing slowly—idling. (This sounds like perpetual motion, but it isn't. The recycling isn't 100% efficient, so the plant will eventually exhaust its resources.) Just as an idling engine can rev up to full speed more quickly than a cold one, an idling CAM plant can resume full growth in 24–48 h after a rain. Agaves have visible new roots just 5 h after a rain. A dormant nonsucculent shrub takes a couple of weeks to resume maximum metabolic activity. Therefore, succulents can take rapid and maximum advantage of the soil moisture that quickly evaporates after a summer rain. The combination of shallow roots and CAM-idling with its rapid response enables succulents to respond to and benefit from less than a quarter-inch (6 mm) of rain.

### 8.4.4 Protection

Stored water in an arid environment requires protection from thirsty animals. Most succulent plants are spiny, bitter, or toxic, often all three. Some unarmed, nontoxic species are restricted to inaccessible locations. Smooth prickly pear (*Opuntia phaeacantha* var. *laevis*) and live-forever (*Dudleya* spp.) grow on vertical cliffs or within the canopies of armored plants. Still others rely on camouflage; Arizona night blooming cereus (*Peniocereus greggii*) closely resembles the dry stems of the shrubs in which it grows.

These adaptations are all deterrents that are never completely effective. Evolution is a continuous process in which some animals develop new inheritable behaviors to avoid spines or new metabolic pathways to neutralize the toxins of certain species. In response the plants are continually improving their defenses. For example, pack rats can handle even the spiniest chollas and rarely get stuck. They also eat prickly pear for water and manage to excrete the oxalates that would clog the kidneys of most other animals. Toxin-tolerant insects often incorporate their host plant's toxins into their own tissues for protection against their predators.

### 8.4.5 Drought Tolerance

Drought-tolerant plants often appear to be dead or dying during the dry seasons. They're just bundles of dry sticks with brown or absent foliage, reinforcing the myth that desert organisms are engaged in a perpetual struggle for survival. They're simply waiting for rain in their own way, and are usually not suffering or dying any more than a napping dog is near death (Figure 8.4).

Drought tolerance or drought dormancy refers to desert plants' ability to withstand desiccation. A tomato plant will wilt and die within days after its soil dries out. But many nonsucculent desert plants survive months or even years with no rain. During the dry season the stems of brittlebush (*Encelia farinosa*) and bursage (*Ambrosia*) are so dehydrated that they can be used as kindling wood, yet they are alive. Drought tolerant plants often shed leaves during dry periods and enter a deep dormancy analogous to *torpor* (a drastic lowering of metabolism) in animals. Dropping leaves reduces the surface area of the plant and thus reduces transpiration. Some plants that usually retain their leaves through droughts have resinous or waxy coatings that retard water loss (e.g., creosote bush [*Larrea tridentata*]).

The roots of desert shrubs and trees are more extensive than those of plants of the same size in wetter climates. They extend laterally two to three times the diameter of the canopy. Most also exploit the soil at greater depth than the roots of succulents. The large expanses of exposed ground between plants in deserts are probably not empty. Dig a hole almost anywhere except in active sand dunes or the most barren desert pavement and you're likely to find roots.

Rooting depth controls opportunities for growth cycles. In contrast to the succulents' shallow-rooted, rapid-response strategy, a substantial rain is required to wet the deeper

**FIGURE 8.4**
Leaf drop is a common drought avoidance feature of arid land plants as represented by ocotillo (*F. splendens*). This plant will have bright green leaves and reddish-orange flowers in wet springs but appears to be dead in mid-summer.

root zone of shrubs and trees. A half-inch is the minimum for even the smaller shrubs; more for larger, deeper-rooted plants. Once a soaking rain has fallen on dormant shrubs such as brittlebush and creosote bush it takes them a couple of weeks to produce new roots and leaves and resume full metabolic activity. The tradeoff between this strategy and that of succulents is that once the deeper soil is wetted it stays moist much longer than the surface layer; the deeper moisture sustains growth of shrubs and trees for several weeks.

Mesquite trees (*Prosopis* spp.) are renowned for having extremely deep roots, the champion reaching nearly 200 ft. (Most flood-plain mesquite, though, die if the water table drops below 40 ft.) These riparian specimens are not drought-tolerating trees—their roots are in the water table. No desert plant is known to use very deep roots as a primary strategy for survival. In fact, the root systems of most trees—including mesquite—are mostly confined to the upper 3 ft of soil. Most rains don't penetrate deeper than this, and at greater depths there is little oxygen to support metabolism for growth.

In contrast to succulents that can take up water only from nearly saturated soil, drought tolerant plants can absorb water from much drier soil. A creosote bush can obtain water from soil that seems dust-dry to the touch. Similarly these plants can continue to photosynthesize with low leaf moisture contents that would be fatal to most plants.

Some plants in this adaptive group are notoriously difficult to cultivate, especially in containers. It seems paradoxical that desert ferns and creosote bushes, among the most drought-tolerant of desert plants, can be kept alive in containers only if they're never allowed to dry out. The reason is that these plants can survive drought only if they dry out slowly and have time to make gradual physiological adjustments. If a potted plant misses a watering, the small soil volume dries out too rapidly to allow the plant to prepare for dormancy, so it dies. Researchers showed that some spike mosses (*Selaginella* spp.) must dehydrate over a 5–7 day period. If they dry more rapidly they lack time to adjust, and if drying takes longer than a week they exhaust their energy reserves and starve to death. *Selaginella lepidophylla* from the Chihuahuan Desert is widely sold as a novelty under the name "resurrection fern." Rehydration and resumption of active life takes only a few hours.

### 8.4.6 Drought Evasion

The stretch of Interstate 40 from Barstow to Needles, California traverses some of the emptiest land in the West. It dashes as straight as it can through 130 miles of dry valleys that are almost devoid of human settlements. The vegetation is simple, mostly widely scattered creosote bushes. It's difficult to tell if you're driving through the Mohave or Sonoran desert. The small, rocky mountain ranges interrupting the valleys beckon to true desert lovers, but the drive is just plain bleak to the clueless. The exits on this freeway average 10 miles apart and connect to two-lane roads that shoot straight over the distant horizon with no visible destinations. You rarely see a vehicle on any of them.

Frequent travelers on this freeway become accustomed to its monotony until they think they know what to expect. The creosote bush may turn if there's been a rain; ocotillo (*Fouquieria splendens*) always flowers in April; most of the time it's just brown gravel and brown bushes. Then one spring travelers are astonished to discover the ground between the bushes literally carpeted with flowers. It happened in March 1998, when for 3 weeks the freeway bisected a nearly unbroken blanket of desert sunflowers 40 miles long and 10 miles wide (Figure 8.5). At every exit-to-nowhere several cars and trucks had pulled off and people were wandering through the two-foot-deep sea of yellow. Those with a long memory may have recalled that the same thing happened in 1978. They should have wondered where these flowers came from, and where were they during the intervening 20 years.

Those desert sunflowers (*Geraea canescens*) were annual wildflowers, plants that escape unfavorable conditions by "not existing" during such periods. They complete their life cycle during a brief wet season, then die after channeling all of their life energy into producing seeds instead of reserving some for continued survival. Seeds are dormant *propagules* with almost no metabolism and great resistance to environmental extremes. (A propagule is any part of a plant that can separate from the parent and grow a new plant, such as seeds, agave aerial plantlets, and cholla joints.) They wait out adverse environmental conditions, sometimes for decades, and will germinate and grow only when specific requirements are met.

Wildflower spectacles like the one described earlier are rare events. Mass germination and prolific growth depend on rains that are both earlier and more plentiful than normal. The dazzling displays featured in photographic journals and on postcards occur about once a decade in a given place. In the six decades between 1940 and 1998 there have been

**FIGURE 8.5**
Desert bloom following beneficial rainfall in March 1998.

only four drop-everything-and-go-see-it displays in southern Arizona: 1941, 1978, 1979, and 1998. During that period only the displays of 1978 and 1998 were widespread throughout both the Sonoran and Mohave deserts.

There are three groups of annuals in the Sonoran Desert. Winter-spring species are by far the most numerous. The showy wildflowers that attract human attention will germinate only during a narrow window of opportunity in the fall or winter, after summer heat has waned and before winter cold arrives. In most of the Sonoran Desert this temperature window seems to occur between early October and early December for most species. During this window there must be a soaking rain of at least 1 in. (2.5 cm) to induce mass germination. This combination of requirements is survival insurance: An inch of rain in the mild weather of fall will provide enough soil moisture that the resulting seedlings will probably mature and produce seeds even if almost no more rain falls in that season. (Remember that one of the characteristics of deserts is low and *undependable* rainfall.) If the subsequent rainfall is sparse, the plants remain small and may produce only a single flower and a few seeds, but this is enough to ensure a future generation. There is still further insurance: Even under the best conditions not all of the seeds in the soil will germinate; some remain dormant. For example, a percentage of any year's crop of desert lupine seeds will not germinate until they are 10 years old. The mechanisms that regulate this delayed germination are poorly known.

The seedlings produce rosettes of leaves during the mild fall weather, grow more slowly through the winter (staying warm in the daytime by remaining flat against the ground), and bolt into flower in the spring. Since the plants are inconspicuous until they begin the spring bolt, many people mistakenly think that spring rains produce desert wildflower displays.

There is a smaller group of annual species that grow only in response to summer rains. Arizona poppy (*Kallstroemia grandiflora*) and annual devil's claw (*Proboscidea parviflora*) are among the few showy ones. A few opportunistic species will germinate in response to rain at almost any season; most lack showy flowers and are known only to botanists. Several species of buckwheats (*Eriogonum* spp.) germinate in fall or winter and flower the following summer. Finally, the line between annual and perennial is straddled by a number of species that can live more than one season if conditions are favorable. One of our commonest wildflowers is in this group; desert marigold (*Baileya multiradiata*) flowers a few months after germinating and can be annual, but may survive for a few years in Arizona Upland where the dry seasons are usually short and not severe (Figure 8.6).

**FIGURE 8.6**
Desert marigold (*B. multiradiata*). Desert marigold also displays the plant adaptation of having grayish green foliage and many small trichomes on the leaf surfaces.

The annual habit is a very successful strategy for warm-arid climates. There are no annual plants in the Polar Regions or the wet tropics. In the polar zones the growing season is too short to complete a life cycle. In both habitats the intense competition for suitable growing sites favors longevity (once you've got it, you should hang onto it). Annuals become common only in communities that have dry seasons, where the perennials are widely spaced because they must command a large soil area to survive the drier years. In the occasional wetter years both open space and moisture are available to be exploited by plants that can do so rapidly. The more arid the habitat, the greater the proportion of annual species in North America that you will see in the landscape. (The percentage decreases in the extremely arid parts of the Saharan-Arabian region.) Half of the Sonoran Desert's flora is comprised of annual species. In the driest habitats such as the sandy flats near Yuma, Arizona up to 90% of the plants are annuals.

Winter annuals provide most of the color for our famous wildflower shows. Woody perennials and succulents can be individually beautiful, but their adaptive strategies require them to be widely spaced so they usually don't create masses of color. A couple of exceptions are brittlebush when it occurs in pure stands, and extensive woodlands of foothill palo verde (*Cercidium microphyllum*). The most common of the showy winter annuals that contribute to these displays in southern Arizona are Mexican gold poppy (*Eschscholtzia mexicana*), lupine (*Lupinus sparsiflorus* and *Lupinus arizonicus*), and owl clover (*Orthocarpus purpurascens*).

One of the contributing factors to the great number of annual species is *niche* separation. (A niche is an organism's ecological role, e.g., sand verbena [*Abronia villosa*] is a butterfly-pollinated winter annual of sandy soils.) Most species have definite preferences for particular soil textures, and perhaps soil chemistry as well. For example, in the Pinacate region of northwestern Sonora there are places where gravels of volcanic cinder are dissected by drainage channels or wind deposits of fine silt. In wet years purple mat (*Nama demissum*) grows abundantly on the gravel and the related sand bells (*Nama hispidum*) on the silt. I have seen the two species within inches of each other where these soil types meet, but not one plant of either species could be found on the other soil. There are specialists in loose sand such as dune evening primrose (*Oenothera deltoides*) and sand verbena, and others are restricted to rocky soils such as most caterpillar weeds (*Phacelia* spp.). This phenomenon of occupying different physical locations is spatial niche separation.

Another diversity-promoting phenomenon is temporal niche separation: The mix of species at the same location changes from year to year. Seeds of the various species have different germination requirements. The time of the season (which determines temperature) and quantity of the first, germination-triggering, rain determines which species will dominate, or even be present at all in that year. Of the three commonest annuals of southern Arizona listed earlier, any one may occur in a nearly pure stand on a given hillside in different years, and occasionally all three are nearly equally abundant. This interpretation of the cause of these year-to-year variations is a hypothesis based on decades of empirical observation. Much more research is needed to discover the ecological requirements of most species of desert annuals. And of course the Sonoran Desert's two rainy seasons provide two major temporal niches. Summer and winter annuals almost never overlap.

The dramatic wildflower shows are only a small part of the ecological story of desert annuals. For each conspicuous species there are a score of others that either have less colorful flowers or don't grow in large numbers. Every time the desert has a wet fall or winter, it will turn green with annuals, but not always ablaze with other colors. One of the most common winter annuals is desert plantain (*Plantago insularis*). It grows only a few inches (centimeters) tall and bears spikes of tiny greenish flowers, but billions of plants cover many square miles in good years. The tiny seeds are covered with a soluble fiber that forms a sticky mucilage when wet by rain; this aids germination by retaining water around the seed and sticking it to the soil. A related species from India is the commercial source of psyllium fiber (e.g., Metamucil®). The buckwheat family (Polygonaceae) is well-represented. There are more than a score of skeleton weeds (*Eriogonum* spp.) and half as many spiny buckwheats (*Chorizanthe* spp.), most of which go unnoticed except by botanists (see species accounts). Fiddlenecks (*Amsinckia* spp., *Boraginaceae*) may grow in solid masses over many acres, but the tiny yellow flowers don't significantly modify the dominant green of the foliage. These more modest species produce more biomass than the showy wildflowers in most years, and thus form the foundation of a great food pyramid.

Some perennials also evade drought much as annuals do by having underground parts that send up leaves and flowers only during wet years. Coyote gourd (*Cucurbita digitata*) and perennial devil's claw (*Proboscidea althaeifolia*) have fleshy roots that remain dormant in dry years. Desert larkspur (*Delphinium parryi*) is a perennial that has woody rootstocks but also grows only in wetter years. Desert mariposa (*Calochortus kennedyi*) and desert lily (*Hesperocallis undulata*) have bulbs that may remain dormant for several years until a deep soaking rain awakens them. Our desert wildflower displays are in jeopardy from invasive exotic plants. Species such as Russian thistle (*Salsola tragus*), mustards, especially Sahara mustard (*Brassica tournefortii*), red stem filaree (*Erodium cicutarium*), and Lehmann's lovegrass (*Eragrostis lehmanniana*) are more aggressive than most of the native annuals and are crowding them out in many areas where they have become established. Some are still increasing their geographic ranges with every wet winter. Disturbed sites such as sand dunes, washes (naturally disturbed by wind and water, respectively), roadsides, and livestock-grazed lands are particularly vulnerable to invasion by these aliens.

### 8.4.7 Combined Drought Adaptations

These three basic drought coping strategies are not exclusive categories. Ocotillo behaves as if it were a CAM-succulent, drought deciduous shrub, but it is neither CAM nor succulent. The genus *Portulaca* contains species that are succulent annuals. The seeds may wait for a wet spell to germinate, but the resulting plants can tolerate a moderate drought. The semisucculent yuccas have some water storage capacity but rely on deep roots to obtain most of their water. Mesquite trees are often *phreatophytes* (plants with their roots in the water table), but mesquite and some other species can also grow as stunted shrubs on drier sites where ground water is beyond their reach.

## 8.5 Adaptations to Other Desert Conditions

Water scarcity is the most important but not the only environmental challenge to desert organisms. The aridity allows the sun to shine unfiltered through the clear atmosphere continuously from sunrise to sunset. This intense solar radiation produces very high summer temperatures that are lethal to nonadapted plants. At night much of the accumulated heat radiates through the same clear atmosphere and the temperature drops dramatically. Daily fluctuations of 40°F (22°C) are not uncommon when the humidity is very low.

*Microphyllary* (the trait of having small leaves) is primarily an adaptation to avoid overheating; it also reduces water loss. A broader surface has a deeper boundary layer of stagnant air at its surface, which impedes convective heat exchange. A leaf up to 10 mm across can stay below the lethal tissue temperature of about 115°F (46°C) on a calm day with its stomates closed. A larger leaf requires transpiration through open stomates for evaporative cooling. Since the hottest time of year is also the driest, water is not available for transpiration. Large-leafed plants in the desert environment would overheat and be killed. Desert gardeners know that tomatoes will burn in full desert sun even if well watered; their leaves are just too big to stay cool. Desert plants that do have large leaves produce them only during the cool or rainy season or else live in shaded microhabitats. There are a few mysterious exceptions such as Jimson weed (*Datura wrightii*) and desert milkweed (*Asclepias erosa*). Perhaps their large tuberous roots provide enough water for transpiration even when the soil is dry.

Leaf or stem color, orientation, and self-shading are still more ways to adapt to intense light and heat. Desert foliage comes in many shades, but rarely in typical leaf-green. More often leaves are gray-green, blue-green, gray, or even white (Figure 8.7). The light color is usually due to a dense covering of *trichomes* (hairlike scales), but is sometimes from a waxy secretion on the leaf or stem surface. Lighter colors reflect more light (=heat) and thus remain cooler than dark green leaves. Brittlebush and white bursage (*Ambrosia dumosa*) leaves show no green through their trichomes during the dry season, while desert agave (*A. deserti*) is light gray due to its thick, waxy cuticle. Other plants have leaves or stems with vertical orientations; two common examples are jojoba (*Simmondsia chinensis*) and prickly pear cactus (*Opuntia* spp.). This orientation results in the photosynthetic surface facing the sun most directly in morning and late afternoon. Photosynthesis is more efficient during these cooler times of day. Prickly pear pads will burn in summer if their flat surface faces upward. Some cacti create their own shade with a dense armament of spines; teddy bear cholla (*Opuntia bigelovii*) is one of the most striking examples.

**FIGURE 8.7**
Light leaf color provides cooler surfaces to plant bodies in arid climates. The light colored plant in this case is brittlebush (*E. farinosa*).

## 8.6 Pollination Ecology and Seed Dispersal of Desert Plants

Flowers are very useful for identifying plants and providing aesthetic pleasure for humans, but they have a more vital function—they are the sexual reproductive organs of plants. Many plants also have methods of asexual (vegetative) reproduction, which produces offspring that are genetically identical to the parent: (a) root-sprouting in limberbush (*Jatropha cardiophylla*), palo verde, and aspen (*Populus tremuloides*); (b) stolons and rhizomes in agaves, strawberries (*Fragaria* spp.), many grasses; and (c) aerial plantlets (some agaves, mother-of-millions kalanchoe). All of the progeny of asexual reproduction are *clones* of their parent plants. (A clone is a group of organisms that are genetically identical; in the case of flowering plants each clone originates from a single original seed.) Horticulturists have invented additional methods of plant cloning that are valuable in perpetuating superior varieties of plants: cutting, grafting, and tissue culture. The "Kadota" fig is a *cultivar* (contraction for cultivated variety) that has been propagated by cuttings for at least two millennia; it is described under a different name in the writings of Pliny the Younger.

In contrast, sexual reproduction combines half the genes from each of two parents, so sexually produced offspring are different from either of their parents and from one another. This variation is the raw material of natural selection, which in turn results in evolution. A species that cannot reproduce sexually (there are quite a few among both plants and animals) is at greater risk of extinction if its environment changes, because it cannot adapt to new conditions.

Pollination is the transfer of pollen from an anther onto the stigma of a flower. The pollen then grows a tube that penetrates the style down to the ovary; sperm cells swim down the tube and fertilize the ova. Fertilized ova develop into seeds, which are the sexual propagules of flowering plants.

Outcrossing (pollination by pollen from another plant) is evolutionarily advantageous because the offspring are more variable than those from self-pollination. This increases plants' probability of surviving in an ever-changing environment. (But selfing is still sexual reproduction and much better for evolutionary change than vegetative cloning.) Plants have many adaptations that increase the likelihood of outcrossing.

Because plants are rooted in the ground and can't get together to mate, they must employ an agent to transport pollen between plants. From this need has evolved one of the most widespread and complex types of *symbiosis* (interactions) between plants and animals. The pollen-transporting agent is frequently an insect or other flying animal. (Flying animals are more mobile than grounded species, and thus more likely to visit widely separated plants.) In order to get pollinated, a flower must both make its presence known (advertise), and provide an incentive (a reward) for an animal to make repeated visits to flowers of the same species. The advertisements are fragrance and/or conspicuous color. Two kinds of food are the usual reward. Nectar is a sugar solution that provides energy for flight. Flying requires much more energy than terrestrial locomotion. Pollen, besides being the male gamete of a flower, is also rich in the protein essential for maintaining animal tissues and for raising young. In place of nectar some flowers offer oil (fat), another energy food. Others provide fragrances that the pollinator gathers to use for its own reproductive advertisement, and a few fascinating species employ deceit and provide no reward (see the species account on *Pipevine* for an example).

The sugar in nectar and the protein in pollen are expensive to produce, so there is selective pressure to utilize these resources efficiently. It is important that animals other than the pollinators do not eat (steal) the nectar and pollen and that the pollinators transport pollen to other flowers of the same species and deposit it in the right place. Natural selection has produced specialization: most plants with animal-pollinated flowers attract only a few species of animals that have the right size and behavior to reach the reward and pick up pollen. More than 100 million years of coevolution between flowering plants and their pollinators has greatly contributed to the huge number of species in both kingdoms (300,000 flowering plants, 600 hummingbirds, and 15,000 known bees in the world). It also explains why there are so many different shapes and colors of flowers.

Flowers can be classified into several pollination *syndromes* according to their pollinators. (A syndrome is a set of characteristics associated with a specific phenomenon.) This is not the same classification as systematic taxonomy and does not reflect the evolutionary relationships among plants. Species in the same family or even the same genus may attract different pollinators.

The hummingbird pollination syndrome is one of the most easily recognized. Hummingbirds are large compared to most insects, almost unique in their ability to feed while hovering, daytime-active, have no sense of smell, have long narrow beaks and tongues that can probe deep narrow tubes, and have excellent color vision. Hummingbird flowers tend to be long-tubular, nonfragrant, sideways- or down-facing, day-blooming, and brightly colored. Bees and most other animals cannot easily land on a hanging flower, and even if they succeed they cannot reach the nectar at the base of the narrow tube.

There are common misconceptions that all hummingbird flowers are red and that hummingbirds can see only the warm colors of the spectrum. It is true that most hummingbird flowers in the temperate biomes are red, but in the tropics they come in many colors. The predominance of red in temperate hummingbird flowers may be a disincentive to bees. Bees are aggressive pollen collectors in temperate climates. But they cannot see red, so red flowers do not appear conspicuous to them.

Wind-pollinated plants make no investment in attracting animals; their flowers lack fragrance or showy parts. Many people would not recognize them as flowers at all. Prodigious quantities of pollen are released, an infinitesimal proportion of which lands on a receptive stigma of the same species. While this seems inefficient, it is obviously effective judging from the successful groups of plants with this syndrome. Conifers, most riparian trees e.g., willows (*Salix* sp.), sycamores (*Platanus* spp.) oaks, and grasses are all wind-pollinated. Conifers and grasses are the dominant plants in the two biomes that bear their names. Grasses occur in most biomes and comprise the sixth largest family of plants with about 9000 species worldwide.

### 8.6.1 Seed Dispersal

Seeds generally need to be transported some distance from the parent plant in order to find a suitable site for establishment. Some plants have wind-dispersed seeds, which are occasionally blown many miles from their origins. This means of dispersal is common among *pioneer plants*—plants that are adapted to colonizing disturbed habitats. Because of their superior ability to invade newly disturbed ground, pioneer plants comprise many of our agricultural and garden weeds. Moreover, most annual crops are domesticated pioneer plants.

Many plants utilize animals to disperse their seeds in another complex coevolutionary process. Small, brightly colored fruits such as hackberry and boxthorn are offered as food for birds that swallow them whole. Other fruits such as those of hedgehog cacti (*Echinocereus* sp.) are large and the bird feeds on them repeatedly. Some bird fruits are sticky such as mistletoe berries; a few stick to the bird's bill until wiped off on a branch while others are successfully swallowed. The seeds of bird fruits are typically small and hard; they pass through birds' guts undamaged and may be deposited many miles from the parent plant.

Mammal-dispersed fruits tend to be larger, aromatic, not colorful (most nonprimate mammals have poor color vision), and usually have larger seeds than found in fruits birds feed on. The animal often transports the fruits a short distance (compared to the flying distances of many birds) to a safer place before eating the pulp and dropping at least some of the seeds. The seeds of coyote gourds (*Cucurbita* spp.) may be dispersed in this manner. Coyotes swallow the whole fruits of palm trees; they digest the thin pulp and excrete the hard seeds intact. Since seeds contain energy stores to nourish the germinating embryo, seeds themselves are also nutritious food for mammals and birds. Some plants offer their seeds without juicy pulp to attract mammals. Pocket mice and antelope squirrels gather the abundant seeds of foothill palo verdes and bury them as food caches for the dry season. The animals don't eat all that they bury, so some seeds remain in the ground and germinate when the rains come. (Birds that specialize in eating seeds, as opposed to fruits containing seeds, crush and digest the seeds and therefore do not disperse viable propagules.)

Even in the desert some seeds are water-dispersed. Blue palo-verde (*Cercidium floridum*) grows mostly along washes. Flash floods disperse the very hard, waterproof seeds downstream, *scarifying* (abrading the surface) them in the process. In the absence of scarification these seeds must weather in the ground for a few years before the seed coats become permeable and enable germination.

The timing of seed maturation is crucial for many plants. The less time seeds are present before they sprout, the greater is their chance of survival. The tropically derived plants in our region germinate with the summer rains. These species usually flower in spring and their fruits ripen shortly before the arrival of the summer rainy season. Palo verde and saguaro are examples. Other plants produce large quantities of seeds and rely on camouflage or burial in the soil to conceal some of them from hungry animals. Brittlebush, for example, flowers and seeds in spring, but the seeds germinate with fall rains. Annuals do the same.

## 8.7 Invisible Larder

I conducted a wildlife survey in the Lower Colorado River Valley in the 1970s. The site had received almost no biologically effective rainfall for 3 years. Creosote bushes were almost the only plants present; they were widely spaced and had shed most of their leaves. Yet in the kilometer-long by 50-m-wide transect, I trapped one pocket mouse overnight, and in the morning observed a whiptail lizard, a rock wren, and two black-throated sparrows. These are all resident species; not transitory migrants. What were they living on?

The soil seed bank is a phenomenon unique to arid habitats. It provides an unseen (by humans) food source for desert animals as well as survival insurance for plant species. The greater density of seed-eating animals and the abundance of decomposing microbes in the moist soils of wetter regions greatly shortens the viability of seeds. In deserts viable—and nutritious—seeds persist in large numbers through decades of drought. After a wet year there may be 200,000 seeds per square meter (square yard) of soil. Even after several dry years with little or no additional seed production there are still several thousand seeds per square meter, enough to sustain low populations of seed-eaters such as harvester ants, kangaroo rats, and sparrows. The whiptail was foraging for insects that fed on the seeds or plant *detritus* (partially decomposed organic matter) in the soil.

## Further Readings

Bowers, J.E., *A Full Life in a Small Place and Other Essays from a Desert Garden* (Tucson, AZ: University of Arizona Press, 1993).

Dykinga, J.W. and C. Bowden, *The Sonoran Desert* (New York: Harry N. Abrams, 1992).

Hanson, R.B. and J. Hanson, *Southern Arizona Nature Almanac* (Boulder, CO: Pruett Publishing, 1996).

Hartmann, W.K., *Desert Heart: Chronicles of the Sonoran Desert* (Lady Lake, FL: Fisher Books, 1989).

Imes, R., *The Practical Botanist: An Essential Field Guide to Studying, Classifying, and Collecting Plants* (New York: Fireside Books/Simon and Schuster, 1990).

Larson, G., *There's a Hair in My Dirt! A Worm's Story* (New York: Harper-Collins Press, 1998).

Nabhan, G.P., *The Desert Smells Like Rain* (San Francisco, CA: North Point Press, 1982).

Nabhan, G.P., *Gathering the Desert* (Tucson, AZ: University of Arizona Press, 1985).

Nabhan, G.P. and S.L. Buchmann, *The Forgotten Pollinators* (Washington, DC: Island Press, 1996).

Seuss, Dr., *The Lorax* (New York: Random House, 1971*).

* This book can be found in the children's and humor sections of bookshops, it conveys the essence of ecology better than any scientific treatise I have encountered. *The Lorax* should be required reading for every citizen of planet Earth.

# 9

# *Wildlife and Anthropogenic Changes in the Arid Southwest*

**Brian K. Sullivan, David R. Van Haverbeke, and Carol Chambers**

**CONTENTS**



## 9.1 Introduction

Deserts are ecosystems of low rainfall and high temperatures (see Chapter 3). A surprising variety of arid landscapes occur in southwestern North America: the botanically diverse Sonoran Desert receives precipitation in both the winter and summer, while the relatively homogenous Great Basin Desert to the north receives much of its annual rainfall in the form of snow during the cold winter months. Animals need cover, water, and food to survive and reproduce, and in deserts, adequate habitat to satisfy these needs is often limited. Even fully aquatic organisms, such as fish and some amphibians, must adapt to extremes of water temperature, salinity, and flooding that result from rare but intense rainfall events in arid ecosystems.[1,2] Relatively rare aquatic ecosystems in deserts provide habitat to many amphibian and fish species but also draw hundreds of species of birds.[1] Although southwestern desert ecosystems appear sparsely inhabited, even the hottest areas have a diverse wildlife community that includes dozens of species of mammals, birds, reptiles, and even amphibians.[3] Over the past few decades, biologists have come to appreciate that these desert landscapes are especially vulnerable to environmental disturbance wrought by humankind.

## 9.2 Anthropogenic Change

The arid Southwest has been profoundly affected by anthropogenic activities. Efforts to secure adequate water, followed by conversion of land to agricultural use, and more recently, urbanization, have greatly altered much of the region. Human population growth, given the attraction of the "Sunbelt," has been especially dramatic in the four deserts of the Southwest: the Mojave Desert of southeastern California, the Great Basin Desert of Nevada and adjacent states, the Sonoran Desert of Arizona and extreme northern Mexico, and the Chihuahuan Desert of north central Mexico, western Texas, and southern New Mexico (see Chapters 1 and 26). The unique Sonoran Desert has been fragmented because of widespread development,[4,5] especially given rapid growth in the Phoenix and Tucson areas (Figure 9.1). Riparian corridors, vital to so many wildlife species, were often the primary targets of anthropogenic change, given the importance of water to permanent human population centers. Alteration of water sources, including impoundment construction, lowered water tables, and draining of springs, was intimately associated with the expansion of human populations in the Southwest. Introduction of non-native species, especially large grazing mammals, including cattle, horses, and burros, also changed existing habitats and thereby took a toll on native fauna.[6,7] Although direct loss of habitat due to conversion of landscapes to urban and agricultural areas has been a highly visible anthropogenic effect on desert wildlife,[8,9] other activities such as logging, fire suppression, and livestock ranching profoundly altered biotic communities while superficially maintaining natural habitats.

Over the past decade, studies have documented a variety of responses of vertebrates to anthropogenic activities, but some generalities have emerged. Here, we examine anthropogenic effects on wildlife in the arid Southwest with a focus on the two major rivers of the region, the Colorado and Rio Grande, and on the organisms of the Sonoran Desert. These examples highlight the nature of human-related effects on wildlife populations and also indicate the successes and failures of conservation policy. For convenience, we group anthropogenic effects on the fauna of the North American deserts into three basic sources: direct habitat loss, habitat fragmentation, and habitat alteration. We examine each of these in turn while addressing individual examples drawn primarily from the Sonoran Desert before considering the two major watersheds and more detailed examples associated with urbanization.

## 9.3 Habitat Loss

Development, leading to conversion of habitat to urban, rural, and agricultural areas, has occurred throughout the Southwest over the past century. The past few decades have witnessed the expansion of human populations onto landscapes initially converted to agriculture in the early twentieth century. Metropolitan centers in the Southwest typically exhibit extensive sprawl in which outlying areas are developed with little focus on building "up" rather than "out" (Figure 9.2). The widely accepted notion that deserts are largely lifeless wastelands with little if any inherent value no doubt contributed to the lack of concern regarding the loss of these surprisingly diverse landscapes. Habitat loss is the most obvious aspect of anthropogenic effects on wildlife, and in the Southwest, it has

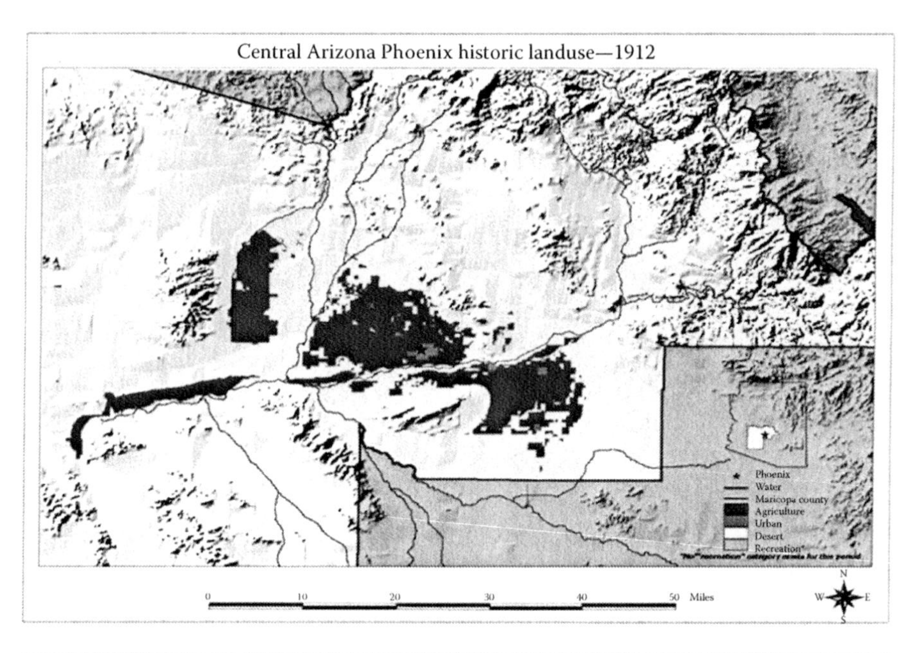


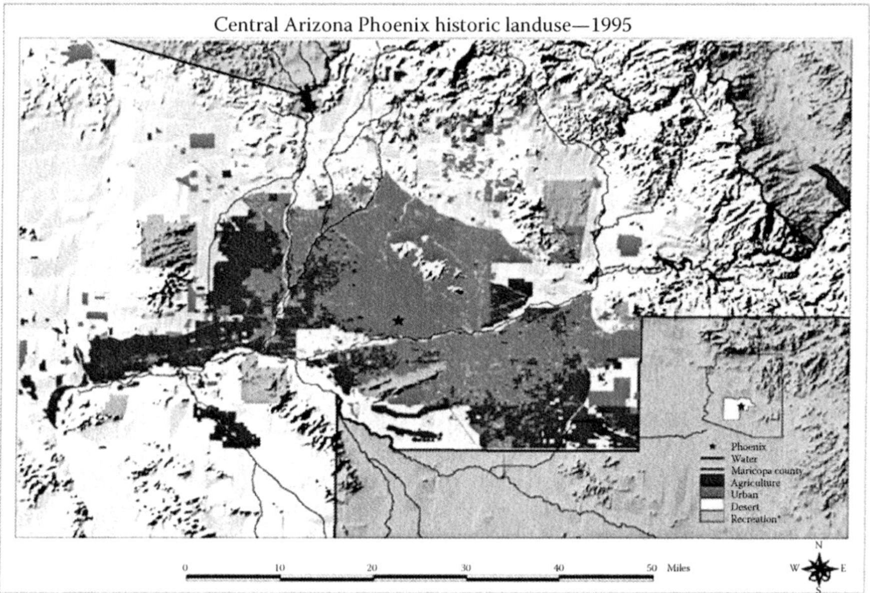


**FIGURE 9.1**
Increase in Phoenix metropolitan area in the twenty-first century. (Modified from Knowles-Yanez, K. et al., Historic land use: Phase I report on generalized land use, Central Arizona-Phoenix Long-Term Ecological Research (CAPLTER), Arizona State University, Tempe, AZ, 1999.)

been especially dramatic in that many organisms historically associated with the most restricted habitat (e.g., riparian corridors) have been significantly affected.

Residential development in urban areas generally replaces native vegetation by adding man-made features (buildings, roads, swimming pools, parks, artificial lights) and often promotes biological invasion of non-native plants and animals, compacts, and

**FIGURE 9.2**
Lookout Mountain, Phoenix Mountains, Maricopa County, Arizona. Note the absence of two habitats, open creosote flats and tree-lined arroyos, and dominance of rocky slope habitat.

**FIGURE 9.3**
Mesopredators such as domestic cats are often predators in urban and suburban settings. A conservative estimate predicts that a single cat kills 56 wild animals each year.

disturbs soils, even altering microclimate.[10,11] With development, the decline of large carnivores due to habitat loss has led to increases in native (striped skunk, gray fox) and introduced (domestic cat) mesopredators that prey on wildlife. For example, pet cats in a small urban subdivision (~100 residences) have been estimated to kill hundreds of rodents, birds, and lizards (Figure 9.3). Cats presumably consume prey uncounted in this estimate; given that activities of feral cats were not considered, these numbers underestimate predation.[12]

Birds are generally sensitive indicators of vegetation change, both composition and structure. Housing density best explained variation in species' richness for birds in Tucson, and housing developments with exotic vegetation did not support native species such as black-throated sparrows, verdins, and northern flickers: high density housing results in habitat loss for these species. Retaining and protecting native vegetation and creating native habitat fragments throughout urban areas can help retain habitat for some wildlife species. However, some species are sensitive to even a small degree of disturbance and will not remain in developed areas.[10] The elf owl and the nectar-feeding lesser long-nosed bat rely on saguaro and other cacti to meet habitat requirements, and development typically eliminates habitat for these species.[13,14] Small mammal communities also change with urbanization. Although suburban neighborhoods with low housing density (0.2 houses/ac) maintained the native nocturnal rodent community found in undeveloped areas, medium density housing (3 houses/ac) did not and in fact encouraged colonization of an exotic species, the house mouse.[15] Lesser long-nosed bats will readily use hummingbird feeders as a food replacement for cactus flowers but these bats apparently do not tolerate artificial lights so remain at the edge of developed areas around Tucson.[16] Some bats use swimming pools as water sources, and Mexican free-tailed bats roost under bridges and forage on insects drawn to artificial lights. These examples highlight the variation among organisms as to whether habitat is best viewed as "altered" or "lost" following development.

Although many wildlife species effectively lose habitat following urbanization, those that respond to the perturbation as habitat alteration experience both pluses and minuses. Cooper's hawk's nest in urban Tucson, selecting native fremont cottonwood (*Populus fremontii*) or non-native eucalyptus (*Eucalyptus* spp.) and aleppo pine (*Pinus halepensis*) for nests. Neither level of disturbance at nest sites nor percent cover of buildings around a nest was a factor in selection.[17–19] Density of nests was higher in the city relative to exurban (outside the city) areas. However, mortality was higher for nestlings in urban areas compared to exurban areas, and the likely cause of death was trichomoniasis caused by a parasitic protozoan. Bird feeders can promote spread of this disease (especially through common species such as doves), and bird feeding is common in urban and suburban areas. Other causes of mortality for urban Cooper's hawks were collisions (e.g., with windows) and organophosphate poisoning used to rid neighborhoods of feral pigeons.

## 9.4 Habitat Fragmentation

Division of contiguous habitat by roads, railroads, canals, dams, and other barriers has affected the population biology of many vertebrates. One of the most dramatic new barriers in the Southwest is a 660 mile long border wall being constructed along the United States and Mexico border.[20] This wall hinders movement of wildlife such as javelina and deer, including some endangered species such as the jaguar (Figure 9.4). Fragmentation affects wildlife movement both among and within habitat patches and can influence genetic structure of populations.* The behavior of even relatively mobile animals, such as mammals and birds, can be altered because of subdivided landscapes resulting from urban development.[23] Roadways not only represent barriers to movement, especially of smaller wildlife species such as amphibians and reptiles, but also represent a significant source of mortality.[24]

Desert mule deer near Tucson, Arizona, primarily selected habitat based on forage quality, slope, and elevation; however, during parts of the year, deer avoided roads, rivers, and canals.[25] For this species, as urbanization increases road density, and habitat becomes more fragmented. Designating wildlife corridors may help mitigate this problem. Although less than 25% of desert mule deer sightings were in a designated corridor for wildlife and remaining locations were in undeveloped lands, some of these lands are scheduled for agricultural development. Deers are likely to increase their use of the designated wildlife corridor if this development occurs.[26]

* See reviews in Kwiatkowski et al.[21] and Dixo et al.[22]

(a)

(b)

**FIGURE 9.4**
The 660-mile-long border fence between Mexico and the United States is a barrier that disrupts animal movement and fragments habitat. Picture (a) indicates the extent to which the border fence can bisect desert ecosystems, while picture (b) illustrates the physical barrier the fence can be to large wildlife species. (From Clark, M., Defenders of Wildlife; San Pedro River Valley fragmented by border wall, previously unpublished.) (Courtesy of Matt Clark.)

Some organisms can withstand and may even benefit from modest levels of habitat fragmentation. For example, prey populations can expand with increased availability of water in areas with low density housing or the construction of golf courses and greenbelts.[21] The negative consequences of habitat fragmentation in these settings may be partially offset by the increased availability of resources required by the native wildlife. In the Sonoran Desert near Tucson, coyotes benefited from living in an area near Saguaro National Park bordering suburban areas. Rodents were captured more frequently in suburban areas than rural ones, but coyotes apparently substituted human-related foods (dog food, bread, fruits, vegetables, table scraps) for rodents and other foods in some areas.[27]

### 9.4.1 Habitat Alteration

Perhaps, the least obvious and more subtle of anthropogenic changes to habitats of the Southwest have had the most far-reaching consequences for wildlife: habitat alteration in the form of introduced species. Urbanization is invariably associated with the introduction of any number of plants and animals including dogs, cats, pigs, horses, and various birds to adjacent habitat resulting in further habitat alteration. Many of these domestic animals, like the cats described earlier, have important effects on native wildlife, but those associated with livestock have received the most attention. Cattle and sheep range far from urban and other obviously developed areas and affect wildlife in significant ways (Figure 9.5). Livestock grazing has led to the establishment of introduced plant species and has been implicated in the conversion of grassland to shrubland in many arid areas of Arizona and New Mexico (Figure 9.6). The ability of cattle to modify habitats has even been used to reduce introduced grasses in some highly disturbed communities, such as the historic desert shrublands of the San Joaquin Valley of California.[28] In southwestern New Mexico, degradation of grassland and increases in shrubland because overgrazing has favored range expansion of the desert grassland whiptail lizard while negatively affecting the grassland specialist Arizona striped whiptail, although this has yet to occur in Arizona.[29]

Livestock ranching and agricultural activities are intimately associated with the construction of impoundments, lowering water tables and draining of springs; all of these

**FIGURE 9.5**
Cattle have greatly altered the desert, grassland, woodland, and even forest biomes of the western United States.

(a)

(b)

**FIGURE 9.6**
Overgrazing by cattle has led to conversion of semidesert grassland biomes to scrublands dominated by cacti ((a) from north of Tucson, Pinal Co., Arizona, June 2009) and mesquite ((b) from west of Rye, Gila Co., Arizona, June 2009).

represent a profound source of habitat alteration, especially prevalent in arid regions.[30] Even if water continues to flow in a river after the construction of a dam, altered flow characteristics and water temperature can significantly modify the habitat for wildlife, especially fish and birds historically associated with riparian and other aquatic communities. Additionally, livestock directly and dramatically influence these fragile riparian systems.[31] Grazing can increase soil compaction, change streambank stability, and enhance establishment of non-native plants in grasslands when coupled with prescribed burns in the presence of non-native plant seedbanks.[32] Livestock can compete directly with wildlife for limited resources such as forage or physically interfere with habitat use.[33,34] Effects of competition are more pronounced if wildlife species competing with grazers are specialists, have small home range sizes and dispersal distances, or rely on habitat features such as forage biomass that are limiting when grazers are present.[35] Removing cattle from

**FIGURE 9.7**
Santa Cruz River, south of Tucson, Pima Co., Arizona (June 2009), showing channelization with the absence of typical riparian vegetation because of lowered water table and ground water pumping (also see Chapter 5).

the San Pedro River in southern Arizona had a profound effect on birds. Abundance of birds doubled over a 4 year period after grazing was eliminated and detection of 26 bird species increased.[36] Although cattle may reduce fuel loads in some biomes, such effects typically do not occur in relatively arid communities.*

Many rivers in the southwest were historically dominated by Fremont cottonwood and willow (*Salix* spp.), but anthropogenic effects typically associated with the cattle industry have changed natural vegetation. When water diversion and ground water pumping cause depletion of alluvial aquifers and the decline of regional groundwaters, perennial stretches of rivers can become ephemeral. This "desertification" along rivers also alters herbaceous vegetation and allows establishment of exotic invasive species such as saltcedar (*Tamarix ramossisima*).[37,38] In some rivers, shallow water tables have declined 200 m and caused total loss of riparian vegetation (Figure 9.7).[39,40] The increase in frequent winter flooding, high rates of stream flow during spring, and removal of livestock allowed some self-repair and reestablishment of native vegetation that benefits wildlife.[37]

One consequence of grazing practices in arid landscapes, especially widespread in the arid Southwest, is increased availability of artificial water sources[41]; ranchers must provide access to water throughout their grazing allotments given the daily requirements of free-ranging cattle by construction of "cattle tanks."[30] In addition, small water developments (guzzlers), created for wildlife beginning in the 1940s, now supply water for wildlife and livestock.[42] Ranchers and those supporting their continuation on public lands have recently recast this activity as environmentally benign, even beneficial to native wildlife.[32] Indeed, water quality in artificial water sources appears to be good with low levels or no evidence of blue-green algal toxins or *Trichomonas*.[42,43] Some wildlife, such as birds (dove, quail) and mammals (deer, rabbits) may benefit from artificial water sources if they are water limited under natural conditions.[44,45] Many of these wildlife are subjected to recreational hunting, and thus this practice is viewed positively by those individuals and in part by state wildlife agencies supported by hunting license fees.

In spite of these examples, artificial water sources represent a dramatic departure from the historic condition for the arid communities of the Southwest and do not benefit all

* See citations in Sullivan.[32]

(a)

(b)

**FIGURE 9.8**
Woodhouse's toads (a) have expanded their range at expense of Arizona toads (b) because of habitat alteration associated with construction of impoundments.

wildlife species. Migratory songbirds use large water bodies (e.g., rivers) rather than small, open-water guzzlers. Migrating birds travel in jet streams at high altitude and probably seek large features such as rivers, lakes, and flowering trees as settling cues for necessary resources.[46] Bats use artificial water sources but larger sources provide water for more species than small guzzlers. Larger sources have more open water and fewer obstacles to flight.[47] Over these sources, bats expend more energy acquiring water because they use more overflights and drinking.[48]

Artificial water sources in arid regions also allow for interactions among species historically separated by habitat differences, such as Woodhouse's and Arizona toads in central Arizona, and Mexican and plains spadefoots in southeastern Arizona; these amphibians may cooccur and even hybridize to a much greater extent than they did in the past (Figure 9.8). Availability of water sources for introduced grazing mammals also allows for the spread of less arid-adapted invasive non-native species such as American bullfrogs and Rio Grande leopard frogs known to negatively affect native amphibians of the Southwest.[49] Other sources of mortality (e.g., various pathogens) may also be increased as a result of these more permanent sources of water.[50] Frogs and toads of the Southwest that use natural temporary rain pools for breeding generally shun artificial water sources that typically support a predatory and pathogenic fauna absent from ephemeral pools and have escaped large-scale declines to date.[51] Nonetheless, by avoiding cattle tanks, green toads of the Chihuahuan Desert were forced to breed in marginal aquatic habitats nearby and experienced high rates of mortality, presumably because water was channeled to the tanks for livestock use.[52] It is increasingly clear that arid land grazing practices have far reaching affects on desert flora and fauna.

### 9.4.2 Habitat Loss and Alteration along the Colorado and Rio Grande Rivers

Forty of the 1061 North American freshwater fishes went extinct in the last century.[53] Evidence suggests that extinction rates for North American freshwater fauna are five times higher than those for terrestrial fauna and that these temperate freshwater systems are being depleted of species as rapidly as tropical forests, some of the most stressed ecosystems on the planet.[54] In the United States, the three leading threats to extinction of freshwater fauna are habitat alteration in the form of nonpoint pollution, interactions with non-native species, and

**FIGURE 9.9**
Glen Canyon Dam on the Colorado River. (Courtesy of U.S. Geological Survey photo, Reston, VA.)

changes in hydrologic regimes associated with impoundment operations.[55] Other authors have stressed that direct modification of aquatic habitats is the most prevalent threat to native fishes in North America.[56,57] Invasion by non-native species, a form of habitat alteration, is recognized as second only to the loss of habitat and landscape fragmentation as a threat to global biodiversity.[58] To date, no North American fish species have been sufficiently recovered to permit removal from the U.S. endangered species list.

The Colorado River drainage encompasses portions of all four North American deserts and has undergone extensive habitat alteration. The Colorado River basin is typically divided into "upper" and "lower" basins, with the division at Lees Ferry, Arizona (see Chapter 4). The upper basin alone has 82 man-made reservoirs, each with a capacity of over 5,000 ac ft (6,030,744 $m^3$), and has 43 diversions to export water from the basin. The largest of these dams include Glen Canyon Dam on the Colorado River, Flaming Gorge Dam on the Green River, the Wayne N. Aspinall Units on the Gunnison River, and Navajo Dam on the San Juan River. These developments have caused significant changes in streamflow, water temperatures, sediment loads, total dissolved-solids, and channel morphologies.[59,60] Glen Canyon Dam adversely affected the aquatic ecosystem of the Colorado River and resulted in permanently fragmented fish populations between the upper and lower basins (Figure 9.9). Cold water releases from reservoirs have disrupted or prohibited mainstream reproduction of native fishes and destabilized rearing habitats because of daily water level fluctuations in the river to meet power demands.[61] Additional threats to water quality and quantity in the upper basin would be posed by the development of an oil shale industry.[62,63] The lower basin is equally, if not more developed.[64,65]

All of these impoundments along the Colorado River have altered large portions of the watersheds. Historically, these were free flowing rivers, largely characterized by high spring runoff and turbid waters with warmer summer water temperatures. Hundreds of miles of these rivers now function as lakes or as altered river habitat generally characterized by year-round cold water temperatures with low turbidities. As a result of these activities, native fish populations have declined from historic levels in the Colorado River basin.[66,67] Just as construction of impoundments and resulting habitat alteration favor some introduced amphibians over native forms, introduced fish often flourish in the unnatural lake habitats associated with arid land reservoirs.

The humpback chub (Figure 9.10), a morphologically unique fish endemic to the Colorado River basin, was listed as endangered in 1967. The species is a member of a relict native fish community, many of which are locally extinct or declining. For example,

(a)

(b)

**FIGURE 9.10**
Humpback chub (a) from the Little Colorado River and razorback sucker (b) from Lake Mohave. (Courtesy of U.S. Fish and Wildlife Service, Washington, DC.)

three of eight native fish species have become extinct in Grand Canyon since the completion of Glen Canyon Dam, including the Colorado pikeminnow, bonytail, and roundtail chub. A fourth, the razorback sucker, may also be extirpated in Grand Canyon.[61] Currently, the humpback chub occupies habitat along an estimated 310 river miles in the Colorado, Green, and Yampa rivers.[68] The largest remaining population of humpback chub occurs in the Grand Canyon, primarily inhabiting the lower 8.4 miles of the Little Colorado River, and the Colorado River in close proximity to the Little Colorado River.[69]

Of all the native endangered fishes of the Colorado River, the Little Colorado River population of humpback chub appears to be faring the best. Despite significant decline documented during the 1990s, the Little Colorado River humpback chub population has recently been documented as increasing (Figure 9.11).[70] This is thought to be the result of a

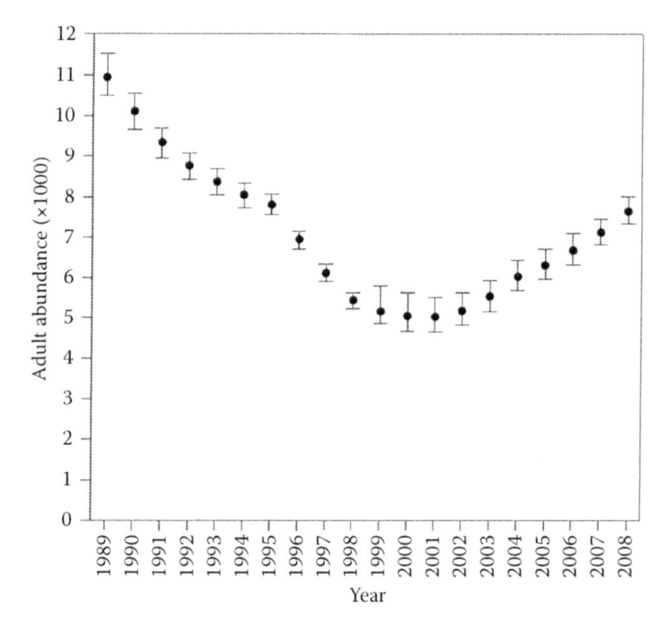


**FIGURE 9.11**
Estimated abundance of the Little Colorado River population of adult humpback chub. (From Coggins, L.G. Jr. and Walters, C.J., Abundance trends and status of the Little Colorado River population of humpback chub: An update considering data from 1989–2008, U.S. Geological Survey Open-File Report 2009-1075, 18 pp., 2009.)

recent warming trend in the temperature of the Colorado River because of drought, a significant effort being launched to remove non-native fish and because of moving the species into habitat previously unoccupied (i.e., translocation efforts).

The razorback sucker was listed as endangered in 1991 as a result of population declines and range contractions (Figure 9.10). As with other southwestern fishes, these problems were largely caused by the establishment of non-native predatory fishes and habitat alterations associated with water development. Historically, this fish was widespread in the Colorado River basin, ranging from Wyoming to the Gulf of California in Mexico. Currently, wild riverine populations only exist in small and isolated patches in the upper Colorado River basin above Lake Powell.[71] The largest remaining lacustrine population occurs in Lake Mohave, although this population has also suffered precipitous population decline, plummeting from the hundreds of thousands to only 44,000 in 1991 and fewer than 3,000 in 2001.[72] As of 2008, fewer than 50 wild adults are thought to be remaining in Lake Mohave.[73] The primary reason for their dramatic decline in Lake Mohave is that natural recruitment is precluded by predation on early life stages by non-native fishes.[74,75] Although biologists have undertaken concerted efforts to bypass the natural recruitment process and to maintain wild genetic diversity via capturing wild razorback larvae and growing them to larger size classes in hatcheries and in isolated off channel habitats, followed by release back into Lake Mohave, the problems with severe non-native predation may be nearly insurmountable.

The Rio Grande River, the largest of the Chihuahuan Desert, has been subjected to alteration by multiple impoundments, channel incision, and water diversions in order to meet the ever growing demands of civilization. The Rio Grande system is operated to reduce flood threats and to supply water for irrigation and municipal and industrial uses via a complex system of dams, ditches, and conveyance channels; as a result, substantial portions periodically dry over the past few decades.[76]

The Rio Grande silvery minnow was declared an endangered species in 1994. Historically, this small fish occurred from northern New Mexico to near the Gulf of Mexico as well as in the Pecos River, a tributary of the Rio Grande. Today, the silvery minnow is thought to only occur in a 168 mile section of the middle Rio Grande between Cochiti Dam to Elephant Butte Dam, which passes through the metropolis of Albuquerque, New Mexico.[77] This is about 7% of its former range, in a stretch of river split into four discrete reaches by three dams. Four other native fishes that have been extirpated in this reach of river include the speckled chub, Rio Grande shiner, phantom shiner, and bluntnose shiner, all thought to have shared similar ecological attributes with the silvery minnow.[78]

Like the Colorado River, the historic Rio Grande was characterized by a diverse native minnow assemblage. However, unlike the humpback chub, the silvery minnow is short-lived (1–2 years), and its population dynamics is likely more susceptible to short-term adverse environmental changes, such as a few years of low river flows. Like most southwestern native riverine fishes, the silvery minnow appears dependant upon annual spring runoff to cue spawning activity, followed by a sustained period of high river discharge that inundates complex shoreline habitat and creates backwaters where growth into larger size classes occurs. These natural attributes are often lacking in today's riverine systems, where artificial floods (dam releases) to improve habitat for native fishes may only last for a day or 2 in order to conserve water. In addition, smaller fish species that can rapidly undergo population growth and decline are sometimes artificially augmented by stocking during periods of low abundance. Sometimes, this appears to be helpful, but more often stocking activities can lead to a wide variety of detrimental genetic issues.[79,80] Finally, rivers that are kept at low volumes via man-made activities can concentrate non-native predator species on native species and lead to decreases in lower trophic level food production (e.g., algal and invertebrate food bases). In Arizona, this pattern of habitat alteration along the many tributaries of the Colorado River has apparently allowed for the expansion of Woodhouse's toad, a species that thrives in many disturbed habitats, relative to the native stream-dwelling Arizona toad that has declined.[81]

### 9.4.3 Habitat Fragmentation and Alteration in Urbanized Sonoran Desert

Anthropogenic effects in urbanized landscapes are receiving greater attention: recent studies of reptiles indicate that many forms have been affected by habitat loss and alteration, leading to local extirpation of many taxa.[82] Reptiles respond individualistically to remaining fragments of desert habitat associated with preserves, and both the size of the preserve as well as floristic and geological structure can be important to viability of populations.[82] Relatively large lizards and snakes appear especially vulnerable to effects of urbanization in the Sonoran Desert, especially in the Phoenix metropolitan area.[83] The common chuckwalla is a large, herbivorous lizard inhabiting the Mohave and Sonoran deserts, and though an extreme habitat specialist restricted to rocky microhabitats, it continues to inhabit the Phoenix Mountain Preserve system. It may be that the more restricted microhabitat preferences of common chuckwallas allow them to persist while other large lizards of the Sonoran Desert preferring open areas (e.g., desert iguanas, long-nosed leopard lizards, regal horned lizards) have declined in these same preserves. This may be a common occurrence for rock dwelling lizards: in many metropolitan regions, rocky slopes may be the least likely area of a preserve to be affected by human activity (Figure 9.12), although off-trail activities (rock-climbing) and collection of common chuckwallas for the pet-trade may contribute to degradation of crevices.[84] In contrast to the habitat specialist common chuckwallas, birds persisting in urban areas generally had broader environmental tolerance than rural congeners.[85]

Other studies suggest that squamate reptiles can be negatively affected by habitat alteration, even when preserves of natural habitat are set aside. Some lizards are more attuned to variation in habitat quality (thermoregulation sites, percentage of ground cover) than preserve size per se.[82] Habitat diversity within preserves is associated with lizard diversity; presumably, increased habitat variation allows for a larger number of specialized forms to

(a)

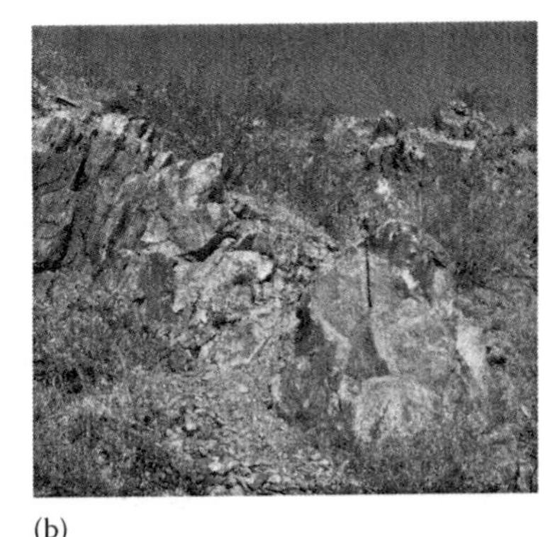

(b)

**FIGURE 9.12**
Chuckwallas (a) still occur in the Phoenix Mountain Preserves though they are surrounded by a sea of urbanization. Unfortunately, rocks can be permanently damaged by collectors and other off-trail activities leading to the loss of vital crevice refuges (note exposure of underlying light colored rock in (b) center).

(a)

(b)

**FIGURE 9.13**
Gila monsters in the northeastern Phoenix metropolitan area can be observed crossing roadways (a) while searching for prey such as cottontail rabbits (b). (Courtesy of Roger Repp.)

persist (like the common chuckwalla), suggesting that habitat structure rather than vegetation composition is important for reptile species.[82] Similar habitat associations have been detected for Sonoran Desert birds in Tucson: they generally respond to percent cover in areas whether they are urbanized or natural, rather than simply avoiding urban areas.[86]

The Gila monster is a long-lived predatory lizard that feeds on birds (eggs of ground-nesting species) and mammals (rodents, neonatal rabbits), prey species that are increasingly abundant in human-altered landscapes (Figure 9.13).[21] Some small carnivores, such as long-tailed weasels, have smaller ranges in urbanized settings due to greater availability of prey and restrictions on movement due to barriers.* Tiger rattlesnakes in the Tucson region have larger body size and higher reproductive output, and smaller home ranges, near golf courses, apparently due to increased numbers of rodent prey.[87] By contrast, a recent study was unable to document a change in home range size or movement behavior of Gila monsters in response to modest levels of habitat fragmentation and increased availability of prey species in the Phoenix region. Given that Gila monsters have much larger home ranges than common chuckwallas (10–100 times larger) and often make use of human structures as refuges in spite of the availability of natural refuges nearby (e.g., woodrat nests), they are more likely to experience injury or death while crossing roads, which they do with regularity.[21] Surveys of Sonoran Desert snake communities near

* See review in Kwiatkowski et al.[21]

the Phoenix metropolitan area indicate that increased traffic over the past two decades has led to increased mortality.*[88] There may be a trade-off within urbanized populations in which higher prey numbers are coupled with increased mortality due to roadways and other unnatural sources of morbidity.

With respect to activity area (i.e., home range size), a number of birds and mammals have smaller home ranges in areas of increased human activity.† As noted earlier, some have speculated that this is from increased prey availability in urbanized landscapes, and others have suggested that barriers (e.g., roadways) act to constrain home range size. In a number of corvid birds, home range area decreases with increasing urbanization, apparently as a result of increased food resources. A smaller number of studies have documented changes in behavior with increasing urbanization, such as a shift to nocturnal activity for large carnivores in urban settings. For example, coyotes avoided disturbed habitats altogether and limited their movements to corridors of natural habitat and moved less rapidly and less often with increasing urbanization.[89] Like the Gila monsters, some mammalian and avian predators may benefit from increased prey availability while experiencing higher levels of mortality associated with road traffic in urbanized landscapes.

### 9.4.4 Conservation Policy

Many of the investigations of anthropogenic effects on wildlife of the Southwest reviewed earlier were stimulated by federal and state conservation policies. For example, the Heritage Program, administered by the Arizona Game and Fish Department, funded the investigations on Arizona striped whiptails, Arizona toads, common chuckwallas, Gila monsters, and tiger rattlesnakes, in addition to many other valuable studies of urbanization and other affects on southwestern wildlife. A host of conservation policies, including the Endangered Species Act (ESA) and National Environmental Policy Act (NEPA), led to research critical to our present ability to conserve wildlife of North America.[90] However, despite impressive growth in conservation policy, and funding for basic and applied research over the past half century, the loss of biodiversity in the United States has continued.[90] Although there have been a number of impressive successes with respect to endangered species recovery, such as the bald eagle and peregrine falcon, North America has the highest proportion of non-native invasive species making up our current flora and fauna and dozens of species awaiting listing as endangered or threatened. North American conservation policy has not stopped biodiversity loss in part because it is insufficiently valued by society.[91] For example, the costs of setting aside critical habitat typically ignore benefits accruing to recreation or clean water generation. Recent analyses have also revealed that conservation policy can sometimes be as influenced by politics as by biology: scientific evidence can be ignored, and the political orientation of state representatives in part determines the number of species listed as endangered or threatened on a state-by-state basis.[91,92]

State and federal agency personnel have begun moving toward grassroots approaches focused on integrating conservation efforts with local stakeholders (e.g., ranchers, landowners) while managing wildlife. Many view this as an appealing development, but it has been coupled with a movement to eliminate postdecisional appeals (i.e., litigation), a mainstay of nonagency conservationists (e.g., the Tucson based Center for Biological Diversity) attempting to combat perceived inaction or mismanagement on the part of agencies.[93] Ideally, one would hope that all conservation concerns could be addressed at the outset (predecisional), but

* B.K. Sullivan, unpublished data.

† See review in Kwiatkowski et al.[21]

postdecisional litigation has doubtless played a vital role in conservation over the past 30 years. Additionally, depending on one's perspective, this grassroot approach continues the historical trend in which undue influence of local, vested parties can occur.[90,91]

It is natural to expect a tension between local, state, and federal level interests with respect to conservation issues. Mattson and Chambers noted that: "failures of federal conservation policies... are in part due to increasing bureaucratized scientific management with its core doctrine of presumed local democratic responsiveness... and the unique placement of agency experts to identify and solve these problems."[41] State-level management has a long record of deference to special interests, such as those of hunters and ranchers (both can be viewed as "customers" of wildlife management services). Together, both managers and their customers have presented a unified perspective, bound together by their unique economic relationship in which the customers support the agency directly through fees. This has been termed as "ecosystem services" approach to conservation as it places a premium on economic factors (e.g., money generated by grazing organisms) while ignoring aesthetics.[91,94] It even leads to the disturbing prospect that management of introduced species (e.g., cattle) often generates more economic value than native forms. A major goal for those valuing wildlife and natural (i.e., historic) habitats is to champion alternative interests (e.g., wildlife viewing) as equally valuable relative to the traditional utilitarian approaches (ranching, hunting, fishing) of the arid Southwest.

### 9.4.5 Future Directions

In spite of the somewhat bleak picture of conservation policy reviewed earlier, it is clear that preservation of arid land wildlife can succeed. Indeed, one obvious conclusion from the investigations reviewed earlier is that setting aside large areas encompassing the range of variation in historic habitats, while eliminating non-native species, would suffice to preserve much of the wildlife of the arid Southwest. As revealed by studies of Sonoran Desert birds and reptiles, habitat variability and adequate linkages (corridors) between reserves and larger natural areas to avoid untoward population genetic effects are clearly optimal.[95] Unfortunately, this is easier said than done; there is little prospect for a return to historic conditions for the major rivers of the Southwest, and removal of introduced species remains controversial, at least with respect to local economically important forms like cattle.

Establishing preserves at the outset, prior to development, is the ideal approach to wildlife conservation. Recent reviews suggest that setting aside large tracts of intact habitat rather than mitigating human effects after the fact is economically prudent.[96] The appealing notion of integrated landscapes (i.e., "conservation construction" in which housing developments are seamlessly merged with habitat remnants) appears to increase the likelihood of detrimental human/wildlife interactions and facilitates spread of introduced species.[97] Any means by which non-native species, including pets, livestock—even live bait for fishing can be eliminated or reduced—should be considered at the outset. The costs associated with attempts to return altered habitat to a natural state are often extreme, though this is the only option remaining for desert fish.[98,99]

Mitigation of habitat damage associated with introduced species remains a challenge for conservation biologists and land managers.[100] Given widespread population growth throughout the Southwest, some have proposed that cattle ranches, in spite of known negative affects on wildlife, can at least serve as buffers against loss of land to development.[94] It has even been suggested that "... over the next 50 years, we will need more ranches, not less."[101,102] Supporters of the ranching industry in the Southwest argue that the "new ranch" breaks with historical practices and adopts ecologically sound principles in which cattle are mere tools to achieve conservation goals such as biodiversity.[94] Nathan Sayre's recent critique of abstractions such as "natural" (used herein to denote "historical") notwithstanding, conservation goals must be implemented relative to some baseline, and for the arid Southwest, the recent evolutionary past is defensible. In the absence of agreement on this issue, one could defend the maintenance of highly artificial reservoirs in Arizona because they achieve the conservation goal of increasing biodiversity. These highly altered habitats support many more species of fish (dozens) today relative to historic aquatic habitats (e.g., Aravaipa Creek, with seven native forms historically), but at the expense of the native forms.[66] Characterization of ranchers as champions of nonequilibrium community ecology in which livestock grazing is recast as a form of disturbance contrasts with efforts of the livestock industry to establish that cattle have simply "replaced" extirpated wildlife.[103] Nonetheless, shifts in perception of aridland grazing in the Southwest cannot obscure the fact that ranching entails enormous effort to transform local biotic communities to a highly altered state with empirically documented negative consequences for arid adapted wildlife.[32]

The aquatic examples outlined earlier portend adverse effects for endangered southwestern fish given the extreme decline documented for so many populations of many species. Predictions of extinction rates have been based largely on the demonstrated relationship between number of species in a given group and habitat area.[104] Principles of metapopulation theory predict that the amount of extinction caused by habitat destruction is an accelerating curve; thus, if a high proportion of habitat is already destroyed, even a very small increase in habitat destruction will dramatically increase extinction risk.[105] Furthermore, there is a limit, termed the extinction threshold, which is the percentage of habitat destruction beyond which a species will ultimately head for extinction.[106,107] In one model, effects of habitat destruction became important with only 28% habitat loss, when some patches became isolated; extinction thresholds occurred at 66% habitat destruction or less.[108] Given sufficient habitat destruction, the equilibrium can be extinction; many rare and endangered species may already be committed to extinction, unless the loss and fragmentation of their habitat are reversed. Global climate change will undoubtedly exacerbate these issues for fish of the arid Southwest as human population growth results in ever increasing demands for water.[109]

Maintenance and restoration of historic aquatic ecosystems (i.e., flow rates, flood regimes, temperature profiles) are perhaps the single most difficult challenge facing conservation biologists and concerned citizens today. Thus far, the section of the Little Colorado River inhabited by humpback chub essentially functions as a free flowing river, characterized by an annual spring runoff, periodic episodes of highly turbid water, and by seasonally warm water temperatures. However, depletion of the aquifer(s) that feed the lower 13 miles of the Little Colorado River and maintain a year-round flow in this stretch of river would seriously threaten this self-sustaining population of endangered fish. For razorback sucker, if large self-sustaining mainstem populations cannot be established, off-channel populations in predator–free habitat are likely the remaining option.[110] For many species of native fishes in the southwest, the construction of physical barriers in small order streams, followed by the removal of non-native fish and the reintroduction of the native fish fauna appears a viable option.

Our review highlights three primary issues with respect to conservation of wildlife of the Southwest. First and foremost, preservation of intact habitat is vital. In situations where habitat alteration and fragmentation have occurred or are inevitable, low density housing with patches of intact habitat may allow for at least some wildlife species to persist. The dramatic consequences of habitat alteration for riparian systems and the associated fish fauna indicate the daunting prospects for recovery "after the fact" of habitat change. Second, mitigation of effects associated with introduced species such as livestock and non-native fish, and non-native plants should be a priority. In part, this issue is the most tractable—in the absence of continued support for these introduced forms, the non-native forms could possibly be controlled. Last, only by continued long-term study as well as experimental field investigations will we gain insight into the conservation and land treatment tactics that will serve us well in the coming decades of continued population growth in the Southwest.

## References


1. Jaeger, E.C., *The North American Deserts* (Stanford, CA: Stanford University Press, 1957).
2. Cornett, J.W., *Wildlife of the North American Deserts* (Palm Springs, CA: Nature Trails Press, 1987).
3. MacMahon, J.A., *Deserts* (New York: Alfred A. Knopf, 1998).
4. Knowles-Yanez, K., C. Moritz, J. Fry, C.L. Redman, M. Bucchin, and P.H. McCartney, Historic land use: Phase I report on generalized land use (Tempe, AZ: Central Arizona-Phoenix Long-Term Ecological Research (CAPLTER), Arizona State University, 1999).
5. Faeth, S.H., P.S. Warren, E. Shochat, and W. Marussich, Trophic dynamics in urban communities, *Bioscience* 55: 399–407, 2005.
6. Smith, C., R. Valdez, J.L. Holechek, P.J. Zwank, and M. Cardenas, Diets of native and non-native ungulates in south-central New Mexico, *The Southwestern Naturalist* 43: 163–169, 1998.
7. Freese, C.H. and D.L. Trauger, Wildlife markets and biodiversity conservation in North America, *Wildlife Society Bulletin* 28: 42–51, 2000.
8. Kareiva, P., S. Watts, R. McDonald, and T. Boucher, Domesticated nature: Shaping landscapes and ecosystems for human welfare, *Science* 316: 1866–1869, 2007.
9. Willis, K.J. and H.J.B. Birks, What is natural? The need for a long-term perspective in biodiversity conservation, *Science* 314: 1261–1265, 2006.
10. Germaine, S.S., S.S. Rosenstock, R.E. Schweinsburg, and W.S. Richardson, Relationships among breeding birds, habitat, and residential development in greater Tucson, Arizona, *Ecological Applications* 8: 680–691, 1998.
11. Roach, W.J., J.B. Heffernan, N.B. Grimm, J.R. Arrowsmith, C. Eisinger, and T. Rychener, Unintended consequences of urbanization for aquatic ecosystems: A case study from the Arizona desert, *BioScience* 58: 715–727, 2008.
12. Crooks, K.R. and M.E. Soulé, Mesopredator release and avifaunal extinctions in a fragmented system, *Nature* 400: 563–566, 1999.
13. Hardy, P.C. and M.L. Morrison, Nest site selection by elf owls in the Sonoran desert, *Wilson Bulletin* 113: 23–32, 2001.
14. Horner, M.A., T.H. Fleming, and C.T. Sahley, Foraging behaviour and energetic of a nectar-feeding bat, *Leptonycteris curasoae* (Chiroptera: Phyllostomidae), *Journal of Zoology* 244: 575–586, 1998.
15. Germaine, S.S., R.E. Schweinsburg, and H.L. Germaine, Effects of residential density on Sonoran desert nocturnal rodents, *Urban Ecosystems* 5: 179–185, 2001.
16. Stone, E.L., G. Jones, and S. Harris, Street lighting disturbs commuting bats, *Current Biology* 24: 1–5, 2009.

17. Boal, C.W. and R.W. Mannan, Nest-site selection by Cooper's hawks in an urban environment, *Journal of Wildlife Management* 62: 864–871, 1998.
18. Boal, C.W. and R.W. Mannan, Comparative breeding ecology of Cooper's hawks in urban and exurban areas of southeastern Arizona, *Journal of Wildlife Management* 63: 77–84, 1999.
19. Mannan, R.W., R.N. Mannan, C.A. Schmidt, W.A. Estes-Zumpf, and C.W. Boal, Influence of natal experience on nest-site selection by urban-nesting Cooper's hawks, *Journal of Wildlife Management* 71: 64–68, 2005.
20. Flesch, D., C.W. Epps, J.W. Cain III, M. Clark, P.R. Krausman, and J.R. Morgart, Potential effects of the United States–Mexico border fence on wildlife, *Conservation Biology* 24: 171–181, 2009.
21. Kwiatkowski, M.A., G.W. Schuett, R.A. Repp, E. Nowak, and B.K. Sullivan, Does urbanization influence the spatial ecology of Gila monsters in the Sonoran Desert? *Journal of Experimental Biology* 276: 350–357, 2008.
22. Dixo, M., J.P. Metzger, J.S. Morgante, and K.R. Zamudio, Habitat fragmentation reduces genetic diversity and connectivity among toad populations in the Brazilian Atlantic Coastal Forest, *Biological Conservation* 142: 1560–1569, 2009.
23. Kokko, H. and A. Lopez-Sepulcre, From individual dispersal to species ranges: Perspectives for a changing world, *Science* 313: 789–791, 2006.
24. Glista, D.J., T.L. DeVault, and J.A. DeWoody, Vertebrate road mortality predominantly impacts amphibians, *Herpetological Conservation and Biology* 3: 77–87, 2008.
25. Marshal, J.P., V.C. Bleich, P.R. Krausman, M.L. Reed, and N.G. Andrew, Factors affecting habitat use and distribution of desert mule deer in an arid environment, *Wildlife Society Bulletin* 34: 609–619, 2006.
26. Tull, J.C. and P.R. Krausman, Use of a wildlife corridor by desert mule deer, *The Southwestern Naturalist* 46: 81–86, 2001.
27. McClure, M.F., N.S. Smith, and W.W. Shaw, Diets of coyotes near the boundary of Saguaro National Monument and Tucson, Arizona, *The Southwestern Naturalist* 40: 101–104, 1995.
28. Germano, D.J., G.B. Rathbun, and L.H. Saslaw, Managing exotic grasses and conserving declining species, *Wildlife Society Bulletin* 29: 551–559, 2001.
29. Sullivan, B.K., P.S. Hamilton, and M.A. Kwiatkowski, The Arizona striped whiptail: Past and present, in G.J. Gottfried, B.S. Gebow, L.G. Eskew, and C.B. Edminster, eds., *Connecting Mountain Islands and Desert Seas: Biodiversity and Management of the Madrean Archipelago II, Proceedings RMRS-P-36*, USDA, Forest Service, Rocky Mountain Research Station, Fort Collins, CO, 2005, pp. 145–148.
30. Fensham, R.J. and R.J. Fairfax, Water-remoteness for grazing relief in Australian arid-lands, *Biological Conservation* 141: 1447–1460, 2008.
31. Belsky, A.J., A. Matzke, and S. Uselman, Survey of livestock influences on stream and riparian ecosystems in western United States, *Journal of Soil and Water Conservation* 54: 419–431, 1999.
32. Sullivan, B.K., The greening of public lands grazing in the arid Southwestern U.S.A., *Conservation Biology* 23: 1047–1049, 2009.
33. Szaro, R.C. and J.N. Rinne, Ecosystem approach to management of southwestern riparian communities, *Transactions of the 53rd North American Wildlife and Natural Resource Conference*, Louisville, KY, 1988, pp. 502–511.
34. Kie, J.G., V.C. Bleich, A.L. Medina, J.D. Yoakum, and J.W. Thomas, Managing rangelands for wildlife, in T.A. Bookhout, ed., *Research and Management Techniques for Wildlife and Wildlife Habitats* (Bethesda, MD: The Wildlife Society, 1996).
35. Holechek, J.L., R.D. Pieper, and C.H. Herbel, *Range Management: Principles and Practices*, 3rd edn. (Upper Saddle River, NJ: Prentice Hall, Inc., 1998).
36. Krueper, D., J. Bart, and T.D. Rich, Response of vegetation and breeding birds to the removal of cattle on the San Pedro River, Arizona (U.S.A.), *Conservation Biology* 17: 607–615, 2003.
37. Stromberg, J.C., R. Tiller, and B. Richter, Effects of groundwater decline on riparian vegetation of semiarid regions: The San Pedro, Arizona, *Ecological Applications* 6: 113–131, 1996.

38. Stromberg, J., Dynamics of Fremont cottonwood (*Populus fremontii*) and saltcedar (*Tamarix chinensis*) populations along the San Pedro River, Arizona, *Journal of Arid Environments* 40: 133–155, 1998.
39. Bryan, K., Change in plant association by change in ground water level, *Ecology* 9: 474–478, 1928.
40. Judd, J.B., J.M. Laughlin, H.R. Guenther, and R. Handergrade, The lethal decline of mesquite on the Casa Grande National Monument, *Great Basin Naturalist* 31: 153–159, 1971.
41. Mattson, D.J. and N. Chambers, Human-provided waters for desert wildlife: What is the problem? *Policy Science* 42: 113–135, 2009.
42. Rosenstock, S.S., V.C. Bleich, M.J. Rabe, and C. Reggiardo, Water quality at wildlife water sources in the Sonoran Desert, United States, *Rangeland Ecology and Management* 58: 623–637, 2005.
43. Bleich, V.C., N.G. Andrew, M.J. Martin, G.P. Mulcahy, A.M. Pauli, and S.S. Rosenstock, Quality of water available to wildlife in desert environments: Comparisons among anthropogenic and natural sources, *Wildlife Society Bulletin* 34: 627–632, 2006.
44. Lynn, J.C., C.L. Chambers, and S.S. Rosenstock, Use of wildlife water developments by birds in southwest Arizona during migration, *Wildlife Society Bulletin* 34: 592–601, 2006.
45. O'Brien, C.S., R.B. Waddell, S.S. Rosenstock, and M.J. Rabe, Wildlife use of water catchments in southwestern Arizona, *Wildlife Society Bulletin* 34: 582–591, 2006.
46. McGrath, L.J., C. van Riper III, and J.J. Fontaine, Flower power: Tree flowering phenology as a settlement cue for migrating birds, *Journal of Animal Ecology* 78: 22–30, 2009.
47. Rabe, M.J. and S.S. Rosenstock, Influence of water size and type on bat captures in the lower Sonoran Desert, *Western North American Naturalist* 65: 87–90, 2005.
48. Tuttle, S.R., C.L. Chambers, and T.C. Theimer, Potential effects of livestock water-trough modifications on bats in northern Arizona, *Wildlife Society Bulletin* 34: 602–608, 2006.
49. Maret, T.J., J.D. Snyder, and J.P. Collins, Altered drying regime controls distribution of endangered salamanders and introduced predators, *Biological Conservation* 127: 129–138, 2006.
50. Lanoo, M., ed., *Amphibian Declines: The Conservation Status of United States Species* (Los Angeles, CA: University of California Press, 2005).
51. Sullivan, B.K., Southwestern desert bufonids, in M. Lanoo, ed., *Amphibian Declines: The Conservation Status of United States Species* (Los Angeles, CA: University of California Press, 2005, pp. 237–240).
52. Griffis-Kyle, K.L., *Bufo debilis* (Green Toad): Breeding habitat selection, *Herpetological Review* 40: 199–200, 1989.
53. Miller, R.R., J.D. Williams, and J.E. Williams, Extinctions of North American fishes during the past century, *Fisheries* 14: 22–38, 1989.
54. Ricciardi, A. and J.B. Rasmussen, Extinction rates of North American freshwater fauna, *Conservation Biology* 13: 1220–1222, 1999.
55. Richter, B.D., D.P. Braun, M.A. Mendelson, and L.L. Master, Threats to imperiled freshwater fauna, *Conservation Biology* 11: 1081–1093, 1997.
56. Deacon, J., Endangered and threatened fishes of the west, *Great Basin Naturalist Memoirs* 3: 41–64, 1979.
57. Williams, J.E., J.E. Johnson, D.A. Hendrickson, S. Contreras-Balderas, J.D. Williams, M. Navarro-Mendoza, D.E. McAllister, and J.E. Deacon, Fishes of North America endangered, threatened, or of special concern, *Fisheries* 14: 2–20, 1989.
58. Simberloff, D., Introduced species, in W.A. Nierenburg, ed., *Encyclopedia of Environmental Biology*, Vol. 2 (San Diego, CA: Academic Press, 1995, pp. 323–336).
59. Andrews, E.D., Downstream effects of Flaming Gorge reservoir on the Green River, Colorado and Utah, *Geological Society of American Bulletin* 97: 1012–1023, 1986.
60. Van Steeter, M.M. and J. Pitlick, Geomorphology and endangered fish habitats of the upper Colorado River: 1. Historic changes in streamflow, sediment load, and channel morphology, *Water Resources Research* 34: 287–302, 1998.

61. Minckley, W.L., Native fishes of the Grand Canyon region: An obituary? in G.R. Marzolf, ed., *Colorado River Ecology and Dam Management* (Washington, DC: National Academy Press, 1991, pp. 124–177).
62. Leenheer, J.A. and T.I. Noyes, Effects of organic wastes on water quality from processing oil shale from the Green River formation, Colorado, Utah, and Wyoming, U.S. Geological Survey Professional Paper 1338, 56 pp., 1986.
63. Lindskov, K.L. and B.A. Kimball, Water resources and potential hydrologic effects of oil-shale development in southeastern Uinta Basin, Utah and Colorado, U.S. Geological Survey Professional Paper 1307, 32 pp., 1984.
64. Fradkin, P.L., *A River No More: The Colorado River and the West* (Tucson, AZ: University of Arizona Press, 1981).
65. Reisner, M., *Cadillac Desert: The American West and Its Disappearing Water* (New York: Viking Press, 1986).
66. Minckley, W.L., *Fishes of Arizona* (Phoenix, AZ: Arizona Game and Fish Department, 1973).
67. Minckley, W.L. and J.E. Deacon, Southwestern fishes and the enigma of "Endangered Species," *Science* 159: 1424–1432, 1968.
68. U.S. Fish and Wildlife Service, Humpback chub (*Gila cypha*) recovery goals: Amendment and supplement to the Humpback Chub Recovery Plan (2007 Revisions) (Denver, CO: U.S. Fish and Wildlife Service, Mountain-Prairie Region (6), 2007).
69. Douglas, M.E. and P.C. Marsh, Population estimates/population movements of *Gila cypha*, an endangered Cyprinid fish in the Grand Canyon region of Arizona, *Copeia* 1996: 15–28, 1996.
70. Coggins, L.G. Jr. and C.J. Walters, Abundance trends and status of the Little Colorado River population of humpback chub: An update considering data from 1989–2008, U.S. Geological Survey Open-File Report 2009-1075, 18 pp., 2009.
71. Modde, T., K.P. Burnham, and E.J. Wick, Population status of the razorback sucker in the middle Green River, *Conservation Biology* 10: 110–119, 1996.
72. Marsh, P.C., C.A. Pacey, and B.R. Kesner, Decline of the razorback sucker in Lake Mohave, Colorado River, Arizona and Nevada, *Transactions of the American Fisheries Society* 132: 1251–1256, 2003.
73. Marsh, P.C., Personal communication, 2008.
74. Minckley, W.L., Status of the razorback sucker, *Xyrauchen texanus* (Abbott), in the lower Colorado River basin, *Southwestern Naturalist* 28: 165–187, 1983.
75. Carpenter, J. and G.A. Mueller, Small nonnative fishes as predators of larval razorback suckers, *Southwestern Naturalist* 53: 236–242, 2008.
76. Chernoff, B., R.R. Miller, and C.R. Gilbert, *Notropis orca* and *Notropis simus*, cyprinid fishes from the American southwest, with description of a new subspecies, *Occasional Papers of the Museum of Zoology* (University of Michigan) 698: 1–49, 1982.
77. Bestgen, K.R. and S.P. Platania, Status and conservation of the Rio Grande silvery minnow, *Hybognathus amarus*, *Southwestern Naturalist* 36: 225–232, 1991.
78. U.S. Fish and Wildlife Service, Rio Grande silvery minnow (*Hybognathus amarus*) recovery plan (Albuquerque, NM: U.S. Fish and Wildlife Service, 2007, xii + 175 pp.).
79. Ford, M.J., Selection in captivity during supportive breeding may reduce fitness in the wild, *Conservation Biology* 16: 815–825, 2002.
80. Ryman, N. and L. Laikre, Effects of supportive breeding on the genetically effective population size, *Conservation Biology* 5: 325–329, 1991.
81. Schwaner, T.D. and B.K. Sullivan, Fifty years of hybridization: Introgression between the Arizona toad (*Bufo microscaphus*) and Woodhouse's toad (*Bufo woodhousii*) along Beaver Dam Wash in Utah, *Herpetological Conservation and Biology* 4: 198–206, 2009.
82. Sullivan, B.K. and K.O. Sullivan, Common chuckwalla (*Sauromalus ater*) populations in the Phoenix metropolitan area: Stability in urban preserves, *Herpetological Conservation and Biology* 3: 149–154, 2008.
83. Sullivan, B.K. and M. Flowers, Large iguanid lizards of urban mountain preserves in northern Phoenix, Arizona, *Herpetological Natural History* 6: 13–22, 1998.

84. Goode, M.J., W.C. Horace, M.J. Sredl, and J.M. Howland, Habitat destruction by collectors associated with decreased abundance of rock-dwelling lizards, *Biological Conservation* 125: 47–54, 2005.
85. Bonier, F., P.R. Martin, and J.C. Wingfield, Urban birds have broader environmental tolerance, *Biology Letters* 3: 670–673, 2007.
86. Turner, W.R., Interactions among spatial scales constrain species distributions in fragmented urban landscapes, *Ecology and Society* 11: 6, 2006.
87. Goode, M., Personal communication, 2009.
88. Jones, T.R., Personal communication, 2008.
89. Atwood, T.C., H.P. Weeks, and T.M. Gehring, Spatial ecology of Coyotes along a suburban-to-rural gradient, *Journal of Wildlife Management* 68: 1000–1009, 2004.
90. Noss, R., E. Fleishman, D.A. Dellasala, J.M. Fitzgerald, M.R. Gross, M.B. Main, F. Nagle, S.L. O'Malley, and J. Rosales, Priorities for improving the scientific foundation of conservation policy in North America, *Conservation Biology* 23: 825–833, 2009.
91. Redford, K.H. and W.M. Adams, Payment for ecosystem services and the challenge of saving nature, *Conservation Biology* 23: 785–787, 2009.
92. Harllee, B., M. Kim, and M. Nieswiadomy, Political influence on historical ESA listings by state: A count data analysis, *Public Choice* 140: 21–42, 2009.
93. Manring, N.J., The politics of accountability in national forest planning, *Administration and Society* 37: 57–88, 2005.
94. Sayre, N.F., Bad abstractions: Response to Sullivan, Conserv*ation Biology* 23: 1050–1051, 2009.
95. Beier, P. and R.F. Noss, Do habitat corridors provide connectivity? *Conservation Biology* 12: 1241–1252, 1998.
96. Langpap, C. and J. Wu, Predicting the effect of land-use policies on wildlife habitat abundance, *Canadian Journal of Agricultural Economics* 56: 195–217, 2008.
97. Hostetler, M. and D. Drake, Conservation subdivisions: A wildlife perspective, *Landscape and Urban Planning* 90: 95–101, 2009.
98. Lovich, J.E. and D. Bainbridge, Anthropogenic degradation of the southern California desert ecosystem and prospects for natural recovery and restoration, *Environmental Management* 24: 309–326, 1999.
99. Markovchick-Nicholls, L., H.M. Regan, D.H. Deutschman, A. Widyanata, B. Martin, L. Noreke, and T.A. Hunt, Relationships between human disturbance and wildlife land use in urban habitat fragments, *Conservation Biology* 22: 99–109, 2007.
100. Stohlgren, T.J., L.D. Schell, and F.V. Heuvel, How grazing and soil quality affect native and exotic plant diversity in Rocky Mountain grasslands, *Ecological Applications* 9: 45–64, 1999.
101. White, C., The 21st century ranch, *Conservation Biology* 22: 1380–1381, 2008.
102. Brunson, M.W. and L. Huntsinger, Ranching as a conservation strategy: Can old ranchers save the new west? *Rangeland Ecology & Management* 61: 137–147, 2008.
103. List, R., G. Ceballos, C. Curtin, P.J.P. Gogan, J. Pacheco, and J. Truett, Historic distribution and challenges to bison recovery in the northern Chihuahuan desert, *Conservation Biology* 21: 1487–1494, 2007.
104. Raven, P.H. and J.A. McNeely, Biological extinction: Its scope and meaning for us, in L.D. Guruswamy and J.A. McNeely, eds., *Protection of Global Biodiversity: Converging Strategies* (Durham, NC: Duke University Press, 1998, pp. 13–32).
105. Tilman, D., R.M. May, C.L. Lehman, and M.A. Nowak, Habitat destruction and the extinction debt, *Nature* 371: 65–66, 1994.
106. Hanski, I., Habitat destruction and metapopulation dynamics, in S.T.A. Pickett, R.S. Ostfeld, M. Shackak, and G.E. Likens, eds., *The Ecological Basis of Conservation* (New York: Chapman & Hall, 1997, pp. 217–227).
107. Hanski, I.A., A. Moilanen, and M. Gyllenberg, Minimum viable metapopulation size, *American Naturalist* 147: 527–541, 1996.

108. Bascompte, J. and R.V. Solé, Habitat fragmentation and extinction thresholds in spatially explicit models, *Journal of Animal Ecology* 65: 465–473, 1996.
109. Rockstrom, J., W. Steffen, K. Noone, A. Persson, F.S. Chapin, E.F. Lambin, T.M. Lenton et al., A safe operating space for humanity, *Nature* 461: 472–475, 2009.
110. Minckley, W.L., P.C. Marsh, J.E. Deacon, T.E. Dowling, P.W. Hedrick, W.J. Matthews, and G. Mueller, A conservation plan for native fishes of the lower Colorado River, *Bioscience* 53: 219–234, 2003.

# 10

# *Healing the Wounds: An Example from the Sky Islands*

**Dave Foreman, Rurik List, Barbara Dugelby, Jack Humphrey, Robert Howard, and Andy Holdsworth**

## CONTENTS



> One of the penalties of an ecological education is that one lives alone in a world of wounds. … An ecologist must either harden his shell or make believe that the consequences of science are none of his business, or he must be the doctor who sees the marks of death in a community that believes itself well and does not want to be told otherwise.
>
> **Aldo Leopold,** *Round River: From the Journals of Aldo Leopold*, 1972*

Aldo Leopold came to understand land health and ecological wounds from his experience in New Mexico and Arizona from 1909 to 1924 and trips to the Sierra Madre in Chihuahua in the mid-1930s. In 1937, he wrote

> For it is ironical that Chihuahua, with a history and a terrain so strikingly similar to southern New Mexico and Arizona, should present so lovely a picture of ecological health, whereas our own states, plastered as they are with National Forests, National

* See Leopold[1] and also Ehrlich.[2]

Parks and all the other trappings of conservation, are so badly damaged that only tourists and others ecologically color-blind, can look upon them without a feeling of sadness and regret.[3]

## 10.1 Introduction

Far before his time in his ability to wisely read the story of the land, Leopold understood that free Apaches kept settlement out of the northern Sierra Madre Occidental well into the twentieth century. Without livestock grazing and with healthy populations of mountain lions and wolves, mountain ecosystems in Mexico were ecologically healthy, whereas similar mountain ecosystems in the United States were deeply wounded.[3] Unfortunately, since Leopold's time, the mountain vastness of northern Mexico has been as carelessly exploited as the southwestern United States.

In recent years, ecological and historical researchers have greatly improved our understanding of the ecological wounds in the Sky Islands region (see Chapter 7). Even in the best-protected areas, such as national parks and wilderness areas ungrazed by domestic livestock, preexisting wounds may continue to suppurate.[4] For example, without wolves, natural fire, and recovered riparian forests (*bosques*), even the large Gila Wilderness Area is not a healthy landscape; in fact, without restoration, its health may continue to decline.

Efforts to protect the land and create a sustainable human society in the Sky Islands region will come to naught without understanding these wounds and their underlying causes and then attempting to heal them. More than 60 years ago, Aldo Leopold[3] worried that "our own conservation program for the [Sky Islands] region has been in a sense a post-mortem cure." Medicine for the land, or ecological restoration, has advanced much in the last 60 years (or so we trust). Perhaps, we can raise this Lazarus of a landscape to robust good health (Figure 10.1). It is, at the very least, our duty as conservationists to try.

The human history of the Sky Islands region is a litany of anthropogenic wounds to terrestrial and aquatic communities. Even the earliest humans in the region, the Clovis culture of big game hunters, around 13,000 years ago (calendar years or 11,000 uncalibrated radiocarbon years ago) wounded the land by causing the Pleistocene megafauna extinction, in which 33 out of 45 genera of large mammals in North America became extinct.[5] Martin and Burney[6] identify 27 species of mammals larger than 100 lb that became extinct in the western United States and northern Mexico alone at that time. The overwhelming evidence points to human hunting as the major cause. Among the animals lost in the Sky Islands region were mammoths, mastodons, camels, horses, tapirs, shrub oxen, musk oxen, llamas, peccaries, bison, mountain goats, mountain deer, giant ground sloths, glyptodonts, dire wolves, saber-toothed cats, shortfaced bears, American lions, American cheetahs, and giant condors.[5,7] Some authorities, including Paul Martin of the University of Arizona, believe that the plant communities of the region are still in disequilibrium from this loss—an example of a long-festering ecological wound precipitated by the cessation of top-down regulation[6] (see Chapter 6).

With the arrival of Europeans in the Sky Islands region less than 200 years ago (300 years ago for the Santa Cruz Valley), the land again suffered deep and debilitating wounds. Of these ecological wounds, we have identified six as major. Each of these has more than one cause, and several of the causes contribute to more than one wound. The overall impact of these wounds is greater than their sum.

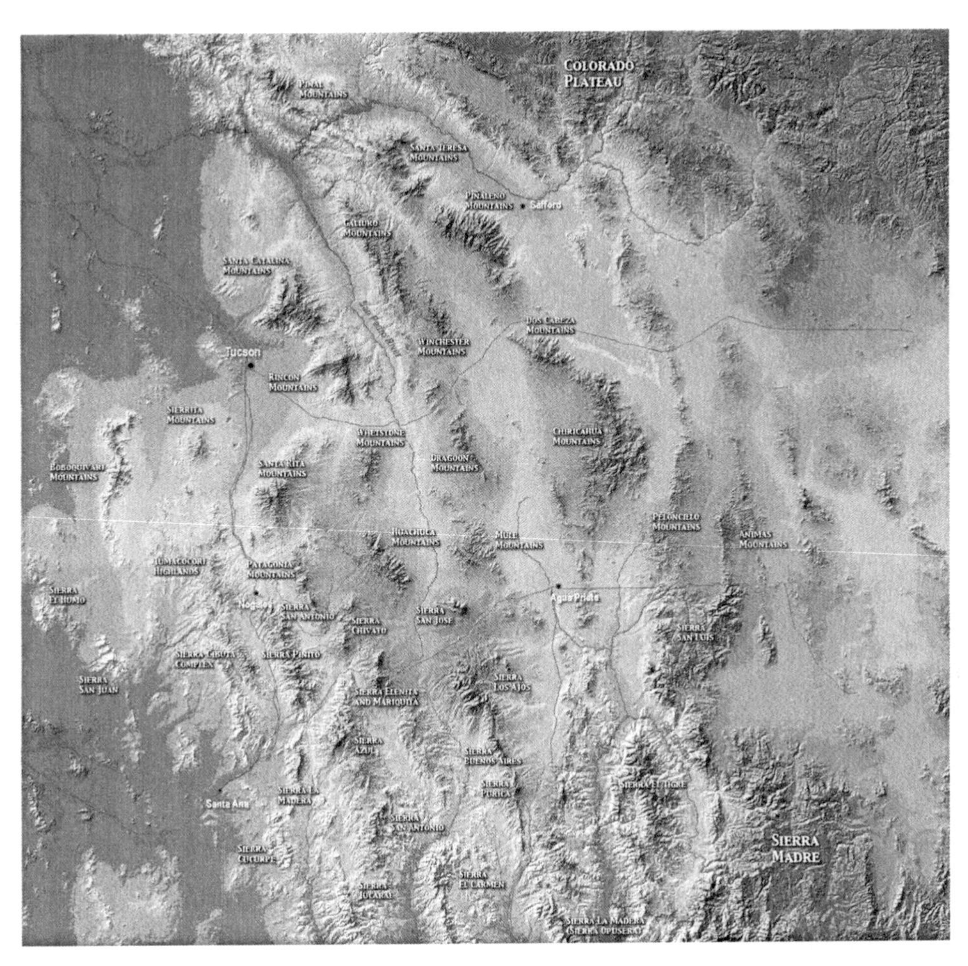


**FIGURE 10.1**
Diagram of Sky Island habitats in the southwest United States and northern Mexico. The habitats are on the higher elevations of the mountain ranges of this region. (Courtesy of Sky Jacobs.)

We will first discuss the major wounds and then we will present the goals and objectives of the Sky Islands Wildlands Network Conservation Plan, which is designed to heal the wounds.

## 10.2 Wounds to the Land

The six major wounds in the Sky Islands/northern Sierra Madre Occidental landscape are as follows:

- Many species of native animals—especially carnivores, large ungulates, and keystone rodents—have been extirpated or greatly reduced in numbers.
- Watersheds, stream channels, and riparian forests have been damaged almost beyond measure.

- Over a century of fire suppression has eliminated a natural disturbance regime vital to the integrity and function of forest, woodland, and grassland ecosystems.
- The region has been fragmented by roads, dams, and other works of civilization, potentially isolating wide-ranging species in nonviable habitat islands.
- Aggressive and disruptive exotic species, both plants and animals, have invaded or been purposefully introduced, threatening ecosystem integrity and the survival of individual species.
- Beginning in the 1870s with cutting for mine timbers, railroad ties, and firewood and continuing to the present day with industrial logging operations, all forest types in the region have been degraded.

Other ecological wounds have occurred as well, but these six are the most pervasive and destructive.

### 10.2.1 Wound 1: Loss of Important Species

Causes: During the preceding 200 years or so, native animals—carnivores, large ungulates, keystone rodents, and other species—have been extirpated or greatly reduced in numbers by (1) trapping, (2) market hunting, (3) competition from domestic livestock, (4) diseases introduced by settlers and domestic livestock, (5) livestock fencing, (6) predator and rodent control, (7) trophy and fur hunting, and (8) transformation of natural habitats for different human uses.

One species, the imperial woodpecker, and two (perhaps three) subspecies are extinct because of hunting, poisoning, trapping, and habitat destruction: Merriam's elk, the Mexican grizzly, and likely the Arizona river otter. In addition, desert bighorn sheep, Rocky Mountain bighorn sheep, pronghorn, and even javelina, mule deer, and Coues white-tailed deer were nearly extirpated around 1900. The bison was probably extirpated, although a handful of survivors may have persisted in northwestern Chihuahua. Except for 20 or so individuals reintroduced recently to the Apache National Forest of Arizona, the Mexican wolf has been extirpated in the wild, although a few individuals may remain in remote areas of the Sierra Madre. Breeding populations of jaguars, ocelots, and jaguarundis were reduced or eliminated in the United States. Mountain lions and black bears also declined sharply. Two keystone rodents—beavers and prairie dogs—suffered tremendous declines (Figure 10.2). Thick-billed parrots and aplomado falcons were extirpated from Arizona and New Mexico. The Tarahumara frog disappeared from the United States by the early 1980s.[8]

American trappers entered the Sky Islands region (then part of newly independent Mexico) in the 1820s.[9] Beavers were abundant in the Gila, Rio Grande, and Little Colorado watersheds. By the 1840s, beavers were functionally extinct in the Sky Islands region, as they were throughout what is now the western United States.[10,11] Market and hide hunters killed off the southern herd of bison in the 1870s.[12] In the Sky Islands, mining camps sprang up in the 1870s, drawing market hunters who slaughtered pronghorn, deer, javelina, bighorn sheep, turkey, and even thick-billed parrots to feed the miners. Authorities on the thick-billed parrot believe that hunting may have been the main cause for its disappearance from the United States.[13] The largest subspecies of elk, Merriam's, was abundant in the Mogollon Highlands (now the Gila and Apache National Forests). This subspecies may have ranged south through the Sky Islands ranges and valleys into Mexico, but reports are inconsistent.[14] They were completely exterminated by hunters: the last few individuals were shot on Fly's Peak in the Chiricahuas in 1906.[12]

**FIGURE 10.2**
Beavers were largely trapped out of wetland habitats in the southwest helping lead to degraded riparian habitats (From Rurik List—beaver picture, Previously unpublished. With permission.) (Courtesy of Rurik List.)

Cattle and sheep ranchers moved into the Sky Islands area in the 1880s, and many encouraged the slaughter of wild ungulates, seeing them as competitors with cattle and sheep for forage. Domestic sheep transmitted diseases to both desert and Rocky Mountain bighorns, causing their near-extinction. Livestock fencing has disrupted the movement of pronghorn to seasonal water sources, leading to their rapid decline and agonizingly slow recovery. Botteri's and rufous-winged sparrows declined sharply because cattle grazing in southern Arizona severely damaged their grassland habitat.[15]

With their natural prey gone, Mexican wolves, Mexican grizzlies, mountain lions, and jaguars turned to cattle and sheep. In the United States, the Department of Agriculture's Predatory Animal and Rodent Control agency used traps, guns, and poison to try to completely exterminate predators, including bobcats, ocelots, and coyotes.[16] By the mid-1930s, grizzlies were extirpated and wolves were functionally extirpated from New Mexico and Arizona.[17,18] Mountain lion populations were greatly reduced. Prairie dogs were functionally exterminated as a result of a taxpayer-sponsored, government poisoning program that continues today. Many ranchers disliked prairie dogs because of the mistaken belief that they damage the range. The black-footed ferret was lost from the region because of the massive decline of prairie dogs.[19] Prairie dogs and predators also fell victim to so-called varmint hunters. Jaguars and ocelots in the United States were shot on sight as valuable trophies or for their fur.

In Mexico, where cattle ranching moved into the mountains later, Mexican wolves, Mexican grizzlies, jaguars, and prairie dogs survived longer.[3] The introduction of the 1080 compound (a powerful "predicide") in the 1950s was the major cause for the decline of wolf populations. The grizzly was a victim of the 1080 campaign against wolves.[20] With their numbers dramatically reduced, traps and guns took care of the surviving individuals. By 1980, the grizzly and wolf were functionally extinct even in Mexico. Large prairie dog towns remain in Chihuahua, although poisoning and conversion of their habitat to irrigated potato fields threaten them. Trophy and fur hunting of jaguars greatly reduced their populations in northern Mexico; they are still heavily hunted as livestock killers.[21]

Subsistence hunting before the 1950s and logging of the forest in the Sierra Madre Occidental of Mexico thereafter was responsible for the extinction of the imperial woodpecker[22] as well as for the decline of the thick-billed parrot and military macaw.

### 10.2.2 Wound 2: Watershed, Stream, and Riparian Damage

Causes: Watersheds, stream channels, and riparian forest (bosques) have been severely damaged by (1) trapping-out of beavers, (2) livestock grazing, (3) water diversions, (4) groundwater pumping, (5) fuelwood cutting, (6) agricultural clearing, and (7) watershed damage from a variety of human activities.

In the arid Sky Islands region, water is generally the limiting resource. Some 80% of vertebrate species in the region are dependent on riparian areas for at least part of their life cycle; over half of these cannot survive without access to riparian areas.[23] In Arizona and New Mexico, more than a hundred federally and state-listed species are associated with cottonwood (*Populus*)–willow (*Salix*) bosques.[23] Over half of the Threatened and Endangered species in the U.S. portion of the Sky Islands region became so because of riparian losses.[24] Arizona and New Mexico have lost 90% of presettlement riparian ecosystems.[25] The Nature Conservancy lists the Fremont cottonwood (*Populus fremontii*)–Goodding willow (*Salix gooddingii*) riparian community as highly imperiled.

In 1870, the total number of cattle in the Arizona Territory was only 5000. By 1891, the population of cattle in the territory had grown to an estimated 1.5 million. In 1870, the cattle population in 17 western states was estimated to be 4–5 million head; by 1890, that had grown to 26.5 million. During this period, great numbers of sheep also grazed the Sky Islands region, and herds of goats were common in some Sky Island ranges.[26] In this grossly overstocked range, thunderstorms carried away the topsoil in sheets, and gully washers turned placid streams into dry arroyos with 40 ft sheer banks. Arizona rancher H.C. Hooker described the San Pedro.

San Pedro river valley in 1870 as "having an abundance of timber with large beds of sacaton and grama grasses. The river bed was shallow and grassy with its banks with luxuriant growth of vegetation." He gave a different description 30 years later, saying that "the river had cut 10 to 40 feet below its banks (Figure 10.3) with its trees and underbrush gone, with the mesas grazed by thousands of horses and cattle."[27] Botanist Tourney[28] wrote, "There are valleys [in the Sky Islands region] over which one can ride for several miles without finding mature grasses sufficient for herbarium specimens without searching under bushes or in similar places." Before 1891, for example, the Santa Rita Mountains south of Tucson had 25,000 cattle and horses and 5,000 sheep grazing in them.[26] Drought struck Arizona and New Mexico in 1891–1893, killing 50%–75% of the total cattle population. "Witnesses stated that a person could stand at one carcass and throw rocks to others nearby."*

Since the cattle crash 100 years ago, herds have built back up in the Sky Islands region. Some desert grasslands were transformed into creosote bush (*Larrea tridentata*) desert by the overgrazing/drought/soil erosion "triple-whammy"; thoughtful observers like rancher Jim Winder believe that some of these areas can never be restored. In naturally occurring, periodic droughts, livestock grazing is even more destructive than otherwise, as cattle will eat everything they can before dying—after which vegetative recovery is nearly impossible. In much of the Sky Islands region, in spite of the improvement from near desertified conditions at the turn of the century, millions of acres of grazing lands remain in only poor or fair condition. Riparian areas are considered by many authorities to be in their worst condition ever. Aldo Leopold[3] wrote, "I sometimes wonder whether semi-arid mountains can be grazed at all without ultimate deterioration." His question remains unanswered.

During early settlement, bosques were heavily cut for fuelwood, fence posts, and mine timbers.[26] This cutting of mesquite (*Prosopis velutina*), cottonwood, willow, and other tree

* Ferguson and Ferguson, "Sacred Cows."

**FIGURE 10.3**
Degraded desert riparian wash showing accelerated streambank erosion.

species degraded wildlife habitat and led to greater erosion of channels. Agricultural clearing along the Gila, San Francisco, Mimbres, San Simon, San Pedro, and Santa Cruz rivers eliminated or degraded the most productive and extensive bosques. Water diversion for irrigation and later for mining, the downcutting of arroyos (lowered streambeds in arroyos intercept ground water at a greater depth, thus drawing the water table down), and groundwater pumping for agriculture, mining, and urban use have lowered the water table, resulting in dried-up *cienegas* (wet meadows), dewatered rivers, and dying bosques. This loss of habitat and degradation of ecological resilience has encouraged the spread of exotic species and the elimination of sensitive native species. Watersheds were damaged not only by livestock grazing, but also by the widespread clearcutting of piñon (*Pinus*), juniper (*Juniperus*), and oak (*Quercus*) woodlands for mining timbers and fuelwood.[26]

In the northern Sierra Madre Occidental of Chihuahua and Sonora, cattle freely graze riparian areas. Especially in the lowlands, where there is little tree cover outside the riparian areas, cattle have limited the growth of new trees, so when the old cottonwoods, sycamores (*Platanus wrightii*), walnuts (*Juglands major*), and other riparian trees die, no young trees replace them. Cattle do similar damage in Arizona and New Mexico.

Another problem in the riparian areas in Mexico is that the river bottoms are often turned into access roads for timber exploitation. Related to this exploitation is the practice of throwing sawdust and other byproducts from the lumberyards into the rivers, which adversely changes the water quality, in turn affecting native fish and other freshwater species.

Too few have heeded Leopold's warning[3]: "Somehow the watercourse is to dry country what the face is to human beauty. Mutilate it and the whole is gone."

### 10.2.3 Wound 3: Elimination of Natural Fire

Causes: A natural disturbance regime vital to the health of forest, woodland, and grassland ecosystems in the Sky Islands region has been largely eliminated by over a century of (1) livestock grazing and (2) fire suppression.

Most ecosystems in the Sky Islands region coevolved with frequent fire. Only the most arid Chihuahuan and Sonoran desert communities in the region are not adapted to regular fire. Before about 1900, most montane forests burned in accordance with the

**FIGURE 10.4**
Wildfire in overstocked pine forests in the southwest have long-lasting impacts to upland habitats. This condition is the result of natural wildfire suppression and other land management actions.

2–7 year wet-dry cycles associated with the El Niño-Southern Oscillation.[29–31] Primitive understandings of the ecological role of natural fire in these ecosystems led the Forest Service and other land managers to aggressively try to put out fires from about 1906 on. In addition to fighting fires, the Forest Service deliberately encouraged overgrazing by cattle and sheep to eliminate grass that carried the natural, cool, ground fires. Increasing numbers of scientists recognized fire's important role by the 1960s, but such ideas were heresy to many foresters and ranchers (see Chapter 11).

The reduction in fire frequency combined with overgrazing by cattle and sheep has allowed woody plants to out-compete grasses (competition from grasses was as significant as fire in keeping pine and juniper stands from becoming too dense and extensive). Consequently, snakeweed (*Gutierrezia sarothrae*), creosote bush, prickly pear (*Opuntia* sp.), cholla (*Opuntia* sp.), catclaw (*Acacia greggii*), mesquite, and piñon–juniper woodland have invaded and replaced grasslands. This has changed the balance of natural ungulates that graze and browse. Forested areas have been extensively degraded by the combination of fire control and overgrazing. By eliminating frequent, cool, ground fires in forests, land managers have allowed the fuel load to build up, thereby creating conditions for destructive conflagrations and crown fires (Figure 10.4).[11,24,26,32–34]

The control of natural fires has decreased their frequency, which has allowed enough time for seedlings to develop into trees large enough to withstand the occasional light surface fires. This has also led to the expansion of forests over grasslands.

### 10.2.4 Wound 4: Fragmentation of Wildlife Habitat

Causes: Wildlife habitat in the region has been fragmented by (1) highways, roads, and vehicle ways; (2) dams, irrigation diversions, and dewatering of streams; (3) destruction and conversion of natural habitat; and (4) other works of civilization, such as urban and ranchette development.[35–37] Fragmentation has severed historic wildlife migration routes and has potentially isolated wide-ranging species in nonviable habitat islands. Expanding human populations and development continue to increase fragmentation (Figure 10.5).

**FIGURE 10.5**
Habitat fragmentation from residential development and in the foreground is a major pipeline that bisects many miles of Arizona landscapes from woodlands, interior chaparral, and into the Sonoran Desert.

At certain scales, isolation of habitats can contribute to native biodiversity. At the landscape or regional scale, the higher elevations of the Sky Island ranges are naturally isolated,[38] permitting genetic divergence and speciation. However, native species using stream and riparian habitats and wide-ranging species such as carnivores, large ungulates, and migratory birds need natural connectivity in the landscape. This natural connectivity has been severed during the last century. Soulé and Terborgh[39] remind us that "connectivity is not just another goal of conservation: it is the natural state of things."

Coolidge Dam on the Gila River, Presa de la Angostura on the Rio Bavispe, and Presa del Novillo on the Rio Yaqui; smaller dams on headwater streams of the Gila, San Francisco, Santa Cruz, Janos, and other rivers; irrigation diversion dams; and dewatered and degraded stretches of once-perennial streams have fragmented the habitat for native fish, amphibians, and aquatic invertebrates. Habitat loss and degradation of bosques have harmed riparian-dependent birds and other species. Habitat for wide-ranging species such as wolf, mountain lion, jaguar, pronghorn, and bighorn has been fragmented by roads, agriculture, and urban, suburban, and ranchette development.

Interstate Highways 10 and 19 are formidable barriers to many kinds of wildlife. Increased traffic on and the proposed widening of Mexico Highway 2 will make it a significant barrier too. Even two-laned paved roads cause many deaths of animals trying to cross. Dirt roads fragment the landscape for wolves, jaguars, and other species vulnerable to opportunistic poaching. For example, at least five released Mexican wolves were shot alongside roads in the Apache National Forest in 1998. Even dirt tracks can fragment the landscape for slow-moving desert tortoises and snakes, especially when many off-road vehicle enthusiasts deliberately run over reptiles for thrills.

In Mexico, public access to private ranches is more open than in the United States, and the access to ejidos (community lands) is practically uncontrolled. Under this situation, roads are a permanent source of poaching. Although the northern Sierra Madre Occidental does not have the industrial and agricultural infrastructure of the southwestern United States, the landscape in Mexico is becoming increasingly fragmented because of growing economic pressure in the region and conversion of natural vegetation to agriculture, often for export products to the U.S. market—all exacerbated by free trade agreements like NAFTA.

### 10.2.5 Wound 5: Invasion of Exotic Species

Causes: Aggressive and disruptive exotic species, both plants and animals, have (1) invaded; (2) escaped from cultivation; or (3) been deliberately introduced, threatening ecosystems and the survival of individual native species.

Conservation biologists now recognize exotic species as a leading cause of extinction, second only to habitat destruction.[40,41] In the Sky Islands region, non-native plants and animals (primarily in aquatic, riparian, and mesic communities) are a major cause of endangerment of native species. Some of these destructive invaders were deliberate introductions; some escaped from cultivation; others hitchhiked in. Most do well in disturbed habitats (Box 10.1).

Tamarisk (salt cedar) (*Tamarix ramossisima*), a native of the Middle East, was planted ornamentally in the late 1800s. It spread through cattle-damaged riparian areas and benefits from dams and flood-control levees, which prevent natural cycles of drying and flooding with which native species evolved. Tamarisk is now a major competitor of native cottonwoods, willows, and other riparian trees (Figure 10.6). It provides very little habitat or food for native species, although it does provide critical interim nesting habitat for the endangered southwestern willow flycatcher in a few areas where native vegetation has been lost. As a phreatophyte, tamarisk sucks up large amounts of water through its roots and transpires this moisture into the air, thereby drying up springs and streams upon which native species depend. Other destructive invader plants include Russian thistle (tumbleweed) (*Salsola iberica*), sweet resin bush (*Europs subcarnosus*),[42] vinca (*Vinca minor*), Bermuda grass (*Cynodon dactylon*), buffel grass (*Cenchrus ciliaris*), Johnson grass (*Sorghum halpense*), and lovegrasses (*Eragrostis* sp.). Warshall[38] reports that over 60 non-native plants have been naturalized in the region. Bowers and McLaughlin[43] report 65 alien plants in the Huachuca Mountains alone.

Rainbow trout (not native to the Southwest) and European brown trout have been deliberately stocked in the high country streams of the Sky Islands region, where they threaten

**FIGURE 10.6**
Salt cedar invasion along a reach of the Gila River in central Arizona. Salt cedar has largely replaced native riparian vegetation resulting in poor habitat quality.

## BOX 10.1 INVASIVE SPECIES

### INVASION OF EXOTIC SPECIES

Biological invasion is considered to be one of the symptoms of global environmental change. Exotic species (from other continents or geographically distant floras) and some native species can become invasive in habitats to which they are not part of the local biota. Invasive species are normally defined as *a non-native plant or animal or other organism whose introduction causes or is likely to cause economic or environmental harm or harm to human health.*

Invasive species displace native species when introduced to habitats where they did not evolve as part of a functionally organized community. Their success is often linked to the lack of natural enemies from their origin ecosystem(s) that are not present in the new habitats to keep them in check. In some cases, aggressive invasive species can literally transform the invaded habitat changing its ecological structure and function. A good example of this would be downy bromegrass (*Bromus tectorum*) in the Great Basin desert where it serves as a fire source and after fire becomes the dominate plant in what was a sagebrush steppe plant community. In the warm deserts, red bromegrass (*B. rubens*) and/or buffelgrass (*Pennisetum cilare*) provide fine fuel, where in the native state, such fuel is rare. After a fire in the invaded sites, these exotic grasses can play a similar role in transforming the Sonoran desert from a community dominated by shrubs and cacti to one largely devoid of these growth forms. The transformed habitat does not support functional groups of native organisms and may exhibit changes in abiotic processes such as modified runoff and stream flow patterns and accelerated soil erosion.

Some exotic plant species were intentionally introduced, such as saltcedar (*T. ramossisima*), Russian olive (*Eleaganus angustifolia*), and buffelgrass for examples. Many potentially exotic species exist as landscape plants in urban and human residential settings in rural areas. Some of these exotic species invade wildlands such as fountain grass (*Pennisetum setaceum*), and some native species are invasive to urban landscapes, such as the shrub, desert broom (*Baccharis sarathroides*). Some invasive animals in the southwestern deserts include crayfish, a fish named gizzard shad, bull frogs, and New Zealand mud snails. These animals invade aquatic habitats and can greatly interfere with native fishes and other native aquatic species. There are invasive organisms, that are unicellular, some vector diseases, and golden algae, produces a toxin to fish.

The majority of introductions were accidental. About 10% of the introduced plants adapt their new habitat and spread (naturalize) while about 1% of the introduced species become ecosystem transformers. Transformer invasive species literally change the structure and function of the habitat they have invaded. Without their natural enemies from their original habitats, invasive species tend to out-compete native species for environmental resources (water and soil nutrients) and have life cycles favoring their growth compared to native species. The ecological traits of invasive plants that make them competitive are largely physiological and reproductive rather than morphological. Invasive plant's effects on reduced ecological biodiversity constitutes a critical concern to managers of desert ecosystems and associated biotic communities of North America.

**John H. Brock**
*Professor Emeritus, Arizona State University Polytechnic*

native Gila and Apache trout and, in the case of rainbows, breed with them, thereby diluting the gene pool. Bass, catfish, sunfish, other game fish, and bullfrogs have been deliberately planted in the Sky Islands region's warm-water streams and reservoirs where they are direct threats to native fish and frogs. Bait fish and crayfish also have spread and threaten aquatic natives. Bullfrogs are the primary threat to native frogs. Rosen et al. state, "In the American Southwest, the native fish fauna is ... facing extinction due primarily to introduced predators and competitors."[44,45] Fifteen non-native fish species are established.[46] Among invertebrates, feral and domesticated honeybees aggressively compete for food with native bees, which may be vital to the pollination of native plants.[47]

In parts of the Sierra Madre, the larger Texas white-tailed deer has been introduced in the range of the smaller Coues white-tailed, with potentially disastrous consequences for the native subspecies through interbreeding. The size difference between the subspecies is such that a female Coues can die while giving birth to a Texan hybrid.[48] European wild boars have been introduced in the Sierra Madre Occidental, competing with the smaller white-collared peccary and damaging the fragile soil of the arid forests of the region.[49]

### 10.2.6 Wound 6: Degradation of Forests and Woodlands

Causes: Degradation of forests is closely related to some of the wounds already discussed, especially Wound 3, elimination of natural fires. Beginning in the 1870s with (1) cutting for mine timbers, railroad ties, and firewood and continuing to the present day with (2) industrial saw timber operations, all forest types in the region have been degraded (Figure 10.7).

Bahre[26] reports that more than 30 mining centers operated in the Arizona portion of the Sky Islands in the late 1800 s. Wood was the sole fuel for the mines and for all other uses. Madrean evergreen woodlands, mesquite bosques, and riparian woodlands were heavily exploited. Bahre also reports that significant saw timber logging occurred in the Graham, Chiricahua, Huachuca, Santa Rita, and Santa Catalina mountains during the late 1800s. A sawmill was located in the Santa Ritas as early as 1857. "Nearly 30 percent of the ponderosa pine (*Pinus ponderosa*) and mixed-conifer forest in the Chiricahuas had been logged by eleven different sawmill operations before 1900."[26]

**FIGURE 10.7**
Timber harvest, especially historic tree cutting, can reduce habitat quality by changing vegetation strata.

Bahre[26] summarizes the early impact on forests:

> None of the sky island evergreen woodlands and forests was pristine before they were set aside as forest reserves and national forests. By 1900, nearly all had been affected to some degree or another by mining, logging, fuelwood cutting, and grazing. At present, we have little idea what these woodlands and forests would be like had they not been logged or grazed, had the fire regimes not been manipulated, or had Forest Service management not occurred.

After World War II, commercial saw timber operations increased on the Gila and Apache National Forests, as they did throughout the National Forest System.[50] Current overstocking of forests was created purposely by the USFS and industry to maximize tree growth for fiber production. They wanted to eliminate old-growth forests and replace them with what they believed were "more efficient young forests".

Old-growth ponderosa pine forests are listed as one of the 21 most endangered ecosystems in the United States (Figure 10.8).[23] For all Arizona and New Mexico National Forests, the Southwest Forest Alliance reports, "About 90 percent of the old-growth has been liquidated, including 98 percent of the old-growth ponderosa pine." Wallace Covington, forestry professor at Northern Arizona University, says, "I've made it clear for 20 years there's been a population crash of old-growth trees—leave the damn things alone." He also writes, "The cumulative effect of old-growth logging, non-native species introductions, overgrazing, predator control, and fire exclusion has been ecosystem simplification so great that Southwestern forest ecosystems are at risk of catastrophic losses of biological diversity."[11,24,51]

Seventy-three percent of the natural forest ecosystems of Chihuahua and Sonora have been severely altered.[52] From the original 23 million acres occupied by old-growth pine-oak forests in Mexico, only 0.6% (41,000 ac) remains.[22] This in turn has led to the decline of species dependent on the old-growth forest, like the extinct imperial woodpecker and the endangered thick-billed parrot and Mexican spotted owl.[53] Nearly all the Sierra Madre Occidental has been logged at some point, and because of this, the present vegetation may be different from the original cover. For example, small oak forests surround large (over 100 ft high) conifer trees, reminders of the forest that once was.

**FIGURE 10.8**
Old growth ponderosa pine is critical habitat for several threatened and/or endangered wildlife species, like the southwest spotted owl and goshawk.

## 10.3 Healing the Wounds

In 1992, Noss* wrote

> A conservation strategy is more likely to succeed if it has clearly defined and scientifically justifiable goals and objectives. Goal setting must be the first step in the conservation process, preceding biological, technical, and political questions of how best to design and manage such systems. Primary goals for ecosystem management should be comprehensive and idealistic so that conservation programs have a vision toward which to strive over the decades. A series of increasingly specific objectives and action plans should follow these goals and be reviewed regularly to assure consistency with primary goals and objectives.

The goals of the Sky Islands Wildlands Network Conservation Plan are based on its mission of healing the ecological wounds of the region. Healing-the-wounds goal-setting also directs the selection of focal species. We have tried to select focal species whose viability or recovery is tied to our six goals. Each of our established six goals is tied to healing a major wound:

Goal 1. Recover all large carnivores and ungulates and other species native to the region.

Goal 2. Restore watersheds, streams, and riparian forests.

Goal 3. Restore a natural fire disturbance regime.

Goal 4. Protect and restore landscape connectivity for wide-ranging species native to the region.

Goal 5. Eliminate or control exotic species.

Goal 6. Protect all remaining native forests and woodlands and restore natural forest conditions.

Objectives are how goals are implemented. Given our goals and approach, we outline our objectives here.

### 10.3.1 Objectives for Goal 1: Recover Native Species

1. Maintain the viability of focal species; this requires large core reserves and landscape connectivity as well as redundancy in the system, owing to probable but unpredictable natural and anthropogenic changes in the future.
2. Protect, recover, or reintroduce all missing or reduced-in-number large and mid-sized carnivores native to the region. These include Mexican wolf, jaguar, ocelot, jaguarundi, river otter, and black-footed ferret.
3. Protect, recover, or reintroduce missing or reduced-in-number ungulates, keystone rodents, and other native species. These include bison, bighorn, elk, beaver, prairie dog, aplomado falcon, thick-billed parrot, southwestern willow flycatcher, and Chiricahua leopard frog.

* Noss, "Wildlands project."

### 10.3.2 Objectives for Goal 2: Protect and Restore Riparian Areas

4. Identify and protect all riparian forest patches, no matter how small.[54]
5. Restore watersheds and watercourses so they can support focal species and maintain regional ecosystem integrity. This restoration program should include removal (or much better management) of exotic species, including cattle, from riparian areas, planting of riparian trees and shrubs, restoration of natural populations of beavers,[11] erosion control structures, and so on.[55]
6. Purchase private lands and bid on federal and state grazing allotments in riparian areas.

### 10.3.3 Objectives for Goal 3: Restore Natural Fire

7. Implement a comprehensive program to restore natural fire to the landscape, while respecting the special requirements of management in wilderness areas (see Chapters 11 and 12).
8. Modify or end domestic livestock grazing so that its role in disrupting natural fire cycles is eliminated or greatly reduced.[51]

### 10.3.4 Objectives for Goal 4: Restore and Protect Connectivity

9. Identify riparian linkages and areas important for wildlife movement.
10. Develop management standards and legal protection for such "corridor" areas.

### 10.3.5 Objectives for Goal 5: Control Exotic Species

11. Implement a comprehensive program to control and mitigate exotic species, including plants and animals such as tamarisk, bullfrogs, rainbow trout, and bass.

### 10.3.6 Objectives for Goal 6: Restore and Protect Native Forests

12. Protect all native forests (old-growth and other generally intact forests) and restore large areas of previously logged or degraded forests so that they recover old-growth characteristics.[24,55] Wilderness and wilderness recovery area designation should be proposed for most of these areas (see Chapter 12).
13. Implement ecological grazing management that allows for restoration of natural forest conditions and processes.[33,55]

These goals and objectives are "clearly defined and scientifically justified" and are based on "a vision toward which to strive over the decades."* However, while the goals and objectives of a conservation plan should be bold, even audacious, they should also be achievable. Ideally, objectives should "specify results to be achieved, specific criteria to measure degree to which results are achieved, time frame for achieving results, [and] target group."[56] For the SIWN Conservation Plan, specific implementation steps address these points. Action plans will be developed for each implementation step.

* Noss, "The Wildlands Project."

We believe that a healing-the-wound approach is an excellent way to analyze conservation problems and to accomplish visionary but achievable goals across a landscape. Healing the wounds is also a powerful metaphor that can move conservationists to action and can inspire the public. Healing ecological wounds can change people from conquerors to plain citizens of the land community.[35] Unless we heal the wounds, we will have a continent "wiped clean of old-growth forests and large carnivores"; we will "live in a continent of weeds."[57]

## References


1. Leopold, A., The Round River—A parable, in *Round River: From the Journals of Aldo Leopold*, pp. 158–165 (New York: Oxford University Press, 1972).
2. Ehrlich, P.R., *A World of Wounds: Ecologists and the Human Dilemma* (Oldendorf/Luhe, Germany: Ecology Institute, 1997).
3. Leopold, A., Conservationist in Mexico, *American Forests* 43: 118–120, 1937.
4. Sydoriak, C.A., C.D. Allen, and B.F. Jacobs, Would ecological landscape restoration make the Bandelier wilderness more or less of a wilderness?, in D.N. Cole, S.F. McCool, W.T. Borrie, and J. O'Loughlin eds. *Wilderness Science in a Time of Change Conference. Volume 5: Wilderness Ecosystems, Threats, and Management*, pp. 209–215 (Missoula, MT., Fort Collins, CO: U.S.D.A., Forest Service, Rocky Mountain Research Station, RMRS-P-15, 2000).
5. Martin, P.S. and R.G. Klein, eds., *Quaternary Extinctions: A Prehistoric Revolution* (Tucson, AZ: The University of Arizona Press, 1984).
6. Martin, P. and D. Burney, Bring back the elephants! *Wild Earth* 9: 57–64, 1999.
7. Ward, P.D., *The Call of Distant Mammoths: Why the Ice Age Mammals Disappeared* (New York: Copernicus, 1999).
8. Sredl, M.J. and J.M. Howland, Conservation and management of Madrean populations of the Chiricahua leopard frog, in F.L. DeBano, P.F. Ffolliott, A. Ortega-Rubio, G.J. Gottfried, R.H. Hamre, and C.B. Edminster, tech. cords., *Biodiversity and Management of the Madrean Archipelago: The Sky Islands of Southwestern United States and Northwestern Mexico*, General Technical Report RM-GTR-264, pp. 379–385 (Fort Collins, CO: U.S. Department of Agriculture, Forest Service, Rocky Mountain Forest and Range Experiment Station, 1994).
9. Hafen, L.R. and C.C. Rister, *Western America: The Exploration, Settlement, and Development of the Region beyond the Mississippi* (Englewood Cliffs, NJ: Prentice-Hall, 1950).
10. Beck, W.A., *New Mexico: A History of Four Centuries* (Norman, OK: University of Oklahoma Press, 1962).
11. Pollock, M.M. and K. Suckling, *Beaver in the American Southwest* (Flagstaff, AZ: The Southwest Forest Alliance, 1998).
12. Matthiessen, P., *Wildlife in America* (New York: Viking, 1987).
13. Snyder, N.F.R., S. Koenig, and T.B. Johnson, Ecological relationships of the thick-billed parrot with the pine forests of southeastern Arizona, in F.L. DeBano, P.F. Ffolliott, A. Ortega-Rubio, G.J. Gottfried, R.H. Hamre, and C.B. Edminster, tech. coords., *Biodiversity and Management of the Madrean Archipelago: The Sky Islands of Southwestern United States and Northwestern Mexico*, General Technical Report RM-GTR-264, p. 288 (Fort Collins, CO: U.S. Department of Agriculture, Forest Service, Rocky Mountain Forest and Range Experiment Station, 1994).
14. Bailey, V., *Mammals of the Southwestern United States* (New York: Dover Publications, 1971).
15. Rising, J.D., *Guide to the Identification and Natural History of the Sparrows of the United States and Canada* (San Diego, CA: Academic Press, 1996).
16. Dunlap, T.R., *Saving America's Wildlife: Ecology and the American Mind, 1850–1990* (Princeton, NJ: Princeton University Press, 1988).

17. Brown, D.E., D.M. Gish, R.T. McBride, G.L. Nunley, and J.F. Scudday, *The Wolf in the Southwest: The Making of an Endangered Species* (Tucson, AZ: The University of Arizona Press, 1984).
18. Brown, D.E., *The Grizzly in the Southwest* (Norman, OK: University of Oklahoma Press, 1985).
19. Miller, B., R.P. Reading, and S. Forrest, *Prairie Night: Black-Footed Ferrets and the Recovery of Endangered Species*, pp. 22–26 (Washington, DC: Smithsonian Institution Press, 1996).
20. McBride, R.T., The Mexican wolf (*Canis lupus baileyi*): A historical review and observations in its status and distribution, Report to U.S. Fish and Wildlife Service, Albuquerque, NM, 1980.
21. Lopez Gonzales, C.A., Personal communication, 1999.
22. Lammertink, J.M., J.A. Rojas Tomé, F.M. Casillas Orona, and R.L. Otto, Situacion y conservacion de los bosques antiguos de pino-encino de la Sierra Madre Occidental y sus aves endémicas, (Seccion Mexicana, Mexico Consejo Internacional para la Preservacion de las Aves, 1997).
23. Noss, R.F. and R.L. Peters, *Endangered Ecosystems of the United States: A Status Report and Plan for Action* (Washington, DC: Defenders of Wildlife, 1995).
24. Suckling, K., *Forests Forever! A Plan to Restore Ecological and Economic Integrity to the Southwest's National Forests and Forest Dependent Communities* (Flagstaff, AZ: The Southwest Forest Alliance, 1996).
25. Noss, R.F., E.T. LaRoe III, and J.M. Scott, Endangered ecosystems of the United States: A preliminary assessment of loss and degradation, Biological Report 28 (Washington, DC: U.S. Department of the Interior, 1995).
26. Bahre, C.J., Late 19th century human impacts on the woodlands and forests of southeastern Arizona's Sky Islands, *Desert Plants*, 14: 8–21, June 1998.
27. Johnson, S., Learning to miss what we never knew, *The Home Range*, Predator Project, Summer, 1997.
28. Tourney, J.W., *Overstocking the Range* (Tucson, AZ: University of Arizona Agricultural Experiment Station Bulletin 2, 1891).
29. Swetnam, T.W. and J.L. Betancourt, Fire-southern oscillation relations in the southwestern United States, *Science* 249: 1017–1020, 1990.
30. Swetnam, T.W. and J.L. Betancourt, Mesoscale disturbance and ecological response to decadal climatic variability in the American southwest, *Journal of Climate* 11: 3128–3147, 1998.
31. Swetnam, T.W. and C.H. Baisan, Fire histories of montane forests in the Madrean borderlands, in P.F. Ffolliott, L.F. DeBano, M.B. Baker, G.J. Gottfried, G. Solis-Garza, C.B. Edminster, D.G. Neary, L.S. Allen, and R.H. Hamre, tech. cords., *Effects of Fire on Madrean Province Ecosystems*, General Technical Report, RM-GTR-289, pp. 15–36 (Fort Collins, CO: U.S. Department of Agriculture, Forest Service, 1996).
32. Humphrey, R.R., *The Desert Grassland* (Tucson, AZ: The University of Arizona Press, 1958).
33. Morgan, D. and K. Suckling, *Grazing is the Major Cause of Forest Health Problems in Southwestern Forests* (Flagstaff, AZ: The Southwest Forest Alliance, 1995).
34. Fule, P.Z. and W.W. Covington, Comparisons of fire regimes and stand structures in unharvested Petran and Madrean Pine Forests, in F.L. DeBano, P.F. Ffolliott, A. Ortega-Rubio, G.J. Gottfried, R.H. Hamre, and C.B. Edminster, tech. coords., *Biodiversity and Management of the Madrean Archipelago: The Sky Islands of Southwestern United States and Northwestern Mexico*, Rocky Mountain Forest and Range Experiment Station, General Technical Report RM-GTR-264, pp. 408–415 (Fort Collins, CO: U.S. Department of Agriculture, Forest Service, 1994).
35. Leopold, A., *A Sand County Almanac* (Oxford, U.K.: Oxford University Press, 1949).
36. Fisher, R.F., M.J. Jenkins, and W.F. Fisher, Fire and the prairie forest mosaic of Devils Tower National Monument, *American Midland Naturalist* 117: 250–257, 1987.
37. Houston, M.A., *Biological Diversity: The Coexistence of Species on Changing Landscape* (Cambridge, U.K.: Cambridge University Press, 1994).
38. Warshall, P., The Madrean sky island archipelago: A planetary overview, in F.L. DeBano, P.F. Folliott, A. Ortega-Rubio, G.J. Gottfried, R.H. Hamre, and C.B. Edminster, tech. coords., *Biodiversity and Management of the Madrean Archipelago: The Sky Islands of Southwestern United*

*States and Northwestern Mexico*, Rocky Mountain Forest and Range Experiment Station, General Technical Report RM-GTR-264, pp. 379–385 (Fort Collins, CO: U.S. Department of Agriculture, Forest Service, 1994).

39. Soulé, M.E. and J. Terborgh, The policy and science of regional conservation, in Soulé, M.E. and J. Terborgh, eds., *Continental Conservation: Scientific Foundations of Regional Reserve Network*, p. 12 (Washington, DC: Island Press, 1999).
40. Wilcove, D.S., D. Rothstein, J. Dubow, A. Philips, and E. Losos, Quantifying threats to imperiled species in the United States, *BioScience* 48: 607–615, 1998.
41. The Wilderness Society, *The Wilderness Act Handbook* (Washington, DC: The Wilderness Society, 1998).
42. Pierson, E.A. and J.K. McAuliffe, Characteristics and consequences of invasion by sweet resin bush into the arid southwestern United States, in F.L. DeBano, P.F. Ffolliott, A. Ortega-Rubio, G.J. Gottfried, R.H. Hamre, and C.B. Edminster, tech. coords., *Biodiversity and Management of the Madrean Archipelago: The Sky Islands of Southwestern United States and Northwestern Mexico*, Rocky Mountain Forest and Range Experiment Station, General Technical Report RM-GTR-264, pp. 219–230 (Fort Collins, CO: U.S. Department of Agriculture, Forest Service, 1994).
43. Bowers, J.E. and S.P. McLaughlin, Flora of the Huachuca Mountains, Cochise County, Arizona, in F.L. DeBano, P.F. Ffolliott, A. Ortega-Rubio, G.J. Gottfried, R.H. Hamre, and C.B. Edminster, tech. coords., *Biodiversity and Management of the Madrean Archipelago: The Sky Islands of Southwestern United States and Northwestern Mexico*, Rocky Mountain Forest and Range Experiment Station, General Technical Report RM-GTR-264F, pp. 135–143 (Fort Collins, CO: U.S. Department of Agriculture, Forest Service, 1994).
44. Rosen, P.C., C.R. Schwalbe, D.A. Parizek Jr., P.A. Holm, and C.H. Lowe, Introduced aquatic vertebrates in the Chihuahua region: Effects on declining native ranid frogs, in F.L. DeBano, P.F. Ffolliott, A. Ortega-Rubio, G.J. Gottfried, R.H. Hamre, and C.B. Edminster, tech. coords., *Biodiversity and Management of the Madrean Archipelago: The Sky Islands of Southwestern United States and Northwestern Mexico*, Rocky Mountain Forest and Range Experiment Station, General Technical Report RM-GTR-264, pp. 251–261 (Fort Collins, CO: U.S. Department of Agriculture, Forest Service, 1994).
45. Rinne, J.N., Sky Island aquatic resources: Habitats and refugia for native fishes, in F.L. DeBano, P.F. Ffolliott, A. Ortega-Rubio, G.J. Gottfried, R.H. Hamre, and C.B. Edminster, tech. coords., *Biodiversity and Management of the Madrean Archipelago: The Sky Islands of Southwestern United States and Northwestern Mexico*, Rocky Mountain Forest and Range Experiment Station, General Technical Report RM-GTR-264, pp. 351–360 (Fort Collins, CO: U.S. Department of Agriculture, Forest Service, 1994).
46. Warshall, P., Southwestern sky island ecosystems, in E.T. LaRoe, G.S. Farris, C.E. Pucket, P.D. Doran, and M.J. Mac, eds., *Our Living Resources: A Report to the National on the Distribution, Abundance, and Health of US Plants, Animals, and Ecosystems* (Washington, DC: U.S. Department of Interior, National Biological Service, 1995).
47. Buchmann, S.L., Diversity and importance of native bees from the Arizona/Mexico Madrean archipelago, in F.L. DeBano, P.F. Ffolliott, A. Ortega-Rubio, G.J. Gottfried, R.H. Hamre, and C.B. Edminster, tech. cords., *Biodiversity and Management of the Madrean Archipelago: The Sky Islands of Southwestern United States and Northwestern Mexico*, Rocky Mountain Forest and Range Experiment Station, General Technical Report RM-GTR-264, pp. 301–310 (Fort Collins, CO: U.S. Department of Agriculture, Forest Service, 1994).
48. Weber, M. and C. Galindo-Leal, Istocia en vandao cola blanca: Informe de un caso reincidente, *Veterinaria México* 23: 79–81, 1992.
49. Galindo-Leal, C. and M. Weber, El venado de la Sierra Madre Occidental: Ecología, manajo y conservacilin (Mexico: EDICUSA-CONABIO, 1998).
50. Clary, D.A., *Timber and the Forest Service* (Lawrence, KS: University Press of Kansas, 1987).
51. Suckling, K., Fire & forest ecosystem health in the American southwest: A brief primer, (Flagstaff, AZ: Southwest Forest Alliance, 1996).

52. Flores-Villela, O. and P. Gerez Fernández, Patrimonio vivo de México: Un dignóstico de la diversidad biológica (Mexico: Conservation International, 1989).
53. Lammertink, J.M. and R.L. Otto, Report on fieldwork in the Rio Bavispe/Sierra Tabaco area of northern Sonora in November–December 1996 (1997).
54. Skagen, S.K., C.P. Melcher, W.H. Howe, and F.I. Knopf, Comparative use of riparian corridors and oases by migrating birds in southeast Arizona, *Conservation Biology* 12: 896–909, 1998.
55. Simberloff, D.J., D. Doak, M. Groom, S. Trombulak, A. Dobson, S. Gatewood, M.E. Soulé, M. Gilpin, C. Martinez del Rio, and L. Mills, Regional and continental restoration, in M.E. Soulé and J. Terborgh, eds., *Continental Conservation: Scientific Foundations of Regional Reserve Networks* (Washington, DC: Island Press, 1999).
56. Arthur Carhart National Wilderness Training Center, A unified national strategic plan for wilderness education: Framework for development (Arthur Carhart National Wilderness Training Center, Missoula, MT, April 1999).
57. Terborgh, J. and M.E. Soulé, Why we need megareserves: Large-scale reserve networks and how to design them, in M.E. Soule and J. Terbourg, eds., *Continental Conservation: Scientific Foundations of Regional Reserve Networks*, p. 199 (Washington, DC: Island Press, 1999).

# 11

## *Built to Burn*

**Stephen J. Pyne**

**CONTENTS**



### 11.1 Introduction

Earth, air, water, fire—these are truly the basics of the Southwest's elemental landscapes. But among them fire is the oddity because it is not a substance but a reaction, and while the others shape the character of life, fire is a creation of life. In peculiar but powerful ways, it is biologically constructed.

The Southwest is built to burn. The fundamental rhythms of fire are set by a cadence of wetting and burning. It has to be wet enough to grow combustibles and dry enough to ready them to actually combust (see Chapter 4). The Southwest's deserts thus burn after wet winters that fluff up the landscape with grasses and forbs; its mountains burn amid droughts that leach away moisture from forests (see Chapters 7 and 8). Add to this the need for a spark to initiate the reaction. The Southwest has plenty in the form of lightning.[1] In this regard, it isn't the number of flashes that matter but their relative dryness. Early monsoon storms often have their rain evaporate before it strikes the ground while lightning suffers no such loss. Look at a map of lightning flashes, and you will find the Southwest well on the margins. Look at a map of lightning-kindled fires, and you will find it at the epicenter (Figure 11.1).[2]

To these factors knead in its rugged terrain. There is, it seems, always some place dry and some place where lightning is temporarily segregated from rain. It all makes for a complex geography of fire, full of niches, of quirky topography and odd cadences, amid broad regional rhythms that adds up to an abundance of fires, mostly small, but occasionally large, that can reside amid tiny terrains or ramify across the region. Year in and year out, something is always available to burn, and something nearly always does.

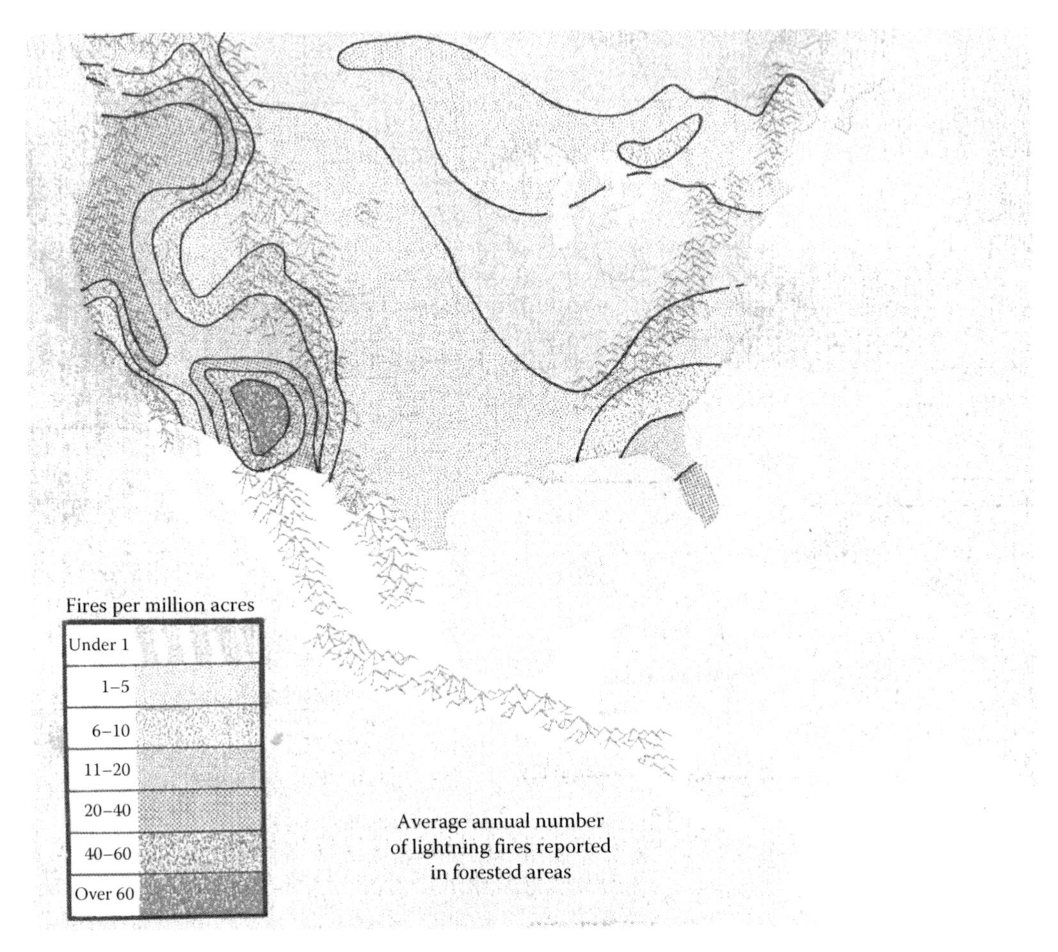


**FIGURE 11.1**
The geography of lightning fires. Note that the Southwest is the national epicenter, with a secondary focus in Florida. (From Schroeder, M.J. and Buck, C., *Fire Weather*, Government Printing Office, Washington, DC, Agriculture Handbook 360, 1970.)

## 11.2 Humans and Fire in North America

This natural matrix, however, has evolved with people present. They carried fire with them and used it to make their world more habitable. They used fire in hearths to cook, heat, light, to work wood and stone, to produce smoke to ward off insects. They broadcast-burned the landscape to help them hunt, freshen spring fodder, and assist with foraging. They burned for wood rats along Colorado River tules and for deer amid pine steppes. In this enterprise, they had biotic allies; the extinction of megafauna encouraged more browse and pasture for fire, which further leveraged the power of the torch. They burned pinyon (*Pinus edulis*)-dominated landscapes to help harvest pine nuts. They burned small plots for gardens. They used fire to signal, fire for ambush and war, fire for ceremony. And they littered the landscapes they inhabited or traversed with campfires and the odd spark. They kept a land ever ready to burn ever simmering with fire.[3]

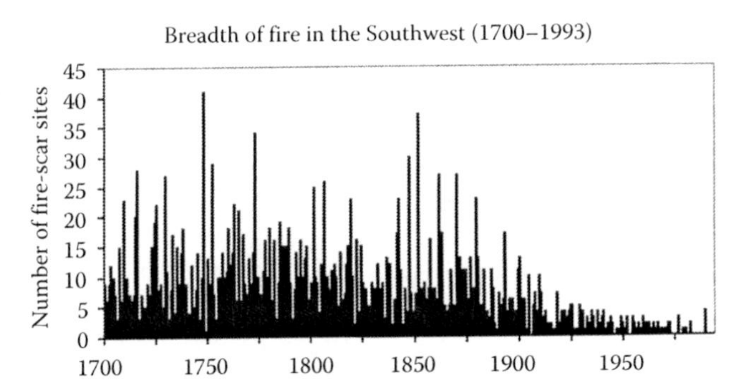


**FIGURE 11.2**
Until the end of the nineteenth century, fires ebbed and flowed with climatic tides. Then, overgrazing, the removal of indigenous burners, and active fire suppression—all a product of settlement powered by industrialization—caused a full-blown recession. The graph shows the breadth of sites holding fire-scarred trees. (Data from Laboratory of Tree-Ring Research, University of Arizona.)

This pyric geography bent when Europeans and their American offspring arrived, and by the 1870s it was breaking. The newcomers reintroduced megafauna in the form of sheep, cattle, and horses, which slow-combusted with their metabolism what open burning had fast-combusted. Once overgrazing set in, they stripped the primary fuels that had carried flame (see Chapter 10). The newcomers also removed the aboriginal fire-starters through war, relocation, and introduced disease; a prominent source of chronic ignition vanished with them. Beginning in the 1890s, they then set aside vast chunks of the land as forest reserves that had as a primary charge to remove fire of all sorts. This change in fire's regime changed also the regimes of earth, water, and air. As Aldo Leopold wrote in 1924, "When the cattle came, the grass went, the fires diminished, and erosion began".[4]

Fire receded, then collapsed, paradoxically creating an ecological insurgency that has grown uncontrollably over the past few decades (Figure 11.2). Fire's ecological power was as great withdrawn as it was applied. Initially this was not obvious: observers saw only an ebb of flame, which for them measured the success of their ecological stewardship. They did not appreciate that humanity's role as keep of the flame applied to landscapes as well as to hearths. They converted their technology and combustion economy to one based on burning fossil biomass, and used those internal-combustion fire engines to help hold the line against free-burning flame.

And for decades the consequences remained unseen. Then, stoked by combustibles that were no longer routinely flushed away by frequent burning, big fires returned (Figure 11.3). But both feral fire and lost fire were a phenomenon of distant wildlands. Then people decided to move their cities against and into those wildlands. The result has created a visible fire crisis to match the invisible one sequestered in the woods and brush.

The big picture is easy enough to grasp. For several decades Americans have been recolonizing their once rural lands. Satellite photos of settlement in Breckenridge, Colorado, look surprisingly like those from Rhondonia in southwestern Amazonia. The American newcomers, however, do not live off the land, only on it. They do not graze, prune, plow, slash, plant, or burn. They come from cities and carve small exurban enclaves out of abandoned farmland or platted ranchettes (see Chapter 10). In the eastern United States, the outcome dapples the countryside with patches of subdivisions and woods, cloying perhaps but not intrinsically volatile. They are routinely blasted by wind and water, with vast damages—the ice storms of 2003, for example, acting like a kind of frozen

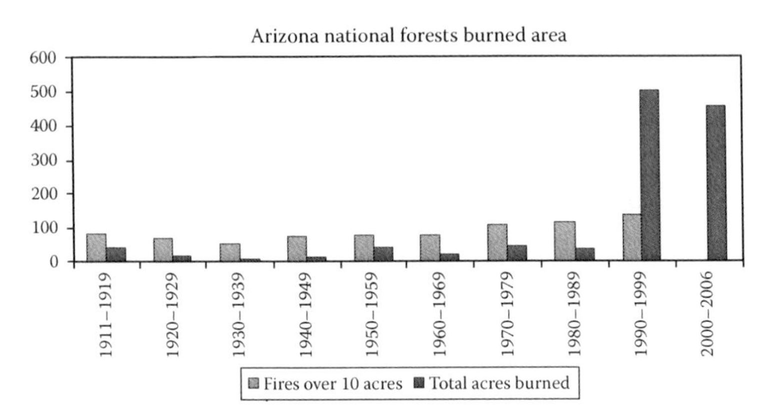


**FIGURE 11.3**
After declining to minimum in the 1930s, both large fires and burned area have increased, dramatically in the past two decades. The reasons are several, including drought, availability of wildland fuels, changes in land use, and reforms in fire policy and practice that encourage more burning. (Data from U.S. Forest Service, Region 3.)

fire, and hurricanes battering barrier islands. In the West, the resulting landscape quilt stitches houses to fire-prone public wildland. Such places are primed to burn.[5]

The urban and the wild—their compound is a kind of environmental nitroglycerine, and when shaken by drought, wind, or spark, they explode. Fire is not alone: sprawl interbreeds with whatever indigenous hazards exist; but fire is the most visible. Over the past two decades the number of structures burned in this intermixed zone (or "wildland/urban interface," as officialdom prefers to call it) has escalated, the irrational exuberance of homeowners having helped the NASDAQ Nineties to create a bull market for burning. The subprime loans that fueled Wall Street's conflagration in 2008 expressed themselves equally on the land. Even more, those enclaves have projected a vast fire-protectorate of urban-centered values across the countryside (e.g., exurbanites particularly loath smoke). In the nineteenth century, Bernhard Fernow denounced America's rural fire scene as one of "bad habits and loose morals." Today we might restate him to read one of "bad habits and loose money."*

## 11.3 Wildfires

Since 1990 the issue has dominated the national discourse on fire. California looms over the national statistics: It stands by itself in the economics of fire losses and costs. With 85% of the houses burned in the United States since 1990, California is to fire what Florida is to hurricanes. When politicians and pundits speak of America's "fire problem," this is usually what they mean, and it is why fire matters just now to the public at large, even though its practical domain lies within a single state (Figures 11.4 and 11.5).[5]

Still, the Southwest is in the thick of it. Its mix of public and private lands makes "sprawl," a perhaps less useful descriptor than "splash." The recolonizing supernova is blasting exurban enclaves from the Sonoran Desert to the Mogollon Rim to the flanks of Mount

* Fernow, quoted in Rodgers III.[6]

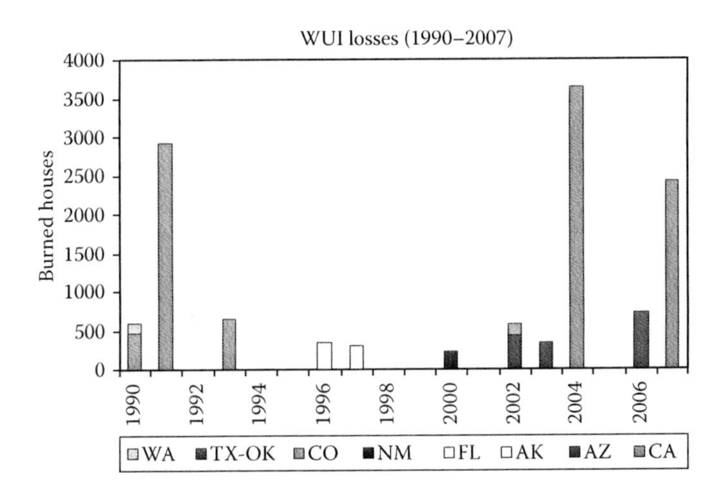


**FIGURE 11.4**
National losses from burned houses from 1990 to 2007. (From Cohen, J.D., *For. Hist. Today*, Fall, 20, 2008.)

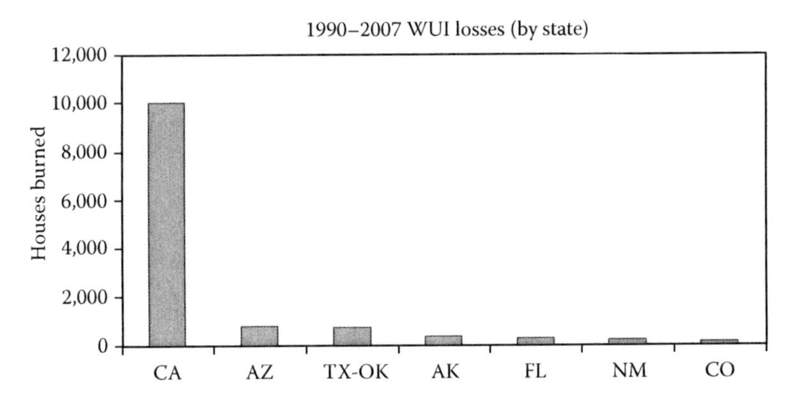


**FIGURE 11.5**
WUI losses by state. Note the dominance of California. (From Cohen, J.D., *For. Hist. Today*, Fall, 20, 2008.)

Humphreys. Some such clusters are at risk from fire, and some, not. Desert fires—flame tearing through exotic grasses and ephemerals flushed with winter rains—may cause fright and prove annoying, and its smoke may smudge paint or damage drapes, but such fires are unlikely to rip through cinder-block or stucco houses with tile roofing. Similarly, away from the advancing frontier, the thickening settlement deepens the zone of protection, which becomes a landscape of fire exclusion. The hazard lies where the two meet and abrade.

The hazard thus changes not only in space but in time: It is the advancing edge—the flaming front of settlement, as it were—that holds the greatest threats. This zone moves with the thrust of construction, but also with the evolution of fuel buildup among the vegetation. Moreover, transitional eras are always the most dangerous: The land is neither fish nor fowl, but held in perpetual, unstable suspension between two states of being. This is unarguably the condition of the intermix zone, as such lands leap from a rural frying pan into an exurban fire. The intermix fire happens in those places and at those times when the peri-urban confronts the quasi-wild.

In the Southwest, this means those sites where dense patches of natural combustibles meet equally dense patches of human-erected combustibles. Log or wood-frame houses are, after all, a reconstituted forest and burn according to similar principles. This collision can occur in chaparral, desert, and grassland, but granted the availability of land and the relative attractiveness of second- or retirement-home sites, it happens mostly in montane forests, particularly those dominated by ponderosa pine (*Pinus ponderosa*).

## 11.4 Wildfire Management

This is a dumb problem to have because technical solutions exist. They begin with the house itself. Banning wood-shingle roofs, attention to simple yard maintenance around structures, installing hydrants, the application of some basic codes for construction and zoning—such measures would eliminate the worst of the situation. The resulting landscape would be less that of a cabin in the woods than of woods in fragmented city. Such places require the techniques—modified, adapted—of urban fire protection.*

A similar logic applies also to the flame-threatened fringe of communities, where an evolving compromise is pointing to thinning-driven fuelbreaks as a means of protection. Is this sensible? It is, if it addresses the particulars of how a house actually burns. Conduction burns structures when flame makes contact. Clearing away the space immediately around a house will break the continuity that allows this transfer. Radiation kindles by immersing combustible material in heat. The distance needed to shield a structure depends on how intense the source flame is and how readily the object ignites. (The intensity varies with the square of the distance, such that small changes can yield big differences, which is why it is so hard to find the right seating distance around a campfire.) Crown fire experiments in Canada recommend 100–200 ft as a minimum distance, which is probably a maximum anywhere. This refers to a tree-enveloping sheet of flame blasting its heat against a wooden structure. Radiant heat from smaller caches of fuel will shrink the zone of danger. Planting weakly flammable vegetation, and eliminating flaky decoration or needle-drenched roofs will expand the zone of safety. Similarly, the proximity of structures one to another matters. As lots fill up with big houses, detached garages, and assorted sheds, one structure can radiate against its too-proximate others, spreading the fire directly from one erected lumber pile to another.[7]

The greater problem is convection, and more broadly, wind, because it carries sparks. Over short stretches, ember showers can saturate a site with new starts. But firebrands can also travel long distances, which argues for protection not only at the house but over broad areas, and they blow about well after the flaming front has passed, which means someone has to be there to swat them out. Studies in both the United States and Australia have shown that many houses have burned not in the thermal wave of a fire tsunami, but later, from small flames that crept to a flammable deck or porch or from windblown sparks that found tiny points of tinder. Had someone been on the scene they could have stopped them with a squirt gun and a whisk broom.

All parties agree that house protection begins with the house itself. A wooden roof is lethal (this particular threat has been known for more than 10,000 years). Other features like eaves can either cache or discourage combustibles and sparks. The arrangement of houses matters, the fire flaming from one to another, or from roof to roof without resort of ground vegetation, leading to the curious spectacle of burned houses amid green

* http://www.firewise.org (accessed July 15, 2009).

(a)

(b)

**FIGURE 11.6**
Burned houses (a and b) at Summerhaven, Arizona, showing structures completely consumed while surrounding (dense) forest is unburned; the fires spread along the surface, or between houses, not through the coniferous canopy. (From Cohen, J.D., An examination of the Summerhaven, Arizona home destruction related to the local wildland fire behavior during the June 2003 Aspen fire, Intermountain Fire Sciences Laboratory, unpublished report, 2003.)

landscapes. Investigations into the Aspen fire that gnawed into Summerhaven on the top of Mt. Lemmon in Arizona suggest that most conifer crown burning was the result of ignition from burning houses, not vice versa, while between houses, fire spread on surface litter (Figure 11.6). In wildlands, the zone of protection must extend outward, an aura known as defensible space. This near-landscaping need not be stripped, only sculpted to dampen fire's ability to creep into, radiate toward, or hurl embers at the house. All parties agree that this is properly the duty of a homeowner, not only to himself but to his neighbors. The shouting begins when defensible space is expanded to the community itself, particularly when a hamlet abuts against public land because it effectively extends the influence of private landholdings into the public domain and becomes subject to national politics.[8]

One proposed solution is to adopt fuelbreaks as a kind of fire levee that can keep the flow of wildfire from overspilling into communities. On this subject, the United States has considerable experience, with mixed lessons for community protection. The core reason for ambiguity is that large fires are large events; they can swallow whole swathes of landscapes. Slivers of thinned fuels—the fuelbreak as moat—will not halt the big fire that most threatens a reserve or hamlet. Fuelbreaks work best when they are built into the design of landscapes, not retrofitted. They function nicely in pine or teak plantations, for example, when constructed as part of the original layout. They work poorly when imposed on mature forests. They are, moreover, temporary features. They reduce an immediate hazard, but cannot hold forever. Broad corridors (and roads) slashed through Oregon's Tillamook Burn helped break the continuity of fire-killed snags, but only until the cycle of reburns ceased and the mountains were replanted. Plowed and fired fuelbreaks through the Nebraska Sandhills helped shatter the near-annual flow of prairie fires, but overgrew after the pine plantations ripened. Fuelbreaks require maintenance, and once the crisis has passed, reluctance at the expense and labor of annually weeding, cutting, and burning overwhelms the project. (Clearing fuelbreaks was one of the early uses for which Agent Orange was developed.) Fuelbreaks are, that is, transient devices that work best at the onset of a project for lands of considerable value.[5]

Several grand experiments have attempted to install truly massive arrays of fuelbreaks; interestingly, all have been in California. The Ponderosa Way was a 650-mile-long fuelbreak that spanned the entire west slope of the Sierra Nevada. Built with the bottomless labor of the Civilian Conservation Corps (CCC), is sought to segregate permanently the lower-elevation chaparral from the higher-level conifers, a Maginot Line of fire protection (and a weird counterpoint to the New Deal's Shelterbelt tree-planting scheme on the Plains). When the CCC camps left, the Way went with them. Later, several experiments in "conflagration control" designed broader fuelbreaks, along ridgelines, both in the dense-conifered Sierras and the chaparral-clothed mountains of Southern California. The conifer model involved selective thinning (not scalping). These proved expensive, however, and failed during the extreme events that they were intended to staunch. After the 1970 fires, a network of fuelbreaks was constructed along the mountains of the Los Angeles Basin, swinging from ridge to ridge like a Great Wall. While they have their value for access, the control of minor fires, and firefighter safety, they cannot alone halt a major fire. They still have to work with suppression forces, and an all-out conflagration will fling sparks across the barrier as readily as over rock outcrops.[9]

What the intermix scene demands is a much broader scope for defensible space, though it not be to the same standards as adjacent to a house. What the scene needs, considered on a landscape scale, is not a fuelbreak but a fire greenbelt. It needs something on the scale of a golf course, not a moat. The width and character of such a greenbelt will depend on the properties of large fires in that setting, but probably anything less than a mile will prove doubtful, and a mile and a half is a more reasonable scale. The purpose is to break the momentum of a crown fire and the saltation of spotting, the process by which wind-blown sparks rekindle new fires well in advance of the nominal front. The scheme is less a seawall than a series of speed bumps. No one enters a residential neighborhood from an interstate freeway—no one brakes from 75 to 25 mph—within 200 ft. The exit occurs in a graduated series of slowing speeds. So it should be with fire greenbelts.[10]

The intensity of the landscaping will increase as it approaches the village. There is no reason to nuke the woods: the purpose is not to stop fire cold, by paving a surrounding lagoon of asphalt, but to force the flames out of the canopy and onto ground and then, by offering only lightly textured combustibles, to tame the fire into something controllable. There will be fire; there will be a need for firefighters; there may well be some houses lost, the outcome of poor housekeeping or bad luck. But firefighters could stand against such flames, and the community will enjoy a reasonable degree of protection.

Or they could if there were enough of them. If anything has become crystal clear over the past decade it is that professional firefighters cannot do the job adequately. They are too few in numbers and too scattered to mass together at critical points during the fire eruptions. It will always be impossible to have enough, along with engines and apparatus, to muster during the first outbreaks or on extended tours. The only force that can protect homes is, finally, homeowners. The Community Fireguard programs devised in Australia show how to prepare a local populace to stay and defend, or if they choose to leave (as many should), when and how to do so. The current American practice—mass evacuations, partial protection of structures, maximum risk to firefighters—is too idiotic to continue indefinitely. But for any fire protection to work, the structure must be defensible, and the surrounding landscape in a form that will allow a fire militia to hold against an approaching firefront.*

The greenbelts could well become recreational sites, wildland parks, suitable for picnics and nature walks; they could be regularly maintained by burning, at least in some locales;

* http://www.cfa.vic.gov.au/residents/programs/cfg.htm (accessed July 15, 2009).

in places where the existing forest is a shambles, they might well improve biotic health and biodiversity (see Chapters 10 and 12). They would offer an opportunity for constructive landscaping, almost certainly an ecological improvement over tangles of pine-jungles and a soil paved in conifer needles. Such a program will be neither cheap nor simple. Probably, though, we could come to some consensus, community by community, about how to do it.

All sides, however, will quickly look beyond that penumbral border. The deeper problem will not end at the hamlet's shadow. Sooner or later we will have to pursue it into the backcountry; not everywhere, but in enough critical sites to matter. Advocates of a changing-the-combustibility strategy will see in the fire greenbelts a demonstration of how that projection might be done, and why. Critics will worry that, once launched, those fidgeting hands and conniving minds will not exercise the same level of care and planning and will blast recklessly into the land beyond. They may become corridors for human traffic of all kinds, like highways into the Amazon. Trailer parks, trophy homes, and casinos may not be far behind. The concept of a fire greenbelt, that is, may prove a tough nut, not because the projects are intrinsically problematic but because of what, to various imaginations, they represent for the future.

It is unlikely that the intermix scene is simply a contemporary fad. Americans are creating a new kind of landscape, a postmodern pastoral, something neither urban nor wild nor rural (see Chapter 18). Like strip malls, these landscapes are destined to become permanent features that won't give way to "real" architecture. They are what they are and can be done well or poorly. Such communities need reasonable standards for fire codes, not simply the jeers and *Schadenfreude* of the chattering classes, eager to gloat over the spectacle of NASDAQ millionaires trying to protect log-plated trophy homes with garden sprinklers. And unless we want our intermix milieu to look like the parking lots of big box retailers and woody trailer slums, we need to think about landscaping; settings that are fire-safe but bio-friendly. We need an esthetic for ourselves in the scene. All of which argues that, once the transitional phase has passed, a new kind of fire will exist, more or less permanently, and that the fire community should think about what kind of institutions and practices are suitable to cope with it.

The trick is to remember that not all biomass is fuel, and that all fuels are parts of a biota. Fire is more than a tool, like a candle, or a "process," like a flood: It is an ecological catalyst, a kind of biotic defibrillator. A flood or earthquake can occur without a molecule of life present. A fire can't. We are not hammering and sawing fuelbeds; we're massaging ecosystems. We are creating habitats; for fire, for ourselves, for the other creatures with whom we share the site. If we get fire right, we will probably get much of the rest of our stewardship right as well.

To its credit the fire community early identified the intermix issue and is succeeding in taming it. This seems counterintuitive: The burning houses and evacuated towns crowd TV screens every summer and fall. But the public has heard the message, new communities are incorporating fire safety into their design, and the crisis, stoked by a long drought and a bull market, is cresting. The most stubborn problems involve those communities created early in the movement and under the worst circumstances; it is this backlog that is most vulnerable. (While California in the early 1990s legislated against shake-shingle roofs in new construction, the Cedar fire of 2003 took those Scripps Ranch houses not yet retrofitted to the new code, and the Sayre fire of 2008 blasted through trailer parks.)

If we do what we already know needs doing, and are lucky, the intermix fire wave could pass within the next 5 years, and it is plausible that, within another decade, the intermix fire scene will be sufficiently domesticated to no longer pose a startling challenge. It will take its place beside urban, wildland, and rural fire, overlapping with them all but with its own distinctive character. It will become a nuisance rather than a nemesis. And we will inhabit a landscape that better reconciles how we live with nature with how we say we would like to live, in a place not simply infested with threats, bristling with protective countermeasures, and occasionally aflame with feral fire, but one aglow with promise.

## 11.5 Conclusion

That promise might look like this. Instead of hardened physical countermeasures, we would search out biological controls that will help dampen the hazard in ecologically friendly ways. Instead of DMZs between the wild and the urban, we might occupy a shared habitat, a site suitable for fire as for elk and wolves, hummingbirds and chipmunks. In place of either walking away from fire or attempting to beat it into submission, we reclaim our heritage as keeper of the flame and see that the right kind of fire gets to the right places at the right times. Not a likely outcome, but an honorable and sensible one, and the only kind of truce imaginable in what will otherwise become an endless war on nature—and worse, one of our own contriving.

## References

1. Schroeder, M.J. and C. Buck, *Fire Weather* (Washington, DC: Government Printing Office, Agriculture Handbook 360, 1970).
2. Krammes, J.S., Tech. Coord., Effects of fire management of southwestern natural resources (U.S. Forest Service, General Technical Report, RM-191, 1990).
3. Pyne, S.J., *Fire in America: A Cultural History of Wildland and Rural Fire* (Seattle, WA: University of Washington Press, 1997).
4. Leopold, A., Grass, brush, timber and fire in southern Arizona, *Journal of Forestry*, 22: 1–10, 1924.
5. Cohen, J.D., The wildland-urban interface fire problem: A consequence of the fire exclusion paradigm, *Forest History Today*, Fall: 20–26, 2008.
6. Rodgers III, D., *Bernard Eduard Fernow: A Story of North American Forestry* (New York: Hafner, 1968), p. 167.
7. Cohen, J.D., Relating flame radiation to home ignition using modeling and experimental crown fires, *Canadian Journal of Forest Research*, 34: 1616–1626, 2004.
8. Cohen, J.D., An examination of the Summerhaven, Arizona home destruction related to the local wildland fire behavior during the June 2003 Aspen fire (Intermountain Fire Sciences Laboratory, unpublished report, 2003).
9. Pyne, S.J., P. Andrews, and R. Laven, *Introduction to Wildland Fire*, 2nd edn. (New York: Wiley, 1996).
10. Pyne, S.J., *Tending Fire. Coping with America's Wildland Fires* (Washington, DC: Island Press, 2004).

# 12

# *Restoring Ecosystem Health in Frequent-Fire Forests of the American West**

**William Wallace Covington**

**CONTENTS**



## 12.1 Introduction

We are at a fork in the road in the American West. Down one fork lies burned-out, depauperate forest landscapes—landscapes that will be a liability for current and future generations. Down the other fork lies healthy, diverse, sustaining forest landscapes—landscapes that will bring multiple benefits for generations to come. Our present inaction is taking us down the path to unhealthy forest landscapes that are costly to manage. Scientifically based forest restoration treatments, including thinning and prescribed burning, will set us on the path to healthy forested landscapes—landscapes like the early settlers and explorers saw in the late 1800s.

Knowing what we now know, it would be grossly negligent for us not to move forward with large-scale, restoration-based fuel treatments in the dry forests of the western United States. Inaction is now the greatest threat to the long-term sustainability of these ecosystems. It is time for ecologists, natural resource professionals, and others with relevant expertise to bring coherent, objective facts, and informed recommendations to the public and to national, regional, and local decision-makers.

This chapter presents my thinking on how to reverse the trend of increasing catastrophic wildfires in the dry forests of the West. I contend that we can help these forests recover their self-regulatory mechanisms, conserve biological diversity, and improve human habitat values by implementing science-based ecological restoration treatments.

## 12.2 Clear Thinking Is Essential

What is needed today is clear thinking. Fuzzy thinking can be a major threat to marshalling the nation's resources to address the critical problem in time to prevent catastrophic losses that will affect future generations. Elementary logic suggests that clear problem definition, explicitly stated premises, and collection and analysis of relevant facts are essential and should lead to the development, implementation, monitoring, and ongoing evaluation of a range of feasible solutions. Logic also cautions us against being misled by logical dodges, faulty premises, and faulty arguments. For example, recent debates about limiting restoration-based fuel treatments to the urban-wildland interface zone, whether to set limits to the size of trees that can be thinned, and whether or not to utilize thinned trees are rife with game-playing that exploits the full range of false logic, misleading facts, and obfuscation. It is time for that to stop. Such unethical, position-based "negotiation" and inflammatory rhetoric only increases the likelihood of continued ecosystem-scale destruction of the western forests.

## 12.3 To Whom Should We Be Listening?

In the public debate about what to do about declining ecosystem health, it is sometimes difficult to figure out how much credence to give to various pronouncements. David L. Sackett, in his foreword to William A. Silverman's book, *Where's the Evidence?: Debates in Modern Medicine*,[1] suggests that we can more clearly judge proponents and critics by the amount of personal risk they have in the situation at-hand. He suggests that the ideas and feelings of proponents or critics deserve our attention when those people live where events are played out on a daily basis. He cautions about giving credence to those who have not even bothered to visit the front lines. Analyzing propositions and criticisms from such individuals requires judgment as to whether they know what they are talking about and, if so, whether what they propose or criticize is informed by knowledge, or is even remotely feasible.

Sackett also suggests that we can judge proponents and critics by the way they handle themselves in public debates. He writes, "Those who focus on ideas rather than their advocates, and treat those with whom they disagree as worthy individuals who just might be right, deserve our most careful and serious study." He suggests that those who cannot disparage an idea without devaluing the person who proposes it, deserve one of three fates: "...simply being ignored, being employed for slightly drunken after-dinner light entertainment..., or serving as subjects in studies of the psychopathology of academe."

Ultimately, special attention must be given to those who have to live with the outcome of the decisions being made, whether those decisions are about medical treatment for humans or medical treatment for ecosystems. It is all too easy to engage in ideologically based arguments from afar. When a person lives with the land, the outcomes are not ideological or theoretical—they are real, they are tangible, and they make a huge difference in the quality of the land and the quality of all life, including that of humans.

## 12.4 What We Know and What It Means

In her authoritative book, *Thinking Like a Mountain: Aldo Leopold and the Evolution of an Ecological Attitude Toward Deer, Wolves, and Forests*, Susan Flader[2] writes:

Ecology, he [Leopold] realized, was among the most complex of the sciences and might therefore be the last to achieve the state of predictable reactions. Yet, committed as he was to deep-digging ecological research, he was equally convinced that the ecologist had a responsibility to "step beyond 'science' in the narrow sense" and offer modes of guidance for meeting ecological problems that were not yet fully understood.... [Leopold] "offered a rueful definition of 'conservation' as 'as series of ecological predictions made by laymen because ecologists have failed to offer any'." He was pleading for ecological predictions by ecologists, whether or not the time was ripe. "If we wait, he warned, there will not be enough health land left to even define health." (p. 207)

We have come to understand quite a bit about ponderosa pine (*Pinus ponderosa*) forests since Leopold penned those words in 1933. Throughout this 70 year period, warnings from ecologists have become more strident as symptoms of forest ecosystem disease have increased. These reports include, but are not limited to, notes about significant increases of woody plant populations in Arizona and New Mexico,[3] increasing forest insect and disease infestation,[4] crashes of native biological diversity,[5] stagnation of nutrient cycling,[6] and the increasing size and severity of crownfires (Figure 12.1).[7]

Ecologists, the public, and public officials now recognize that there is a serious problem—a crisis in our western forests. What is not agreed upon is how best to solve the

**FIGURE 12.1**
Permanent photo point from Pearson Natural Area Ecological Restoration Study Area, 1993. There were 22.8 trees per acre in this area in 1876, and 1253.5 trees per acre in 1993. In 1993, there was 28.3 tons per acre of fuel, and 120 lb per acre of herbaceous production. (From Covington, W.W., *Ecol. Restor.*, 21, 7, 2003. Copyright by the Board of Regents of the University of Wisconsin System. Reproduced by the permission of the University of Wisconsin Press.)

problem. I submit that there are four general statements on which to begin a dialog about this situation. They are as follows: (1) we are faced with a complex, but understandable problem; (2) we believe that ecological restoration provides a solution to this problem; (3) there are benefits and challenges to implementing ecological restoration on dry forests in the West; and (4) we need to act swiftly but with great care and with the best available knowledge and forethought of the consequences of both our actions and inactions.

### 12.4.1 A Complex, but Understandable Problem

We now know from historical ecology studies that frequent fires were typical of pre-European settlement ponderosa pine and dry mixed conifer forests in the western United States. These studies also indicate that the current overcrowded stands of trees do not sustain the diversity of wildlife and plants that existed a century ago. Moreover, today's large, catastrophic, stand-replacing fires, which are natural in chaparral, lodgepole pine (*Pinus contorta*), and spruce-fir (*Picea-Abies*) forests, are a major ecological threat to the integrity and sustainability of frequent-fire forest types. In addition, research dating from the 1940s to the present indicates that prescribed burning and mechanical thinning in combination with raking heavy fuels from the base of old-growth trees can rapidly restore ecosystem health.[8]

From an ecological perspective, we know that the areas that support these dry forest ecosystems have had, and probably always will have, periodic droughts. The problem is not droughty conditions, but the unprecedented high levels of fuel in the forests. The problem is about too few old-growth trees and far too many younger trees. We also recognize that crownfires in these systems are symptomatic, like unnatural epidemics of pine beetles and other insect infestations, of failing ecosystem health. This failing health can also be seen in the loss of native biodiversity, the decline of watershed functions, and increased erosion and sedimentation.

From a social perspective, we know that small-scale fuel reduction projects (40 ac stands or a quarter-mile strip around a town) are not the answer. The problem concerns landscape-scale, overgrown forested ecosystems that are no longer sustainable and which represent a danger to present and future generations. We also recognize that there is increasing pressure to build homes in the urban-wildland zone, and that these buildings and their occupants are often in the path of dangerous wildfires.

### 12.4.2 Ecological Restoration Represents a Solution

Ecological restoration of dry western forests is pretty straightforward. It involves the following: (1) retaining trees that predate European settlement, (2) retaining post-European settlement trees that are needed to reestablish pre-settlement stand structure, (3) thinning and removing most of the excess trees, (4) raking heavy fuels from the bases of the remaining trees, (5) conducting prescribed burns to emulate the natural disturbance regime, (6) seeding or planting native species, (7) controlling exotic plants where necessary, and (8) restoring meadows, seeps, and springs (Figure 12.2). Naturally, this work should be done in a systematic and scientifically rigorous fashion.

In terms of removing trees, ecological restoration is about thinning (cutting selected trees) that will release larger old-growth and open the understory for native species as opposed to logging (cutting trees for the highest yield of a commodity). Restoration activities should always be focused on working at the landscape or watershed scale in

**FIGURE 12.2**
Same permanent photo point from Pearson Natural Area Ecological Restoration Study Area, 1997. Four years after thinning to 60 trees per acre, raking and prescribed burning, there was 6 tons per acre of fuel and 400 lb per acre of herbaceous production, thus effectively moving the fuel load from the trees and shrubs to the ground layer where fires burn cooler and are more easily controlled. (From Covington, W.W., *Ecol. Restor.*, 21, 7, 2003. Copyright by the Board of Regents of the University of Wisconsin System. Reproduced by the permission of the University of Wisconsin Press.)

order to preserve and enhance critical habitat for humans, other animals, and plants. In dealing with the forests of the American West, restorationists need to think about landscapes that typically cover 100,000–1,000,000 ac—large pieces of land that include not only wildlands but also human communities.

For those living in the urban-wildland zone, the evidence seems pretty clear. They can protect their properties by building with fire-resistant materials, planting fire-resistant landscapes, thinning excess trees close to their homes, and avoiding building in or near forest sites with high fuel loads.

### 12.4.3 Benefits and Challenges of Restoring Western Dry Forests

The benefits of restoring the dry western forests are many; the challenges are great. The benefits include: (1) improved watershed function and sustainability; (2) enhanced native plant and animal biodiversity, including threatened and endangered species; (3) elimination of unnatural forest insect and disease outbreaks; (4) enhanced natural beauty of the land; (5) improved resource values for humans, now and into the future; and (6) economic benefits from jobs and goods created by restoration activities.

Several things could hamper restoration efforts, not the least of which is short-term thinking. Specifically, we know that the restoration of these forests will be expensive in the short term, although its benefits will accrue as time progresses. Counterproductive, position-based negotiating stances employed by some are now standing in the way of moving forward with implementing restoration treatments. This kind of situation results in political maneuvering about setting iron-clad management prescriptions, such as one-size-fits-all diameter caps, that can interfere with cost-effective, ecologically sound restoration efforts. Finally, restoration efforts will be hampered if we fail to make sure that trees are being removed principally for the purpose of restoring natural forest patterns and processes, not economic gain.

### 12.4.4 Acting Swiftly, but Prudently

We need to act swiftly, but with great care, to solve this forest crisis. If we do not, even greater problems will be left for our children and grandchildren. The best way to do this is to follow a scientifically rigorous, environmentally responsible, and socially and politically sound approach. Collaborative approaches where local interests are well represented are most likely to result in projects which meet these criteria.

We must work to understand the underlying disease and treat it at all levels. Piecemeal solutions will only treat the symptoms and not the causes. We also must work quickly to engage all the scientific and political will possible because time is not on our side. According to a U.S. General Accounting Office report,[9] the country is already spending more than 90% of its fire suppression monies fighting fires in ponderosa pine and dry mixed conifer forests in the West. Scientific evidence supports the prediction that if we do nothing, the number, size, severity, and costs of wildfires in these dry western forests will only increase. This problem could be further compounded if large-scale infestations of pine beetle occur along with catastrophic fires, as they already have in northern Arizona.

## 12.5 Recommendations

With these ideas in mind, I would like to make the following recommendations:

*Design treatments starting with solid science, set standards for effectiveness, and measure progress*:

Research indicates that alternative fuel reduction treatments have strikingly different consequences not just for fire behavior but also for biodiversity, wildlife habitat, tree vigor, and forest health. Treatment design should be based on maintaining overall ecosystem health and reducing catastrophic fire. Science-based guidelines should be developed and become the foundation for treatments. In addition, they should be the criteria for evaluating the effectiveness of treatments. Guidelines will help managers design, implement, and monitor restoration treatments, and provide a base of certainty to those distrustful of land management agencies. The standard should be clear—if a treatment does not permit the safe reintroduction of fire and simultaneously facilitate the restoration of the forest, it is not a solution.

*Reduce conflict by using an adaptive management framework to design, implement, and improve treatments*:

We can wait no longer. Solutions to catastrophic wildfire must be tested and refined in a "learning while doing" mode.[10] Two of the barriers preventing the implementation of landscape scale treatments are the unrealistic desire for scientific certainty and a fear that once an action is selected it becomes a permanent precedent for future management. In the first place, scientific certainty will never exist. Second, the past century of forest management demonstrates the need for applied research and active adaptation of management approaches using current knowledge. We should expand our environmental review process to provide approval of a series of iterative treatments provided they are science based, actively monitored, and committed to building from lessons learned and new information.

*Rebuild public trust in land management agencies by continuing to support a broad variety of partnership approaches for planning and implementing restoration-based fuel treatments*:

The lack of trust that exists between some members of the public and land management agencies is the genesis for obstructionist actions. A fundamental way to rebuild trust is to develop meaningful collaborations between the agencies, communities, and the public. There are emerging models of various forms of collaborative partnerships (for example, the Greater Flagstaff Forest Partnership in Arizona, the Applegate Partnership in Oregon, the Four Corners Sustainable Forests Partnership in Arizona, Colorado, New Mexico, and Utah) working to reduce the threat of fire while restoring the forest for its full suite of values. Success of such collaborations depends on meaningful community collaboration, human and financial resources, and adequate scientific support to make well-informed management recommendations. Congress, federal agencies, universities, and non-governmental organizations must support these communities to help them achieve success.

The time to act is now. Ecological restoration provides the solid foundation for helping these damaged ecosystems recover, and adaptive management is the only viable approach for dealing with the forest health crisis. Together they will allow us to move forward at the pace and at the scale that we must, learning as we go.

## References

1. Sackett, D.L., Foreword, in W.A. Silverman, ed., *Where's the Evidence?: Debates in Modern Medicine* (New York: Oxford University Press, 1998).
2. Flader, S., *Thinking Like a Mountain: Aldo Leopold and the Evolution of an Ecological Attitude Toward Deer, Wolves, and Forests* (Columbia, MO: University of Missouri Press, 1974).
3. Leopold, A., Grass, brush, timber, and fire in southern Arizona, *Journal of Forestry* 22: 1–10, 1924.
4. Weaver, H., Fire as an ecological factor in the southwestern pine forests, *Journal of Forestry* 49: 93–98, 1995.
5. Cooper, C.F., Changes in vegetation, structure, and growth of southwestern pine forests since white settlement, *Ecological Monographs* 30: 129–164, 1960.
6. Covington, W.W. and S.S. Sackett, The effects of a prescribed burn in Southwestern ponderosa pine on organic matter and nutrients in woody debris and forest floor, *Forest Science* 30: 183–192, 1984.
7. Covington, W.W., R.L. Everett, R.W. Steele, L.I. Irwin, T.A. Daer, and A.N.D. Auclair, Historical and anticipated changes in forest ecosystems of the Inland West of the United States, *Sustainable Forestry* 2: 13–63, 1994.
8. Covington, W.W., P.Z. Fulé, M.M. Moore, S.C. Hart, T.E. Kolb, J.N. Mast, S.S. Sackett, and M.R. Wagner, Restoring ecosystem health in ponderosa pine forests of the Southwest, *Journal of Forestry* 95: 23–29, 1997.
9. General Accounting Office, Western forests: Catastrophic fires threaten resources and communities, Report GAO T-RCED-98-273, 23 pp. (Washington, DC: United States Government Printing Office, 1998).
10. Walters, C.J. and C.S. Holling, Large-scale management experiments and learning by doing, *Ecology* 71: 2060–2068, 1990.
11. Covington, W.W., Restoring ecosystems health in frequent-fire forests of the American West, *Ecological Restoration* 21: 7–11, 2003.

# Part III

# Desert Planning

## Richard A. Malloy

The scarcity of water and extreme temperature that shaped the southwestern desert environment have created a fragile balance between life and death for anything trying to sustain life. Therefore, the idea of planning under these conditions is to find manners to protect life and provide the means for civilizations to grow without adverse harm to this population. The question is "how can you do this?" Ancient civilizations of the southwest (Hohokam, Anasazi) and elsewhere around the world have disappeared without a clear understanding of what caused the collapse of these communities. Was the force of nature the cause or was it a lack of planning or understanding of sustainable measures needed to survive under these conditions of scarcity? For modern civilizations, it is important to understand the past to bring continuity to measures to develop the future in desert communities.

To that end, we begin with Frederick R. Steiner's chapter, "Ecological Planning Method." This chapter provides the reader with an easy-to-follow guide to the steps of the ecological planning process. The method outlined by Steiner is based on Ian McHarg's landmark work in landscape ecological planning. The ecological planning method outlined here attempts to bring together the best available scientific data and social science methodologies that can lead to the development of a landscape plan. This plan will involve the community and takes into account multiple environmental constraints and opportunities identified through the planning process.

Chapter 14, "Phoenix as Everycity: A Closer Look at Sprawl in the Desert," by Sandy Bahr, Renée Guillory, and Chad Campbell presents an overview on urban sprawl and urban growth management as it applies to the city of Phoenix, Arizona. The Southwest is growing rapidly, and the result of this growth is not always in the best interest of the established residents of the city. Sprawl often bypasses existing urban infrastructure to accommodate new development on the fringe. The effects of urban sprawl have a negative impact on the quality of life for urban residents, including urban environmental quality, traffic congestion, longer commuting times, access to recreational opportunities, and added financial burdens to pay for the infrastructure required to deal with the consequences of these developments. The authors discuss several options for communities to consider going forward to mitigate the adverse impacts of sprawl, which include a wide variety of tools for cities to use as planning tools for sustainable growth management.

Sharon B. Megdal and Joanna B. Nadeau take on the very important task of providing us in Chapter 15, "Water Planning for Growing Southwest Communities," with an overview of water planning concerns for Southwest communities. The authors begin with some macro planning concerns for the region and some of the major water engineering projects along the Colorado and Rio Grande Rivers. This chapter addresses the critical need for communities to develop water planning goals and conservation measures for drought management and potential shortfalls in the local or regional water portfolio. The involvement of critical stakeholders of water users is identified as a critical element of the decision-making process concerning water issues.

The American West is home to a large amount of the country's natural resources. In Chapter 16, "Removable and Place-Based Economies," Kim Sorvig evokes Aldo Leopold's "land ethic" in response to the concern about natural resources that can be extracted and transported to support distant and disconnected areas. Sorvig talks about amenity communities: Those geared to developing a quality of life for the residents pose an alternative economic base for places that can leverage the place-based qualities the land has to offer. Extractive industries such as mining or mineral operations can have an impact on the community. As the markets for these resources fluctuate or dry up, so does the means to support a viable community.

Poor and disadvantage communities have many barriers in front of them to reach a dignified quality of life. This struggle can be exacerbated by public planning or community development that encourages a disproportionate burden of risk from environmental toxins from industries or facilities that pose health or safety risks to the local residents. In Chapter 17, "Environmental Injustice in the Urban Southwest," Bolin et al. present a study of south Phoenix using census data and other public records on toxic materials to identify the adverse impacts posed by environmental toxins, particularly on the poor and minority communities. These disadvantaged groups are often without a voice within the political arena and have trouble getting issues of concern put on the agenda for consideration.

Connecting communities with nature should be a real consideration for development projects, but this is not always the reality. Geoffrey Frasz presents an interesting critique of three communities in the Las Vegas, Nevada area in Chapter 18, "Dwelling in Expanded Biotic Communities." Frasz, a philosopher, provides a theoretical foundation for what a community should strive to achieve for its residents. He then outlines the good points found in master-planned communities and the elements that leave the resident disconnected from the natural environment and other people. He also makes some interesting comments on how people can more effectively dwell within a biotic community.

Richard A. Malloy evokes a stirring call to open a dialogue on development in Chapter 19, "Dialogue on Development." Communities need development to sustain the needs of its residents. It would behoove everyone to come together in areas of common concern, setting aside critical lands for protection, effective water planning to support new development, and sensible measure of using existing public infrastructure to avoid undo costs to the public. Development should not drive cities in a race to the bottom on public giveaways of tax incentives, sales tax revenues, or development impact fees. Rather, developers should have a responsive government planning partner in areas where development is not controversial or out of accord with community interests as a whole.

# 13

# *Ecological Planning Method**

**Frederick R. Steiner**

**CONTENTS**



## 13.1 Introduction

What is meant by ecological planning? Planning is a process that uses scientific and technical information for considering and reaching consensus on a range of choices. Ecology is the study of the relationship of all living things, including people, to their biological and physical environments. Ecological planning then may be defined as the use of biophysical and sociocultural information to suggest opportunities and constraints for decision making about the use of the landscape. Or, as defined by Ian McHarg, the approach "whereby a region is understood as a biophysical and social process comprehensible through the operation of laws and time. This can be reinterpreted as having explicit opportunities and constraints for any particular human use. A survey will reveal the most fit locations and processes."[1]

McHarg has summarized a framework for ecological planning in the following way:

> All systems aspire to survival and success. This state can be described as synthropic-fitness-health. Its antithesis is entropic-misfitness-morbidity. To achieve the first state

* Adapted from *The Living Landscape*, by Frederick Steiner. Copyright © 2008 Frederick Steiner. Reproduced with permission from Island Press, Washington, DC.

> requires systems to find the fittest environment, adapt it and themselves. Fitness of an environment for a system is defined as that requiring the minimum of work and adaptation. Fitness and fitting are indications of health and the process of fitness is health giving. The quest for fitness is entitled adaptation. Of all the instrumentalities available for man for successful adaptation, cultural adaptation in general and planning in particular, appear to be the most direct and efficacious for maintaining and enhancing human health and well-being.[2]

Arthur Johnson explained the central principle of this theory in the following way:

> The fittest environment for any organism, artifact, natural and social ecosystem, is that environment which provides the [energy] needed to sustain the health or well-being of the organism/artifact/ecosystem. Such an approach is not limited by scale. It may be applied to locating plants within a garden as well as to the development of a nation.[3]

The ecological planning method is primarily a procedure for studying the biophysical and sociocultural systems of a place to reveal where specific land uses may be best practiced. As Ian McHarg summarized repeatedly in his writings and in many public presentations:

> The method defines the best areas for a potential land use at the convergence of all or most of the factors deemed propitious for the use in the absence of all or most detrimental conditions. Areas meeting this standard are deemed intrinsically suitable for the land use under consideration.

## 13.2 Steps in the Ecological Planning Method

As presented in Figure 13.1, there are 11 interacting steps. An issue or group of related issues is identified by a community—that is, some collection of people—in step 1. These issues are problematic or present an opportunity to the people or the environment of an area. A goal(s) is then established in step 2 to address the problem(s). Next, in steps 3 and 4, inventories and analyses of biophysical and sociocultural processes are conducted, first at a larger level, such as a drainage basin or an appropriate regional unit of government, and second at a more specific level, such as a watershed or a local government.

In step 5, detailed studies are made that link the inventory and analysis information to the problem(s) and goal(s). Suitability analyses are one such type of detailed study. Step 6 involves the development of concepts and options. A landscape plan is then derived from these concepts in step 7. Throughout the process, a systematic educational and citizen involvement effort occurs. Such involvement is important in each step but especially so in step 8 when the plan is explained to the affected public. In step 9, detailed designs are made that are specific at the individual land-user or site level.

These designs and the plan are implemented in step 10. In step 11, the plan is administered. The heavier arrows in Figure 13.1 indicate the flow from step 1 to step 11. Smaller arrows between each step suggest a feedback system whereby each step can modify the previous step and, in turn, change from the subsequent step. Additional arrows indicate other possible modifications through the process. For instance, detailed studies of a planning area (step 5) may lead to the identification of new problems or opportunities or the amendment of goals (steps 1 and 2). Detailed designs (step 9) may change the landscape plan, and so on. Once the process is complete and the plan is being administered and monitored (step 11), the view of

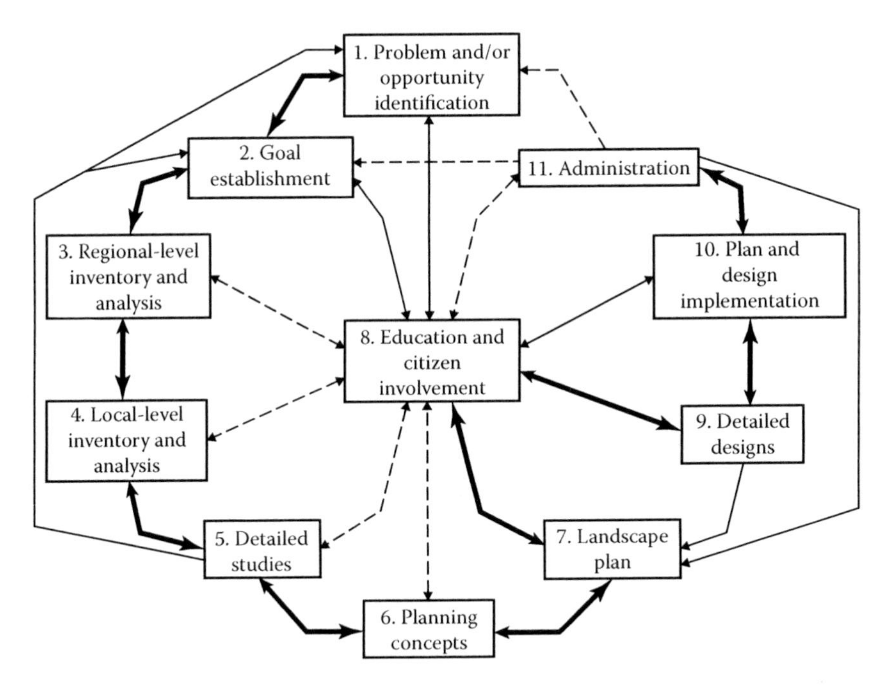


**FIGURE 13.1**
Ecological planning model. (Adapted from Steiner, F., *The Living Landscape*, pp. 3–24. Copyright 2008 Frederick Steiner. Reproduced by permission of Island Press, Washington, DC.)

the problems and opportunities facing the region and the goals to address these problems and opportunities may be altered, as is indicated by the dashed lines in Figure 13.1.

This process is adapted from the conventional planning process and its many variations* as well as those suggested specifically for landscape planning.[10–14] Unlike some of these other planning processes, design plays an important role in this method. Each step in the process contributes to and is affected by a plan and implementing measures, which may be the official controls of the planning area. The plan and implementing measures may be viewed as the results of the process, although products may be generated from each step. The approach to ecological planning developed by McHarg at the University of Pennsylvania differs slightly from the one presented here. The Pennsylvania, or McHarg, model places a greater emphasis on inventory, analysis, and synthesis. This one places more emphasis on the establishment of goals, implementation, administration, and public participation, yet does attempt to do so in an ecological manner.

Ecological planning is fundamental for sustainable development. The best-known definition of sustainable development was promulgated by the World Commission on Environment and Development, known as the Bruntland Commission, as that which "meets the needs of the present without compromising the ability of future generations to meet their own needs."[15] A more recent definition was provided by the National Commission on the Environment, which has defined sustainable development as

> a strategy for improving the quality of life while preserving the environmental potential for the future, of living off interest rather than consuming natural capital. Sustainable development mandates that the present generation must not narrow the choices of

* See Hall,[4] Roberts,[5] McDowell,[6] Moore,[7] Stokes,[8] and Stokes et al.[9]

> future generations but must strive to expand them by passing on an environment and an accumulation of resources that will allow its children to live at least as well as, and preferably better than, people today. Sustainable development is premised on living within the Earth's means.[16]

Scandurra and Budoni have stated the underlying premise for sustainability especially well and succinctly. That is, "The planet cannot be considered as a gigantic source of unlimited raw materials, neither, equally, as a gigantic dump where we can dispose of all waste from our activities."[17] The environment is both a source and a sink, but has limited capacities to provide resources and to assimilate wastes indefinitely.

Beatley and Manning[18] relate sustainable development to ecological planning. They note that "McHargian-style environmental analysis... [has] become a commonplace methodological step in undertaking almost any form of local planning."[19] They note, however, that, although such analyses are "extremely important... a more comprehensive and holistic approach is required."[19] The steps that follow attempt to provide a more comprehensive approach.

### 13.2.1 Step 1: Identification of Planning Problems and Opportunities

Human societies face many social, economic, political, and environmental problems and opportunities. Since a landscape is the interface between social and environmental processes, landscape planning addresses those issues that concern the interrelationship between people and nature. The planet presents many opportunities for people, and there is no shortage of environmental problems.

Problems and opportunities lead to specific planning issues. For instance, suburban development occurs on prime agricultural land, which local officials consider a problem. A number of issues arise, involving land-use conflicts between new suburban residents and farmers such as who will pay the costs of public services for the newly developed areas. Another example is an area with the opportunity for new development because of its scenic beauty and recreational amenities, like an ocean beach or mountain town. A challenge would be how to accommodate the new growth while protecting the natural resources that are attracting people to the place.

### 13.2.2 Step 2: Establishment of Planning Goals

In a democracy, the people of a region establish goals through the political process. Elected representatives will identify a particular issue affecting their region—a steel plant is closing, suburban sprawl threatens agricultural land, or a new power plant is creating a housing boom. After issues have been identified, then goals are established to address the problem. Such goals should provide the basis for the planning process.

Goals articulate an idealized future situation. In the context of this method, it is assumed that once goals have been established, there is a commitment by some group to address the problem or opportunity identified in step 1. Problems and opportunities can be identified at various levels. Local people can recognize a problem or opportunity and then set a goal to address it. As well, issues can be national, international, or global in scope. Problem solving, of which goal setting is a part, may occur at many levels or combinations of levels. Although goal setting is obviously dependent on the cultural-political system, the people affected by a goal should be involved in its establishment.

Goal-oriented planning has long been advocated by many community planners. Such an approach has been summarized by Herbert Gans:

> The basic idea behind goal-oriented planning is simple; that planners must begin with the goals of the community—and of its people—and then develop those programs which constitute the best means for achieving the community's goals, taking care that the consequences of these programs do not result in undesirable behavioral or cost consequences.[20]

There are some good examples of goal-oriented planning, such as Oregon's mandatory land-use law.* However, although locally generated goals are the ideal, too often goals are established by a higher level of government. Many federal and state laws have mandated planning goals for local government, often resulting in the creation of new administrative regions to respond to a particular federal program. These regional agencies must respond to wide-ranging issues that generate specific goals for water and air quality, resource management, energy conservation, transportation, and housing. No matter at what level of government goals are established, information must be collected to help elected representatives resolve underlying issues. Many goals require an understanding of biophysical processes.

### 13.2.3 Step 3: Landscape Analysis, Regional Level

This step and the next one involve interrelated scale levels. The method addresses three scale levels: region, locality, and specific site, with an emphasis on the local. The use of different scales is consistent with the concept of levels-of-organization used by ecologists. According to this concept, each level of organization has special properties. Novikoff observed, "What were wholes on one level become parts on a higher one."† Watersheds have been identified as one level of organization to provide boundaries for landscape and ecosystem analysis. Drainage basins and watersheds have often been advocated as useful levels of analysis for landscape planning and natural resource management.[26–34] Dunne and Leopold provide a useful explanation of watersheds and drainage basins for ecological planning. They state that drainage basin

> is synonymous with watershed in American usage and with catchment in most other countries. The boundary of a drainage basin is known as the drainage divide in the United States and as the watershed in other countries. Thus the term watershed can mean an area or a line. The drainage basin can vary in size from that of the Amazon River to one of a few square meters drainage into the head of a gully. Any number of drainage basins can be defined in a landscape depending on the location of the drainage outlet on some watercourse.[35]

Essentially, drainage basins and watersheds are the same thing (catchment areas), but in practical use, especially in the United States, drainage basins generally are used to refer to larger regions and watersheds to more specific areas. Lowrance et al.,[36] who have developed a hierarchical approach for agricultural planning, refer to watersheds as the landscape system, or ecologic level, and the larger unit as the regional system, or macroeconomic level. In the Lowrance et al. hierarchy, the two smallest units are the farm system, or microeconomic level, and field system, or agronomic level. The analysis at the

* See, for instance, Pease,[21] Eber,[22] DeGrove,[23] and Kelly.[24]

† Novikoff, 1945, as quoted in Quinby.[25]

regional drainage-basin level provides insight into how the landscape functions at the more specific local scale.

Drainage basins and watersheds, however, are seldom practical boundaries for American planners. Political boundaries frequently do not neatly conform with river catchments, and planners commonly work for political entities. There are certainly many examples of plans that are based on drainage basins, such as water quality and erosion control plans. Several federal agencies, such as the U.S. Forest Service (USFS) and the U.S. Natural Resources Conservation Service (NRCS, formerly known as the Soil Conservation Service or SCS), regularly use watersheds as a planning unit. Planners who work for cities or counties are less likely to be hydrologically bound.

### 13.2.4 Step 4: Landscape Analysis, Local Level

During step 4, processes taking place in the more specific planning area are studied. The major aim of local-level analysis is to obtain insight about the natural processes and the human plans and activities. Such processes can be viewed as the elements of a system, with the landscape a visual expression of the system.

This step in the ecological planning process, like the previous one, involves the collection of information concerning the appropriate physical, biological, and social elements that constitute the planning area. Since cost and time are important factors in many planning processes, existing published and mapped information is the easiest and fastest to gather. If budget and time allow, then the inventory and analysis step may be best accomplished by an interdisciplinary team collecting new information. In either case, this step is an interdisciplinary collection effort that involves search, accumulation, field checking, and mapping of data.

Ian McHarg and his colleagues developed a layer-cake model (Figure 13.2) that provides a central group of biophysical elements for the inventory, what the ancient Greeks called the "chorography" of the place. Categories include the earth, the surface terrain, groundwater, surface water, soils, climate, vegetation, wildlife, and people (Table 13.1). UNESCO, in its Man and the Biosphere Programme, has developed a more exhaustive list of possible inventory elements (Table 13.2).

Land classification systems are valuable at this stage for analysis because they may allow the planner to aggregate specific information into general groupings. Such systems are based on inventoried data and on needs for analysis. Many government agencies in the United States and elsewhere have developed land classification systems that are helpful. The NRCS, USFS, the U.S. Fish and Wildlife Service, and the U.S. Geological Survey (USGS) are agencies that have been notably active in land classification systems. However, there is not a consistency of data sources even in the United States. In urban areas, a planner may be overwhelmed with data for inventory and analysis. In remote rural areas, on the other hand, even a NRCS soil survey may not exist or it may be old and unusable. An even larger problem is that there is little or no consistency in scale or in the terminology used among agencies. A recommendation of the National Agricultural Lands Study* was that a statistical protocol for federal agencies concerning land resource information be developed and led by the Office of Federal Statistical Policy and Standards. One helpful system that has been developed for land classification is the USGS Land Use and Land Cover Classification System (Table 13.3).

The ability of the landscape planner and ecosystem manager to inventory biophysical processes may be uneven, but it is far better than their capability to assess human

* National Agricultural Lands Study Final Report, 1981.

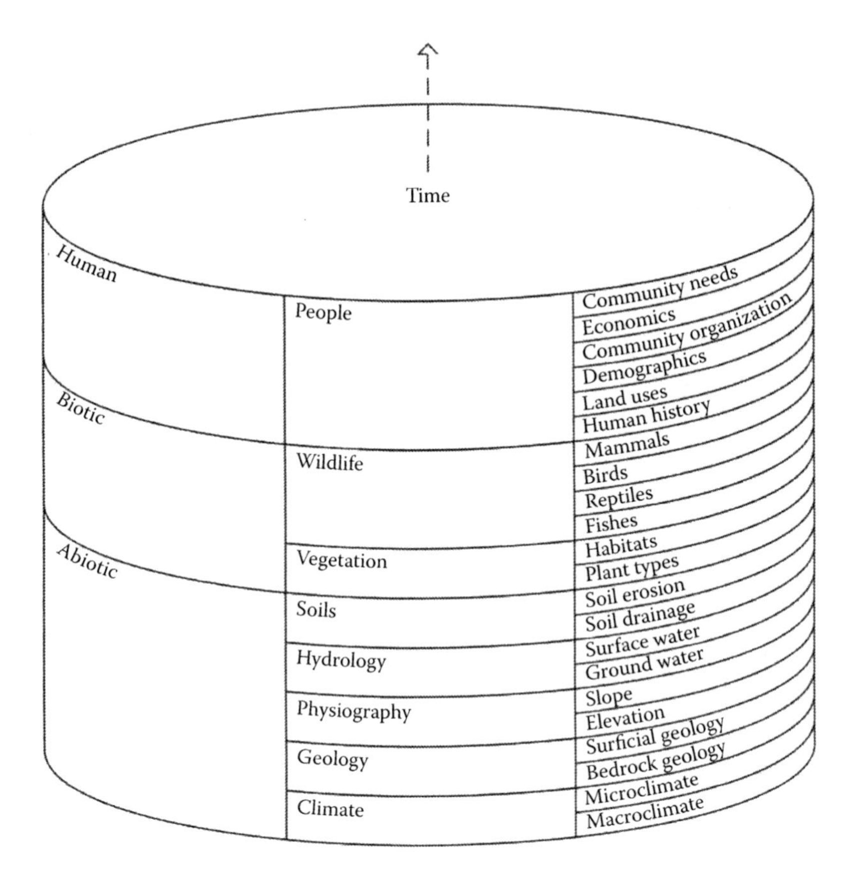


**FIGURE 13.2**
Layer-cake model.

ecosystems. An understanding of human ecology may provide a key to sociocultural inventory and analysis. Since humans are living things, *human ecology* may be thought of as an expansion of ecology, of how humans interact with each other and their environments. Interaction then is used as both a basic concept and an explanatory device. As Gerald Young,[37–40] who has illustrated the pandisciplinary scope of human ecology, noted:

> In human ecology, the way people interact with each other and with the environment is definitive of a number of basic relationships. Interaction provides a measure of belonging, it affects identity versus alienation, including alienation from the environment. The system of obligation, responsibility and liability is defined through interaction. The process has become definitive of the public interest as opposed to private interests which prosper in the spirit of independence.[41]

### 13.2.5 Step 5: Detailed Studies

Detailed studies link the inventory and analysis information to the problem(s) and goal(s). One example of such studies is suitability analysis. As explained by McHarg,[42] *suitability analyses* can be used to determine the fitness of a specific place for a variety of land uses based on thorough ecological inventories and on the values of land users. The basic purpose of the detailed studies is to gain an understanding about the complex relationships

**TABLE 13.1**

Baseline Natural Resources

| Data Necessary for Ecological Planning |
|---|
| The following natural resource factors are likely to be of significance in planning. Clearly the region under study will determine the relevant factors, but many are likely to occur in all studies |
| Climate. Temperature, humidity, precipitation, wind velocity, wind direction, wind duration, first and last frosts, snow, frost, fog, inversions, hurricanes, tornadoes, tsunamis, typhoons, Chinook winds |
| Geology. Rocks, ages, formations, plans, sections, properties, seismic activity, earthquakes, rock slides, mud slides, subsidence |
| Surficial Geology. Kames, kettles, eskers, moraines, drift and till |
| Groundwater Hydrology. Geological formations interpreted as aquifers with well locations, well logs, water quantity and quality, water table |
| Physiography. Physiographic regions, subregions, features, contours, sections, slopes, aspect, insolation, digital terrain model(s) |
| Surficial Hydrology. Oceans, lakes, deltas, rivers, streams, creeks, marshes, swamps, wetlands, stream orders, density, discharges, gauges, water quality, floodplains |
| Soils. Soil associations, soil series, properties, depth to seasonal high water table, depth to bedrock, shrink-swell, compressive strength, cation and anion exchange, acidity-alkalinity |
| Vegetation. Associations, communities, species, composition, distribution, age and conditions, visual quality, species number, rare and endangered species, fire history, successional history |
| Wildlife. Habitats, animal populations, census data, rare and endangered species, scientific and educational value |
| Human. Ethnographic history, settlement patterns, existing land use, existing infrastructure, economic activities, population characteristics |

between human values, environmental opportunities and constraints, and the issues being addressed. To accomplish this, it is crucial to link the studies to the local situation. As a result, a variety of scales may be used to explore linkages.

A simplified suitability analysis process is provided in Figure 13.3. There are several techniques that may be used to accomplish suitability analysis. Again, it was McHarg who popularized the "overlay technique."[42] This technique involves maps of inventory information superimposed on one another to identify areas that provide, first, opportunities for particular land uses and, second, constraints.[43] MacDougall[44] has criticized the accuracy of map overlays and made suggestions on how map overlays may be made more accurate.

Although there has been a general tendency away from hand-drawn overlays, there are still occasions when they may be useful. For instance, they may be helpful for small study sites within a larger region or for certain scales of project planning. It is important to realize the limitations of hand-drawn overlays. As an example, after more than three or four overlays, they may become opaque; there are the accuracy questions raised by MacDougall[44] and others that are especially acute with hand-drawn maps; and there are limitations for weighting various values represented by map units. Computer technology may help to overcome these limitations.

Numerous computer program systems, called geographic information systems (GIS), have been developed that replace the technique of hand-drawn overlays. Some of these programs are intended to model only positions of environmental processes or phenomena, while others are designed as comprehensive information storage, retrieval, and evaluation systems. These systems are intended to improve efficiency and economy in information handling, especially for large or complex planning projects.

**TABLE 13.2**

UNESCO Total Environmental Checklist: Components and Processes

| | |
|---|---|
| *Natural Environment—Components* | |
| Soil | Energy resources |
| Water | Fauna |
| Atmosphere | Flora |
| Mineral resources | Microorganisms |
| *Natural Environment—Processes* | |
| Biogeochemical cycles | Fluctuations in animal and plant growth |
| Irradiation | Changes in soil fertility, salinity, alkalinity |
| Photosynthesis | Host/parasite interactions and epidemic processes |
| Animal and plant growth | |
| *Human Population—Demographic Aspects* | |
| Population structure: | Population size |
| • Age | Population density |
| • Ethnicity | Fertility and mortality rates |
| • Economic | Health statistics |
| • Educational | |
| • Occupational | |
| *Human Activities and the Use of Machines* | |
| Migratory movements | Mining |
| Daily mobility | Industrial activities |
| Decision making | Commercial activities |
| Exercise and distribution of authority | Military activities |
| Administration | Transportation |
| Administration | Recreational activities |
| Farming | Crime rates |
| *Societal Groupings* | |
| Government groupings | Information media |
| Industrial groupings | Law-keeping media |
| Commercial groupings | Health services |
| Political groupings | Community groupings |
| Religious groupings | Family groupings |
| Educational groupings | |
| *Products of Labor* | |
| The built environment: | Food |
| • Buildings | Pharmaceutical products |
| • Roads | Machines |
| • Railways | Other commodities |
| • Parks | |
| *Culture* | |
| Values | Technology |
| Beliefs | Literature |
| Attitudes | Literature |
| Knowledge | Economic system |
| Information | |

*Source:* Boyden, S., *An Integrative Ecological Approach to the Study of Human Settlements*, MAB Technical Notes 12, UNESCO, Paris, France, 1979.

**TABLE 13.3**

U.S. Geological Survey Land Use and Land-Cover Classification System for Use with Remote Sensor Data

| Level I | Level II |
|---|---|
| 1. Urban or built-up land | 11 Residential |
| | 12 Commercial and services |
| | 13 Industrial |
| | 14 Transportation, communication, and services |
| | 15 Industrial and commercial complexes |
| | 16 Mixed urban or built-up land |
| | 17 Other urban or built-up land |
| 2. Agricultural land | 21 Cropland and pasture |
| | 22 Orchids, groves, vineyards, nurseries, and ornamental horticulture |
| | 23 Confined feeding operations |
| | 24 Other agricultural |
| 3. Rangeland | 31 Herbaceous rangeland |
| | 32 Shrub and brush rangeland |
| | 33 Mixed rangeland |
| 4. Forest land | 41 Deciduous forest lad |
| | 42 Evergreen forest land |
| | 43 Mixed forest land |
| 5. Water | 51 Streams and canals |
| | 52 Lakes |
| | 53 Reservoirs |
| | 54 Bays and estuaries |
| 6. Wetland | 61 Forested wetland |
| | 62 Nonforested wetland |
| 7. Barren land | 71 Dry salt flats |
| | 72 Beaches |
| | 73 Sandy areas other than beaches |
| | 74 Bare exposed rocks |
| | 75 Strip mines, quarries, and gravel pits |
| | 76 Transitional areas |
| | 77 Mixed barren land |
| 8. Tundra | 81 Shrub and brush tundra |
| | 82 Herbaceous tundra |
| | 83 Bare ground |
| | 84 Mixed tundra |
| 9. Perennial snow ice | 91 Perennial snow ice |
| | 92 Glaciers |

*Source:* Anderson, J.R. et al., A land use and land cover classification system for use with remote sensor data, U.S. Geological Survey, Professional Paper 964, 1976.

### 13.2.6 Step 6: Planning Area Concepts, Options, and Choices

This step involves the development of concepts for the planning area. These concepts can be viewed as options for the future based on the suitabilities for the use(s) that give a general conceptual model or scenario of how problems may be solved. This model should be presented in such a way that the goals will be achieved. Often more than one scenario has to be made. These concepts are based on a logical and imaginative combination of the information gathered through the inventory and analysis steps. The conceptual model

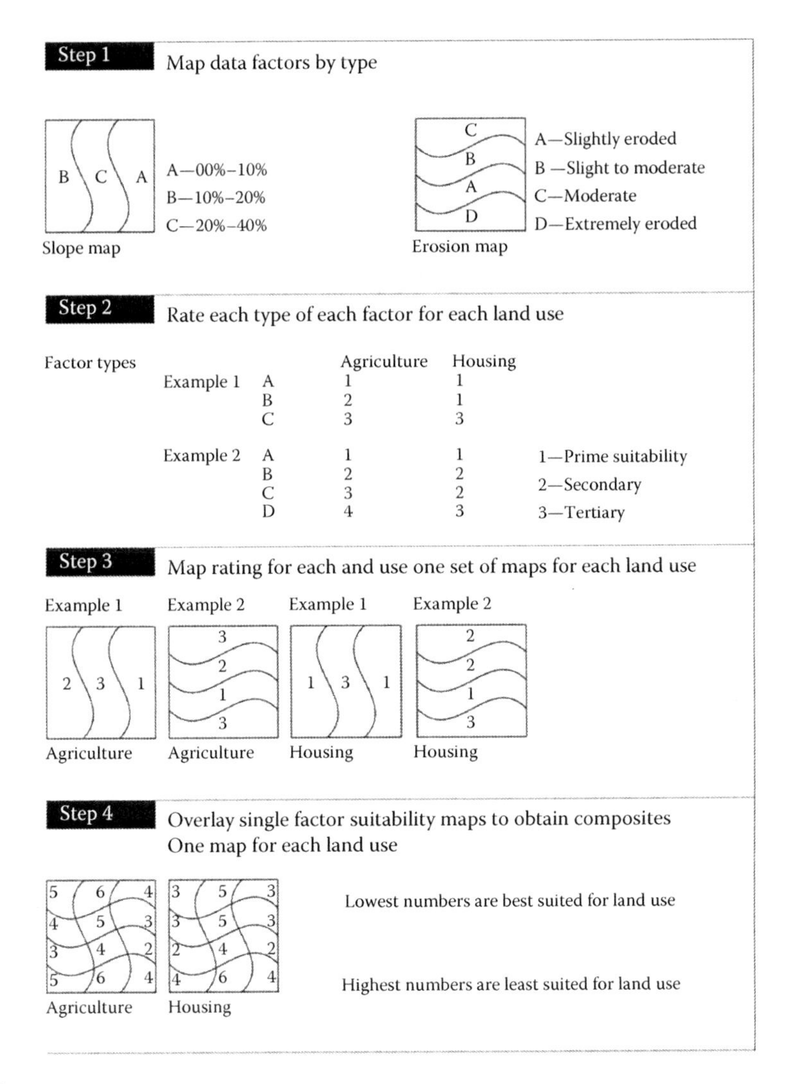


**FIGURE 13.3**
Suitability analysis procedure.

shows allocations of uses and actions. The scenarios set possible directions for future management of the area and therefore should be viewed as a basis for discussion where choices are made by the community about its future.

Choices should be based on the goals of the planning effort. For example, if it is the goal to protect agricultural land yet allow some low density housing to develop, different organizations of the environment for those two land uses should be developed. Different schemes for realizing the desired preferences also need to be explored.

The Dutch have devised an interesting approach to developing planning options for their agricultural land reallocation projects. Four land-use options are developed, each with the preferred scheme for a certain point of view. Optional land-use schemes of the area are made for nature and landscape, agriculture, recreation, and urbanization. These schemes are constructed by groups of citizens working with government scientists and planners. To illustrate, for the nature and landscape scheme, landscape architects and ecologists from the *Staatsbosbeheer* (Dutch Forest Service) work with citizen environmental action groups. For agriculture, local extension agents and soil scientists work with farm commodity organizations and farmer cooperatives. Similar coalitions are formed for recreation and urbanization. What John Friedmann[45] calls a dialogue process begins at the point where each of the individual schemes is constructed. The groups come together for mutual learning so that a consensus of opinion is reached through debate and discussion.

Various options for implementation also need to be explored, which must relate to the goal of the planning effort. If, for example, the planning is being conducted for a jurisdiction trying to protect its agricultural land resources, then it is necessary not only to identify lands that should be protected but also the implementation options that might be employed to achieve the farmland protection goal.

### 13.2.7 Step 7: Landscape Plan

The preferred concepts and options are brought together in a landscape plan. The plan gives a strategy for development at the local scale. The plan provides flexible guidelines for policymakers, land managers, and land users about how to conserve, rehabilitate, or develop an area. In such a plan, enough freedom should be left so that local officials and land users can adjust their practices to new economic demands or social changes.

This step represents a key decision-making point in the planning process. Responsible officials, such as county commissioners or city council members, are often required by law to adopt a plan. The rules for adoption and forms that the plans may take vary widely. Commonly, in the United States, planning commissions recommend a plan for adoption to the legislative body after a series of public hearings. Such plans are called *comprehensive plans* in much of the United States, *general plans* in Arizona, California, and Utah. In some states like Oregon, there are specific, detailed elements that local governments are required to include in such plans. Other states permit much flexibility to local officials for the contents of these plans. On public lands, various federal agencies, including the USFS, the U.S. National Park Service, and the U.S. Bureau of Land Management, have specific statutory requirement for land management plans.

The term *landscape plan* is used here to emphasize that such plans should incorporate natural and social considerations. A landscape plan is more than a land-use plan because it addresses the overlap and integration of land uses. A landscape plan may involve the formal recognition of previous elements in the planning process, such as the adoption of policy goals. The plan should include both written statements about policies and implementation strategies as well as a map showing the spatial organization of the landscape.

### 13.2.8 Step 8: Continued Citizen Involvement and Community Education

In step 8, the plan is explained to the affected public through education and information dissemination. Actually, such interaction occurs throughout the planning process, beginning with the identification of issues. Public involvement is especially crucial as the landscape plan is developed, because it is important to ensure that the goals established by the community will be achieved in the plan.

The success of a plan depends largely on how much people affected by the plan have been involved in its determination. There are numerous examples of both government agencies and private businesses suddenly announcing a plan for a project that will dramatically impact people without consulting those individuals first. The result is predictable—the people will rise in opposition against the project. The alternative is to involve people in the planning process, soliciting their ideas and incorporating those ideas into the plan. Doing so may require a longer time to develop a plan, but local citizens will be more likely to support it than to oppose it and will often monitor its execution.

### 13.2.9 Step 9: Design Explorations

To design is to give form and to arrange elements spatially. By making specific designs based on the landscape plan, planners can help decision makers visualize the consequences of their policies. Carrying policies through to arranging the physical environment gives meaning to the process by actually conceiving change in the spatial organization of a place. Designs represent a synthesis of all the previous planning studies. During the design step, the short-term benefits for the land users or individual citizen have to be combined with the long-term economic and ecological goals for the whole area.

Since the middle 1980s, several architects have called for a return to traditional principles in community design. These "neotraditionals" or "new urbanists" include Peter Calthorpe, Elizabeth Plater-Zyberk, Andres Duany, Elizabeth Moule, and Stefanos Polyzoides. Meanwhile, other architects and landscape architects have advocated more ecological, more sustainable design, including John Lyle, Robert Thayer, Sim Van Der Ryn, Carol Franklin, Colin Franklin, Leslie Jones Sauer, Rolf Sauer, and Pliny Fisk. Michael and Judith Corbett helped merge these two strains in the Ahwahnee Principles[46] (Table 13.4). Ecological design, according to David Orr, is "the capacity to understand the ecological context in which humans live, to recognize limits, and to get the scale of things right."[47] Or, as Sim Van Der Ryn and Stuart Cowan note, ecological design seeks to "make nature visible."[48] These principles provide clear guidance for ecological design,* although some designers and some planners might object to the placement of design within the planning process. In an ecological perspective, such placement helps to connect design with more comprehensive social actions and policies.

### 13.2.10 Step 10: Plan and Design Implementation

Implementation is the employment of various strategies, tactics, and procedures to realize the goals and policies adopted in the landscape plan. The Ahwahnee Principles provide guidelines for implementation (Table 13.4). On the local level, several different mechanisms have been developed to control the use of land and other resources. These techniques include voluntary covenants, easements, land purchase, transfer of development rights, zoning, utility extension policies, and performance standards. The preference selected should be appropriate for the region. For instance, in urban areas like King County, Washington, and Suffolk County, New York, traditional zoning has not been effective to protect farmland. The citizens of these counties have elected to tax themselves to purchase development easements from farmers. In more rural counties like Whitman County, Washington, and Black Hawk County, Iowa, local leaders have found traditional zoning effective.

One implementation technique especially well suited for ecological planning is performance standards. Like many other planning implementation measures, *performance standards* is a general term that has been defined and applied in several different ways. Basically, performance standards, or criteria, are established and must be met before a

---

* See also Beatley and Manning.[18,19]

**TABLE 13.4**

Awahnee Principles

*Preamble:*

Existing patterns of urban and suburban development seriously impair our quality of life. The symptoms are more congestion and air pollution resulting from our increased dependence on automobiles, the loss of precious open space, the need for costly improvements to roads and public services, the inequitable distribution of economic resources, and the loss of a sense of community. By drawing upon the best from the past and the present, we can plan communities that will more successfully serve the needs of those who live and work within them. Such planning should adhere to certain fundamental principles.

*Community Principles:*

1. All planning should be in the form of complete and integrated communities containing housing, shops, work places, schools, parks, and civic facilities essential to the daily life of the residents.
2. Community size should be designed so that housing, jobs, daily needs, and other activities are within easy walking distance of each other.
3. As many activities as possible should be located within easy walking distance of transit stops.
4. A community should contain a diversity of housing types to enable citizens from a wide range of economic levels and age groups to live within its boundaries.
5. Businesses within the community should provide a range of job types for the community's residents.
6. The location and character of the community should be consistent with a larger transit network.
7. The community should have a center focus that combines commercial, civic, cultural, and recreational uses.
8. The community should contain an ample supply of specialized open space in the form of squares, greens, and parks whose frequent use is encouraged through placement and design.
9. Public spaces should be designed to encourage the attention and presence of people at all hours of the day and night.
10. Each community or cluster of communities should have a well-defined edge, such as agricultural greenbelts or wildlife corridors, permanently protected from development.
11. Streets, pedestrian paths, and bike paths should contribute to a system of fully-connected and interesting routes to all destinations. Their design should encourage pedestrian and bicycle use by being small and spatially defined by buildings, trees and lighting and by discouraging high-speed traffic.

*Regional Principles:*

1. The regional structure should be integrated within a larger transportation network built around transit rather than freeways.
2. Regions should be bounded by and provide a continuous system of greenbelt/wildlife corridors to be determined by natural conditions.
3. Regional institutions and services (government, stadiums, museums, etc.) should be located within the urban core.

*Implementation Strategies:*

1. The general plan should be updated to incorporate the above principles.
2. Rather than allowing for developer-initiated, piecemeal development, a local government should initiate the planning of new and changing communities within its jurisdiction through an open planning process.
3. Prior to any development, a specific plan should be used to define communities where new growth, infill, or redevelopment would be allowed to occur. With the adoption of specific plans, complying projects can proceed with minimal delay.
4. Plans should be developed through an open process and in the process should be provided illustrated models of the proposed design.

*Source:* Calthorpe, P. et al., The Ahwahnee principles, in *Creating Sustainable Places Symposium*, A.B. Morris, ed., Herberger Center for Design Excellence, Arizona State University, Tempe, AZ, pp. 3–6, 1998.

certain use will be permitted. These criteria are usually a combination of economic, environmental, and social factors. This technique lends itself to ecological planning because criteria for specific land uses can be based on suitability analysis.

### 13.2.11 Step 11: Administration

In this final step, the plan is administered. *Administration* involves monitoring and evaluating how the plan is implemented on an ongoing basis. Amendments or adjustments to the plan will no doubt be necessary because of changing conditions or new information. To achieve the goals established for the process, planners should pay especial attention to the design of regulation review procedures and of the management of the decision-making process.

Administration may be accomplished by a commission comprising citizens with or without the support of a professional staff. Citizens should play an important role in administering local planning through commissions and review boards that oversee local ordinances. To a large degree, the success of citizens' boards and commissions depends on the extent of their involvement in the development of the plans that they manage. Again, Oregon provides an excellent example of the use of citizens to administer a plan. The Land Conservation and Development Commission, comprising seven members who are appointed by the governor and supported by its professional staff, is responsible for overseeing the implementation of the state land-use planning law. Another group of citizens, 1000 Friends of Oregon, monitors the administration of the law. The support that the law has from the public is evidenced in the defeat of several attempts to abolish mandatory statewide land-use planning in Oregon. However, as Department of Land Conservation and Development staff member Ron Eber observes, "It is a myth that planning is easy in Oregon—it is a battle every day!"* For example, in the early 1990s, a counter force to 1000 Friends of Oregon was organized. "Oregons in Action" is a property rights group, which is opposed to the progressive statewide planning program.

## 13.3 Working Plans

A method is necessary as an organizational framework for landscape planners. Also, a relatively standard method presents the opportunity to compare and analyze case studies. To adequately fulfill their responsibilities to protect the public health, safety, and welfare, actions of planners should be based on a knowledge of what has and has not worked in other settings and situations. A large body of case study results can provide an empirical foundation for planners. A common method is helpful for both practicing planners and scholars who should probe and criticize the nuances of such a method in order to expand and improve its utility.

The approach suggested here should be viewed as a working method. The pioneering forester Gifford Pinchot advocated a conservation approach to the planning of the national forests. His approach was both utilitarian and protectionist, and he believed "wise use and preservation of all forest resources were compatible."[49] To implement this philosophy, Pinchot in his position as chief of the U.S. Forest Service required "working plans." Such plans recognized the dynamic, living nature of forests. In the same vein, the methods used to develop plans should be viewed as a living process. However, this is not meant to imply

* Personal communication, 1999.

that there should be no structure to planning methods. Rather, working planning methods should be viewed as something analogous to a jazz composition: not a fixed score but a palette that invites improvisation.

The method offered here has a landscape ecological, specifically human ecological, bias. As noted by the geographer Donald W. Meinig, "Environment sustains us as creatures; landscape displays us as cultures."[50] As an artifact of culture, landscapes are an appropriate focus of planners faced with land-use and environmental management issues. Ecology provides insight into landscape patterns, processes, and interactions. An understanding of ecology reveals how we interact with each other and our natural and built environments. What we know of such relationships is still relatively little but expanding all the time. As Ilya Prigogine and Isabelle Stengers have observed, "Nature speaks in a thousand voices, and we have only begun to listen."[51]

## References

1. McHarg, I.L., Ecology and design, in *Ecological Design and Planning*, G.F. Thompson and F.R. Steiner, eds. (New York: John Wiley & Sons, 1997), 321 pp.
2. McHarg, I.L., Human ecological planning at Pennsylvania, *Landscape Planning* 8: 112–113, 1981.
3. McHarg, I.L., Human ecological planning at Pennsylvania, *Landscape Planning* 8: 107, 1981.
4. Hall, P., *Urban and Regional Planning* (New York: John Wiley & Sons, 1975).
5. Roberts, J.C., Principles of land use planning, in *Planning the Uses and Management of Land*, M.T. Beatty, G.W. Petersen, and L.D. Swindale, eds. (Madison, WI: American Society of Agronomy, Crop Science Society of America, and Soil Science Academy of America, 1979), pp. 47–63.
6. McDowell, B.D., Approaches to planning, in *The Practice of State and Regional Planning*, F.S. So, I. Hand, and B.D. McDowell, eds. (Chicago, IL: American Planning Association, 1986), pp. 3–22.
7. Moore, T., Planning without preliminaries, *Journal of American Planning Association* 54(4): 525–528, 1988.
8. Stokes, S.N., A.E. Watson, G.P. Keller, and J.T. Keller, *Saving America's Countryside* (Baltimore, MD: Johns Hopkins University Press, 1989).
9. Stokes, S.N., A.E. Watson, and S. Mastran, *Saving America's Countryside: A Guide to Rural Conservation*, 2nd edn. (Baltimore, MD: Johns Hopkins University Press, 1997).
10. Lovejoy, D., ed., *Land Use and Landscape Planning* (New York: Barnes & Noble, 1972).
11. Fabos, J.G., *Planning the Total Landscape* (Boulder, CO: Westview Press, 1979).
12. Zube, E.H., *Environmental Evaluation* (Monterey, CA: Brooks/Cole, 1980).
13. Marsh, W.M., *Landscape Planning* (Reading, MA: Addison-Wesley, 1983).
14. Duchhart, I., *Manual on Environment and Urban Development* (Nairobi, Kenya: Ministry of Local Government and Physical Planning, 1989).
15. World Commission on Environment and Development, *Our Common Future* (Oxford, U.K.: Oxford University Press, 1987).
16. National Commission on the Environment, *Choosing a Sustainable Future* (Washington, DC: Island Press, 1993).
17. Scandurra, E. and A. Budoni, For critical revision of the concept of sustainable development: Ten years after the Brutland report, Paper presented to the *20th Annual Meeting of the Northeast Regional Science Association*, Boston, MA, May 30–June 1, 1997, 2 pp.
18. Beatley, T. and K. Manning, *The Ecology of Place: Planning for Environment, Economy, and the Community* (Washington, DC: Island Press, 1997).
19. Beatley, T. and K. Manning, *The Ecology of Place: Planning for Environment, Economy, and the Community* (Washington, DC: Island Press, 1997, p. 86).
20. Gans, H.J., *People and Plans* (New York: Basic Books, 1968), 53 pp.

21. Pease, J.R., Oregon's land conservation and development program, in *Planning for the Conservation and Development of Land Resources*, F.R. Steiner and H.N. van Lier, eds. (Amsterdam, the Netherlands: Elsevier Scientific Publishing, 1984), pp. 253–271.
22. Eber, R., Oregon's agricultural land protection, in *Protecting Farmlands*, F.R. Steiner and J.E. Theilacker, eds. (Westport, CT: AVI Publishing Company, 1984), pp. 161–171.
23. DeGrove, J.M., *The New State Frontier for Managing Land-Use Policy: Planning and Growth Management in the States* (Cambridge, MA: Lincoln Institute of Land Policy, 1992).
24. Kelly, E.D., *Managing Community Growth: Policies, Techniques and Impacts* (Westport, CT: Praeger, 1993).
25. Quinby, P.A., The contribution of ecological sciences to the development of landscape ecology: A brief history, *Landscape Research* 13(3): 9–11, 1988.
26. Doornkamp, J.C., The physical basis for planning in the Third World IV: Regional planning, *Third World Planning Review* 4(2): 11–118, 1982.
27. Young, G.L., F.R. Steiner, K. Brooks, and K. Stuckmeyer, Determining the regional context for landscape planning, *Landscape Planning* 10(4): 269–296, 1983.
28. Steiner, F.R., Resource suitability: Methods for analyses, *Environmental Management* 7(5): 401–420, 1983.
29. Dickert, T. and R.B. Olshansky, Evaluating erosion susceptibility for land-use planning in coastal watersheds, *Coastal Zone Management* 13(3/4): 309–333, 1986.
30. Easter, W.K., J.A. Dixon, and M.M. Hufschmidt, eds., *Watershed Resource Management* (Boulder, CO: Westview Press, 1986).
31. Fox, J., Two roles for natural scientists in the management of tropical watersheds: Examples from Nepal and Indonesia, *Environmental Professional* 9: 59–66, 1987.
32. Erickson, D.L., Rural land-use and land-cover change: Implications for local planning in the River Raisin watershed (USA), *Land Use Policy* 21(19): 377–378, 1995.
33. Smith, T., B. Trushinski, J. Willis, and G. Lemon, *The Laurel Creek Watershed Study* (Waterloo, Ontario, Canada: Grand River Conservation Authority, 1997).
34. Golley, F.B., *A Primer for Environmental Literacy* (New Haven, CT: Yale University Press, 1998).
35. Dunne, T. and L.B. Leopold, *Water in Environmental Planning* (New York: W.H. Freeman, 1978), 495 pp.
36. Lowrance, R., P.F. Hendrix, and E.P. Odum, A hierarchical approach to sustainable agriculture, *American Journal of Alternative Agriculture* 1(4): 169–173, 1986.
37. Young, G.L., Human ecology as an interdisciplinary concept: A critical inquiry, *Advances in Ecological Research* 8: 1–105, 1974.
38. Young, G.L., *Human Ecology as an Interdisciplinary Domain: An Epistemological Bibliography* (Monticello, IL: Vance Bibliographies, 1978).
39. Young, G.L., *Origins of Human Ecology* (Stroudsburg, PA: Hutchinson Ross Publishing, 1983).
40. Young, G.L., A conceptual framework for an interdisciplinary human ecology, *Acta Oecologiae Hominis* 1: 1–136, 1989.
41. Young, G.L., Environmental law: Perspectives from human ecology, *Environmental Law* 6(2): 294, 1976.
42. McHarg, I.L., *Design with Nature* (Garden City, NY: Doubleday, The Natural History Press, 1969), 197 pp.
43. Johnson, A.H., J. Berger, and I.L. McHarg, A case study in ecological planning: The Woodlands, Texas, in *Planning the Uses and Management of Land*, M.T. Beatty, G.W. Peterson, and L.D. Swindale, eds. (Madison, WI: American Society of Agronomy, Crop Science Society of America, and Soil Science Society of America, 1979), pp. 935–955.
44. MacDougall, E.B., The accuracy of map overlays, *Landscape Planning* 2: 23–30, 1975.
45. Friedmann, J., *Retracking America* (Garden City, New York: Anchor Press/Doubleday, 1973).
46. Local Government Commission, The Ahwahnee principles, Sacramento, CA, 1991.
47. Orr, D.W., *Earth in Mind: On Education, Environment, and the Human Prospect* (Washington, DC: Island Press, 1994), 2 pp.
48. Van der Ryn, S. and S. Cowen, *Ecological Design* (Washington, DC: Island Press, 1996), 16 pp.

49. Wilkinson, C.F. and H.M. Anderson, Land and resource planning in national forests, *Oregon Law Review* 64(1 and 2): 22, 1985.
50. Meinig, D.W., Introduction, in *The Interpretation of Ordinary Landscapes*, D.W. Meinig, ed. (New York: Oxford University Press, 1979), 3 pp.
51. Prigogine, I. and I. Stengers, *Order Out of Chaos* (Toronto, Ontario, Canada: Bantam Books, 1984), 77 pp.
52. Boyden, S., *An Integrative Ecological Approach to the Study of Human Settlements*, MAB Technical Notes 12 (Paris, France: UNESCO, 1979).
53. Anderson, J.R., E.E. Hardy, J.T. Roach, R.E. Witmer, A land use and land cover classification system for use with remote sensor data, U.S. Geological Survey, Professional Paper 964, 1976.
54. Calthorpe, P., M. Corbett, A. Duany, E. Plater-Zyberk, S. Polyzoides, E. Moule, J. Corbett, P. Katz, and S. Weissman, The Ahwahnee principles, in *Creating Sustainable Places Symposium*, A.B. Morris, ed. (Tempe, AZ: Herberger Center for Design Excellence, Arizona State University, 1998), pp. 3–6.

# 14

# *Phoenix as Everycity: A Closer Look at Sprawl in the Desert*

**Sandy Bahr, Renée Guillory, and Chad Campbell**

**CONTENTS**



## 14.1 Introduction

Perhaps, this *Arizona Republic* headline says it all: "Growth pattern crippled Phoenix: Half-empty outskirts suffer as once-reliable cycle busts."[1] Phoenix, Arizona, tells a vivid story of urban sprawl and all of its impacts. Sprawling development often occurs far from existing infrastructure—leapfrogging over established areas. In Phoenix, poorly planned, low-density, scattered development consumes open space, empties pocketbooks, and generally serves automobiles better than people. In the recent years, it has also meant a very deep recession as the area suffers the consequences of land speculation and overbuilding in both the residential and commercial sectors.

What is urban, or more accurately, suburban, sprawl? The Anthem development, which opened in 1998, is a clear example of leapfrog sprawl near Phoenix. Anthem is nearly 30 miles north of downtown Phoenix in an area where the land use is regulated by the county. When it was built, there were miles of undeveloped desert between Anthem and the city's established areas. Following a textbook case of sprawl development patterns, Anthem also demonstrates how sprawl inevitably gives way to more development, like it or not. Much of the undeveloped desert between Phoenix and Anthem lives on only in the ubiquitous picture postcards of the area; in reality, this once open north Phoenix corridor is rapidly becoming just another jumble of subdivisions and strip malls.[2]

To the east of Phoenix in northeastern Pinal County is the planned Superstition Vista development. It is being billed as "sustainable" development or even "smart growth," but a closer analysis demonstrates that it is more of the same for the Phoenix area. Superstition Vista would be located on 275 square miles of state trust lands and developed to accommodate hundreds of thousands of houses far from existing infrastructure and transportation all in an area that gets 7–8 in. of rainfall each year. At least two freeways are being considered to serve the new development in this area including State Route 802 and the continuation of U.S. 60. Superstition Vista, much like the sprawl development of the 1980s, 1990s, and the current decade, will rely on the automobile and cheap gasoline, plus eat up thousands of acres of open space. Is it "sustainable" to continue this kind of development in a place that gets less than 10 in. of rainfall each year and where scientists tell us the impacts of climate change are likely to be felt intensely? Hotter and drier projections are in the climate reports for the southwestern United States.[3] This development seems like risky business for more than the environment, though. The first land to be sold relative to this development was affected by the real estate collapse as the parent company of the developer ended up in bankruptcy and missed at least two payments to the Arizona State Land Department.[4]

West of Phoenix there seems to be an unending sea of tile rooftops and examples of sprawling development. The Verrado development, located in the Town of Buckeye, incorporates some "smart growth" concepts, including promoting a "neighborhood" feel with real front porches as well as walkability, but it was still developed far from existing infrastructure and means massive commuting on a crowded I-10 freeway. Probably, one of the most outrageous examples of urban sprawl and the antithesis of sustainability is a water skiing housing development in south Buckeye. Spring Mountain Ski Ranch includes lakes, docks, and a "professional slalom course and ski jump." This is all far from any current development and in one of the driest parts of the Sonoran Desert. The housing bust has left this development with a gate, docks, and water-skiing area, as well as a sign indicating the need for dust control, but no houses (Figure 14.1).

To the south, a "Megapolitan" area often referred to as the Sun Corridor (think of Phoenix and Tucson, amoeba-like, spreading toward each other and becoming one mass) looms large on the landscape and is predicted to be inevitable.[5] A Megapolitan, as defined by Virginia Tech Metropolitan Institute, is "two or more metropolitan areas with anchor principal cities between 50 and 200 miles apart that will have an employment interchange measure of 15% by 2040 based on projection." Signs of this merging of Tucson and Phoenix are evident as development to the north of Tucson and to the south of Phoenix continues and sleepy communities such as Casa Grande, more or less midway between to the two cities, are now known for acres of houses and big box development. Is this Megapolitan concept merely a transition from suburban sprawl?

The collapse of the real estate market has been felt everywhere, but it has been more intense in the Phoenix area than most parts of the country. A report on housing in the Phoenix area published by the W. P. Carey School of Business at Arizona State University stated, "Since the market peaked in 2006 (after price hikes of 74%–81%), the southwest region has fallen the most—59%—with the central and northwest regions close behind."[6]

Sprawl is not unique to Phoenix, of course. All Southwest cities—Tucson, Las Vegas, Albuquerque, and El Paso—have their own challenges with sprawl and its effects. However, Phoenix is an especially good case study on sprawl due to its size, its various attempts to address the problems of sprawl, and its dramatic boom-and-bust growth swings since

**FIGURE 14.1**
Sign for water ski development near Buckeye, the epitome of unsustainability in a desert region.

the mid-twentieth century. Sprawl-related policy debates in Phoenix are ongoing, and we offer here an outline of Phoenix's past, present, and future in hopes that policymakers, once and for all, get serious about addressing sprawl. In the United States, land consumption has far outpaced population growth. Urban areas are expanding at about twice the rate that the population is growing. Each year from 1992 to 1997 and from 1997 to 2001, 2.2 million acres of open space were developed according to the National Resources Inventory Data provided by the U.S. Department of Agriculture, Natural Resource Conservation Service.[7] This was an increase of 1.4 million acres per year over the 1982–1992 time period.[7] Approximately 34 million acres of open space were developed between 1982 and 2001.

Urban sprawl and loss of open space characterize many of the Southwest's desert cities—nowhere is it more evident than in the Phoenix area where desert and farmland were being developed at more than acre an hour for most of the past three decades, according to a report by the Morrison Institute for Public Policy at Arizona State University. This report states that in the Phoenix area, an average of 23 square miles of desert and farmland were converted to urban use annually between 1975 and 1995.[8] This trend in land conversion showed no sign of letting up until the recent economic downturn. With a total of 347.2 square miles of open space in the path of current development—and expected to fall under development by 2025—Maricopa County's wildlife and natural habitat is listed as the fourth most endangered in the nation due to sprawl.[9]

## 14.2 Phoenix Metropolitan Area: A Case Study on Sprawl in the Desert

Maricopa County is home to Phoenix, Arizona's largest city, which contains 60% of Arizona's population and is characterized by the vast and biologically diverse Sonoran Desert.[10] Even in the face of a 28.5% increase in population in the Phoenix area between 1990 and 2007, officials in Phoenix and the surrounding communities have not managed growth with comprehensive planning; instead, land speculation and development have negatively affected the quality of life for all residents (Figure 14.2).[11]

**FIGURE 14.2**
Generic sprawl development around Shadow Mountain.

The result? Phoenix's modernist groove—so well established in the 1950s and 1960s—gave way to generic sprawl development. A typical development scenario in the Phoenix area includes the construction of big box stores surrounded by acres of asphalt, strip malls, and more and more cookie-cutter subdivisions. This type of development encourages traffic congestion and clogged freeways, even in normal daily activities.

The consequences of rapid and generally poorly planned growth include poor air quality, burdens on infrastructure (including schools), rising costs of services and infrastructure, increases in paved areas and the rising urban heat island effect, and fragmentation of wildlife habitat—in short, a plummeting quality of life. In fact, the Maricopa Association of Governments (MAG) report View of the Valley in 2040 found that 45% of area residents surveyed said that they would move immediately if they could. Many of the reasons cited were related to the way the area was growing and the impacts of that growth—too many people, traffic, and pollution/air quality, among others.[12]

Arizona, overall, is ranked 36th for livability in the "Most Livable State Award 2008," which uses 43 factors to determine livability. Arizona received a lower ranking due to a number of sprawl-related factors, including high student-to-teacher ratios in classrooms, high crime rates, and high freeway fatality rates.[13] Can fixing sprawl directly address all of these issues? No. But Arizona would rank much higher on any livability list if we built neighborhoods, not developments; planned transportation systems, not freeways; and ensured that schools are adequately funded and could plan better for enrollment changes. Let us look at some of these problems in more detail.

## 14.3 Problems Associated with Sprawl

### 14.3.1 Lack of Transportation Choices and Associated Costs

At least partly as a result of sprawl, there has been an enormous increase in automobile travel. The need for increased driving in sprawling areas requires a massive and expensive

**FIGURE 14.3**
Freeways are not really transportation systems: they are habitat for cars.

network of highways. As sprawl intensified between 1990 and 1999, the number of miles Americans drove increased 24% and the amount of roadway per person grew by 17% in the 23 metro areas with the greatest increase in road building (Figure 14.3).[14]

According to MAG, residents of Maricopa County travel a combined 67 million miles on an average weekday. This number is expected to grow by 140% over the next 40 years.[15] With rising gas prices, now well over 3 dollars per gallon, vehicle miles traveled did decrease some, including in 2008 when they were at over $2/gal. In View of the Valley in 2040, the MAG projects that the regional costs for transportation will be between $45 and $50 billion[12]—and if recent transportation plans are any indication; most of the dollars will be devoted to roads. A small percentage of the federal transportation dollars is used for transit, and all of the Highway User Revenue Funds (which consists primarily of gas tax), plus about two-thirds of a half-cent sales tax, are used for roads. The Regional Transportation Tax passed in 2004 extended the sales tax for another 20 years and directed about a third of the dollars to transit. While this was a vast improvement, it still guarantees a transportation system heavily weighted toward roads and freeways.[10]

In Phoenix, 89% of the workforce drove or carpooled to work with only 3% using public transportation in 2007.* To discover the reason, one needs to look no further than transportation spending. While the Phoenix area spends millions of dollars each year to

* http://www.clrsearch.com/RSS/Demographics/AZ/Phoenix/Drive_Time (accessed July 23, 2009).

**FIGURE 14.4**
Once the destination for people with lung ailments needing a rest cure, Phoenix now has many "bad air days."

build new and wider roads, ostensibly to accommodate increased traffic, building new and wider roads does not relieve traffic congestion, according to a study by the Surface Transportation Policy Project. In this study, researchers found that high road-building areas actually showed more traffic congestion than low road-building areas. Newer and wider roads generate more traffic, a phenomenon known as "induced traffic."[14] According to the 2011 urban mobility report by the Texas Transportation Institute, Phoenix ranked 15th worst in terms of annual delay per traveler and 12th worst in wasted fuel per traveler.[16]

In addition to costs associated with just building more and more freeways, sprawl drives up transportation costs for individual families as well. Families living in an area that is sprawling significantly spend thousands of dollars more per year than those living in areas where they can use mass transit, bike, or walk to work, shop, etc.[17] Families in Phoenix spend 18.2% of their family budget—about $6826 per year—on transportation (including purchase, operation, and maintenance of automobiles), resulting in a total household expenditure on transportation in Phoenix of $7.6 billion (Figure 14.4).[17]

Because sprawl does not accommodate a mix of transportation choices, when the population increases, there is a disproportionate increase in vehicle miles traveled. Sprawl development also disturbs large areas of land that in turn contributes to the particulate pollution. Phoenix is a nonattainment area, meaning that it does not meet the federal health-based standards for both ozone and coarse particulates, referred to as $PM_{10}$ as they are 10 μm in size or smaller.

The Phoenix area failed to meet its deadline for reducing $PM_{10}$ pollution by the end of 2006 and therefore had to develop a special plan to reduce particulates by 5% per year or risk losing federal highway dollars, not to mention the health impacts. Maricopa County experienced 59 PM10 violation days in 2007 and 16 in 2008.* The MAG submitted a 2012 Five Percent Plan for PM10 for the Maricopa County Nonattainment Area in May 2012. Coarse particulates, $PM_{10}$, are generated by construction-related activities, vehicular travel, driving on unpaved lots, road shoulders, and roads. Other sources of $PM_{10}$ pollution are created by off-road vehicles, agriculture, and leaf blowers.

According to the Arizona Department of Transportation, on the average weekday, residents of Maricopa County make 13 million vehicle trips covering approximately

* http://arizonaindicators.org/sustainability/particulate-matter

67 million miles—and this number is projected to increase significantly in the next 40 years.[18] Both the traffic and the increase in vehicle miles traveled contribute significantly to air quality problems—despite lower emissions from newer vehicles, a large percentage of the air pollution still comes from vehicles.

Approximately 34% of the volatile organic compounds and 65% of the oxides of nitrogen that contribute to the formation of ozone pollution come from mobile sources.[19] The American Lung Association of Arizona gave Maricopa County, which includes the City of Phoenix and 24 other municipalities, an "F" for ozone pollution and a "C" for particulate pollution in its State of the Air Report for 2012.[20] In 2000, noted national researcher C. Arden Pope III from Brigham Young University told the governor's Brown Cloud Summit that poor air quality in the Phoenix area is reducing a person's life expectancy by 1%.[21] Beyond the significant impact poor air quality has on public health, it also has enormous pocketbook costs. In the Phoenix area, recent research indicates that there is a strong correlation between asthma-related emergency room visits for children and poor air quality.[21] Hospitalization costs associated with asthma in Arizona are $650 million.[22]

Sprawl development and lack of planning for pedestrian and bicycle transportation have other downright dangerous health implications. According to another Surface Transportation Policy Project study that covered 2002–2003, Phoenix was one of the 10 most hazardous metro areas for walking with a slight improvement in the 2007–2008 period when it ranked as 16th most hazardous for walking.[*,23] The Phoenix area averages about 2.44 pedestrian deaths per 100,000 people, despite the fact that only about 2.1% of the population walks to work.[23]

Finally, setting aside the economic costs and impacts on physical health, let us not forget the impact of traffic congestion and sprawl on the nonquantifiable issues. How many hours of being stuck in traffic are too many? According to the U.S. Census, people in Phoenix spend just over 4 days per year commuting to work.[24] These 4.3 days per year spent in a car waiting to get to and from work is an unfortunate loss of time that could otherwise be spent with family, friends, and other more relaxing and reinvigorating pursuits. Unless Maricopa County pursues more transit options and more sensible planning in the very near future, commute times will continue to get longer as the Valley continues to grow.

### 14.3.2 Sprawl Burdens Infrastructure and Inner City Residents

Beyond roads, sprawl places a significant burden on other infrastructure as well. Fire response times increase dangerously; communities cannot build enough new stations to accommodate growth, so firefighters have to drive farther and farther to the scenes of accidents or emergencies. This, too, has a cost.

Schools are also overburdened. In the winter of 1998, schools in Peoria (a north Phoenix suburb), overwhelmed by a surge in population growth, closed their doors to new students. Rapid, uncontrolled growth—in Peoria and many other Phoenix-area cities—has also resulted in a proliferation of portable classrooms that often eat up playground space. According to the Arizona Education Association, Arizona ranks second highest in both growth of student population and in student–teacher ratio.[25]

Sprawl often leads to situations in which residents that live in the central city and inner suburbs subsidize development at the fringes. As growth moves to the edges of the metropolitan area, farther away from the existing infrastructure, it costs cities and

* Ernst, M. and Shoup, L., *Dangerous by Design—Solving the Epidemic of Preventable Pedestrian Deaths (and Making Great Neighborhoods)*. Surface Transportation Policy Project, 2009.

counties more to deliver public services. These costs, which are borne equally by all of the taxpayers, result in an entire city or county picking up the tab for new and costlier services that only serve new, sprawling neighborhoods.[26] Some government agencies in Arizona have performed detailed analyses on the costs of growth and have confirmed national findings that sprawl development often costs communities much more to service than development on smaller parcels within established communities. For example, another Phoenix area city, Scottsdale, commissioned two independent fiscal impact studies and determined that sprawl costs nearly twice as much as development in established neighborhoods and three to four times as much as compact development across all expenditure categories analyzed (fire, general government, municipal services, planning and development, police, and community services). Studies performed for both nearby Mesa and Gilbert found similar results.[26] Collectively, these studies helped municipalities implement higher impact fees per dwelling unit, but, in the case of Mesa, political will could not be mustered to charge impact fees at actual cost, which means that even cities armed with the best information about the costs of sprawl can still lose money to finance new infrastructure needed by leapfrog developments. (It is important to note that because of the idiosyncrasies of Arizona law, no community can charge impact fees for schools without facing a lawsuit brought by developers. So one of the largest infrastructure costs associated with sprawl cannot be addressed by any municipality.)[27] University of Arizona Professor Ignacio San Martín argues that within the decade "cities could go broke trying to provide services for subdivisions that don't pay for themselves."[28]

Perhaps one of the most ironic results relating to the costs of sprawl in some areas is that the more affluent households at the fringes are subsidized by the poorer households located in the central city. For example, Myron Orfield in *Metropolitics* found that households in the central cities of Minneapolis and St. Paul pay $6 million more each year in fees for sewer capital and operational expenses than they receive in services. In the central cities, where there are high numbers of low-income residents, those residents subsidize the new sewer systems by paying rates $19–$25 per year above the actual cost of delivery to their own neighborhoods.[29] When dollars are focused on the suburbs, the inner and older parts of cities are often neglected. A 1998 study by Subhrait Guhathakurta concluded that suburban areas received the majority of planned capital expenditure funds and that the spending per household in these suburban areas was 77% higher than for the average Phoenix household.[30]

### 14.3.3 Sprawl Eats Up Open Space and Fragments Wildlife Habitat

Areas with the highest growth rates also happen to have high numbers of imperiled species, according to a report from the National Wildlife Federation.[31] The fastest growing metropolitan areas with more than 1 million people in the contiguous 48 states are home to approximately 29% of the nation's imperiled species. Maricopa County has 22 endangered species and is one of the fastest growing counties in the United States.[31] Species such as the Sonoran desert tortoise, while not listed as endangered, its numbers are declining. A key reason includes loss of habitat including from development and urbanization (Figure 14.5).

### 14.3.4 Sprawl Hurts Tourism

As noted earlier, sprawl consumes natural open space, a key attraction in Arizona, which also has a negative economic impact on communities. Tourism is the second largest

**FIGURE 14.5**
Sonoran desert tortoise is a candidate species for listing under the Endangered Species Act. Development and urbanization are the key factors in the species' decline.

industry in the state. In 2007, according to the Arizona Office of Tourism, spending associated with travel in Arizona helped generate 171,500 jobs and $5.1 billion.[32] Many of those visitors engaged in sightseeing; Arizona "... is well above the U.S. average on sightseeing, national/state parks, hike/bike, and looking at real estate."[32]

From 1998 to 2002, winter visitor numbers fell in the Phoenix area. Some of the reasons people cited for not returning to Phoenix included the problems associated with increasing urbanization—traffic congestion, poor air quality, and the "disappearing desert."[33]

## 14.4 New Challenge to Planning and Limiting Sprawl in Phoenix

Property rights are a sacred covenant in the Southwest. A 2006 ballot measure has made it more of a challenge to plan and curb the excesses of sprawl through land-use planning. Proposition 207 expanded the definition of "takings" to include any regulation that a property owner believes has diminished the value of his or her property. The measure does not require the property owner to prove the claim with appraisals, and it gives the property owner the right to sue the government if he or she is not paid for the claim. The regulations can also be waived regardless of the effect on neighboring properties.

The measure was modeled after a law passed in Oregon—Measure 37, which was later repealed when it became clear that it limited the ability of local government to regulate land use and protect open space. The full implications of Proposition 207 have yet to be realized. Other Southwest cities may face future challenges imposed by ballot measures or legislative initiatives that limit the ability of local planners to deal with land-use regulations that may restrict some uses on land, even if these regulations have collective good for the community to deal with responsible growth and development. Currently, most cities, including Phoenix, use a waiver to accommodate new development, but little is being done to implement new protective land-use regulations.

## 14.5 Sustainable Solutions to Sprawl

Sprawl has many negative impacts on communities with very little positive other than the profits gained through land development or home sales. No matter the definition of sustainability, few would consider sprawl-style development sustainable. To be sustainable, it is critical that future generations not be saddled with the true costs of today's development. With urban sprawl, the costs are deferred to taxpayers as well as to future generations. "The Sonoran Desert can support a surprisingly large number of people, but there are limits to its life-sustaining capacity," states Todd Bostwick, City of Phoenix Archaeologist.[34]

After many years of discussion and consideration of sprawl by community planners, developers, and the public at large, the debate on how to address the concerns of sprawl has no clear answer or solution. Over time, a series of measures to address sprawl and its negative impacts have received attention from positive applications in other areas. The following section represents some key tools for communities to address appropriate growth strategies for their residents. The solutions best suited for a given community will depend on the local land uses, political will, citizen involvement, and financial resources available to implement the conservation measure (see Chapter 13).

### 14.5.1 Smart Growth?

A solid alternative to urban sprawl might include slower and better planned growth and making more conscious decisions about where development should go. Consideration of what is sustainable in a desert city is essential. Concentrating development where schools, roads, and sewer lines are already in place and reinvesting in older communities instead of abandoning them is critical to limiting sprawl. To curb sprawl, communities must revitalize downtowns and place homes near transit centers or within walking distance of shops, restaurants, and offices, preserving open space and channeling growth away from critical open space and sensitive habitat. Cutting the subsidies that feed sprawl and reinvesting in existing communities can help rein in urban sprawl as well. Today, these types of actions are often packaged as "Smart Growth." According to Smart Growth America, there are 10 principles of Smart Growth[34]:

- Provide a variety of transportation choices.
- Mix land uses.
- Create a range of housing opportunities and choices.
- Create walkable neighborhoods.
- Encourage community and stakeholder collaboration.
- Foster distinctive, attractive communities with a strong sense of place.
- Make development decisions predictable, fair, and cost effective.
- Preserve open space, farmland, natural beauty, and critical environmental areas.
- Strengthen and direct development toward existing communities.
- Take advantage of compact building design and efficient infrastructure design.

The term "Smart Growth," however, has been used to describe a number of developments that are far from the city centers. These developments, such as the Verrado development in the West Valley of Phoenix, incorporate one or two of these "smart growth" principles but continue to promote urban sprawl by leapfrogging far from the city centers. Overall, "Smart Growth" has achieved mixed results throughout the country. A report issued in May 2009 by the Lincoln Institute of Land Policy indicated that the states that adopted "smart growth policies achieved success in areas such as protecting open space and expanding transportation choices, but no state was able to make gains in all the major objectives of smart growth."[35] The highest ranking states were Oregon, Colorado, and New Jersey, states that ranked well due to "land use, urbanization, and concentration" factors.

In Phoenix, the "Smart Growth" concepts have yet to take hold, but some individual projects do incorporate most of them. They can be found in the downtown area along the light rail corridor. The Roosevelt Square development incorporates retail and residential, townhomes and apartments, as well as transportation choices as it is conveniently located on the light rail line and bus routes, and is also in area that is walkable (Figure 14.6).

Arizona has had some so-called Smart Growth provisions in statute since 1998. These laws mandate inclusion of the various elements in the general plans for cities as well as sending the plans to a vote of the public in the larger cities, but the real-world impacts have been limited. A smart growth scorecard was also implemented. However, the impacts cannot yet be assessed as it was just adopted in 2008, and the first completed scorecards were submitted in early 2009. The scorecard awards discretionary state funds according to a number of factors, including how the community does its planning, what and how much open space they protect, and how they conserve resources, among several other factors.

**FIGURE 14.6**
Roosevelt Square in downtown Phoenix includes mixed-use development and is conveniently located on the light rail line.

### 14.5.2 Urban Growth Boundaries

Tools such as "urban growth boundaries" (UGBs) should be considered and have proved effective where well implemented. UGBs are basically a boundary a city or town establishes around its urban development. Outside that boundary, development is limited, and the land is protected for conservation, farming, watershed values, etc. The first UGB in the United States was drawn around Lexington, Kentucky, in 1958.[36] In 1973, the state of Oregon passed legislation requiring all cities to include UGBs in their comprehensive land-use plans. Similar requirements were passed in Washington State in 1989. Tennessee also requires cities to establish UGBs. In addition to these states, several cities have adopted growth boundaries, including Livermore and Pleasanton in the bay area of California; Minneapolis, Minnesota; and Miami-Dade County, Florida.*

While UGBs alone do not address all of the problems associated with poorly planned growth and urban sprawl, they do provide an important tool. Communities that have implemented UGBs have been able to focus more on revitalization of their urban centers, save tax dollars by using public facilities more efficiently, and develop in a way that gives their communities more accessible public transit.†

### 14.5.3 Open Space Planning

Planning and preservation of open space is also an important tool that is being used to help deter sprawl. Rural preservation programs can limit the loss of farmland, where appropriate, and also protect important open space and wildlife habitat. Maryland has worked on a combination of measures to deter sprawl including directing resources to growth areas with a preservation program. Maryland also directs growth to priority funding areas that must meet guidelines for minimum density and adequate sewer and water. In Montgomery County, Maryland, a comprehensive preservation program has saved over 93,000 ac of working farmland and open space. Florida is also looking to preserve open space and has designated over $300 million per year (to support bond acquisition) for preservation. New Jersey has approved $1 billion for a 10-year bond program to preserve open space and farmland.[37]

Conservation easements are another tool that can help deter urban sprawl and protect open space. These easements generally limit development on a parcel of private land but allow continued agricultural activities such as livestock grazing or farming. According to The Nature Conservancy, nearly 86,000 ac in Arizona are under conservation easements. However, 57% of the private land outside incorporated areas has already been subdivided for development, so future applicability of conservation easements in Arizona may be limited, especially for deterring urban sprawl.[38]

### 14.5.4 Urban Revival

Community revitalization—investment in downtowns and inner suburbs—is an important component of deterring sprawl (see Chapter 25). Providing for affordable housing must be a component of revitalization. Plans and tools that simply push out low-income people and gentrify those neighborhoods, leaving limited-income families with no alternatives, are not the way to revitalize downtowns and curb urban sprawl. Offering assistance to existing neighborhoods so families can stay put as their neighborhood improves

* http://www.ccei.udel.edu/files/urbangrowthboundary.pdf (accessed July 23, 2009).
† www.GreenbeltAlliance.org, (accessed July 23, 2009).

is one way to help ensure that redevelopment does not just push out lower-income people, and, for new neighborhoods, inclusionary zonings (zoning that requires a certain portion be devoted to affordable housing) can also help accomplish it.

The City of Rockville, Maryland, has required subdivisions of 50 or more dwellings to include moderately priced units as part of their development. In Florida, the state has an active affordable housing program that takes $0.20 out of a $100 real estate transfer tax and sets it aside for affordable housing—that amounts to about $200,000 per year.[37]

The state of California passed the California Community Redevelopment Law, which requires private developers to set aside 15% and public agencies to set aside 30% of units for affordable housing in all redevelopment areas.[39] Minnesota has a voluntary "Innovative and Inclusionary Housing Program" that provides gap financing and regulatory relief for builders who set aside 10%–15% of their developments for lower-income families.[40]

An important component of community revitalization is promoting infill development. This can be done via incentives such as waiving development impact fees, but a critical component for infill is ensuring that the proper amenities and transportation are available. For example, Phoenix has opened its first stretch of light rail. Much of the route within the city itself has seen an increase in mixed-use and higher density development, and ridership on the light rail line itself has far exceeded projections by both federal and local agencies. The light rail line opened in December of 2008, and by February 2009, the average weekday ridership was 35,277, over a third greater than the 26,000 riders Valley METRO (the Maricopa County Public Transportation Agency) had forecasted.[41] Projects such as this indicate that if further transit options are effectively integrated into land-use planning for the Greater Phoenix Area, they would be wildly successful as well.

A number of these "smart growth" ideas have caught on. According to a 2007 survey done for the National Association of Realtors®, "three-fourths of Americans believe public transportation and smarter development will do more to cure traffic than building new roads." In that same survey, 81% said that they wanted to redevelop older areas rather than building new, and 70% indicated that they were concerned about "loss of open land such as fields, forests, and deserts."[42]

### 14.5.5 Effective and Integrated Transportation and Land-Use Planning

With better transportation planning and support of alternative transportation choices like commuter trains, regular bus service, light rail, and walking and bike paths, vital urban centers can be developed and encouraged and residents can spend less time behind the wheel (see Chapter 27). New light rail systems have helped contribute to an increase in transit use in many communities, including some in the Southwest. Cities such as Denver, Dallas, and Salt Lake City and at least 10 other places have added new lines or extensions in the last 5 years (Figure 14.7).

Other large urban areas should develop real regional planning entities similar to the Metro in Portland, Oregon. The Metro is a regional elected body that serves 3 counties and 25 cities. Twenty-seven municipalities in the Phoenix area are addressed by the MAG, including 24 cities and towns, 2 tribal nations, and Maricopa County itself. MAG lacks any real authority, however, and tends to look out for the parochial interests of individual cities and serve more as a pass-through for federal highway dollars rather than provide for regional land-use and transportation planning. Clearly, regional planning for growth and transportation is imperative. What one of these 24 cities does impacts all others in the nearby metropolitan area.

FIGURE 14.7
Bicycling around Phoenix can offer an alternative to commuting by automobile, but bikes must share the wide streets with fast-paced traffic.

Regional planning needs to extend to open space preservation as well. Pima County (in southern Arizona) is attempting to plan preservation on a regional level via the Sonoran Desert Conservation Plan. This plan identifies important habitat for wildlife, areas for preservation, wildlife corridors, and also where development is appropriate. MAG has put together a decent and relatively comprehensive regional open space plan, but there has been no follow through or implementation of the proposal. It looks good on paper but will never be implemented without some regional guidance.

Preservation in many communities in the west also means identifying ways to protect state trust lands. These are the lands granted to states at statehood to generate revenues for the trust beneficiaries—primarily the public schools. Protecting key lands is a challenge as the mandate is to maximize revenue. That often can mean development, at least in the short term. Phoenix, Tucson, and Flagstaff all have state trust lands around their urban fringes. To adequately preserve these areas, a constitutional amendment is required, a change in Arizona's enabling act, and leadership and the political will from elected officials.

## 14.6 Conclusion

Communities should be encouraged to do a better job of recovering the costs of growth and to limit the burden of those who can least afford it with the costs of paying for sprawl development. There should be different levels of impact fees depending on where the development is located, and communities should consider waiving them altogether in downtown areas. Lawmakers should work to ensure that development is truly paying for itself and that, in particular, developers pony up for the cost of schools and other infrastructure needs. In Arizona, the legislature should explicitly authorize cities to impose school impact fees and to recover the full costs for development, including costs associated with parks and protecting open space.

Communities should evaluate and then implement measures that require inclusionary zoning so that the current economic segregation patterns are curtailed. Cities also need to look at allowing more mixed-use zoning and lessening some of the zoning restrictions that require wide streets, huge lots, and, ultimately, more sprawl.

Better transportation planning and investment in mass transit, so that people have real choices, are another imperative for better planning—not to mention protecting air quality and the overall quality of life. Communities must look at better and safer methods for accommodating both bicycle and pedestrian transportation—for example, developing bike paths away from the roads themselves and allowing for narrower streets that accommodate pedestrian travel and slow down automobile traffic.

If Phoenix and other desert communities are to address urban sprawl, live sustainably, and better protect and live in the desert, it will require significant leadership from elected officials, business interests, neighborhoods, and environmental and civic organizations. In lieu of leadership from elected officials, it may require additional ballot measures in the states that allow them. As these areas grow into this century, it will require a focus on quality rather than quantity and an end to the growth-is-good-for-growth's-sake mentality. It means an end to measuring value by how much concrete is laid and instead measuring value by the quality of the air, the safety of streets and highways, the time not spent in traffic, and the vibrancy of our communities.

Phoenix demonstrates all too well the downside of denying the environment in developing communities as well as the significant costs associated with sprawl. But Phoenix is also showing us that it is not too late to change, to revitalize our downtown neighborhoods, to invest in mass transit, and to reconsider sprawl. Perhaps, the 2008 real estate crash that has hit deeper and harder in Phoenix and its surrounding suburbs is helping to drive home these lessons and giving policymakers some room to consider more sustainable alternatives to sprawl. The City of Phoenix lays claim to several planning awards, including the title of "best managed city."* However, in our opinion, Phoenix is currently not on a sustainable path—environmentally or economically—in building neighborhoods and communities that adequately connect the residents to sensible transportation, housing, and employment options, which allow these people to recreate and live without the burdens posed by urban sprawl, but it could be.

## References

1. Reagor, C., Growth pattern crippled Phoenix: Half-empty outskirts suffer as once-reliable cycle busts, *The Arizona Republic*, February 15, 2009.
2. Pilkington, E., Paving Arizona: The state is urbanizing the desert at a blistering pace as millions grab a slice of the American dream, *The Guardian*, Guardian International section, February 23, 2007, p. 29.

* http://www.cityofphoenix.gov/citygovernment/awards/index.html (accessed August 10, 2011).

3. Christensen, J. H., B. Hewitson, A. Busuioc, A. Chen, X. Gao, I. Held, R. Jones et al., Regional climate projections, in *Climate Change 2007: The Physical Science Basis. Contribution of Working Group I to the Fourth Assessment Report of the Intergovernmental Panel on Climate Change*, Solomon, S., D. Qin, M. Manning, Z. Chen, M. Marquis, K. B. Averyt, M. Tignor, and H. L. Miller, Eds. (Cambridge, England: Cambridge University Press, 2007).
4. Dougherty, J., The growth machine is broken, *High Country News*, April 27, 2009.
5. Megapolitan: Arizona's sun corridor (Tempe, AZ: Morrison Institute for Public Policy, Arizona State University, May 2008).
6. The view from the bottom: Phoenix real estate market, July 16, 2009 in Knowledge@W.P. Carey.
7. Natural Resources Conservation Service, http://www.nrcs.usda.gov/programs/commplanning/ (accessed August 10, 2009).
8. Hits and misses: Fast growth in Metropolitan Phoenix (Tempe, AZ: Morrison Institute for Public Policy, Arizona State University, September 2000).
9. Ewing, R., J. Kostyack, D. Chen, D. Stein, and M. Eanst, Endangered by sprawl, 2005: How runaway development threatens America's wildlife (Washington, DC: National Wildlife Federation, Smart Growth America, and NatureServe, 2005).
10. Regional transportation plan (Phoenix, AZ: Maricopa Association of Governments, November 25, 2003), http://www.azmag.gov/Documents/RTP_2003-Regional-Transportation-Plan.pdf (accessed July 23, 2009).
11. See U.S. Census Bureau, http://www.census.gov (accessed July 23, 2009).
12. View of the Valley in 2040: What are we leaving our grandchildren? Phoenix, AZ: Maricopa Association of Governments, http://www.mag.maricopa.gov/detail.cms?item=4081 (accessed July 23, 2009, undated).
13. *Most Livable State Award 2008* (Lawrence, KS: Morgan Quitno Press, 2008).
14. Easing the burden: A companion analysis of the Texas Transportation Institute's congestion study (Washington, DC: Surface Transportation Policy Project, May 2001), http://www.transact.org/PDFs/etb_report.pdf (accessed July 23, 2009).
15. Transportation Policy Committee, Did you know? (Phoenix, AZ: Maricopa Association of Governments), http://www.letskeepmoving.com/The_Problem/Did_You_Know_/did_you_know_.html (accessed July 23, 2009).
16. Schrank, D., T. Lomax, and B. Eisele, The 2011 urban mobility report (College Station, TX: Texas Transportation Institute, 2011).
17. Driven to spend: The impact of sprawl on household transportation expenses (Washington, DC: Surface Transportation Policy Project, November, 2000).
18. Transportation Policy Committee, Did you know? (Phoenix, AZ: Arizona Department of Transportation).
19. 2008 PM10 periodic emissions inventory for the Maricopa County, Nonattainment area revised, June 2011 (Phoenix, Arizona: Maricopa County Air Quality Department, June 2011).
20. State of the air: 2012 (Washington, DC: American Lung Association, 2012).
21. Pitzl, M.J., See Valley air cuts life expectancy, *The Arizona Republic*, July 12, 2000.
22. Arizona comprehensive asthma control plan. Phoenix, AZ: Arizona Department of Health Services, http://www.azasthma.org/Common/Files/Asthma%20Plan.pdf (accessed July 23, 2009, undated).
23. Mean streets: How far have we come? (Washington, DC: Surface Transportation Policy Project, November, 2004).
24. See American Community Survey (Washington, DC: U.S. Census Bureau, 2002).
25. See Arizona Education Association, http://www.arizonaea.org/profiles.php?page=135 for guidance on class size in Arizona (accessed July 23, 2009).
26. San Martín, I., Town of Gilbert: Cost of community services (Tempe, AZ: Herberger Center for Design Excellence, Arizona State University, April 1998); Tischler and Associates' (1996) studies conducted for the City of Scottsdale, Average land use prototype analysis, fiscal impact analysis, and water, water resources, and sewer development fees; James Duncan and Associates' (1997) analysis conducted for the City of Mesa (related media published in the *Tribune Newspapers*

on August 27, 1997). For national statistics on the costs of growth, see also N. Newman, Fiscal impacts of alternative land development patterns: The costs of current development versus compact growth (Detroit, MI: Southeast Michigan Council of Governments Final Report, Eds. R. Burchell and N. Newman, June 1997), Issues and facts: Urban service areas for New Mexico (Albuqueque:1000 Friends of New Mexico, No. 2, Fall 1997), and, Alternatives for future growth in California's Central Valley: The bottom line (Washington, DC: American Farmland Trust, Summary Report, October 1995).

27. Arizona Revised Statutes 9-463.05.
28. Yantis, J., Valley, state make leaps in growth—Experts fear increased numbers could mean LA-like nightmares. *Tribune Newspapers*, December 31, 1997.
29. Orfield, M., *Metropolitics* (Washington, DC: Brookings Institute, 1997).
30. Guhathakurta, S., Who pays for growth in the city of Phoenix? An equity-based perspective on suburbanization, *Urban Affairs Annual Review* 33(5), 813–838, 1998.
31. Endangered by sprawl (Washington, DC: National Wildlife Federation, 2005).
32. See Arizona 2007 tourism facts year-end summary. Phoenix, AZ: Office of Tourism, http://www.azot.gov/documents/2007_AOT_Tourism_Facts.pdf (accessed July 23, 2009).
33. See Snowbird business down in East Valley, *The Arizona Republic*, February 2, 2000; Snowbird flock is shrinking, *The Arizona Republic*, July 20, 2001.
34. Sustainability for Arizona (Tempe, AZ: Morrison Institute for Public Policy, Arizona State University, November 2007).
35. Smart growth policies: An evaluation of programs and outcomes (Cambridge, MA: Lincoln Institute of Land Policy, May 2009).
36. Nelson, A. C. and J. B. Duncan, *Growth Management: Principles and Practice* (Chicago, IL: American Planning Association, 1996).
37. Meck, S. et al., Planning communities for the 21st century (Chicago, IL: American Planning Association, December 1999).
38. Conservation easements. Phoenix, AZ: The Nature Conservancy of Arizona, July 2007, http://www.nature.org/wherewework/northamerica/states/arizona/files/new_ce_az.pdf (accessed July 23, 2009).
39. Community redevelopment law: Your guide to California's redevelopment law. Sacramento, CA: MHA, McDonough Holland & Allen PC: Attorneys at Law, http://www.mhalaw.com/mha/practices/publicLawPubs/MHA_CRL_2009_updated.pdf (accessed July 23, 2009).
40. Inclusionary housing initiative. St. Paul, MN: Alliance for Metropolitan Stability Minneapolis, http://www.policylink.org/pdfs/EDTK/IZ/Alliance.pdf (accessed July 23, 2009).
41. Light-rail ridership rises in February, *East Valley Tribune*, March 16, 2009.
42. National Association of Realtors, *2007 Housing Opportunity Pulse Survey* by Public Opinion Strategies, Alexandria, VA. Key findings from a national survey of 1000 adults conducted October 5, http://www.cityofphoenix.gov/citygovernment/awards/index.html 7, 9–10, 2007.

# 15

# *Water Planning for Growing Southwestern Communities*

**Sharon B. Megdal and Joanna B. Nadeau**

**CONTENTS**



## 15.1 Introduction

The Southwest is known for its diversity of topography, ecology, economic activity, and cultures. Communities large and small have been growing rapidly. Population changes reflect the net inflow of people into communities as well as the net growth of the existing population. The relative importance of these two components of growth will vary across regions within a state and across states. Planning for growth is a challenging endeavor. It requires vision and careful consideration of alternatives. Failure to plan will not stop growth, but effective planning will influence the pattern of land development that results from increasing populations in Southwestern urban centers and changing economic activities.

Water is essential to life and living communities. Yet the water supplies are not always where the people and their associated economic activities locate, as is apparent in the widespread reliance on imported supplies to meet the needs of large Western cities, such as Las Vegas, Phoenix, and San Diego. The Southwest is home to dams and constructed water projects that are technological marvels (Figure 15.1). It is also home to rivers that only flow during storm events and degraded riparian areas. Federal funding has been essential to the growth of the Southwest. Witness water delivery projects such as the Salt River Project, Central Arizona Project (CAP), and Central Valley Project. But the engineering and calculus of water planning have changed. Rather than dams, there is focus on desalination technology. Location of wastewater treatment facilities must now consider the potential for reuse of the water rather than only the disposal of the outflows. The natural environment is being recognized more and more as a water using sector, in addition to the traditional three: agricultural, municipal, and industrial.

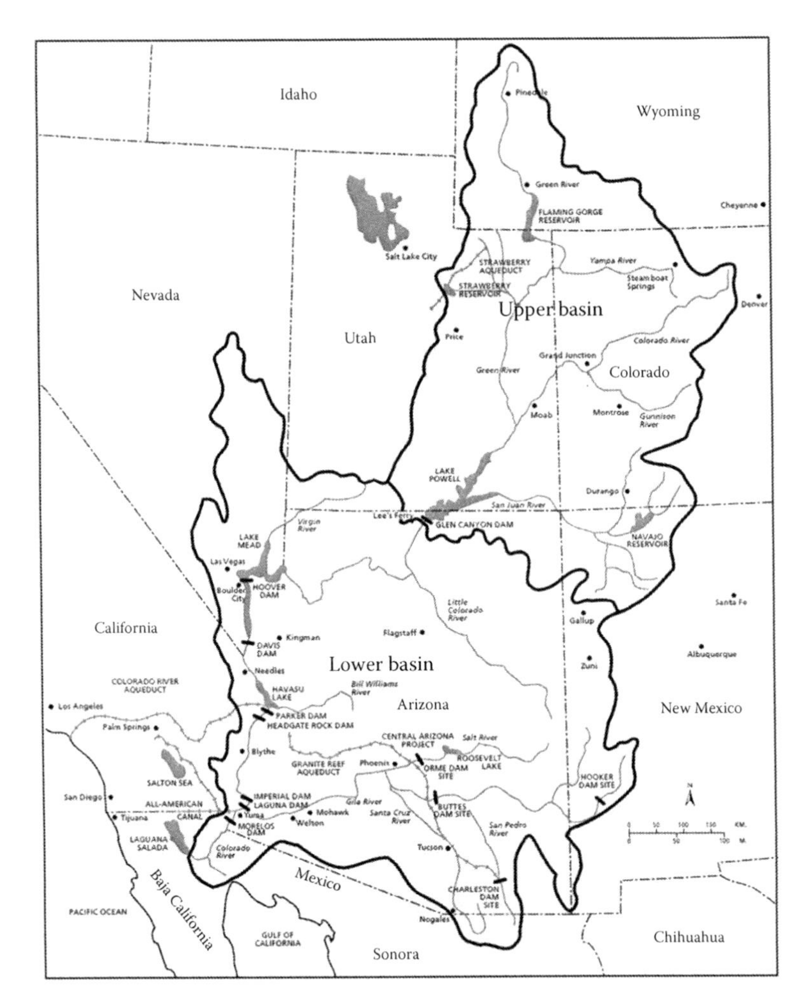


**FIGURE 15.1**
Map of Western water projects. (From Central Arizona Project. With permission.) (Courtesy of The Pacific Institute.)

Water supply planning has in the past been the exclusive domain of water agencies, districts, and suppliers. The job of the water purveyors and suppliers has been to ensure sufficient supplies for existing and future populations. Many communities across the Southwest are looking to a water future characterized by significant unknowns. Concerns abound regarding long-term overdrafting of aquifers, or pumping in excess of the supply, and overallocated river systems where total allocations exceed average flows (see Chapter 4). For example, studies suggest that the Colorado River's annual flow historically varied widely around the average of 14.3 million acre feet/year, despite the fact that the allocations total 16.5 million acre feet/year (Table 15.1).*

As we learn more about the history of droughts and the implications of climate change, concerns exist regarding overly optimistic projections for surface water flows and their inherent variability. The public may recognize the importance of good water

* See Gelt[1] and Colorado River Compact 1922 for a discussion of Colorado River allocations and Wong et al.[2] For a discussion of Rio Grande allocations see Woodhouse et al.[3] for the most recent information about annual flows in the Colorado River.

**TABLE 15.1**

Colorado River Allocations

| State | Annual Allocation (Million Acre-Feet per Year) |
|---|---|
| Upper basin | |
| Colorado | 3.88 MAFY |
| Utah | 1.73 MAFY |
| Wyoming | 1.05 MAFY |
| New Mexico | 0.85 MAFY |
| Lower basin | |
| Nevada | 0.30 MAFY |
| Arizona | 2.85 MAFY |
| California | 4.40 MAFY |
| Mexico | 1.50 MAFY |
| Total | 16.5 MAFY |

stewardship at the home, but is aware that water saved at one household may enable the next home to be built.[4]

While the connection of population growth to water availability is obvious, there is often a significant disconnect between land-use planning and water planning. Different agencies and individuals split or share responsibilities, as in Arizona where water regulations are handed down from the state level for municipalities to implement alongside local comprehensive land-use plans. There is an intermingling of the private and public sector in water provision. As communities in the Southwest find themselves growing into their known water supplies or finding out that certain water supplies cannot be counted on as reliably as previously thought, water planning has taken on a more visible and important function. This essay considers key issues related to planning water supply portfolios and water management strategies for growing Southwestern cities.

## 15.2 Traditional and Nontraditional Water Supplies

As discussed elsewhere in this volume, traditional sources of water have been groundwater and surface water, the latter often diverted for delivery through human-made systems (see Chapters 4, 21, and 26). Groundwater and surface water are connected, as any student of the hydrologic cycle knows, but they are often managed as separate systems. Water quality regulations are often established and enforced by different agencies than those that oversee water quantity (water supplies), although water quantity and quality are obviously connected as well. The quality of water is a key determinant of the water available for use. Seawater had not been considered a water supply available for community use until the technology to remove salt became economically feasible. The outflow of wastewater treatment plants—called *effluent*—had been considered a nuisance or something to be disposed of rather than a valuable water resource. Like desalinated seawater or brackish water, effluent treated to high standards is now considered an important water resource.[5] Arizona laws consider discharged effluent to be appropriable as surface water.* In New Mexico, state and regional

* 1989 John F. Long water rights case.

water plans encourage municipal water rights holders to reuse effluent to augment return flows to streams, as this flow can help meet the allotments of downstream users.* Long used for industrial and agricultural and turf irrigation, there is widespread recognition of the value of this water resource, with less agreement on the uses of it. Effluent, or as is more politically correct, *reclaimed water*, is often regulated separately from other sources of water. For example, in Arizona, it is distinct from surface water, even when flowing in a river after discharge, and from groundwater. The Arizona Department of Water Resources, which oversees the state's groundwater regulations, does not have the same authority over effluent.

In the category of nontraditional sources, I include desalinated water, highly treated reclaimed water, and conserved water, that is, water saved as a result of efficiencies in use. I label these supplies as nontraditional because the first two have not been common components of community potable water portfolios in the Southwestern United States. Even though conservation has long been a part of the water ethic of the Southwest, many communities find that they have not emphasized the importance of conserved water as an indirect water supply. Reduction in community per capita water use typically means that a given water supply can serve more users.† An interesting example of new, nontraditional, initiatives is illustrated by the city of El Paso, which has included reclaimed wastewater and conserved water as components of their water budget and is now using desalinated brackish water to help meet water demand.‡ In New Mexico, most municipal water right permits include a condition imposed by the State Engineer that the utility achieve a specific conservation goal.§

Desalinated seawater has become an economically viable option for communities (Figure 15.2). In Mexico and California, seawater desalination programs are in various stages of development.¶ Inland states are also dealing with saline water issues as a response

**FIGURE 15.2**
Desalination plant. (From U.S. Bureau of Reclamation home page for the San Juan-Chama Project Colorado and New Mexico, http://www.usbr.gov/projects/Project.jsp?proj_Name=San%20Juan-Chama%20Project, accessed on July 15, 2011.) (Courtesy of the US Bureau of Reclamation.)

* New Mexico State Water Plan and Regional Water Plan Template, *Middle Rio Grande Regional Water Plan*, A-27.
† For an example of how community per capita water use affects the population that can be served, see Megdal.[6]
‡ El Paso Water Utilities has been using reclaimed water since 1963. Its most recent endeavor, a joint project with Fort Bliss is a desalinization plant capable of producing 27.5 million gallons of fresh water daily. See website for more information: http://www.epwu.org/water/water_resources.html (accessed July 28, 2009).
§ See Longworth[7] and Office of the State Engineer.[8]
¶ See, for example, Rodgers.[9]

to shortages of freshwater. Arizona, for example, has interest in seawater desalination due to the potential to either pipe water inland or, more likely, enable exchanges to be made, whereby more Colorado River is delivered into Central Arizona in exchange for funding for desalinated water. No such long-term agreements are yet in place, and it may take years for them to be formulated.

Desalting technology also has the potential to treat brackish water sources to potable quality. Operation of the Yuma Desalting Plant is of interest to the CAP, the Metropolitan Water District of Southern California, and the Southern Nevada Water Authority. Though built to treat return flows from the Wellton-Mohawk Irrigation and Drainage District, the plant could treat high-salt groundwater, as well. Economic and environmental considerations, such as brine disposal, are key to operating this plant on a continuing basis. As in Arizona, in New Mexico, where at least two-thirds of all groundwater is considered brackish, communities are increasingly considering desalination as part of their future plans.*

Technology is a key consideration but clearly not the only consideration related to the use of reclaimed water (highly treated wastewater) for potable use. The "Yuck Factor" is clearly a key consideration.[13,14] While some communities have met significant resistance to connecting reclaimed water to the potable water system, if only indirectly through recharge, other communities have not. Orange County, California's Groundwater Replenishment Project, and Cloudcroft, New Mexico's reclaimed water treatment both involve systems that, albeit in different ways, mix highly treated reclaimed water into their potable water systems.[5] Reclaimed water has long been used for outdoor watering. The purple pipes associated with this use are common features of golf courses, school grounds, and parks. They are also becoming part of communities plumbed with dual systems to individual properties. Industrial uses and exchanges of effluent with farmers help to supplement water supplies. And even where effluent is not used to supplement other supplies, effluent-dominated riparian areas and created wetlands occur widely across the Southwest, such as at the Tres Rios Wetlands in Arizona and the Prado Wetlands in California.

A way of gaining more water for potable needs as communities grow is through conservation and/or diversification of individual water portfolios. Conservation, which has been long practiced in the Southwest, reduces per capita consumption of water. Lower per capita community water use means a given supply can be spread over more users, thereby forestalling the need to secure additional supplies. There is significant opportunity to water conservation, as evidenced by the conservation programs introduced in Las Vegas, Nevada, and San Diego, California.† Although conservation is not something people oppose, some who are concerned about the rapid pace of growth in their communities question the value of conserving water if the conserved water is used to support growing populations rather than ecosystems degraded by water pumping and diversions.‡

Increased capture of rainwater or stormwater is another means of increasing the diversity of water supplies available to a community. However, laws in several states recognize that the hydrologic cycle prevails, and water taken from one location in the hydrologic cycle means that there is less water for another location. Large-scale capture of rainwater means less water flowing downstream in one form or another, and small-scale actions of individual households can add up. Nevertheless, rainwater harvesting, whether passive or active, is a means of substituting rainwater for pumped water delivered through a

* See, for example, Sandia National Laboratories,[10] McGavock and Cullom,[11] and Hill.[12]

† See, as but one example in a series about San Diego's experience in implementing voluntary water conservation Lee and Gardner.[15]

‡ For a discussion of a way of connecting concerns with the environment and conservation, see Schwarz and Megdal.[16]

groundwater-dependent community water systems (see Chapter 21). Outdoor watering does not require highly treated water. Therefore, as communities find themselves faced with expensive treatment options for potable water systems, separating indoor and outdoor water systems make sense. Because the cost of retrofitting plumbing on older buildings can be high, this may be more appropriate in areas of new construction. In 2008, Tucson, Arizona, became the first city in the United States to require new commercial buildings to rely on harvested rainwater for half of their outdoor watering needs, and both New Mexico and Texas promote rainwater harvesting initiatives.* To the extent that the rainwater is used to displace water delivered through the potable system, more use of rainwater will reduce demands on the engineered systems communities use to deliver water to their customers. It should be noted that the rules and regulations pertaining to rainwater systems vary significantly by state.† For example, in Colorado, rainwater harvesting is only allowed in rural areas.‡

Gray water systems provide the opportunity for individual water users to recycle their water and thereby reduce their demands on the potable water system. Widespread installation of gray water systems must be factored into the design and operation of wastewater collection and treatment systems. Existing wastewater collection systems rely on outflows from washing machines and dishwashers. These flows provide water used to move waste solids through the wastewater collection system to the treatment plants. In Arizona, gray water systems require a valve so that users can either send the gray water outflows to the sewage lines or divert them for on-premise use.§ This customer choice means that the flows to the wastewater treatment plant will not be as large or predictable as historically has been the case. This may require changes to the long-range water resource plans of community water systems. Lower demand on the potable system means less need for new water sources in the future, reduced flows to the treatment facilities, and less reclaimed water available to the community water system for turf facility or other uses. More widespread use of treated effluent, particularly in planned communities and large developments, will likely require smaller and more strategically located treatment plants than the large-scale treatment plants built in the past.

## 15.3 Investments in Water Infrastructure

The discussion of treatment technology should highlight the fact that identifying and paying for wet water supplies are not the only challenges associated with providing water to growing urban areas. The infrastructure investments associated with treatment and delivery of the water supplies can be huge. Although there is limited discussion of new dams,¶ transporting water from the Colorado River to the Navajo Nation or desalinated water

* See O'Dell,[17] The Texas Manual on Rainwater Harvesting, prepared by the Texas Water Development Board, 2005, and A waterwise guide to rainwater harvesting by the New Mexico Office of the State Engineer.

† See the student research paper by Lien.[18] Available from S.B. Megdal by request.

‡ See Johnson.[19]

§ It will be interesting to measure, if possible, the changes in outdoor water use that result from installation of gray water systems over time to see if the ready availability of flows from dishwashers and washing machines is significant enough to lead to increased water use for landscape irrigation in semi-arid or arid areas. Data are not often available to measure the effectiveness of water conservation approaches. See Megdal.[20]

¶ Dams are still under consideration. See Schwarzenegger ...[21] and Schultz.[22]

from coastal areas inland, or even Mississippi River water to the West,[23] would require huge investments. Quantifying the costs of long-term and large investments is challenging, as is determining how to pay for them. Fiscal systems that depend on growth to pay its own way—a principle that most agree is sound—can be subject to huge swings in revenues as the economy cycles. The economic downtown that began in late 2007 is a stark reminder of this reality. Contingency planning for expected swings must be a part of infrastructure financial and investment plans. Once begun, construction of a large treatment plant cannot easily be curtailed. In addition to the investments required for new supplies and associated infrastructure, investments to maintain the aging water infrastructure of existing systems must be factored into the capital investment plans of community water systems.

Infrastructure may not only be dams, reservoirs, canals, pipes, and treatment plants. Underground storage and recovery facilities have become an increasingly important component of water systems. They are being used to address water treatment considerations as well as differences in the availability of surface water across locations and/or time (Figure 15.3). We see innovative uses of underground storage in numerous areas, including many in Arizona and California. While in some locations there remains a disconnect between where water is stored and where it is recovered, in Arizona, groundwater recharge and recovery laws have enabled innovative approaches to groundwater management.[24] New Mexico is exploring the possibilities of groundwater banking as legislation for such was passed in 1999 and has a demonstration site within the city of Albuquerque and

**FIGURE 15.3**
Underground recharge facility. (From U.S. Bureau of Reclamation home page for the San Juan-Chama Project Colorado and New Mexico, http://www.usbr.gov/projects/Project.jsp?proj_Name=San%20Juan-Chama%20 Project, accessed on July 15, 2011.) (Courtesy of the Central Arizona Project.)

several other projects in development.[25] Arizona, California, and Nevada have engaged in interstate water storage or banking arrangements that have been important to addressing water supply needs across time and place.

Considerations of infrastructure to deliver water should include energy infrastructure. It takes significant amounts of energy to deliver water. The CAP, which pumps diverted Colorado River water uphill and as far as 336 miles, is the single largest consumer of electricity in Arizona.[26] Similarly, the San Juan-Chama Project uses a series of channels and tunnels to divert 110,000 ac ft annually, nearly 40 miles from the San Juan River Basin southeast across the continental divide to the Rio Grande River Basin.* The approximately 5000 ac ft a year received by Santa Fe from this project must be lifted nearly 1500 ft to reach municipal distribution systems.[28] Discussions of seawater or brackish water desalination quickly involve questions about the availability, cost, and carbon emissions associated with removing the salt from the water. Likewise, advanced treatment of wastewater for reuse requires significant energy.

Finally, water research should be considered another form of water infrastructure. Research and development are essential to understanding and evaluating alternatives in order to make wise resource allocation decisions. The private and public sectors, with the involvement of universities, will need to invest significantly in advancing our state of knowledge.

## 15.4 Regulatory Considerations

It is important to separate the matter of water to supply existing populations from that of water availability for growing populations. Regulations, such as the Assured Water Supply Program in Arizona, are designed to ensure that water supplies are legally, physically, and continuously available to support growing populations for 100 years. In New Mexico, similar planning regulations require a 40 year scope. Although, in the case of Arizona, these Assured Water Supply requirements do not apply to all parts of the state, they do apply to heavily populated Central Arizona, home to more than 80% of Arizona's population. These regulations allow groundwater pumping to depths as low as 1000 ft below land surface, and the groundwater replenishment requirements can be met in a very flexible manner,[29] but the requirement for physically available water ensures that there is water to serve existing populations.† Where water availability becomes a key concern is in identifying the water supplies to support future populations, and even then the challenges are many years into the future for some communities. Water planners must look to the long term because the solutions to these challenges will likely take many years to gain approval and implement.

The limitations on water availability for communities may be legal and economic rather than physical. Laws vary considerably across states. Legal water rights determine who has more senior rights to surface water in the Southwest. Agricultural users were often the first users and therefore hold the more senior rights. They may be willing partners in transactions to provide water to cities and towns, yet the terms of the transactions could take years to develop and take effect.[30] The law will determine who has the right to drill what kind of wells, and the rights and regulations may vary with the type of water user. Exempt, also

* See U.S. Bureau of Reclamation homepage....[27]

† This statement assumes surface water supplies associated with the Assured Water Supply Requirements are reliably available over the period of time either through surface water delivery or through recovery of stored water.

called domestic, wells can frustrate water planning for rural areas as they are outside the authority of the state to regulate. In many cases, state laws do not govern water use by Indian Nations, yet Indian Nations may be partners in water leases with the approval of the U.S. government. Therefore, there may be existing traditional water resources—surface water and groundwater—that can be redirected to cities and towns. Economic and legal considerations, as well as infrastructure investment requirements, tend to determine the feasibility of these voluntary transactions. These opportunities must be considered along with opportunities to invest in technology that enables use of water supplies that heretofore have not been used to meet potable water demands, what I call "nontraditional" sources.

Regulations affect water planning in many ways. Perhaps the most obvious set of regulations pertains to water quality. The federal 1972 Clean Drinking Water Act controls discharges of pollutants into waters of the United States.[31] The 1974 Safe Drinking Water Act requires the U.S. Environmental Protection Agency to set standards for drinking water quality and oversee the implementation and compliance with the standards.* The standards cover many constituents, including constituents found naturally in water, such as arsenic. Not having standards for some constituents does not mean the public is not concerned about them. Trace pharmaceuticals found in wastewater supplies are mentioned by many when the safety of water supplies is addressed.[32] Safe Drinking Act water quality regulations pertain to public water systems, where EPA defines a public water system as "a system for the provision to the public of water for human consumption through pipes or other constructed conveyances, if such system has at least fifteen service connections or regularly serves at least twenty-five individuals."† Most states have established agencies that focus on these and other water quality regulations, such as those relating to use of reclaimed water and gray water systems. These agencies are often separated from agencies overseeing surface water and groundwater management and distinct from those who oversee water company planning, policies, pricing, and profits.

Federal regulatory considerations come into play through the Endangered Species Act (ESA) and Clean Water Act (CWA), both of which increasingly involve watershed-scale ecosystem protection. Concerns about the delta smelt population in the San Francisco Bay-Delta region have affected water supplies for the Metropolitan Water District of Southern California, which serves approximately 17 million people, and other water users. Proposed solutions to the Delta conflict may cost as much as $7.5 billion.‡ Similarly, it has taken years for regional state and federal parties to formulate, and it will take many millions of dollars to implement the Lower Colorado River Multi-Species Conservation Program. In the Middle Rio Grande River Basin, the Silvery Minnow was placed on the endangered species list in 1994 and occupies an estimated 7% of its historical range. The U.S. Fish and Wildlife service has estimated that it may take up to 25 years to be able to reclassify the fish as threatened with a cost of nearly 115 million dollars.§ It has been suggested that for full recovery of the Silvery Minnow and restoration of its habitat, the Cochiti Dam (located approximately 50 miles north of Albuquerque on the Rio Grande) would need to be removed or reengineered.[35] Even where the ESA may not have direct applicability, concerns about species and their habitat influence water supply planning. For example, concerns about the Cienega de Santa Clara in Mexico have influenced decisions regarding the operation of the Yuma Desalting Plant along the Colorado River (Figure 15.4).

* http://en.wikipedia.org/wiki/Safe_Drinking_Water_Act (accessed March 28, 2009).
† http://www.epa.gov/safewater/pws/pwsdef2.htm (accessed March 28, 2009).
‡ For an example proposal, see *San Jose Mercury News*.[33]
§ See Rio Grande Silvery Minnow Draft Revised Recovery Plan.[34]

**FIGURE 15.4**
Cienega de Santa Clara in Mexico. (From U.S. Bureau of Reclamation home page for the San Juan-Chama Project Colorado and New Mexico, http://www.usbr.gov/projects/Project.jsp?proj_Name=San%20Juan-Chama%20 Project, accessed on July 15, 2011.) (Courtesy of the US Bureau of Reclamation.)

Who regulates and/or oversees these four Ps of water provision—planning, policies, pricing, and profits—has implications for water system decision making. Many water providers are governed by local governing bodies, such as a city or town council. The local elected officials who govern the city govern the municipally owned water system. Some are districts governed by elected boards distinct from city or town councils. Then there are water systems that are owned by private water companies. These companies are most often regulated by a statewide body, often called a public utility commission or corporation commission. These commissions typically base their decisions on factors or standards that can be quite different from those of local jurisdictions. For example, a city or town may allow a water company to collect fees in advance of building or using a treatment facility. A commission may require a private water company's plant to be "used and useful" before allowing any costs of the plant to be recovered. These approaches may pertain to infrastructure investment or investment in new water supplies as well. This asymmetry in private versus public water company oversight and regulation becomes more important as collaborative approaches to securing water supplies and infrastructure increase.

## 15.5 Uncertainties in Water Planning

The regulatory framework, along with other economic considerations, establishes important parameters for water planning. Yet any planning for the future is fraught with uncertainties. Key uncertainties include the rate of population growth and regulatory provisions, which often change over time. Consideration of alternative scenarios is essential to water planning. Uncertainties regarding where responsibility resides for water provision and water supply acquisition due to changing ownership over time can be important but difficult to address in water plans. Expansion of a service area by acquisition of other providers occurs as discrete events. The press has given much attention to the issue of privatization of water provision but little to the trend in some areas of the United States, such as Arizona, that, as urban areas grow, there is "municipalization" of private utilities as city-operated utilities absorb the operations of smaller private systems or new municipal utilities are formed to take over the operations of private systems.

Another unique situation that leads to uncertainty of water planning is shared groundwater with Mexico. A variety of aquifers underlie the international boundary. These shared resources have been increasingly utilized by urban populations in the border region, where growth rates have frequently outpaced national averages. Yet there is no binational agreement related to their shared management and planning across the border. Sister cities such as El Paso-Ciudad Juárez and Nogales-Nogales pump groundwater destined for municipal supply from shared aquifers, and while there may exist some informal communication on local usage, planners at the municipal, state, and federal levels on both sides of the border lack a binational framework, which would allow for conjunctive planning. In light of this situation, the U.S. Congress passed the Transboundary Aquifer Assessment Act of 2006 with the goal of assessing priority transboundary aquifers in the border regions of Texas, New Mexico, and Arizona.* As binational assessment activities are carried out, uncertainties regarding aquifer properties will be reduced, allowing for more informed future planning strategies.

What is likely of paramount concern to water planning in semiarid regions of the Southwest are uncertainties associated with physical/natural systems, such as those associated with drought and climate change. Some say that water-scarce regions are always in a drought. It is interesting to look at a definition or two of drought. The *Pocket Oxford American Dictionary, Second Edition*, defines drought as "a very long period of abnormally low rainfall, leading to a shortage of water."[36] An online source, the Free Dictionary by Farlex, defines drought as "a long period of abnormally low rainfall, especially one that adversely affects growing or living conditions."† Both suggests that drought involves a long period of abnormally low rainfall, which would mean that rainfall would have to be below the normally low amounts in many parts of the West in order for drought to be declared. Both also connect the low rainfall to effects on water availability and, therefore, living conditions. In any case, as communities have grown or are projected to grow into their known water supplies, planning for future water supplies under conditions of uncertainties about the length and severity of drought has become of greater importance. Water table declines, of concern in nondrought conditions, are exacerbated in times of drought. Reduced surface water flows, including reduced baseflows caused by reduced precipitation or groundwater pumping, have to be considered. Drought planning is a critically important component of water planning, and many municipal providers have drafted drought plans.‡ Yet in Arizona, it was not until 2004 that a Statewide Operational Drought Plan was adopted. Drought preparedness has become an important component of water systems planning.§

Important to many urban centers of the Southwest are the shortage sharing regulations for the Colorado River. It was not until late 2007 that then Secretary of the Interior Dirk Kempthorne approved terms for sharing a declared shortage on the Colorado River.[37] The terms were negotiated by the seven Colorado River Basin states (Arizona, California, Colorado, Nevada, New Mexico, Utah, and Wyoming) and presented to the Secretary. Mexico, which also shares Colorado River water, is not included in this shortage sharing framework. The criteria

* For text of the enrolled bill, see http://www.govtrack.us/congress/bill.xpd?bill=s109-214 (accessed April 30, 2009).

† http://www.thefreedictionary.com/drought (accessed March 28, 2009).

‡ Phoenix drafted in 1993 and revised in 2000 http://phoenix.gov/WATER/drtmain.html (accessed April 29, 2009), New Mexico drafted in 2002 and updated annually http://www.ose.state.nm.us/DroughtTaskForce/droughtplans.html (accessed April 29, 2009).

§ http://www.azwater.gov/dwr/Drought/ADPPlan.html (accessed March 28, 2009).

**FIGURE 15.5**
Lake Mead. (From U.S. Bureau of Reclamation home page for the San Juan-Chama Project Colorado and New Mexico, http://www.usbr.gov/projects/Project.jsp?proj_Name=San%20Juan-Chama%20Project, accessed on July 15, 2011.) (Courtesy of the US Bureau of Reclamation.)

for declaring and sharing a shortage are important for all involved, but their importance for the CAP, which delivers 1.5 million acre feet of water to three counties in Arizona, cannot be overstated. The CAP, which had to agree to lowest priority of Colorado River users in order to secure federal funding for the project, could suffer cutback of its entire allocation in times of drought, adversely affecting the water supplies of Central Arizona cities, such as Tucson and Phoenix.* Instead, under what are considered the most likely drought scenarios, this agreement called for more frequent and therefore less severe cutbacks through 2026.†

These guidelines do not cover all scenarios. River modeling is a complex task. It is difficult to model things that have never occurred before. Recent research has focused on the question of whether the Colorado River flows could diminish such that Lake Mead, the storage reservoir for the Lower Basin States (Arizona, California, and Nevada), would run dry (Figure 15.5). There is disagreement on the probabilities of this occurring, but scientists have placed the probability at greater than zero.‡

What is most challenging for planners is that paradigms for water modeling and therefore water management have changed. In an oft-cited article in *Science*, seven scientists concluded that approaches to data gathering and modeling must be adapted to incorporate fundamental changes in climate.[42] They conclude that, in the face of the enormous challenges of renewing decaying infrastructure and building new infrastructure, we must "update the analytic strategies used for planning such grand investments under an uncertain and changing climate." The lack of "stationarity" cited by the scientists essentially means that water planners must consider new/more scenarios with greater unknowns.§ If the scientists can change and explain their enhanced approach to modeling, the water planners can better understand

* Central Arizona Project.[38]

† Arizona began a system banking CAP water for future times of shortage in 1997. Therefore, planning for drought on the Colorado River began well before the official shortage sharing guidelines were established.

‡ See Barnett and Pierce,[39] McKinnon,[40] and Myers.[41]

§ The concept of stationarity is similar to what Alan Greenspan stated to Congress in the Fall of 2008 regarding the economic models on which he based a lifetime of economic thinking. He stated that the economic meltdown had left him in a "state of shocked disbelief." He cited "a flaw in the model ... that defines how the world works." See The Associated Press.[43]

how to incorporate these changing circumstances in their water plans and thereby create more robust or resilient strategies. What might seem like a workable strategy in some circumstances, such as transferring water from agricultural to municipal use, may not seem as appealing in a world where agricultural water use is a needed buffer to absorb reductions in water availability. One example of an expanded modeling approach is King County, Washington's use of adaptive management in water planning to address risks from climate change.* An example of modified water planning to/that reflect uncertainties is Tucson Water's emphasis on scenario planning as the focal point of its long-range water plan.†

## 15.6 Planning by and for Whom and What?

Is it realistic to think that water planning for the urban areas of the Southwest is only about the needs of the growing populations? Water resources support multiple parts of the economy. Is agricultural activity in a state dispensable? Food availability and security are public concerns. Agricultural land is a form of open, green space. Should environmental needs for water be factored into urban water planning? It has been demonstrated that property values in urban areas vary directly with the proximity to riparian areas.[46] Large sums of money are being spent on environmental restoration, indicating the public values environmental amenities. What should be the geographic area of water plans? How should land-use plans and water plans be connected?

These are obvious questions without obvious answers. Those operating and managing water systems will often say that their responsibility is to make sure water flows out of the pipe or tap. It is not their job to determine the nature of land uses, be it agricultural, commercial, industrial, natural open space, or residential. Communication and cooperation are keys to developing good plans. The physical landscape of a community will depend on the availability of water. Therefore, land-use planners must understand the framework for water planning and management. The water experts must take the time and invest the resources in fostering an understanding of water planning and management by those in public policy decision-making positions, the business community, and the community at large. Good water planning is necessary for good community planning.

While no one would argue with recommendations for more and better communication and understanding, one would be a Pollyanna to think that disputes can be avoided. Solutions to disputes or divergent interests will likely emanate from creative problem solving and development of voluntary agreements to address competing water needs. Changes to law may be necessary, but such changes will require consultation and negotiation with the affected parties. Therefore, we must look to resolving matters through negotiations and voluntary agreements. The seven-state agreement on Colorado River shortage sharing is an outcome of lengthy negotiation. Water sharing agreements between agricultural users and urban water authorities have also been achieved.‡ The purchase of instream flow rights to support riparian areas is yet another. These agreements can take a long time to develop and even longer to implement. They can involve significant monetary resources and always require leadership.

As water availability relative to projected demand is becoming a shared challenge of many communities, more recognize that they are not in it alone. State water planning exercises

* King County Climate Plan.[44]
† Tucson Water, *Water Plan: 2000–2050*.[45]
‡ http://www.sdcwa.org/sites/default/files/files/publications/watertransfer-fs.pdf (accessed March 29, 2009).

**FIGURE 15.6**
Stakeholder meeting. (From U.S. Bureau of Reclamation home page for the San Juan-Chama Project Colorado and New Mexico, http://www.usbr.gov/projects/Project.jsp?proj_Name=San%20Juan-Chama%20Project, accessed on July 15, 2011.) (Courtesy of the Central Arizona Project.)

underscore this, such as that of California, as do state visioning exercises, such as Envision, Utah. In Arizona, on the other hand, a statewide water plan is not required of the Arizona Department of Water Resources. In addition, Arizona shares a border with Mexico, shares Colorado River water with 6 other states, and is home to 22 Indian Nations and Tribes. Indian water rights settlements have quantified certain water rights, but some Indian water rights claims remain unsettled. Surface water rights remain unresolved by the courts. Effective and comprehensive water planning involves a complex web of players and jurisdictions.

Although separated by oceans, there are lessons to be learned from other countries, especially those who have dealt with drought and adverse water conditions. Australia is an example. There, the federal government has required State Natural Resources Management Plans. In developing them, they must engage a broad set of stakeholders including community groups and volunteers, local government bodies, industry sector, landholders and other natural resource users, the federal government, state government and state agencies, and regional natural resource boards and groups.* In Australia, they are investing heavily into seawater desalination. As a representative of the Perth Water Utility, Perth, Australia, stated, their water security is through diversity of water sources.† Areas have been faced with restrictions residents of the Southwest would consider extreme, such as a ban on outdoor watering. Perhaps it is a testament to our excellent water planning that, despite extended drought, communities have not had to impose such severe restrictions. Yet where voluntary cutbacks have not been successful, consideration of mandatory restrictions is the next step.‡

Engaging the full range of stakeholders is a crucial part of water planning. Among the associated challenges to meaningful stakeholder engagement are the following: facilitating sufficient understanding of complex issues; obtaining the long-term commitment of the players to participate; developing a customized approach to planning because one size does not fit all; implementing appropriate communications mechanisms; and, perhaps most importantly, funding the costs involved (Figure 15.6).

---

* Slide presented by Dr. Jennifer McKay of the University of South Australia, Seminar at the University of Arizona, February 20, 2009, Tucson, Arizona.

† Comment in presentation made at the *Colorado River Water Users Association Annual Conference*.[47]

‡ See Lee.[48,49]

## 15.7 Conclusion

Water planning has never been easy, but it is becoming more difficult due to relatively scarce supplies, growing populations, extended drought, and climate change. As communities grow, issues become more diverse and complex. The public is very aware of issues related to the safety of their water supplies. Many Southwestern areas are pushing the envelope of physical constraints. In the twentieth century, the challenges were largely engineering, and they were overcome through the building of dams and water delivery systems. Water delivery to areas like Los Angeles, Phoenix, and Las Vegas meant that they could grow. The early twenty-first century presents different challenges. Rivers are fully or over-allocated. Water tables are declining. Dams are rarely considered an option, even if there were water supplies to capture and deliver. Citizens recognize the value of open spaces and riparian areas. There are concerns about the security of our food supplies. Planning to meet the water needs of Southwestern urban areas is not only the job of water planners and managers. Water planning is a community exercise, where the size of the community likely differs from the boundaries of a single water provider or even a collection of them.

Water planning will have to incorporate public values regarding uses of water, such as reclaimed wastewater and water conservation. It will have to acknowledge that individuals may have more control of their water use through rainwater harvesting or on-site water recycling (gray water systems). The geographic area of relevance may be wider than in prior periods. It will have to incorporate a changed paradigm for scientific water modeling. In summary, good water planning will be more difficult but also more necessary than ever before.

## References

1. Gelt, J., Sharing Colorado river water, *Arroyo*, August 1997.
2. Wong, C.M., Williams, C.E., Pittock, J., Coller, U., and Schelle, P., World's top 10 rivers at risk. Gland, Switzerland: WWF International, 2007, available at http://assets.panda.org/downloads/worldstop10riversatriskfinalmarch13_1.pdf (accessed April 29, 2009).
3. Woodhouse, C.A., Gray, S.T., and Meko, D.M., Updated streamflow reconstructions for the Upper Colorado River basin, *Water Resources Research* 42: W05415, 2006.
4. Davis, T., UA idea: Tucsonans save water; funds go to restore our rivers, *Arizona Daily Star*, July 16, 2008.
5. Grenoble, P.B., Toilet to tap: Once again, *Water Efficiency: The Journal for Water Conservation, Professionals* 4(1): 10–15, January–February 2009, http://www.waterefficiency.net/WE/Articles/Toilet_to_Tap_Once_Again_5038.aspx (accessed April 29, 2009).
6. Megdal, S.B., Water resource availability for the Tucson region, http://ag.arizona.edu/azwater/publications.php?rcd_id=12 (accessed July 15, 2009).
7. Longworth, J., New Mexico courts uphold water conservation, Presentation at *WaterSmart Innovations Conference* (Las Vegas, NV: 2008), http://www.watersmartinnovations.com/PDFs/Thursday/Napa%20A/1100-%20John%20Longworth-%20Conservation%20Policy,%20Enforcement%20and%20Administration-%20New%20Mexico%20Courts%20Uphold%20Water%20Conservation%20Regulations.pdf (accessed July 31, 2009).
8. Office of the State Engineer, New Mexico Water Conservation Program, Policy Development, http://www.ose.state.nm.us/wucp_policy.html (accessed July 15, 2011).
9. Rodgers, T., Desalination plant plans OK'd, *San Diego Union Tribune*, August 7, 2008.

10. Desalination of saline and brackish water becoming more affordable, Sandia National Laboratories, *News Release*, March 19, 2009.
11. McGavock, E.H. and Cullom, C.C., Desalination of brackish groundwater in Arizona, *Proceedings of the 2008 Meeting of the American Institute of Professional Geologists, Arizona Hydrological Society, and 3rd International Professional Geology Conference* (Flagstaff, AZ: September 20–24, 2008).
12. Hill, K., Researchers develop low-cost, low-energy desalination process, *New Mexico State University News Release*, May 29, 2007.
13. Nancarrow, B.E., Leviston, Z., Po, M., Porter, N.B., and Tucker, D.I., What drives communities' decisions and behaviours in the reuse of wastewater, *Water Science and Technology* 57(4): 485–491, 2008.
14. Friederici, P., Facing the yuck factor: How has the West embraced water recycling? Very (gulp) cautiously, *High Country News*, 39, September 17, 2007.
15. Lee, M. and Gardner, M., Agencies get aggressive in efforts to curb waste, *San Diego Union Tribune*, September 8, 2008.
16. Schwarz, A. and Megdal, S.B., Conserve to enhance: Voluntary municipal water conservation to support environmental restoration, *Journal of the American Water Works Association* 100(1): 42–53, 2008.
17. O'Dell, R., City green step is nation's 1st, *Arizona Daily Star*, Tucson/Region Section, October 15, 2008.
18. Lien, A., Legal implications of rainwater harvesting for existing surface water right holders: Does Arizona have a problem? (Tucson, AZ: University of Arizona, unpublished manuscript).
19. Johnson, K., It's now legal to catch a raindrop in Colorado, *New York Times*, Page A1, June 28, 2009.
20. Megdal, S.B., Evolution and evaluation of the active management area management plans, http://ag.arizona.edu/azwater/publications.php?rcd_id=57 (accessed July 15, 2007).
21. Office of the Governor. Schwarzenegger proposes major water infrastructure upgrades to meet California's future needs as part of California's strategic growth plan, *Press Release*, January 9, 2007, http://gov.ca.gov/index.php?/press-release/5083/ (accessed March 29, 2009).
22. Schultz, E.J., Visalian in state Cabinet sees opportunity in water crisis, *Fresno Bee*, March 22, 2009, http://www.fresnobee.com/local/story/1278068.html (accessed March 29, 2009).
23. Brean, H., Mulroy advice for Obama: Tap Mississippi floodwaters, *Las Vegas Review-Journal*, January 12, 2009, http://www.lvrj.com/news/37431714.html (accessed July 15, 2011).
24. Megdal, S.B., Arizona's recharge and recovery programs, in B.G. Colby and K.L. Jacobs, eds., *Arizona Water Policy: Management Innovations in an Urbanizing, Arid Region* (Washington, DC: RFF Press, pp. 188–203, 2007).
25. Moore, S.J., An Overview of the Bear Canyon Recharge Demonstration Project, in *Proceedings from the 53rd Annual New Mexico Water Conference*, Albuquerque, NM, Water Resources Research Institute Report Number 347, 2008.
26. Scott, C.A., Varady, R.G., Browning-Aiken, A., and Sprouse, T.W., Linking water and energy along the Arizona–Sonora border, *Southwest Hydrology* 6(5): 26–31, September/October 2007.
27. U.S. Bureau of Reclamation homepage for the San Juan-Chama Project Colorado and New Mexico, http://www.usbr.gov/projects/Project.jsp?proj_Name=San%20Juan-Chama%20Project (accessed July 15, 2011).
28. Kevin F., Rio Grande water users rely on San Juan-Chama project water, *Waterline*, Fall, 2001.
29. Avery, C., Consoli, C., Glennon, R., and Megdal, S.B., Good intentions, unintended consequences: The Central Arizona groundwater replenishment district, *Arizona Law Review* 49(2): 339–359, 2007.
30. Council for Agricultural Science and Technology (CAST), *Water, People, and the Future: Water Availability for Agriculture in the United States* (Ames, IA: CAST, Issue Paper 44, 2009).
31. Smith, K.L. and Graf, C.G., Protecting the supply: Arizona's water quality challenges, in B.G. Colby and K.L. Jacobs, eds., *Arizona Water Policy: Management Innovations in an Urbanizing, Arid Region* (Washington, DC: RFF Press, pp. 121–136, 2007).

32. National Science Foundation, Pharmaceuticals in drinking water guide, http://www.nsf.org/consumer/newsroom/pdf/pharmaceuticals_water.pdf (accessed July 31, 2009).
33. Delta water plan will cost billions, *San Jose Mercury News*, June 21, 2008.
34. Rio Grande Silvery Minnow Draft Revised Recovery Plan, U.S. Fish and Wildlife Service http://www.fws.gov/southwest/es/Documents/R2ES/Rio_Grande_Silvery_Minnow_DRAFT_Recovery_Plan_Jan-2007.pdf (accessed April 29, 2009).
35. Cowley, D.E., Strategies for ecological restoration of the middle Rio Grande in New Mexico and recovery of the endangered Rio Grande Silvery Minnow, *Reviews in Fishery Science* 14: 169–186, 2006.
36. *Pocket Oxford American Dictionary*, 2nd edn. (New York: Oxford University Press, p. 243, 2008).
37. Boxall, B. and Powers, A., Colorado River water deal is reached—The Interior secretary calls it an 'agreement to share adversity,' *Los Angeles Times*, December 14, 2007, Nation Section, http://articles.latimes.com/2007/dec/14/nation/na-colorado14 (accessed March 28, 2009).
38. Central Arizona Project, *Colorado River Shortage*, Issue Brief, January 19, 2011, http://www.cap-az.com/Portals/1/Documents/Shortage-Issue-Brief-Jan-19.pdf (accessed July 15, 2011).
39. Barnett, T.P. and Pierce, D.W., When will Lake Mead go dry?, *Journal of the American Water Resources Association* 44: W03201, 2008.
40. McKinnon, S., Climate-change realities could ruin water planning, *Arizona Republic*, February 1, 2008.
41. Myers, A.L., Lake Mead fares better in new study, *Las Vegas Review-Journal*, November 19, 2008.
42. Milly, P.C.D., Betancourt, J., Falkenmark, M., Hirsch, R.M., Kundzewicz, Z.W., Lettenmaier, D.P., and Stouffer, R.J., Stationarity is dead: Whither water management?, *Science*, 319(5863): 573–574, 2008.
43. The Associated Press, Greenspan: More bad news due, *Arizona Daily Star*, October 24, 2008.
44. King County Climate Plan, February 2007 found on line at: http://www.kingcounty.gov/environment/waterandland/drinking-water.aspx (accessed July 30, 2009).
45. Tucson Water, *Water Plan: 2000–2050*, (Tucson, AZ: Tucson Water, http://www.ci.tucson.az.us/water/waterplan.htm) (accessed September 8, 2009).
46. Colby, B. and Wishart, S., *Riparian Areas Generate Property Value Premium for Landowners* (Tucson, AZ: Department of Agricultural and Resource Economics, University of Arizona), http://www.cals.arizona.edu/arec/pubs/riparianreportweb.pdf (accessed March 29, 2009).
47. Warburton, A., *Colorado River Water Users Association Annual Conference* (Las Vegas, NV, December 15, 2009).
48. Lee, M., Rationing proposal may pit super savers vs. wasters, *San Diego Union Tribune*, February 7, 2009.
49. Lee, M., Mayor Sanders releases plan for major water-use cuts, *San Diego Union Tribune*, March 20, 2009.

# 16

## *Removable and Place-Based Economies: Alternative Futures for America's Deserts*

**Kim Sorvig**

**CONTENTS**



### 16.1 Ghost Towns and Their Beneficiaries

Two telling photographs, from very different desert locations, introduce Wescoat and Johnston's *Political Economies of Landscape Change*.[1] The contrasting images form a sad and eloquent summary of removable-resource regions. The resource in question in these visual case studies is, unsurprisingly, water (Figure 16.1a and b). Figure 16.1a shows a ghost town called Roan Creek in the upper Colorado River drainage basin. From this vicinity, water has been removed, and along with it, human life. Wescoat's photo shows the abandoned Roan Creek Community Center, its surroundings eroded by footsteps of a vanished population, its backdrop a towering mesa that must have been a landmark even to casual visitors. This was a Place, but exporting its resources has led to its abandonment.

Figure 16.1b shows the beneficiary of the removal of Roan Creek's water: Las Vegas, Nevada. Here, a city-block-sized fountain forms the foreground for an entirely artificial place: a mishmash of architectural borrowings, simulated boulders, and imported palm trees. If the casino landscape ever had any genuine landmarks or real sense of place, those have been overwritten by Place-on-Steroids. Las Vegas, emblematic of most desert cities, is built and maintained with resources—water, gasoline and asphalt, hydroelectricity, cement, transplanted trees—removed from Roan Creek and a thousand other places like it.

(a)

(b)

**FIGURE 16.1**
(a) The ghost town of Roan Creek, abandoned when its water was removed to supply Las Vegas. (b) Las Vegas' illusory landscapes are created almost entirely from imported resources. (From Wescoat, J.L. Jr. and Johnston, D.M., *Political Economics of Landscape Change*, Springer, Dordrecht, the Netherlands, 2008.)

Dried-up small towns and artificial oases are opposite facets of a single problem. Both fail the test that Aldo Leopold used to define his land ethic. "A thing is right," he wrote in his 1949 *Sand County Almanac*, "when it tends to preserve the integrity, stability and beauty of the biotic community. It is wrong when it tends otherwise." Leopold suggests that economic value can—and must, especially in deserts—be based on the inherent and complex worth of place.

## 16.2 Sustaining the Desert?

All human development simultaneously depends upon and alters the ecosystem in which it occurs. Thus, sustaining a place and sustaining a human population (especially a large one with modern consumption habits) may be incompatible goals. Nowhere is this thrown into sharper relief than in desert regions. *Sustaining* the desert means maintaining conditions that many humans consider inimical. *Developing* the desert conventionally means replacing predevelopment ecosystems with landscapes dependent on imported and exported resources. The importation and exportation of resources is a critical question but often underemphasized in discussions of desert settlement and, indeed, in envisioning sustainability in general.

A desert, by definition, is "a region rendered barren by environmental extremes" unpopulated, unproductive, lacking useful vegetation, water, and other essential resources.* Deserts are defined by their *inability* to support human life, or their *undesirability* as places to live. Yet today, North American deserts are popular sites for development, their spacious, iconic landscapes attracting both tourism—since at least the 1800s—and urbanized settlement.

Modern desert development relies heavily on imported water and food; exports, especially of minerals, support these imports as an economic exchange. Conventional development of regions with limited carrying capacity requires exportive economics—removing resources

* *American Heritage Dictionary*, 1976.

from one region to support another. Even in purely economic terms, this conventional relationship between desert and nondesert is hardly sustainable. Exporting raw resources and importing finished goods puts a region at a serious disadvantage, as the 13 American colonies soon realized even in a setting far from desert. As for the environmental results, as early as the 1950s, author Peter Matthiessen warned that the American West and the Southwest more particularly, were slated to become a "national sacrifice zone."[2] His warning was born out by a headlong increase in mining and drilling lease activity across the American West during the Bush–Cheney administration.

## 16.3 Resources versus Places

To be sustainable, desert development must go well beyond "green" technologies, although these will clearly be important.[3] In particular, desert regions must resolve the paradox of resources versus places. Put succinctly, that paradox is this: *we value raw materials enough to destroy the living land in extracting them.* In nondesert regions, society may be able to ignore the conflict between place and resource. In the desert, the paradox is unavoidably and starkly visible.

Desert places and resources have been fertile ground for conflict since European development began. Today, deserts worldwide are threatened by exponentially growing demands, driven by declining mineral supplies and changing demographics. These trends contain the seeds of new conflict—and in the desert, dormant seeds can lie hidden for years, and then sprout with sudden ferocity.

## 16.4 Wholesale History

A history of desert settlement has been detailed in previous chapters (see Chapter 26). When evaluating prospects for desert sustainability, historical habits of resource use (and changes in those patterns) become critical.

The American West was acquired wholesale, that is, in large chunks and at a steep discount. Whether taken directly from native cultures or bartered among colonial powers, the land transfers were vast: the Louisiana purchase, the Alaska "folly," the Guadalupe-Hidalgo treaty. Although not exactly *terra incognita* to white settlers, much about these regions was unknown. Land was subdivided and property rights were assigned using a grid of anonymous square tracts, a land-use system that military and imperial regimes throughout history have found practical for occupied regions (Figure 16.2)* Land claims—and government giveaways to the railroad or other industries—were by the square mile. Since so little was actually known about the characteristics of any given tract of land, profiting from the mineral and agricultural claims was considered a gamble.

* For a variety of interpretations of the significance of grid settlement planning, including its prevalence under expansionist empires including Chinese, Japanese, and Roman, see Smith.[4] A grid is an ideal system for rapidly identifying locations in unfamiliar territory; in some ways, the gridding of land might be added to Jared Diamond's list (*Guns, Germs, and Steel*) of factors giving "Western" cultures the ability to conquer other societies, whose way-finding methods commonly required intimate first-hand knowledge of place.

**FIGURE 16.2**
The West's ubiquitous square-mile grid shows a culture treating unique places as interchangeable units of land.

These conditions shaped attitudes toward the land: an unpredictable adversary in its existing form, and an unlimited resource if transformed and shipped elsewhere.

Early uses of the Western deserts focused on removable resources: mining, timber, and large-scale ranching.[5] Local settlements were a limited market; the profit lay in export. Local use of resources usually fosters a degree of stewardship, but exportation demands ever-increasing scale, cheap production, and minimal involvement with what today would be termed "environmental" effects. In sparsely inhabited deserts especially, local residential interests were too few and far between to balance exportive schemes.

With enough space, early surface interests (ranching and timber) were able to coexist with subsurface ones (mining and oil drilling), which were small, low-tech, even pick-and-shovel operations compared to today's massive ventures. The doctrine of "split estate" gave (and still gives) a miner or oil-wildcatter the ability to override any objections from surface owners.* When ranches were large and mines or well-fields relatively small, this system was apparently a workable compromise. This is far from true when the same laws are applied to modern settlement patterns.

Laws governing that other contentious Western resource—water—also favor commercial use, giving the first surface-water claimant complete precedence over any subsequent ones.† Unlike Eastern U.S. and European water law, this is not a system that recognizes the irreducible water needs of residents who must share water to survive. The Western water-law system puts Roan Creek and Las Vegas in an inevitable fight to the death.

Although it would take several dissertations to fully separate fact from fiction on these topics, the land-use and water laws that govern all of America's deserts clearly favored the *removal* of resources; settlements endured despite these laws, rather than with their protection. Increasing residential numbers and density, industrial-scale extraction methods, and economic pressure on declining supplies of resources are amplifying this situation toward an uncertain outcome.

---

* This theory, which one expert has referred to as "part of the Balkanization of land-ownership rights," is enshrined in the Mining Act of 1872, under which subsurface interests are given utterly dominant rights, and in practice, get resource without paying significant royalties to the government. Reform was passed in the House (HR 2262 of 2007 and HR 699, 2008) but defeated in the Senate.

† See for example http://wwa.colorado.edu/western_water_law/ (accessed August 12, 2011), the University of Colorado's excellent summary of these issues. Many other websites on the topic are also readily searchable.

Desert land-uses now are vastly different than in the 1800s, yet the laws remain nearly unchanged. In particular, the balance of export and local use has changed, with explosive implications for sustainable development in North America's deserts, and throughout the American West.

## 16.5 Amenity Migrants and Resource Refugees

Today, more than 13.5 million people live in the four main desert Southwest states (Arizona, Nevada, New Mexico, and Utah). Of these, almost 83% have moved there since the Second World War; in the single decade between 1990 and 2000, Arizona's population grew over 66%.* The trend is even more pronounced in Southwestern cities: about 3.5 million live in El Paso, Albuquerque, Las Vegas, Phoenix, and Tucson today; almost 99% of them arrived since World War II (see Malloy, this volume).

Few of these settlers make a living directly from the land; most newcomers, even in rural areas, own parcels closer in size to large suburban lots than to the vast spreads claimed by historic owners. More often than not they own only surface rights, the minerals beneath them split off and held by others. Many "subsurface" or mineral owners are absentee or corporate; their identities, in states like New Mexico, are deliberately kept secret and "proprietary," preventing surface owners from knowing who has dominant rights over their homes. Small lots offer little room to accommodate extractive industry. Secrecy and philosophical differences inflame the inevitable clashes.

The West's postwar settlers include large cadres who have been termed "amenity migrants"[6] because they choose their homes primarily based on quality of life. Many are retirees, free to live wherever they can afford. Employed amenity migrants tend to have jobs that are portable, digital, and/or entrepreneurial. Amenity migrants are not tourists, but choose to live in places, including deserts, that also attract tourism by their beauty, climate, cultural associations, and relatively low densities of development. Amenity migration has brought a new and influential demographic into the desert heart of extractive-resource country.

Amenity migration entails many contradictions. Although not tourists, they are often mistaken for tourists by established residents (and even some academics). Some amenity migrants are back-to-the-land types, knowledgeable and committed about sustainability. Their jobs (and/or retirement income) have positive economic impacts, while their genuinely land-based lifestyles have relatively few environmental impacts. Other amenity migrants, however, are looking for high-style living in the country. This is a resource-consumptive lifestyle that contributes to rising global demand for resources.

Amenity residents of both kinds hold values that conflict with local resource extraction. They set high value on healthy and scenic lands, and look unfavorably on industrial land-uses in their communities. Yet a host of rather recent technologies supports amenity migrants: cars allow them to live far from sources of supply; air conditioners keep climate at bay; communication systems make remote living (and working) safer and simpler than

* U.S. Census, compiled per state 1790–2007 by InformationPlease, online at http://www.infoplease.com/ (accessed August 12, 2011). The page specific to demographics is http://www.infoplease.com/ipa/A0004986.html (accessed August 12, 2011).

at any previous period. Most studies of the amenity migration phenomenon note these contradictions; some treat the migrants as outright hypocrites.

Poorly managed, amenity-driven population growth, like tourism, can be destructive to existing interests and to the sought-after amenities themselves. Nonetheless, this demographic can create healthy local economies that are viable alternatives to more conventional Western industries.* In doing so, amenity migrants are on a collision course with the resource-extractive users who have dominated the desert Southwest for so long. Many established domestic mines and well-fields are becoming depleted (e.g., "peak oil") while foreign ones are politically unstable. At the same time, recent arrivals who have come to the desert for its amenities are resisting the "resource sacrifice zone" concept—even as their numbers and lifestyles contribute, in varying degrees, to rising worldwide resource demand.

Oil, gas, mining, timber, and ranching interests are not just threatened, but outraged by place-protective public activism, and are fighting back.† The previously mentioned Western laws (most from the late 1800s) are all on industry's side. Split estate allows oil corporations or uranium mines to take whatever surface land they want, including people's homes, often without recourse or recompense. Rocketing mineral demand (much of it from developing nations, e.g., China and India) is pushing new exploration and production into residential areas that until recently would have been off-limits, while amenity migration is pushing residences further into "empty" countryside. New extractive technology exploits previously marginal resource deposits, often near settlements. Bitter conflicts and local disasters are inevitable. These battles are intensified by the American-Dream belief that one's home is sacrosanct and by the amenity migrants' environmental and health concerns, which industry often scorns.

This is important background—in very coarse strokes—to any discussion of whether sustainability can be achieved in the deserts of the United States.

These current trends imply several train-wreck scenarios. If those are to be avoided, one critically important goal is resolving the conflict between removable resources and livable places. A central goal in any discussion of sustainability, it is cast in a more extreme light by desert conditions.

---

* Headwaters Economics, 2000–2009, at least nine published studies of energy economics in the U.S. West, available online from http://headwaterseconomics.org/ (accessed August 12, 2011).

† An example of such a counter-attack by oil and gas producers has been documented in "Split Estate," a film by Debra Anderson, which has been aired on the Discovery Channel and will be released in theaters in 2010. For details, see http://www.splitestate.com/ (accessed August 12, 2011). Similar industry tactics against citizens are recorded in the recently released "Crude," a documentary on irresponsible oil development in Latin America. In their publicity, industry spokespeople focus on portraying surface owners as whining NIMBYs who knowingly bought cheap land (i.e., split estate property), and hypocrites who drive SUVs but don't want the scenery near their mansions littered with oil wells. These are near-quotes from charges leveled by oil industry lobbyists on radio call-in shows against activists trying to prevent unregulated drilling in New Mexico's Galisteo Basin. An archive of news reports and other documentation on this conflict, culminating in the Santa Fe County ordinance referenced in this chapter, can be found at http://drillingsantafe.blogspot.com/ (accessed August 12, 2011). Industry representatives also like to portray secret "proprietary" chemicals used in drilling and production as safe enough to eat. In fact, many such products are extremely toxic; they have killed livestock and even sickened a nurse whose contact was entirely indirect (through contaminated worker clothing after a spill); this occurred in La Plata County, Colorado, and was widely reported. See *High Country News*, http://www.hcn.org/wotr/gas-industry-secrets-and-a-nurses-story (accessed August 12, 2011). Unfortunately, little if any scholarly attention is being given to the medical, environmental, and social impacts of petroleum production itself, or of the public relations methods used by the industry.

## 16.6 Place-Based versus Extractive Communities

Dotting the desert Southwest are towns supported by place—revitalized small towns, resorts, new-economy hotspots modeled on Silicon Valley in the wilderness, retirement communities, and arts colonies. Some are strictly dependent on tourism, but many are not, having attracted permanent entrepreneurial residents.

Precisely because they are place-based, it is harder to generalize about cities like Sedona, Arizona, or Santa Fe, New Mexico, than about communities whose economies are wedded to extractive industry. It is unfair but revealing to contrast an oil town like Farmington, New Mexico, with art/realty/tourism-driven Santa Fe. Both are at the edge of this book's desert region; their economic strategies diverge, one based on removable resources, the other on place.

Farmington[7] was so named for its truck farms and orchards, which once rivaled Colorado and California for superb produce. Today, there is only one "production" orchard in Farmington, plus one research farm funded by private wealth (Figure 16.3). Since the 1920s, oil and gas booms, each followed by a bust, have buffeted Farmington. Reliant on a removable resource whose market value is volatile, Farmington's economy has repeatedly suffered plunging tax revenues due to sudden drops in mineral market prices—similar to what states and nations experienced in the 2009 recession.

Farmington's main drag shows the impact of haphazard extractive development, industrial in scale, pattern, and disregard for anything local. Despite near-heroic revitalization efforts, some pleasant new subdivisions, and remnants of the farm community it once was, Farmington still gives casual visitors an impression dominated by drilling equipment and cheap industrial buildings (some in use, some moldering), obscuring the town's three major rivers and spectacular bluffs.

Farmington began its existence with a place-based economy, and lost it. The city resembles Roan Creek in some ways, but not others. Roan Creek sold its water resources, but could not buy a replacement. Oil profits, while they last, can buy an imported living, but the *processes* of removal and boom-bust are destructive. When the exportive economy crashes, place-based resources must sustain the community—resources that are all too often compromised by each boom.

**FIGURE 16.3**
Despite a rare desert river landscape, Farmington today is visually dominated by extractive industry.

Contrasted with a gritty "Real West," amenity-focused communities are often viewed as interlopers, resented by old-timers and given derisive names like Fanta Se, sometimes well deserved. Yet studies by Headwaters Economics* show that amenity-based economies provide far more revenue and income in the West today than extractive industry. Mining and drilling creates under 3% of real income in the five Western states studied by Headwaters.

Taxes on highly priced energy resources contribute up to 15% of these state's revenues (a fact touted by industry at every opportunity, and often exaggerated by as much as a factor of 3). Yet *nonextractive* industries, most of them place-focused, provide over half the tax revenues and 95% of real income.

Amenity-focused towns clearly exploit their environment and resources, too. The difference is that most work hard (not always successfully, and not necessarily out of altruism) to protect the biosphere, the cultural institutions, and the local landmarks (natural or constructed) that attract people and fuel their economies. This is not to say that all such economies are environmentally harmless, or truly sustainable. Genuinely "green" industries can create lasting economic value while protecting Place. Real estate sales, film location rentals, tourist visitation, arts markets, outdoor recreation businesses, and the like can in theory do the same, though there is a two-edged potential for damage. The Headwaters studies clearly indicate such economies are becoming more prevalent and financially successful than extractive industry as economic drivers in the Western United States.

Although there are many factors that differentiate, say, Farmington and Santa Fe, their histories since the 1920s are instructive. It was in the 1920s that oil and gas first became an industry in the Farmington area. In the same period, Santa Fe was down on its luck, bypassed by the railroad and losing its centuries-old preeminence in New Mexico commerce. A deliberate decision (still controversial) set the city on what would someday be called a New West path—becoming an arts colony (today rivaling New York in gallery sales) that spun off tourism and other place-based industries. Note that a nondestructive amenity-based product—Santa Fe Style arts and architecture—preceded and drove the tourism boom.

Billions in petrochemical wealth have been extracted from Farmington (doubly exported, since much of the resulting profit flowed only to Texas). Yet it is Santa Fe that has seen the more stable prosperity. Where the oil economy experienced at least three major busts in the twentieth century, Santa Fe's place-based economy has seemed nearly recession-proof (Figure 16.4)† including those recessions caused (like those in the 1970s and 2009) by volatile oil prices. Downturns that deter travel or destroy investments affect communities of all types—but place-based resources outlast such fluctuations, while resource removal amplifies instability.

Development using and protecting landscape and cultural resources has great potential resilience and longevity. This is far less true of development that relies on exportation of

* Headwaters Economics, 2000–2009, at least nine published studies of energy economics in the U.S. West, available online from http://headwaterseconomics.org/ (accessed August 12, 2011). In analyzing economic impacts of mining and drilling, Headwaters makes a number of key distinctions, often overlooked in other studies, and blatantly "spun" in industry publicity. These distinctions involve royalties to private versus government landowners; severance taxes; local job creation and personal income (versus out-of-state hiring); and handling of infrastructure necessitated by extractive business (who pays? Industry directly? industry via taxes and fees? Or taxpayers?) Where any of these questions is ignored, the analysis becomes skewed.

† http://www.santafenewmexican.com/Local%20News/Refuge-from-the-crash (accessed August 12, 2011). "Refuge from the Great Depression: As the stock market crash of 1929 destroyed lives and economies throughout the country, Santa Feans largely escaped the bad times."[8]

**FIGURE 16.4**
Faced with declining commerce, Santa Fe adopted place-based architecture and arts as its economic base.

minerals or biological resources clear-cut or factory farmed.[5] "In for the long run" is a well-known investment strategy, and almost always beats short-term returns. Similarly, investing in place is by far the more profitable strategy—in the long run, which is where sustainability will be judged.

## 16.7 Removable versus Place-Based Resources

Implicit in the contrast between economies based on extraction versus place is a distinction between types of "resources." On the one hand, "removable resources" are the focus of much conventional economic theory; on the other, "place-based resources" are undervalued or ignored. Table 16.1 compares these resource types and some of their characteristics.

Place-based resources cannot readily be removed from a region without losing all or most of their economic value. Landscapes and local cultural institutions are obvious examples. Likewise, "ecosystem services" (e.g., cleansing of water by wetlands) have their greatest value in place and on a relatively local scale—relocating wetlands, for example, usually fails to provide equivalent functional wetlands.* Only a few ecosystem functions,

* See Kentula,[9] especially pp. 17–19 and Viani.[10] Both these reports raise serious questions about the possibility of "mitigating" destruction of natural wetlands by constructing replacement wetlands elsewhere. On-site restoration of damaged wetlands offers far better odds of ecological functionality.

**TABLE 16.1**

Comparing Resource Types

| Removable Resources | Place-Based Resources |
|---|---|
| Natural resources as "raw materials" | "Ecosystem services" |
| Products for "export" | (e.g., water and air cleansing by vegetation)<br>Historic and cultural resources, activities, and institutions<br>Landscape amenity resources<br>(landmarks, climate, scenery, outdoor activities)<br>Products for local use and investment<br>"Regional image" products and services |
| Interchangeable enough to be widely valued | Unique to region |
| Other producers can compete directly | Competition based on uniqueness |
| Distance-dependent losses | Proximity-enhanced value |

like the sequestration of the greenhouse gas carbon dioxide, are truly global and place independent. Unlike mechanical functions, ecosystems are inextricable from place.

The distinction between the two types of resources is in some cases a matter of scale. Crops and minerals, for example, have a place-based value that is finite: a given population can only consume so much corn or copper locally, for example. When the same resource is seen as exportable, its production can become an unlimited good—the more that is produced, potentially the more economic value that is created. Treating growth as an unlimited good has been frequently noted as a major contributor to unsustainable development.*

Removable-resource economics relies on several concepts that have troubling results when place is a concern. Exporting a removable resource requires transportation, whether haulage by vehicle or transmission through pipes or wires. Transport has distance-dependent costs, ranging from fuel energy to voltage drop or pipeline leakage.† In addition, geographic separation of production and consumption encourages consumers to ignore limits on the resource. Urban consumption can easily exceed the capacity of invisible hinterlands which become depleted, essentially creating new deserts. The production, transportation, and consumption of economy-of-scale resources also contribute immeasurably to homogenization of landscapes and of architecture.

Distance-based costs are minimized when resources are used locally. In fact, it has become a central principle of green building, as well as of the local food or "locavore" movement, that sustainability requires the shortest possible distance between source and user.‡

Export of resources is most profitable if there is a large market for the same product. Thus, the product must have least-common-denominator characteristics, being interchangeable or generic enough to be universally (or at least widely) valued. Where products are generic, competition is based on volume, price, quick availability, and artificially created branding. Regional place-specific products compete on the basis of uniqueness and distinctive quality, in addition to price and availability.

The theory of economies-of-scale, which applies best to exportive economics, is such gospel that an opposite truth is overlooked: for ecosystem services and green development,

* Probably the first such observations should be credited to Malthus; they have been part of common discourse since the Club of Rome report.

† This type of cost is sometimes known as a "friction cost."

‡ Nelischer[11] is a landscape construction materials expert, author of the *Handbook of Landscape Construction*. One of his primary recommendations for choosing among materials is to favor local ones.

small, local, and distributed systems—the opposite of large-scale centralization—are almost always far more cost-effective. Thus, when removable resources become the main focus of an economy, they can cause much of what is unsustainable and unsatisfying about today's consumer world: failure to live within our means, generic goods and places, and a type of competition that encourages marketing based on fiction.

Another critical issue with extractive resource production is that industries maximize profits by *externalizing costs*, using jobs and tax revenues as a deceptive carrot.* For example the oil industry, like other extractive industries, demands that taxpayers provide the infrastructure (heavy-duty roads, emergency services, etc.) that industry operations require. Many counties or municipalities simply cannot afford to expand such services to meet pressures caused by extractive operations; in that case, residents pay in terms of inadequate and unsafe roads, and underfunded police and fire departments.

Where oil or gas is developed, health care and housing for mostly transient workers is another cost borne by the producing locale. Because drug abuse, particularly methamphetamine use, has a higher than normal incidence among oilfield workers due to the stress, monotony, and danger of such jobs, both rehabilitation and crime may become issues for surrounding communities.† Damage to surface property, excessively sized "well pads," and unplanned dirt roads all diminish the ecological health, productivity, and resale/tax value of the land. Restoration of such damage and mitigation of toxic spills or deliberate waste disposal are theoretically the responsibility of the operator. If the operator walks away, goes bankrupt, or flouts the law, the local jurisdiction and residents are left holding the bill—paying either in dollars or in unmitigated pollution, or both.

Regulation, usually aimed at controlling these externalized costs and preventing irreparable harm to place-based resources, is fiercely resisted by the extractive industries. Despite their operations being, in the words of a major sourcebook on public environmental health, "one of the most serious sources of contamination of soils, waters, and the biosphere,"[12] extractive industries continue to lobby for *deregulation*. The oil industry, for example, has been exempted from at least six major environmental-protection and public right-to-know laws.‡ Antiregulatory policies, such as those adopted by the Bush–Cheney administration, give extractive industries resources from public lands with only a fraction of the resource value being returned to the American people. Nonregulation, along with tax breaks and subsidies, cost the U.S. and state governments millions, if not billions.

Infrastructure, surface protection, and pollution mitigation are all costs of doing business that industry has successfully avoided by externalization. Conventional wisdom holds

---

* See for example the 1995 Union of Concerned Scientists report, "Money Down the Pipeline: The Hidden Subsidies to the Oil Industry" summarized online as "Subsidizing Big Oil," at http://www.ucsusa.org/clean_vehicles/vehicle_impacts/cars_pickups_and_suvs/subsidizing-big-oil.html (accessed August 12, 2011). Public testimony about externalization of costs by the oil and gas industry was presented to the NM Oil Conservation Division "Pit Rule" hearings in spring 2008 by Professor Avraham Shama, UNM School of Management, and is available as public record from OCD.

† See for example http://www.timesrecordnews.com/news/2008/sep/10/ (accessed August 12, 2011). *The Times Record News*, based in Wichita Falls, Texas, specifically notes "Rampant drug and alcohol use among workers, some of whom turn to methamphetamine to get through 12-hour shifts and labor up to 14 days in a row." Similarly, the Denver Post's Mike McPhee (02/04/2009) reported "Colorado 8th in Methamphetamine Use" http://www.denverpost.com/news/ci_11628847 (accessed August 12, 2011), with oilfield workers playing a significant role in achieving this dubious distinction. Many oil-producing states have special (and costly) meth-focused social services directed at oilfield workers.

‡ Natural Resource Defense Council;[13] see also Sumi,[14] online at http://www.earthworksaction.org/oil_and_gas.cfm (accessed August 12, 2011), website of the Oil and Gas Accountability Project, P.O. Box 1102, Durango, Colorado, 81301. Both offer excellent overviews of the issue, available technologies, health and social costs, etc., and include extensive research resource listings.

that these costs are repaid by state severance tax and royalties (to private or government owners). It is usually counties, however, that bear the externalized costs. New Mexico, for example, returns tax revenues to the counties, but only at the whim of the state legislature, which frequently allocates taxes away from producing ("wealthy") counties to fund projects in poorer areas. According to Colorado-based Oil and Gas Accountability Project (OGAP), the actual producing jurisdiction rarely breaks even in the trade-off between externalized local costs and "trickle-back" state funds.*

Local jurisdictions also suffer indirect losses: property values plummet when extractive industry arrives. Amenity migrants, farmers, entrepreneurs, and tourists may shun the area. Established residents flee: unskilled long-term residents can neither get high-paid jobs nor afford boom time prices, while skilled professionals (e.g., doctors) become hard to attract or retain.†Out-migration decreases property tax, business start-ups, and business, income, and sales taxes. Some of the latter are offset by sales and income from oilfield employees, but once the inevitable bust arrives, those revenue streams are gone, while the potentially more stable ones based on residential and place-centered values are weakened or lost.

Because the desert West has such a long history of extractive industry dominance, these issues are at the core of any attempt to find sustainable ways of living in the arid regions of this country. Yet resource removal has been pushed into the background of green building and sustainable development, much as human population control and regional carrying capacity have become almost unmentionable. This could be fatal to sustainability. Admirable though water-harvesting, alternative energy, or New-Urbanist mixed-use planning may be, they cannot create a sustainable future unless the exportation and importation of resources is harnessed, with great care, to the goal of sustaining living places.

## 16.8 Mining, Munificence, and Maintenance

### 16.8.1 Ajo, Arizona

Ajo, Arizona, provides an example of the complex relationships between extractive and local economics.‡ Until 1984, its economy was dominated by a Phelps-Dodge open-pit copper mine (Figure 16.5). Ajo's town center today is graced by an unusually large arcaded plaza and an ornate whitewashed church, an oversized school, and a hospital. Clearly, some mining wealth was carefully invested locally to create these amenities.

In 1984, however, the mine was closed, a victim of international competition and pricing. An attempt to reopen in 2008 was crushed by another market drop. The results are obvious behind the fine facades. Today the Ajo hospital is completely boarded up, while the school and plaza arcade appear to be about half-occupied (Figure 16.6).

Here is a case in which an extractive industry apparently set aside unusually large percentages of boom time profit for civic buildings. Yet those structures themselves became a liability when the bust reduced population and removed funds for maintenance. Adaptive reuse, fundraising and grant-writing, historic preservation assistance, and

* OGAP legal staff, personal communication, November 2007.

† Dr. Jeffrey Neidhart, Farmington Oncology Clinic, personal communication, September 2008.

‡ Personal interview, April 2009, Ajo/ Cornelia Mining Museum staff.

**FIGURE 16.5**
The mine at Ajo, Arizona, idled by international price wars, is the subject of a small museum.

**FIGURE 16.6**
The boarded Ajo hospital, one of many large public buildings funded by the mine; others are only partly occupied.

volunteer maintenance can only go so far—and as population dwindles, so do these resources. While Ajo is far above the squalor of truly negligent company-town operations, it is faced with watching its beautiful but outsized architectural resources decay. Ironically, a town that invested less in large-scale improvements might be better able to reuse and sustain infrastructure after the mining collapse.

Approaching Ajo from the southeast, an astonishing geological formation appears out of the desert. Dead flat on top, monolithic gray brown except for a peculiar white layer, it stretches for miles, completely hiding the townsite from view. This is the spoil dump from the mine. An entire mountain, in effect, has been exported onto the desert.

This degree of landscape destruction also has secondary consequences: signs in Ajo notify the public that, because it is already despoiled, the surroundings are deemed suitable for a new high-voltage electric transmission line. Similarly, the nation's largest "concentrating photovoltaic" power plant has been proposed for a mine-tailings site near Questa, New Mexico—an admirable reuse, at first glance.[15] The fact that it is planned by the mining company that ruined the land (a subsidiary of Chevron Oil) raises knotty questions: conflict of interest, why fossil-fuel corporations control alternative energy futures, and what impact a toxic site will have on green-energy workers.

A final irony here is that abandoned mines can become tourist attractions. Ajo's museum at the pit edge is a step in that direction; Bisbee, Arizona, has a profitable mine-tour industry; there are oil-drilling museums in several states.

### 16.8.2 Yuma, Arizona

Yuma, Arizona, about 150 miles from Ajo, shows that there are also differences *among* extractive economies.* Yuma was founded as a steamboat landing at a river crossing. Its early livelihood was supplying the mining towns that housed most of the county's population. In the 1860s, Yuma became the county's center when those exportive cities went bust. As local markets shrank in importance, however, Yuma's agriculture became exportive. Today Yuma produces about 90% of U.S. lettuce, for example; clearance, drainage, and irrigation to support agriculture have depleted the Colorado River and killed fields with salinity. Agribusiness is often said to "mine the soil." Its boom and bust cycles may be longer than those of mineral extraction, but are still of real concern (Figure 16.7).

**FIGURE 16.7**
Yuma's farm economy outlasted the nearby mines it once served, but "mined the soil," destroying wetlands which have recently been restored.

* Unless otherwise noted, information about Yuma comes from Crowe and Brinckerhoff.[16]

Both mining and agribusiness rely heavily on "removable" water, especially in the desert, bringing the discussion full circle to Roan Creek. In Yuma's case, the fields thrived while the river was dying, its floodplain choked with monocultures of salt cedar and invasive reeds.* The dead river decreased the quality of life both for the city of Yuma, and for the Quechan tribe whose lands occupy the opposite bank. Recently wetland restoration† has become a shared vision for the region—one of an increasing number of initiatives to reverse local impacts of exportive economics.

## 16.9 Exporting the Sky for "Green" Energy

Even "sustainable" and "alternative" resources can suffer when they are seen as export-ready and removable. The current push for wind and solar "farms" is a case in point, and one that specifically targets deserts as remote sources of these resources—which are seen as removable (Figure 16.8).

Small-scale wind and solar electricity have honorable pedigrees in the environmental self-sufficiency movement. Recent interest, however, has focused on "grid-tied" systems. A 2009 clean-energy conference‡ that included Al Gore envisioned expanding the grid to "bring wind and solar energy from remote locations to the nation's cities." Exporting sun and wind from centralized large-scale production plants in the hinterlands "is essential to all that we do to promote renewable fuels," in the words of House Speaker Nancy Pelosi. Further, the participants argued that the transmission grid is so critical that states should be *excluded* from deciding where high-power lines are placed. This issue will directly affect the future of desert states, whose year-round intense sunshine and frequent strong winds are seen as surplus commodities to be relocated from these remote and "undeveloped"

(a)

(b)

**FIGURE 16.8**
Large-scale solar plants cover vast areas with mirrors (a) which concentrate intense reflected heat on towers (b). Risks of local atmospheric heating or wildlife incineration are seldom discussed. (Courtesy of brightsourceenergy.com; photo 105 Bright Source Energy.).

* *Tamarisk*, *Arundo*, and *Phragmites* species.
† Information on these restoration projects is from personal interviews, April 2009, with lead restorationist Fred Phillips and other project participants.[17]
‡ Information on this conference from wire services article by Hebert.[18]

regions to meet growing urban demand. One source estimates current alternative energy "farm" proposals would cover 2.3 million acres.*

The grid, even if new variants are pronounced "smart" by promoters, remains a resource-export technology (see Chapter 29). In the 1920s, as the Rural Electrification project, it was a marvel of engineering. At that time, there were few technologies for clean, small-scale, on-site electric generation. But today's photovoltaic and wind-electric (and recently announced fuel cell) technologies are efficient at small scales, and their main advantage is that they can generate electricity *at the point of use*, bypassing transmission and its many associated problems. Those problems include

- Transmission losses between generation and use due to voltage drop. Step-up and step-down transformers also lose some energy at each transformation. The energy is not actually "lost," but converts to heat.
- Complete lack of energy-storage capacity. Almost all "alternative" energy generating systems incorporate storage, by necessity. The grid only simulates storage, by shunting energy around the grid for slightly later use. This actually increases the overall transmission losses, which are distance-dependent.
- Costs (in materials, money, land, and fuels) of constructing, operating, and maintaining a far-flung linear system. Routed through undeveloped areas, the grid requires access roads, costly both in financial and environmental terms. Because the grid is linear, breaks and failures can occur anywhere along its length, and can be difficult to locate or access for repair. By contrast, on-site generation involves many points of generation, but each is self-contained, almost always served by existing roads, at a known location and of a small size, making diagnosis and access relatively easy.
- Massive site impacts of "farms" covering hundreds of acres with solar panels or 300-foot-tall wind turbines.
- Large differences (two or more orders of magnitude) in voltage for generation, transmission, and use. Centralized generating plants operate at high voltage; long-distance transmission requires even higher voltage; but consumers (except for a few industrial users) require rather low voltages (110 or 220). Dispersed on-site systems generate at voltages near those that are actually used; inversion (from DC to AC) is required primarily because the grid has made AC standard. When a dispersed generator is "grid-tied" to sell energy back to the grid, it is typically generating at low voltages (6–48 V), which are inefficiently coupled into higher transmission voltages.
- Difficulty matching generation to demand. Demand for electricity varies daily and seasonally, but large power plants are not easily or quickly adjustable. Many such plants are oversized in order to meet peak demand, greatly increasing overall wastefulness. Because generation is out of sight and mind, centralization arguably encourages thoughtless consumption. Local generation, by contrast, enforces planning and conservation.
- Vulnerability. Centralized systems are very exposed to disruption by terrorism, sabotage, or natural disaster; widespread disruption of on-site generators would be difficult.

* http://www.solarpowerninja.com/solar-power-government-industry-news/solar-energy-sparks-desert-real-estate-boom/ (accessed March 2, 2010).

There is thus a strong argument to be made that centralized generation and grid distribution are extreme energy-*wasting* systems. The grid entirely misses the primary advantage of "alternative" generation: dispersed on-site location. Centralized control and maintenance is at most a weak counterargument.

The main reason that the grid is still the focus of discussion is that it is owned by large companies protecting their interests. On-site generation would free citizens from paying monthly bills for energy. Tying alternative generation to the grid keeps utility companies in control. In addition, long-distance shunting from cheap-generation areas to expensive-use areas allows speculative trading in electricity as a commodity—gaming the system, the strategy revealed in the collapse of the Enron Corporation.

Desert regions are currently jockeying for position to export alternative energy "surpluses" via new grids. System inefficiencies and the sheer acreage of land impacted by "farms" and grids have all the makings of a desert disaster. Solar and wind generation are "greener" than coal or nuclear generation; that does not change the problems associated with exportive use of any resource.

## 16.10 "Exportive Economics" Elephant

The issue of "exportive" economics remains under the radar of most discussions about development, desert or otherwise. It is the elephant in the room, widely ignored. Thus it may be premature to ask what desert cities and rural residents can do about the problem—the first order of business is to acknowledge the problem.

Once communities recognize that exportive economics must be better managed, the following suggestions and observations may be helpful:

- Any community, rural or urban, that is serious about sustainability and living within regional carrying capacity must consider *transport distances*. Most green building programs, as well as local food and buy-local initiatives, set a radius for preferred procurement. However, many schemes (e.g., modular buildings, smart grids) are promoted as sustainable without evaluating the true costs of transport and installation.
- Decreasing dependence on removable resources requires *reduction of consumption*, conservation, reuse, and recycling, as well as the closely related need for durable rather than disposable design. (Disposal also 'removes' resources.)
- Focusing on *local and on-site resources* can have surprising results. The Architecture and Landscape-Architecture building at the University of Arizona, Tucson, for example, has a 1 ac landscape designed explicitly to "live off the wastes of the building"(Figure 16.9).* Much of the garden was constructed of "urbanite," that is, demolition waste. The entire landscape, formerly a parking lot, is irrigated by roof-harvested rainwater, condensate reclaimed from the HVAC system, and mineral-rich water from backflushing the university's drinking-water wells. "People think the desert lacks water," says Ron Stoltz, chair of the Landscape department; "on this project, we're drowning in it." Similarly, the Yuma wetlands restoration project

* See Sorvig.[19] The quote is from the garden's designer, Christy TenEyck.

**FIGURE 16.9**
The University of Arizona architecture and landscape architecture building's garden is entirely irrigated with surplus water from HVAC and other infrastructure.

recreates wetlands using water pumped out of farm fields. There are limits to such trade-offs, but many of them are to be found in desert regions.

- Requiring extractive industry to *pay upfront for impacts* of their business is another important tactic. Residential and commercial developers are routinely charged "impact fees," but it remains rare for extractive industries, whose impact is often larger and more damaging, to be asked to do so. The 2009 Santa Fe County Oil and Gas Ordinance makes prepaid infrastructure upgrades (stronger road paving, new roads, expanded emergency services) a condition of obtaining a drilling permit.*
- Similarly, it is increasingly common to *require reclamation and revegetation* of extractive sites. The Ajo mine, for example, if ever reopened, would move old and new spoil into the old pit area rather than spread it even further across living landscapes. Antiquated mining laws and industry-sponsored loopholes make these requirements less effective than they need to be.
- Cities and counties have the legal authority to *limit surface disturbance* as a matter of public health, safety, and welfare, even where they are prohibited from directly addressing mineral rights. Santa Fe County's ordinance does this in two ways. First, overlay analysis (GIS mapping of existing conditions, like studies originally

* Santa Fe County's ordinances, including its oil and gas development regulations, are online at: http://www.santafecounty.org/ (accessed August 12, 2011).

envisioned in the 1960s by Ian McHarg) determines areas sensitive to damage by drilling; applications to drill in high-sensitivity zones are presumed unlikely to be granted, although all applications are evaluated case-by-case. Second, total area of surface disturbance and clearing, including roads and drilling pads, is limited. Current drilling technology makes clustered (several wells per pad) directional drilling not only feasible but more profitable than old single-well placements, bringing average pad size down from 4 to 5 ac per well to 0.1 to 0.5 ac.* The Santa Fe law also limits the *total surface area per square mile* that can be developed for wells at a given time. The mineral owners' rights are not "taken," but must be developed in sequence if simultaneous development would require more surface area.

- Because exportive businesses create short-term profit but eventually go bust, communities must *set aside revenues* (severance tax, royalties, etc.) for the future. Norway, for example, carefully invests North Sea oil revenue.† Mitigating extraction impacts is a long-term investment, and varies state-by-state. New Mexico, for example, taxes extraction at fairly high rates, but spends less than any other Western state on oil and gas enforcement, mitigation, or infrastructure impacts.‡ This encourages current spending, at the expense of the future.
- *Resolving "split estate" laws* will require coordinated long-term effort at local, regional, and national levels. Such reform is essential, because split ownership is a demographic disaster waiting to happen—especially in the desert West. Large numbers of citizens live on such lands; many are retirees on fixed incomes whose prime asset is their home. This situation demands equitable reform, in a way that serves the public interest and is sustainable. Both technological and political initiatives will be required and difficult choices.
- *"Adjudication" of water rights* is increasing, in an attempt to avoid the situation where claimed rights exceed actual supplies. This issue, beyond the scope of this chapter, is particularly critical and difficult in arid regions. In effect, the same issue occurs with mineral claims: split estate allocates more rights (surface use plus removal by mining) than can actually coexist simultaneously.

Conflicts among varied claims on the land have led to unusual *coalitions*. A prime example is New Mexico's Quivara Coalition, bringing together two groups long seen as completely at odds: ranchers and environmentalists. Sustaining the land for the long term has provided common ground between what has historically been an extractive industry—cattle ranching—and relative newcomers who value wildlife, quiet outdoor

* Molvar[20] details the technological trends toward footprints (well pads) as small as one-tenth of an acres, citing specific projects where such methods were used—usually voluntarily by companies that recognize profit as a result. The report was reviewed by two senior oil industry geologists.

† http://en.wikipedia.org/wiki/The_Government_Pension_Fund_of_Norway (accessed August 12, 2011) Statens pensjonsfond—Utland, also called Oljefondet, "is a fund into which the surplus wealth produced by Norwegian petroleum income is deposited.... The purpose of the petroleum fund is to invest parts of the large surplus generated by the Norwegian petroleum sector, generated mainly from taxes of companies, but also ... license to explore ... Current revenue from the petroleum sector is estimated to be at its peak period and to decline over the next decades. The Petroleum Fund was established in 1990 after a decision by the Storting to counter the effects of the forthcoming decline in income and to smooth out the disrupting effects of highly fluctuating oil prices."

‡ Headwaters Economics, see Moss.[6]

recreation, and scenic protection.* Similarly, the Yuma East Wetlands project has produced a cooperating web of groups, including the City of Yuma, the adjacent Quechan tribe, and farmers, initially skeptical or opposed to recreating marshes along the Colorado. Ultimately, sustainability requires refocusing on what profits the group and the place—a Leopold-like land ethic—with individual profit as a result rather than the exclusive goal. In short, it requires economics to prioritize place-based benefits rather than removable ones.

More than in any other region, widespread development in deserts is based on importation of some resources and exportation of others. Experts on economic development, construction, sustainability, and public policy have, in many ways, ducked the key question of regional carrying capacity. It is almost un-American to suggest that carrying capacity might mean that some areas cannot and should not be inhabited. However, without careful, realistic life-cycle analysis of exportive systems, both the importing and exporting communities can be destabilized. Place-based development, close-to-source acquisition of resources, and carefully limited removal of resources are essential components of whatever future development occurs in the deserts of the world.

## References

1. Wescoat, J.L. Jr. and D.M. Johnston, *Political Economies of Landscape Change* (Dordrecht, the Netherlands: Springer, 2008).
2. Matthiessen, P., *Wildlife in America* (New York: Viking Press, 1959 and many subsequent editions).
3. Thompson, J.W. and K. Sorvig, *Sustainable Landscape Construction: A Guide to Green Building Outdoors*, 2nd edn. (New York: Island Press, 2007).
4. Smith, M.E., Form and meaning in the earliest cities: A new approach to ancient urban planning, *Journal of Planning History*, 6(1): 3–47, 2007.
5. Velasquez-Manoff, M., Can American West retain its wildness? *Christian Science Monitor*, March 8, 2009.
6. Moss, L., *The Amenity Migrants* (Wallingford, U.K.: CAB International, 2006).
7. Waybourne, M., *Homesteads to Boomtown: A Pictorial History of Farmington New Mexico and Surrounding Areas* (Virginia Beach, VA: Donning Press, 2005).
8. Sharpe, T., Refuge from the Great Depression: As the stock market crash of 1929 destroyed lives and economies throughout the country, Santa Feans largely escaped the bad times, *The New Mexican*, November 16, 2008.
9. Kentula, M. et al., *An Approach to Improving Decision Making in Wetland Restoration and Creation* (Boca Raton, FL: CRC Press, 1993).
10. Viani, L.O., A question of mitigation, *Landscape Architecture*, 96(8): 24, August 2006.
11. Nelischer, M. et al., Quality of an urban community: A framework for understanding the relationship between quality and physical form, *Landscape and Urban Planning*, 39(2–3): 229–241, November 30, 1997.
12. Selinus, O. (ed.), *Essentials of Medical Geology* (London, U.K.: Elsevier, 2004).
13. Natural Resource Defense Council (NRDC), Drilling down: Protecting western communities from the health and environmental effects of oil and gas production, 2007, download from http://www.nrdc.org/land/use/down/down.pdf (accessed August 12, 2011).

* Velasquez-Manoff;[5] See also http://www.quivira.org/ (accessed August 12, 2011).

14. Sumi, L., *Oil and Gas at Your Door? A Landowner's Guide to Oil and Gas Development*, 2nd edn. (Durango, CO: Oil and Gas Accountability Project, 2005).
15. Bryan, S.M., Chevron plans Questa solar plant, Associated Press (*Santa Fe New Mexican*, p. A-6, February 24, 2010).
16. Crowe, R. and S.B. Brinckerhoff, *Early Yuma: A Graphic History of Life on the American Nile* (Yuma, AZ: Yuma County Historical Society, 1976).
17. Sorvig, K., The same river twice, *Landscape Architecture Magazine*, 99(11): 42–53, August 2009.
18. Hebert, H.J., Gore, Clinton headline energy conference, Associated Press, February 23, 2009.
19. Sorvig, K., Drowning in the desert, *Landscape Architecture*, 100(1): 26–36, January 2010.
20. Molvar, E.M., *Drilling Smarter: Using Minimum-Footprint Directional Drilling to Reduce Oil and Gas Impacts in the Intermountain West* (Laramie, WY: Biodiversity Conservation Alliance, 2003).

# 17

# *Environmental Injustice in the Urban Southwest: A Case Study of Phoenix, Arizona*

**Bob Bolin, Sara Grineski, and Edward J. Hackett**

**CONTENTS**



## 17.1 Introduction

This chapter examines the distribution of technological hazards in the Phoenix metropolitan area, focusing on race and class inequalities in facility locations and the issue of environmental justice. Our focus is on identifying a zone-pronounced environmental inequality in Phoenix and examining the historical roots of those inequities. In this study, we consider the development and persistence of environmental inequities in Phoenix, Arizona, a city emblematic of the rapidly urbanizing Southwest. Chronic conditions of poverty, underdevelopment, and environmental degradation in South Phoenix are seen as a consequence of land use, investment, and zoning decisions that extend historically to the early twentieth century. We consider the ways that rapid urban growth and industrialization in the twentieth century has produced and expanded a zone of environmental injustices in the city.

To examine potential environmental justice concerns, we address two questions in what follows. First, what are the current sociospatial patterns of environmental inequities in the Phoenix metropolitan area? Second, what factors in urban development have contributed to these patterns of environmental injustices in the city? Finally, we examine convergences with patterns of environmental inequities in other western Sunbelt cities and discuss strategies for reducing environmental injustices.

As developed in the environmental justice literature, environmental inequities refer to the disproportionate burdening of low-income and people of color communities with industrial hazards, toxic waste sites, and other *locally unwanted land uses* (LULUs) such as freeways, sports stadia, and airports. To address our primary questions, we first highlight prevailing environmental inequities based on a geographic information systems (GIS) analysis of the distribution of four types of point-source environmental hazards. We then sketch the historical geography of their development.

## 17.2 Phoenix, Arizona

The Phoenix metropolitan area, with a current population of more than three million sprawled over 1864 mile$^2$ of former Sonoran Desert and farmland, is the largest of the rapidly growing desert cities of the Southwest (e.g., Albuquerque, El Paso, Las Vegas, Tucson). Phoenix exhibits a dual pattern of industrial concentration and dispersal, with a mix of industrial facilities concentrated near the urban core in combination with newer decentralized industrial nodes in suburban locations. Previous environmental equity research on Phoenix has documented pronounced inequities by race and class, both by the location of toxic release inventory[1] (TRI) facilities and for the volume of toxic atmospheric emissions by those industries.[1] While some large TRI facilities, primarily corporate semiconductor manufacturers, are located in suburbs, a disproportionate number of polluting industries operate in proximity to low income, African American and Latino neighborhoods, producing inequities in the distribution of potential risks from toxic atmospheric emissions.[1] To address key issues, we begin our discussion with a brief review of recent research on environmental inequities in western Sunbelt cities before moving to a discussion of environmental inequity in Phoenix.

## 17.3 Environmental Inequalities in the Western Sunbelt

While initial research on environmental justice focused on the Southeastern United States and cities of the "Rustbelt,"* a number of more recent studies have called attention to environmental inequities in the western Sunbelt.† Befitting its size, economic centrality, and its standing as a "world city,"[11] most research attention has been directed at Los Angeles. Several key studies have detailed both the current prevalence and the historical development of environmental injustices, including the explicit racism in features of industrial development and planning in LA.[7,10,12,13] While cities of the desert Southwest lack LA's history of large-scale, concentrated Fordist industries, they are similar to LA in that they have significant (and growing) Latino populations, expanding post-Fordist industrial sectors, an urban spatial form given sprawl, and persistent air pollution problems.[14–16]

In matters pertaining to environmental quality, water resources, and industrial and transportation pollution, Phoenix exhibits environmental problems shared by other Southwestern cities. The combination of rapid, spatially expansive growth, industrial development, and a heavy dependency on automobiles, conspire to produce chronic air pollution problems of variable severity. Like Albuquerque, El Paso, and Las Vegas,

* See Bullard[2] and Hurley.[3]

† See Bolin,[1] Boer et al.,[4] Clarke and Gerlak,[5] Morello-Frosch et al.,[6] Pastor et al.,[7] Pijawka et al.,[8] Pulido et al.,[9] and Sidawi.[10]

Phoenix is located in a shallow river valley where frequent temperature inversions during the cooler winter months trap and concentrate industrial and transportation-generated pollutants over the city. Unlike other Southwestern cities, Phoenix lacks consistent background atmospheric circulation contributing to the buildup of air pollutants.* Industrial and transportation related emissions are compounded by high particulate levels that derive from the dry conditions and dust produced by construction, gravel pit operations, agriculture, and unpaved roads on the urban periphery. While concerns over criteria air pollutants and Phoenix's frequent "brown cloud" looming over the city are frequently voiced by residents,† for environmental justice movements in the city, the greatest concerns are with the location of industrial polluters in minority neighborhoods.[20] Recent research suggests these concerns are justified.[21]

## 17.4 Phoenix's Riskscape

While most environmental justice research considers the geographic distribution of a single hazard such as hazardous waste handlers (treatment, storage, and disposal facilities or TSDFs), our research investigates the distributions of four types of hazards‡ in the Phoenix metropolitan area. These include industrial facilities emitting toxic substances regulated under the EPA's TRI, manufacturing facilities that produce hazardous wastes (large-quantity generators or LQGs), TSDFs, and toxic contamination sites listed by the federal government under provisions of CERCLA and the national priority list (NPL or Superfund). We document the spatial concentration and compounding of potential risks produced by the agglomeration of these four common types of point-source hazards in neighborhoods with racially and economically marginalized residents. Table 17.1 presents an overview of the number of hazard sites by category included in this study of Phoenix.

To calculate relative hazard burdens of each tract in the metro area, we utilize an approach that assesses the cumulative hazard burdens in census tracts based on .62 mile (1 km) radius buffer zones around each hazard site for each of the four types of hazards. These multiple overlapping hazard zones are summed for each census tract they overlay and the value standardized by the total area of each tract. The resulting cumulative hazard density index (CHDI) calculates the accumulation of all hazard buffers that overlap a given census tract.§ The index provides an aggregate hazard score for each tract which is then correlated with demographic data in order to measure levels of environmental inequity in Phoenix.

* Ellis.[17] See also Grineski et al.[18]

† See, for example, PASS.[19]

‡ The annual TRI provides release data on industrial polluters, including information on the volume, chemical composition, and location of the polluting facilities. Reporting industries must employ at least 10 workers and it manufactures or CERCLA—the Comprehensive Environmental Response, Compensation, and Liability Act processes over 25,000 lb of at least one of the currently listed 600 TRI chemicals, or uses more than 10,000 lb of at least one TRI chemical. See http://www.epa.gov/year2000/toxic.html (accessed August 12, 2011) provides the legal mandates for the investigation and regulation of a wide range of toxic contamination sites. CERCLIS, the Information System maintained under the act, lists heavily contaminated sites (toxic waste dumps, abandoned industrial sites, mines, contaminated federal facilities etc.). See http://rtk.net/cerclissearch.html (accessed August 12, 2011). Large Quantity Generator designations include facilities that produce or accumulate at least 2200 lb of RCRA regulated hazardous waste on a monthly basis. TSDFs comprise a subset of LQGs that treat, store, or dispose of hazardous wastes. See http://d1.rtknet.org/brs/ (accessed August 12, 2011).

§ For a full discussion on this topic, see Bolin et al.[22]

**TABLE 17.1**

Distributions of Hazards by Census Tract

| Type of Hazard | Number of Sites | Number of Tracts with at Least 1 Site | Number of Tracts with No Sites | Number of Tracts with HDI Score > 0 | Number Tracts with 0 HDI Score |
|---|---|---|---|---|---|
| CERCLIS | 412 | 122 | 344 | 305 | 161 |
| LQGs | 117 | 65 | 401 | 231 | 235 |
| TRI | 119 | 58 | 408 | 192 | 274 |
| TSDF | 13 | 11 | 455 | 47 | 419 |
| Any hazard sited in tract | 661 | 156[a] | 310[a] | 361[a] | 105[a] |

[a] Numbers do not sum as some sites are included in more than one hazard category.

To summarize research findings, of the 466 tracts in Maricopa County, 105 (about 22%) tracts are untouched by 1 km buffers around hazardous sites. By comparison, 310 tracts (67%) do not actually contain any of the four types of hazards studied. There are strong, statistically significant relationships between the proximity of a hazard and the income and race of residents. As presented in Table 17.2, tracts with CHDI scores greater than zero had median family incomes that averaged more than $9000 less than "clean" tracts and had significantly higher percentages of Latino, African American, and Indian residents. Taking each hazard type individually, the same pattern holds. Tracts with a hazard score greater than zero for each of the four types of hazards are significantly poorer and house significantly higher proportions of Latino, Black, and Indian residents compared to tracts with a hazard score of zero for the same hazard type. Affected tracts have significantly lower median household incomes and higher proportions of minority residents. This methodology reveals significant social inequalities in the distribution of four major types of technological hazards in Phoenix.

To further specify the areas of highest hazard burden in the city of Phoenix, we next identify those tracts in the 90th percentile of hazard scores, yielding 20 high hazard tracts. Figure 17.1 presents a map of the Phoenix "riskscape" showing major urban features as well as the locations of the highest scoring tracts. These 20 tracts comprise our study area for the discussion to follow. To illustrate the gradients of hazard density values across the metro area, Figure 17.2 presents a contour map of CHDI scores with the most heavily shaded areas denoting the highest hazard scores. Notable in Figure 17.2 are the high concentrations of hazards in older neighborhoods of South Phoenix (discussed later) and in a mixed industrial/residential zone immediately west of the central business district (CBD). For the high hazard study area tracts identified in Figure 17.1, all 20 are within 4.3 mile of Phoenix's CBD covering an area of 49.8 mile$^2$. Taken together, these tracts form a roughly L-shaped zone following rail and highway corridors around the CBD from Sky Harbor airport on the east, across historic minority neighborhoods to the aging suburbs of west central Phoenix. These tracts make up approximately 4% of the urbanized area and have an average CHDI score of 6.4 (maximum is 16.4), compared to a metro- wide CHDI average of .9. Recall too that approximately one quarter of metropolitan tracts, all on the urban periphery, have hazard density scores of zero.

The study area captures an expanding zone of hazardous land uses: from the original core in South Phoenix, it has extended with post-WWII industrial development east into Sky Harbor airport and west into older suburban areas. Seventy-two percent of the housing stock here was built prior to 1970 compared to a metropolitan average of 30%. These 20

**TABLE 17.2**

Mean Sociodemographic Characteristics and Difference of Means t-Tests for Census Tracts with Zero and Nonzero Hazard Density Indices

| | Type of Hazard | | | | |
|---|---|---|---|---|---|
| **Variable** | **CERCLIS** | **LQG** | **TRI** | **TSDF** | **CHDI** |
| White (percent) | | | | | |
| Nonzero | 65.8 | 64.4 | 58.1 | 52.4 | 67.5 |
| Zero | 80.1 | 77.1 | 79.6 | 72.8 | 81.9 |
| t (sig.) | **6.6 (.00)** | **5.8 (.00)** | **9.6 (.00)** | **4.6 (.00)** | **6.3 (.00)** |
| Latino/a (percent) | | | | | |
| Nonzero | 25.5 | 26.2 | 30.9 | 33.1 | 24.0 |
| Without | 14.0 | 16.9 | 14.9 | 20.2 | 12.9 |
| t (sig.) | **6.3 (.00)** | **4.9 (.00)** | **8.1 (.00)** | **3.4 (.00)** | **5.7 (.00)** |
| Black (percent) | | | | | |
| Nonzero | 4.1 | 4.7 | 5.6 | 7.0 | 4.1 |
| Zero | 2.9 | 2.7 | 2.4 | 3.3 | 2.4 |
| t (sig.) | **2.2 (.03)** | **3.7 (.00)** | **5.4 (.00)** | **2.5 (.02)** | **3.4 (.00)** |
| Native (percent) | | | | | |
| Nonzero | 2.0 | 2.1 | 2.6 | 5.1 | 1.8 |
| Zero | 0.7 | 1.0 | 0.8 | 1.2 | 0.6 |
| t (sig.) | **3.4 (.01)** | **2.3 (.02)** | **3.1 (.00)** | 1.7 (.09) | **3.9 (.00)** |
| Income ($) | | | | | |
| Nonzero | 32,649 | 32,347 | 30,544 | 25,716 | 34,292 |
| Zero | 43,444 | 40,440 | 40,473 | 37,622 | 43,616 |
| t (sig.) | **5.8 (.00)** | **4.9 (.00)** | **6.1 (.00)** | 5.6 (.00) | **4.5 (.00)** |

*Source:* Bolin, B. et al., *Environ. Plan. A*, 34, 317, 2002.
t-values significant with $p < .05$ in bold; $n = 466$.

tracts contain 41% of all industrially zoned land in Phoenix although they constitute less than 10% of Phoenix's total urbanized area. Further, this region had been host to a 20% of all metro area industrial expansion since 1970. The placement of Interstate Highways 10 and 17 through portions of South Phoenix and bracketing the CBD has worked to concentrate more recent industrial development in these already burdened areas. A major railroad corridor, new freeways, and an expanding central city airport have each contributed to the production of a region with the highest concentrations of hazardous industries and contamination sites in the metro area.[22,23]

Zoning data illustrate the problem of industrial encroachment on residential portions of the 20 tracts. In metropolitan Phoenix, 1% of residentially zoned areas directly border industrial zoning, in contrast to the 35% of neighborhoods in the study area which do so. The presence of unbuffered industrial activity in the midst of low-income minority neighborhoods has been a persistent feature of South Phoenix since the early twentieth century* and that pattern largely remains today. As Table 17.3 illustrates, the percentage of metropolitan population in the study area dropped from 7% in 1970 to 2.1% in 2000,

* See Luckingham.[24,25]

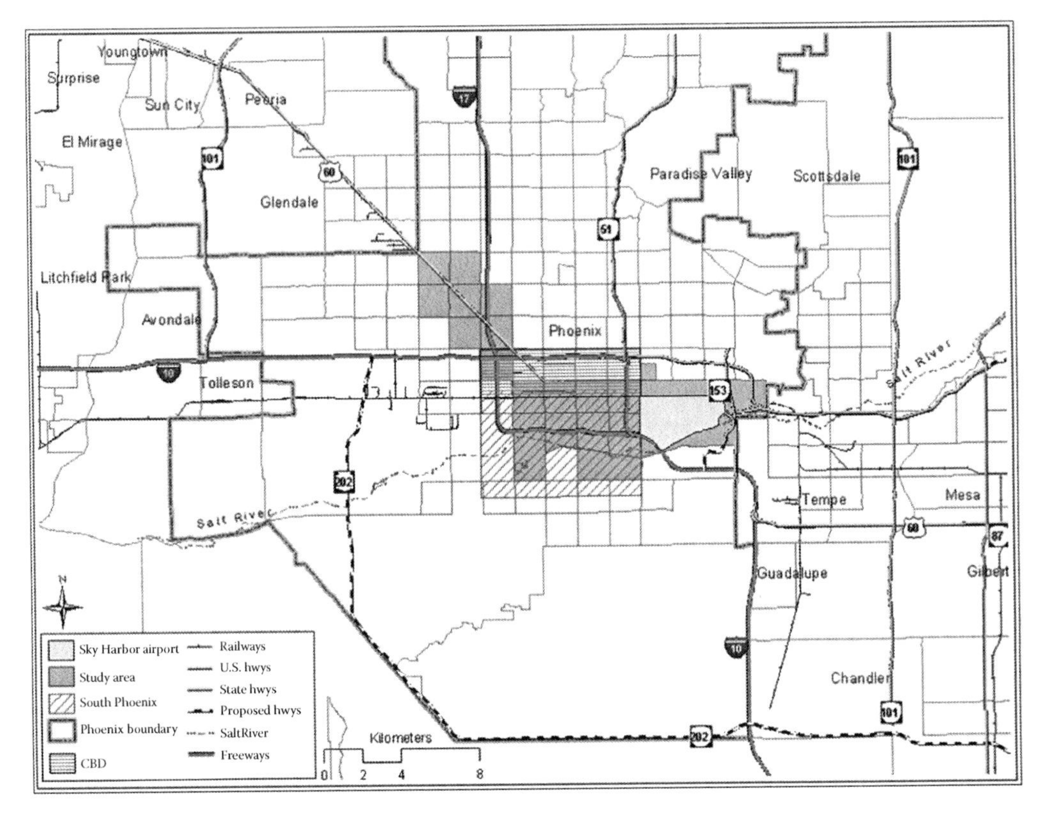


FIGURE 17.1
High hazard census tracts in the Phoenix metropolitan area.

although the population in these tracts has remained relatively steady, currently with a population of 64,590. The study area has become increasingly Latino, more than doubling to 67% in 30 years, and has growing rates of poverty—in some tracts more than five times the city-wide rate of 12%. The "Latinization" of the high hazard area is most pronounced in the western portion of the study area where some tracts have gone from less than 3% Latino to more than 60% over the last 30 years.

## 17.5 Toxic Tracts: Production of Environmental Inequality

The history of this area offers important insights into the sociospatial practices that have produced environmental inequities in Phoenix. This 20-tract "toxic archipelago" has been produced over a period of decades by race-based housing segregation and public disinvestment in conjunction with a variety of zoning, siting, planning, and investment decisions across the metro area. Phoenix's first industrial district, in what is now known as South Phoenix, began developing along the east–west rail corridor south of the CBD in the early twentieth century (see Figure 17.1). The rail corridor functioned both as an anchor to industrial activity and as the physical dividing line between white north Phoenix and Black and Latino South Phoenix.[26] Initial South Phoenix industrial development along the rail corridor included

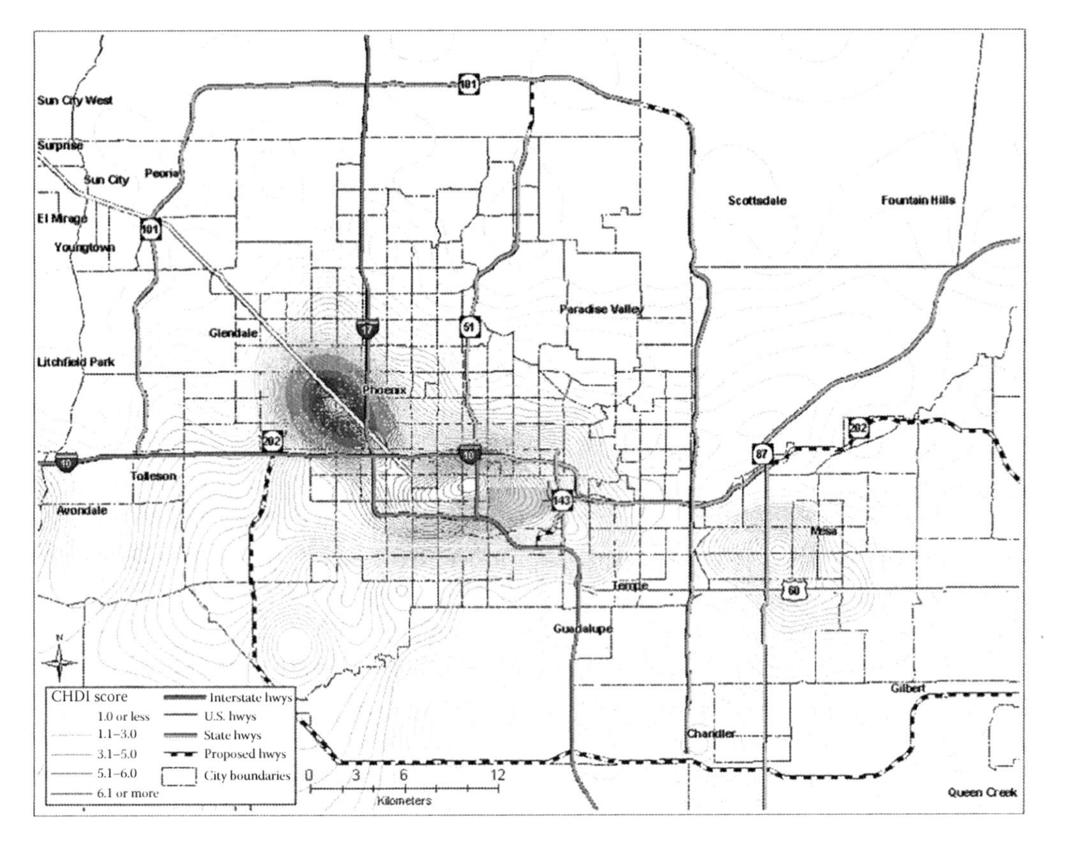


**FIGURE 17.2**
CHDI contours.

brick factories, small foundries and steel mills, petroleum storage facilities, meat packing and dog food plants,* rendering plants, ice factories, and auto parts producers.[27] Most industries located in a patchwork between the railroad and the Salt River in the midst of minority neighborhoods. This industrial district grew steadily, with more than 300 industrial firms locating in this zone by 1921.[28] Plants extended south to the river, where landfills and the city's first sewage plant (1920) were located adjacent to Mexican American settlements. Spur rail lines to connect factories to the main rail line crisscrossed South Phoenix, dispersing industries and warehouses throughout the area by the 1930s, a pattern which remains today.[28]

Early patterns of residential segregation by race and class were established in concert with railroad corridor development.[24] As early as 1910 African Americans were subject to laws enforcing residential, schooling, and employment segregation, practices that persisted into the Civil Rights era of the 1960s. A consequence of racial control and exclusion is the concentration, even today, of much of Phoenix's small Black population in a few census tracts of South Phoenix.[20,29] South Phoenix was also the historic home to Phoenix's *barrios*, where much of the area's Mexican and Mexican American population lived since the

* The meat packing and dog food plants were supported by extensive areas of stock yards adjacent to the rail road tracks, some of which persisted into the 1950s. These facilities, with their heavy production of animal wastes, were placed adjacent to Mexican-American neighborhoods with no apparent concern for the health and well-being of the residents, further stigmatizing the area as socially and environmentally undesirable.[24]

**TABLE 17.3**

Demographic Change in Study Area Tracts

| | 1970 | | 1980 | | 1990 | | 2000 | |
|---|---|---|---|---|---|---|---|---|
| | Total | Percent | Total | Percent | Total | Percent | Total | Percent |
| Total population | 967,522 | 100 | 1,509,052 | 100 | 2,091,151 | 100 | 3,045,615 | 100 |
| | **68,165** | **100** | **57,171** | **100** | **53,670** | **100** | **64,590** | **100** |
| White non-Latino population | 808,955 | 84 | 1,226,362 | 81 | 1,616,280 | 77 | 2,017,231 | 66 |
| | **35,974** | **53** | **25,707** | **45** | **20,761** | **39** | **13,270** | **21** |
| Latino population | 112,225 | 12 | 199,517 | 13 | 335,540 | 16 | 756,601 | 25 |
| | **20,291** | **30** | **22,316** | **39** | **24,520** | **46** | **43,247** | **67** |
| African American population | 31,970 | 3 | 47,833 | 3 | 73,761 | 4 | 113,337 | 4 |
| | **10,537** | **15** | **7,288** | **13** | **6,128** | **11** | **5,038** | **8** |
| Persons in poverty | 112,907 | 12 | 156,768 | 10 | 257,359 | 12 | 355,668 | 12 |
| | **18,682** | **27** | **15,529** | **27** | **17,992** | **34** | **20,480** | **32** |
| Median household income | $30,926 | | $27,679 | | $32,630 | | $45,322 | |
| | **$20,369** | | **$14,024** | | **$14,576** | | **$25,721** | |

Demographic characteristics of the 90th percentile CHDI tracts (bold) compared to metropolitan phoenix 1970–2000.

early twentieth century.[30] Like Blacks, Latinos were segregated by strict, and at times virulent, racist practices, including the vigilantism of an active Ku Klux Klan in the 1920s.[24] And like African Americans, with the decline in nearby agricultural employment in the postwar period, Latinos sought what employment they could in low-wage unskilled jobs in the city. While the last three decades have seen some geographic mobility for African Americans and Latinos outside of historic areas of settlement, South Phoenix itself remains predominantly a Black and Latino low-income area with a concentration of industrial land uses. The hegemonic racism that held sway for much of the twentieth century insured that South Phoenix remained a stigmatized zone of poverty and people of color sequestered on the periphery of the urban core.[31] Referred to as the "the shame of Phoenix" in a 1920 community report, living conditions in South Phoenix were described as "fully as bad as any ¼ in the tenement districts of NY and other large centers of population."* As late as 1947 a *Saturday Evening Post* article noted that conditions in South Phoenix were a match "misery for misery and squalor for squalor with slums anywhere."† The continuing presence into the 1960s of tar paper shacks and ramshackle housing without sewage or running water in the midst of industrial facilities, was a testament to the political power of landlords to ignore building codes, zoning, and human welfare in Phoenix.[32] With the rapid urban expansion of the 1970s, new zoning, planning, and siting decisions, under the guise of urban redevelopment, added new hazardous facilities to the mix of industries already in place in South Phoenix. Among these were hazardous waste handling facilities (TSDFs), mandated by federal Resource Conservation and Recovery Act (RCRA) legislation in the 1970s.[20]

## 17.6 Urban Development, Planning, and Environmental Justice

The high hazard area discussed here reflects a century of urban development stretching from historically segregated and "dumped on" of South Phoenix to aging working class suburbs of west Phoenix. The legacy effects of racial segregation and the not unrelated siting of industries are strongly drawn in South Phoenix. The hazard zone has been shaped by successive waves of transportation infrastructure development, from the late nineteenth century railroads to post-WWII freeways and an expanding airport. While continued economic marginality of central city neighborhoods may be, in part, attributed to suburban expansion and the resource drain on the central city,[33] it is nonetheless a product of decades of planning and investment decisions made by both the public and private sectors.‡ By the 1930s, the race of residents was an intrinsic part of how property values were determined for lending purposes. The presence of the minority population in South Phoenix was considered an investment "hazard," leading to bank redlining, perpetuating economic underdevelopment, and inadequate housing in Black and Latino neighborhoods.[35] Residents in this zone have, for decades, endured pervasive environmental disamenities while persistently failing to receive significant economic benefits from the industrial presence in their neighborhoods. Notably, these burdens no longer go uncontested. In a political and legal environment now shaped by civil rights and environmental justice principles, a variety of recent lawsuits over the permitting of

* As quoted in Kotlanger.[28]
† Quoted in Konig, p. 21.[27]
‡ See, for example Hackworth.[34]

hazardous facilities in South Phoenix have been filed on civil rights grounds, and local environmental justice organizations now frequently deploy the term "environmental racism" at site protests.[20] Because of their historical disenfranchisement, what people of color have lacked is the political power to protect neighborhoods against industrial and transportation encroachment. Once industrial zones and transportation corridors are in place, as they were in Phoenix by the early twentieth century, those with political power will do little or nothing to alter that built landscape to benefit low-income residents.[36] That industries seek vacant land adjacent to both transportation corridors and waste disposal facilities is well documented, insuring that an agglomeration of hazardous sites and other residentially incompatible land uses will tend to develop around an initial transportation corridors, unless zoning and planning begin to actively redirect it elsewhere, something which has not occurred in Phoenix.

The persistence of these environmental burdens, in spite of major changes in federal regulations, knowledge of toxic hazards, and emergence environmental justice principles, underline, with only limited exceptions,* an ongoing official disregard for this region of the city. In the absence of a managerial focus on these environmentally and economically distressed regions, there is no indication that the trends of growing environmental burdens, increasing poverty, and increasing percentage of Latinos in high-hazard areas will be arrested anytime soon. In the absence of federal resources and programs targeting environmental restoration in low-income neighborhoods, urban planners, policy makers, and citizens have restricted options. As Harvey notes,[37] in cities today, "concerns for environmental justice (if they exist at all), are kept strictly subservient to concerns for economic efficiency, continuous growth, and capital accumulation." Harvey's observation is an apt description of the developmental process described here. In Phoenix there has been strong support for industrial expansion and commercial development projects that promise enhanced tax revenues, suburban growth, and private accumulation, irrespective of development's effects on environmental and social conditions in low-income and minority neighborhoods.

## Acknowledgments

Funding for this research was provided by the Central Arizona-Phoenix Long Term Ecological Research (CAP-LTER) program, Global Institute of Sustainability, Arizona State University, Tempe, Arizona.

## References

1. Bolin, B., E. Matranga, E. Hackett, E. Sadalla, D. Pijawka, D. Brewer, and D. Sicotte, Environmental equity in a Sunbelt city: The spatial distribution of toxic hazards in Phoenix, Arizona, *Environmental Hazards* 2(1): 11–24, 2000.
2. See Bullard, R., *Dumping in Dixie* (Boulder, CO: Westview, 1990).

* cf. *A Multi-media Toxic Hazard Reduction Program.*

3. Hurley, A., *Environmental Inequalities: Class, Race, and Industrial Pollution in Gary, Indiana, 1945–1980* (Chapel Hill, NC: University of North Carolina Press, 1995).
4. Boer, J., M. Pastor, J. Sadd, and L. Snyder, Is there environmental racism? The demographics of hazardous waste in Los Angeles County, *Social Science Quarterly* 78(4): 793–810, 1997.
5. Clarke, J. and A. Gerlak, Environmental racism in the Sunbelt? A cross-cultural analysis, *Environmental Management* 22(7): 857–867, 1998.
6. Morello-Frosch, R., D. Pastor, and J. Sadd, Environmental justice and southern California's 'riskscape': The distribution of air toxics exposures and health risks among diverse communities, *Urban Affairs Review* 36(4): 551–578, 2001.
7. Pastor, M., J. Sadd, and J. Hipp, Which came first? Toxic facilities, minority move in and environmental justice, *Journal of Urban Affairs* 23(1): 1–21, 2001.
8. Pijawka, K.D., J. Blair, S. Guhathakurta, S. Lebiednik, and S. Ashur, Environmental equity in central cities: Socioeconomic dimensions and planning strategies, *Journal of Planning Education and Research* 18: 113–123, 1998.
9. Pulido, L., S. Sidawi, and R. Vos, An archaeology of environmental racism in Los Angeles, *Urban Geography* 17(5): 419–439, 1996.
10. Sidawi, S., Planning environmental racism: The construction of the industrial suburban ideal in Los Angeles county in the early twentieth century, *Historical Geography* 25: 83–99, 1997.
11. Keil, R., *Los Angeles: Globalization, Urbanization, and Social Struggles* (New York: Wiley, 1998).
12. Boone, C. and A. Modarres, Creating a toxic neighborhood in Los Angeles county: A historical examination of environmental inequity, *Urban Affairs Review* 35(2): 163–187, 1999.
13. Pulido, L., Rethinking environmental racism: White privilege and urban development in southern California, *Annals of the Association of American Geographers* 90(1): 12–40, 2000.
14. Davis, M., *City of Quartz: Excavating the Future in Los Angeles* (New York: Vintage, 1992).
15. Wiley, P. and R. Gottlieb, *Empires in the Sun: The Rise of the New American West* (New York: Putnam, 1982).
16. Soja, E., *Postmetropolis: Critical Studies of Cities and Regions* (London, U.K.: Blackwell, 2000).
17. Ellis, A., M. Hildebrandt, and H. Fernando, Evidence of lower atmospheric ozone sloshing in an urbanized valley, *Physical Geography* 20(6): 520–536, 1999.
18. Grineski, S., B. Bolin, and C. Boone, Criteria air pollution and marginalized populations: Environmental inequity in metropolitan Phoenix, Arizona, USA, *Social Science Quarterly* 88(2): 535–554, 2007.
19. PASS (Phoenix Area Social Survey), *The Phoenix Area Social Survey: Community and Environment in a Desert Metropolis* (Tempe, AZ: Center for Environmental Studies, Arizona State University, 2003).
20. Sicotte, D.M., Race, class and chemicals: The political ecology of environmental justice in Arizona, PhD dissertation (Tempe, AZ: Arizona State University, 2003).
21. Arizona Department of Environmental Quality, *A Multi-Media Toxic Hazard Reduction Program for South Phoenix* (Phoenix, AZ: Arizona Department of Environmental Quality, 2003).
22. Bolin, B., A. Nelson, E. Hackett, D. Pijawka, S. Smith, E. Sadalla, D. Sicotte, E. Matranga, and M. O'Donnell, The ecology of technological risk in a Sunbelt city, *Environment and Planning A* 34: 317–339, 2002.
23. Sobotta, R., Communities, contours and concerns: Environmental justice and aviation noise, PhD dissertation (Tempe, AZ: Arizona State University, 2002).
24. Luckingham, B., *Minorities in Phoenix* (Tucson, AZ: University of Arizona Press, 1994).
25. Luckingham, B., *Phoenix: A History of a Southwest Metropolis* (Tucson, AZ: University of Arizona Press, 1989).
26. Roberts, S., Minority-group poverty in Phoenix: A socio-economic survey, *Journal of Arizona History* 14(73): 347–362, 1973.
27. Konig, M., Phoenix in the 1950s: Urban growth in the Sunbelt, *Arizona and the West* 24 Spring: 19–38, 1982.
28. Kotlanger, M., Phoenix, Arizona 1920–1940, PhD dissertation (Tempe, AZ: Arizona State University, 1983).

29. Pijawka, D., S. Guhathakurta, S. Lebiednik, J. Blair, and S. Ashur, Environmental equity and stigma in central cities: The case of south Phoenix, in J. Flynn, P. Slovic, H. Kunreuther, eds., *Risk, Media and Stigma: Understanding Public Challenges to Modern Science and Technology* (London, U.K.: Earth Scan, pp. 187–203, 2001).
30. Dimas, P., Progress and a Mexican-American community's struggle for existence: Phoenix's Golden Gate Barrio, Master's thesis (Tempe, AZ: Arizona State University, 1991).
31. Bolin, B., S. Grineski, and T. Collins, The geography of despair, *Human Ecology Review* 12(2): 156–168, 2005.
32. Citron, M., Of slums and slumlords, *Reveille* (1): 8–13, 1966.
33. Guhathakurta, S. and A. Wichert, Who pays for growth in the city of Phoenix? An equity-based perspective on suburbanization, *Urban Affairs Review* 33(6): 813–838, 1998.
34. Hackworth, J., Local planning and economic restructuring: A synthetic interpretation of urban redevelopment, *Journal of Planning Education and Research* 18: 293–306, 1999.
35. Brunk, L., A Federal legacy: Phoenix's cultural geography, *Palo Verde* 4(1): 60–78, 1996.
36. Harvey, D., *Justice, Nature and the Geography of Difference* (London, U.K.: Blackwell, 1996).
37. Harvey, D., *Justice, Nature and Geography of Difference* (London, U.K.: Blackwell, p. 374, 1996).

# 18

## *Dwelling in Expanded Biotic Communities: Steps Toward Reconstructive Postmodern Communities*

**Geoffrey Frasz**

**CONTENTS**



## 18.1 Introduction

One of the things a philosopher can bring to a discussion is how better, sustainable communities can be established either in the American Southwest or, indeed, in any area. Philosophic theory can help identify the conceptual frameworks necessary for strategies that establish the important sense of place for man in the natural world. What is needed is a deeper understanding of what elements are necessary to establish truly environmentally sound communities. This understanding can provide not only a forward-looking vision for newly planned communities but also a means to critique existing communities.* What is presented in this chapter is my personal and philosophical vision of community. This vision can serve as a springboard for thinking about what specific features of existing communities should be praised or faulted. Five essential general features for establishing a successful *reconstructive postmodern* concept of community† are presented. Two different communities in Las Vegas, Nevada, Green Valley and Summerlin, are examined in light of these features. Finally, a planned community outside of Las Vegas, Coyote Springs, is discussed and its proposed features are examined in relation to its contribution to creating a vision for a viable and sustainable community.

* This is what Plato did 2500 years ago in *The Republic*. His account on an ideal state allows him to provide the standard against which various existing states could be judged as just or unjust and to what degree they were.

† I use the term "reconstructive" in the meaning given by David Griffin to distinguish this approach from the "deconstructive" meanings that have come to be attached to the term *postmodern*. See Griffin.[1]

## 18.2 Foundations of Postmodern Communities

An initial difficulty facing any attempt to describe what are and should be postmodern communities results from the fact that the term *postmodern* is itself ambiguous, given the multiple means of the term postmodern. This ambiguity can be found in the distinctions made between "postmodernism" and "postmodernity" where the former refers to changes in the arts, philosophy, and other social and political views, while the latter refers to actual social and political changes in the industrialized world, especially since the 1960s.*

The ambiguity is further compounded by the fact that the term has a descriptive and a normative sense. It first can refer to the significant changes in society in the last 50 years, which supposedly warrant a new term to describe this new state of affairs. Accordingly, the development of postmodern communities would refer to the types of communities that have been developed in post-World War II years that do not share many or most of the features of modern communities that developed in Western countries since, at least, the Industrial Age. The difference between modernism and postmodernism can be seen in the contrast between architectural works by Miles van der Rohe, Le Corbusier, and the Bauhaus Movement on one hand, and the works of Philip Johnson and Michael Graves on the other. The unadorned structures of the modern period reflecting a "form follows function" philosophy is now being replaced with an eclectic approach that draws on the past† and reintroduces the idea of facades on buildings, nonorthogonal angles, and nontraditional surfaces.‡

The alternative, normative meaning to postmodernism refers to a critique of modernism with its focus on progress, rationality, a devaluing of emotional responses, and emphasis on order and structure. The history of the movement from modern to postmodern thinking in urban design can be found in published works by scholars such as Ellin[4] and Frampton.[5] The rationalism implicit in community design, with emphasis on functionality, grid layout, and monotonous order, was now understood as providing a defense against subversive, dangerous, "wild" aspects of the world. The urban area, which was once a place of safety, became, itself a place of danger. Nan Ellin has also described how fear and a desire for safety has shaped the development of communities by invoking "an idealized past, an exoticized other, a fantasy world, group cohesion, or oneself,"[6] where Levittown is being replaced by Celebration, Florida.§ The "New Urbanism" movement itself can be understood through such a critique as having its impulse in a response against the authoritarian control by external power centers, creating communities that are friendly, small scale, and safe but are also, at the same time, gated, restrictive to outsiders, and fixed in a single overarching design theme.

Given the horrific social events of the twentieth century, some have held that the project of modernity needs to be replaced or *deconstructed* so that a better social world can emerge.[7] But I do not share such a radical view. Instead I am arguing that the problems of alienation, domination and control of individuals, objectification of nature, and the consequent postmodern attempts to build communities that address

* This distinctions made by Giddens.[2]
† Las Vegas, Nevada is a prime example of this postmodernism. See Venturi and Brown.[3]
‡ An instance of this in Las Vegas can be found in Frank Gehry's design of the Lou Ruvo Alzheimer's Institute. http://www.dexigner.com/architecture/news-g6875.html (accessed August 10, 2011).
§ http://architecture.about.com/od/communitydesign/g/newurban.htm (accessed August 10, 2011).

these problems require a "reconstructive" approach to community design. I use the term "reconstructive" in the meaning given by David Griffin[1] to distinguish this approach from the "deconstructive" meanings that have come to be attached to the term *postmodern*. Thus I am presenting a postmodern version of community that emerges through a revision of previous modern and traditional concepts. Such as version of community draws upon techniques and technologies that are willing to use any approach that furthers the reestablishment of human-nature connection that was substantially lost in the modern period. In this reconstructive approach I find philosophical foundation the concept of biophilia by Kellert and Wilson.[8] The working out of design features that can be used in reconstructive postmodern communities can be found in works by Kellert et al.[9,10] and also Knowles.[11]

## 18.3 Features of a Postmodern Community

A postmodern vision of community is needed as an alternative to the "modern" conception of the human community. Modern communities can be characterized as collections of people who happen to inhabit a given place. However, there is little common bond between people in modern communities because the communities are created mainly to satisfy immediate human wants and each person tends to focus in an autonomous manner on maximizing short-term monetary exchange value. These communities are distinctively human community, and there are few, if any, perceived links between the human community and the natural world beyond the backyard. This seeming disconnect between humans and nature creates a need to reestablish this link through effective community design and planning. Therefore, the consideration of reconstructive postmodern vision of community concept might be one that helps expands the notion of community to include both human and nonhuman biotic components (see Chapter 25). The adoption of these principal features in community planning and design would lead the residents of this community to recognize that they in fact *dwell in expanded biotic communities* or, in other words, live where the built environment and the natural environment *coexist*.

The *first* feature of reconstructive postmodern communities attempts to establish a new vision of communities as we know them. This new view recognizes that every community is a dynamic, ongoing process, and that no static maintenance of a community is possible. Any attempt to hold back the natural development of a community is to go against its basic nature, as something that—like a baby—comes into being, develops, and then matures. This process is organic and, ultimately, can be flexible as the vision for the new community emerges. Efforts to interfere with natural community development could risk the potential for the community to flourish.

Healthy postmodern communities are characterized by what I call a *mixed* community—one that effectively integrates the human and nonhuman components *by design, not chance* (see Chapter 27). Such mixed communities share a common vision that they are more than just aggregates of persons, but also contain the basis for what Aldo Leopold[12] called the "land," which includes the biotic and abiotic communities, ecosystems, and watersheds that provide the foundation for traditional communities. Thus, I hold that any sustainable postmodern community must involve viewing its members as *persons-in-a-biotic community*.[13] A postmodern version of community based on this enlarged idea of community will recognize that human beings, as persons who take a conscious and decisive role in the development of a particular mixed community, will take on responsibility for all the members of that community. They will respect the process the individuality of all the members of the community, since they contribute to the dynamic processes that make up the community.

The *second*, a successful reconstructive postmodern version of community, will be characterized by attempts by community members to seek a diverse set of insights about the nature of community. This postmodern approach will strive to include various accounts of community that affirm not only the value of persons in community but also the value of persons who recognize that they also live in a biotic community in what I earlier referred to as a mixed community. To maintain the reconstructive postmodern community structure, the most up-to-date ideas of natural and social science should be included in the management of these places, especially the principals of ecology and ecological design. The community would benefit from the use of ecological insights that would identify manners of linking the community within a broader biogeographical framework and use ecological planning to guide decision making by community members. These principles would incorporate wise use of energy sources and pathways, food production, and make use of appropriate level technologies and sustainable development programs. As physics was the paradigmatic science that underlay modern world views, it is to be ecology that plays the same role in postmodern mixed communities.

A *third* feature common to reconstructive postmodern community involves adopting policies and practices that take a long-term approach rather than focus on short-term gain.* Recognizing the need to develop community economic structures and features that reflect and foster this long-term view is an essential part of the development of postmodern communities. Most "modern" communities reflect some theory of political economy concerned primarily with the manipulation of property and value so as to maximize short-term monetary exchange value to the owner.

Postmodern communities will need to reflect a management approach that does not measure wealth solely in terms of money, and the wealth of the community and the homes there derive from the ability of the property to produce what is actually needed for the well-being of all the human and nonhuman members. The problem for creating a successful postmodern community is that in the modern economic models that underlie much of modern community development, the concern for short-term profit making trumps the concern for developing long-term wealth and inherent value of the community and its members. In profit-driven development models, individual investors seek to advance their own personal goals, based on their perception of only when their own self-interests are met, the common good of the community will result.

The *fourth* general feature is that any reconstructive postmodern concept of community must avoid theories of community development and of human nature where certain features of human existence in community are abstracted out and erroneously considered to be the fundamental concrete facts of existence, to the exclusion of other real facts. One such instance is the way any theory radical individualism can abstract individual human existence out of the real connections people have with the world, and then considers these isolated humans as fundamentally real, autonomous individuals. When we forget where food, power, or water comes from, or where waste actually

* This is distinction made by Aristotle in Bk. 1 of the *Politics*, the difference between *chrematistics* and *oikonomia*. The former is the branch of political economy relating to the manipulation of property and value so as to maximize short-term monetary exchange value to the owner. *Oikonomia*, by contrast, is the management of the household so as to increase its use value to all members of the household over the long run. See Aristotle, *The Politics*, Bk. I.

goes, and think of ourselves primarily as urban consumers, we have abstracted out only part of the human experience and view that part as most significant or important. What is forgotten or lost in this abstraction is the actual reality of interconnectedness of life forms—the kinship we have with animals, plants, and soil. This concept is the basis of the term "land" in the context used by Leopold as previously discussed. It is crucial that any postmodern concept of community avoid this error because it comes about from a kind of forgetting. It forgets that most notions of cities are abstractions—separate from the natural setting it originated—and it mistakenly considers that city as an abstraction for a final reality. A critical error involves mistaking the abstract for the concrete perception of the urban environment. In modern models of community certain features of human life are abstracted out, particularly the concept of humans as part of larger, biotic communities. In these accounts, it is forgotten that humans dwell not only with other humans but also within the land. Modern sociological and economic theories that focus on certain abstract aspects of human existence, such as people are fundamentally rational self-interested consumers, treat these abstractions as fundamentally real features of human existence and ignore other equally real features committed to this error.

An example from Las Vegas can help to understand this disconnect between the environment and human behavior. Human life in this area is possible through adaptations to this extreme environment defined by the scarcity of water, yet so much of community design in Las Vegas involves new ways to forget (or abstract out) this fundamental fact. The landscaping that involves grasses and other plants that require heavy water use, artificial lakes, fountains, and other water shows for visitors allow for this kind of forgetting. What is misplaced is the real concrete, the concrete fact that one is living in the desert and the dry streambed that is ignored when building a subdivision will some time become a roaring watercourse. In this case, the physical reality of the landscape is *not* the natural or the most environmentally sound option for this part of the country.

The only way for postmodern accounts of community to avoid committing this error is to maintain a continual recourse to the concrete facts of existence. Doing so is another appropriate task for environmentally based community education, to help community dwellers remember that they live in a much more extensive biotic community, that their lives involve networks of energy and food productions more fundamental than personal networks of shared social interests.

*Fifth* and finally, one of the key features that will set reconstructive postmodern communities apart from modern ones is the degree of communality involved. Much of what passes for community today is actually aggregate clusters of like-minded individuals who engage in collective activities only insofar as those activities further the individual self-interests of each member. Today much of urban living occurs in what are called lifestyle communities such as the over-50 retirement communities that have sprung up across the country, gay communities found in urban areas, and the student-centered communities that group up around colleges and universities. Quite often in these kinds of communities individual interests and the short-term view predominates giving rise to a sense of isolation, fragmentation, and loss of identity. Postmodern concepts of community will seek to replace the concept of isolated, individuals-in-aggregates with the idea of *persons-in-the-community*.

In short, in postmodern communities, the community as a whole takes on the responsibility for its members, and the members take on a responsibility for the community. Part of this responsibility includes a respect for the diverse individuality of the members of the community. In this way a balance can be struck that provides an answer to the problem of community. Respect for persons is balanced against members taking a collective interest in the well-being of the community in which they cannot help but to live and to dwell. This allows us to value communities in terms of how this balance is struck. Diverse, tolerant, open communities that respect the intrinsic worth of individuals will result in greater amounts of complex, intense, harmonious, and novel satisfactions of the entities that make up the community. Plus, the harmony between the individuals will contribute to the overall harmony and balance of the community. Postmodern communities will seek to embody various community features and structures, moral as well as physical that strike a balance between the need for harmony and the importance of the individual.* There will be different degrees in the ways communities will reflect this attempt to balance the individual and the community. But it will be possible to rank communities to the degree that individuals live and act as persons-in-a-community, as described here.

The principles presented here, on features of reconstructing postmodern communities, may seem idealistic or difficult to achieve comprehensively in these modern times; however, the need to move in a direction of creating communities that can sustain and grow responsibly is unprecedented for growing Southwest cities. Many cities have continued to grow endlessly without ever attempting to address the disconnect between people and the environment or communities in nature. The next section will discuss real-life examples of communities in the Las Vegas, Nevada, region and how the principles of reconstructing postmodern communities apply in these examples.

## 18.4 My Critique of Two Existing Las Vegas Communities

In this paper I have presented five features that I consider essential for any new reconstructive postmodern community. This new view of community provides a theoretical foundation for those models of communities that seek to cross the boundaries between human and nonhuman worlds. This view also allows me to understand some of the problems that face my own city of Las Vegas. With this basis of understanding, I will discuss two existing planned communities in the Las Vegas area and comment on areas where these projects fall short of these ideals for this type of community by default, design, or disaster (Figure 18.1).

### 18.4.1 Green Valley

Many residents of Las Vegas live part of the postwar American dream of suburban living in communities of individual homes. One such "community" is known as Green Valley, a slight misnomer since it is neither green nor a valley. This master-planned community is subdivided into more than 30 "neighborhoods," many with tall walls surrounding them and security gates limiting access and fostering isolation. This isolation of the individual is encouraged by other community features. There is little public transportation, forcing residents to use personal automobiles, which in turn requires wide highways and few pedestrian sidewalks. There is no physical or spiritual center of the community, only strip malls at major intersections. The architecture of many of the homes places the garage in the

* See Whitehead's discussion in *Adventures of Ideas*, pp. 290–292 of "order" and "love" as principles that reflect this fundamental balance.

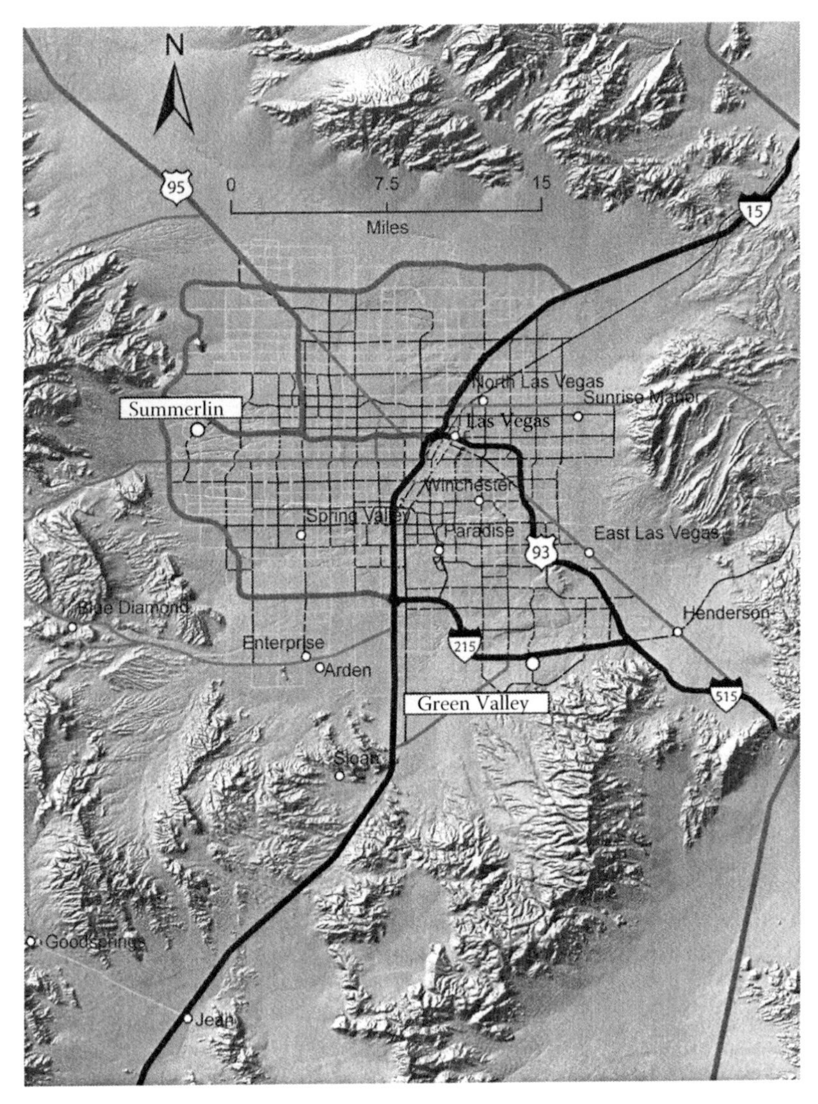


**FIGURE 18.1**
Map of the Las Vegas communities of Summerlin and Green Valley.

front of the house, encouraging residents to leave and return to the development completely sealed in a car, never venturing out into the neighborhood (Figure 18.2). Furthermore, the homes have little room for growth or additional construction, forcing families to move into new constructions when the family grows too large for the original home. This movement adds to the isolation and lack of community cohesion.

A planned community center stood empty soon after its construction because it filled no need for residents and it has become the corporate offices of the developers.[14] The sterility of the former center, with its unused outdoor checkerboard tables by a fountain, surrounded by lifelike statues of people engaged in "typical" acts of community living, illustrates how communities cannot be planned and built overnight and are truly organic entities (Figure 18.3). Effective community is not fostered by building an entire community with all of the structures already in place, but rather the features of a community should

**FIGURE 18.2**
The high walls in Green Valley homes thwart the possibility of a community sense.

**FIGURE 18.3**
Statue art as marketing tool in Green Valley to convey a community feel.

be allowed to emerge slowly to allow specific needs of the community to emerge over time. Otherwise, attempts at community design and planning might lead to alienating community residents by excluding them from decisions affecting the fundamental structure of the community.

While the goal is to create features that foster community feeling and respect for diverse community members, the planned turnover in home ownership has spawned extensive covenants in the terms of ownership of each home to preserve resale values. This is often done in ways that stipulate what may or may not be done to the design of the home, resulting in a bland repetition of landscape and home ornamentation. The legal penalties are enforced, insuring that residents do not attempt to live in a way that might be "different" or disturbing to neighbors, even if this involves retooling homes for solar energy and gray water use. But the reliance on neighborhood associations to enforce such covenants underscores the fact that such groups have no inherent shared vision of a good life that would supersede any fears of diminished property values. The rapid turnover of home ownership also hinders the development of communities of memory,[15] where people have a common history and shared stories of who they are as part of living in a particular place. The turnover in Green Valley prevents people from connecting with anyone other than next-door neighbors and from developing a shared community memory.

Green Valley was built in a desert over shallow waterways. This disruption of the natural hydrologic flow through this built environment illustrates the idea that human communities are all too often created without careful consideration of the larger surrounding biotic community. As more desert land is covered by concrete, the ambient mean temperature of the area increases. This increased heat results in a further demand on air conditioning, which places added demands on energy production. It is possible to design with the desert in mind, as has been done elsewhere in the American Southwest, to build homes that take advantage of natural cooling and shade features. But the home designs have been copies primarily from places in a completely different biogeographic zone such as south coastal California. Consequently, the homes built reflect little concern for designing with nature, including, for example, shaded patios and desert-adapted trees. Furthermore, in an attempt to deny the fact that Green Valley is in a desert, homes are still being built with extensive grass lawns that require constant watering, placing further burdens on the limited local water supply. An additional denial of the desert comes from the way the dry streambeds have been straightened and made into concrete waterways that turn into raging floods during the spring rainstorms (Figure 18.4). The water has no opportunity to slowly sink into the desert soil, replenishing the aquifer, and is moved directly out of the area into the Colorado River. As a result, the rain that used to replenish the local water table is now added through increased surface runoff to the volume of the river whose water itself is primarily sent to California, while Nevada must constantly agitate for more water to support its growth.

**FIGURE 18.4**
A concrete drainage in Green Valley over a desert wash.

The southern boundaries of Green Valley nestle against recreational and wilderness areas including the North McCullough, South McCullough, and Eldorado wilderness areas. Access to these and other recreational facilities in Henderson are desirable features, yet many of the residents who bought homes on the edge of the wilderness areas, supposed for the closeness of their homes to natural areas, are generally not receptive to "outsiders" coming into their neighborhoods to enter the same areas to access the recreational and natural areas.*

Another example of how the human communities like Green Valley are built without much regard for the surrounding biotic community is seen in the problem of the desert tortoise. These indigenous creatures require a large amount of land to support their grazing. But the rapid development in Green Valley displaced many of the tortoises as their habitat was covered over. Until just recently, it was common to gather up all the tortoises and wild burros in the area and move them elsewhere, into places where previous tortoises and burros had already established territories and habitats.

In many ways Green Valley serves as an example of the failure to practice building long-term community wealth. In Green Valley, the residents are limited in their flexibility to build their homes with a design reflecting the natural surroundings and must rely on ready-made development plans for home and neighborhood, often guided by the goal of rapid profit from the sale of these homes. This concern for the short-term profit over the long-term wealth of the community that is found in the quality of the life of the people living there illustrates the inherent problems of community in such developments. It is a common feature for developers in the "communities" of Green Valley and other such master-planned developments to build as large a house as possible on a single site, maximizing the number of properties for sale in a given area. However this leaves little room for homeowners to add new rooms as their family needs change.† The only option is to sell the "starter house" and move to another larger house in a new community where the family has no long-standing connections (Figure 18.5).‡

But even in Las Vegas there are alternatives to Green Valley. And while the problems of community in the southern Nevada area are many, it is possible that postmodern communities, with features described in this chapter can be developed. The problems of isolation in such modern communities as Green Valley result from building in such a way as to disconnect and isolate people not only from each other but from the larger biotic community around them.

* http://www.lasvegassun.com/news/2008/dec/31/residents-say-hikers-not-our-back-yard/ (accessed July 19, 2009). Fortunately not all the local residents share the antipathy to hikers.

† This was not always the case. In many of the early postwar developments such as the well-known Levittown, the developer provided only homes with a similar small, basic plan but left room for homeowners to expand, modify, and create individual homes as would suit their needs. Such homeowners would live for decades in such home, establishing true communities with neighbors. Although Levittown provided affordable houses in what many residents felt to be a congenial community, critics damned its homogeneity, blandness, and racial exclusivity (the initial lease prohibited rental to non-Whites). Today, "Levittown" is used as a term of derogation to describe overly sanitized suburbs consisting largely of tract housing. See Peter Bacon Hale's website Levittown: Transformations of the Postwar Suburb, found in http://tigger.uic.edu/~pbhales/Levittown/oldindex.html (accessed July 19, 2009) for a vivid photographic account of how homes in Levittown have been transformed to fit the needs of community members.

‡ To successfully market a house as a "starter" home requires a buyer who accepts the necessity of eventually moving to another house as a family grows. Such a buyer will be more concerned that property values in the area remain high. To that end, CC&Rs and homeowner associations will be hostile to attempts to create individual homes that do not "fit in" with other homes in the area, no matter what the needs of the individual homeowner are.

**FIGURE 18.5**
The district in Green Valley, an attempt to recreate an urban community.

### 18.4.2 Summerlin

Summerlin is a 22,500 ac master-planned development in the northwest region of Las Vegas. Designed and planned by the Howard Hughes Corporation and named after Hughes' maternal grandmother Jean Amelia Summerlin, it currently has over 95,000 residents. It still has over 9,000 ac to develop and, while there is no expected date of completion, it is expected to eventually have over 160,000 residents.[16] Because of the vast land area in this planned development, Summerlin still has room to grow, making it a more dynamic area than Green Valley, which is contained by engineered features, the major community diversity comes in the designs of the homes.* To its credit, Summerlin planners have been willing to swap land owned by the Hughes Corporation with the Bureau of Land Management (5000 ac in 1988 and 1000 ac in 2002)[16] in order to preserve natural areas adjacent to the Red Rock Natural Conservation Area and provide a buffer to development while still providing nonmotorized access into the area. This willingness to forgo short-term profit for long-term development (thus creating a greater amount of natural wealth) is an example of what needs to be done more often in mixed communities.

Summerlin planners have also attempted to support ecological diversity by building parks and trails through desert washes, allowing for a natural diversity of native plants and animals to flourish amid homes (Figure 18.6). By attempting to build within existing washes the neighborhoods do allow for variety, novelty, and even intensity of natural experiences. This is especially the case when heavy rains introduce a large amount of water into the washes. The challenge facing Summerlin residents is to accept and celebrate the fact that they are living in developments within a mixed community as I have described, instead of attempting to utilize water-intensive landscape with nonnative plants and grasses, landscapes unappealing for indigenous wildlife (Figure 18.7).

Summerlin has tried to locate itself within the broader biogeographical zone of southern Nevada. It does attempt to harmonize neighborhoods in ways that encourage residents to walk, bike, or hike on trails throughout the area (Figure 18.8). In fact, access to various trails is the number one amenity that homeowners seek with buying a home. In April 2002,

* See the Summerlin website for a description of the full range of home designs.

**FIGURE 18.6**
City park in Las Vegas designed to incorporate a desert wash.

**FIGURE 18.7**
Plants along the trail are indigenous to the Mojave Desert region.

a survey of 2000 recent home buyers was cosponsored by the National Association of Home Builders and the National Association of Realtors. The survey asked about the "importance of community amenities," and trails came in second only to highway access. Those surveyed could check any number of the 18 amenities, and 36% picked walking, jogging, or biking trails as either "important" or "very important." Sidewalks, parks, and playgrounds ranked next in importance. Ranking much lower were ball fields, golf courses, and tennis courts.*

Summerlin currently has over 150 miles of trails, some of which act as connector trails to the nearby Red Rock National Conservation Area. Responding to community demand there are street-side village trails that follow highways, bike trails, regional connector trails, and natural

* http://www.americantrails.org/resources/benefits/homebuyers02.html (accessed July 19, 2009).

**FIGURE 18.8**
Desert design of homes with lower walls to be open to the native desert.

**FIGURE 18.9**
A former golf course made into a neighborhood park and walk way.

trails that are built along desert washes (Figure 18.9). Furthermore a development fee was initiated in the 1980s to fund a tortoise habitat, so that desert tortoises could be removed from areas of development and put in a secret location far to the south of Las Vegas.[16*]

Even though Summerlin has golf a course, a program was initiated with the Audubon Society to make the golf courses certified Audubon Cooperative Sanctuaries that provide habitat and resting areas for birds in the Pacific Flyway.[16] Yet Summerlin's golf courses are expensive to maintain, especially in times of severe water shortage.[†]

* Whether the transplanted tortoises would infringe on the habitat of existing tortoises in this area is not known.
† http://www.lvbusinesspress.com/articles/2006/06/26/news/news12.txt (accessed July 19, 2009).

Summerlin is not a completely ecologically sensitive community design. One area that Summerlin (as well as Green Valley) falls short is in the ability to foster a greater sense of communality for all the residents. Most homes still have concrete or cinderblock walls around them, isolating residents from immediate neighbors. Recreation areas in parks can aid in bringing people together, but neighborhood centers are few. There are several county supported libraries, primary and secondary schools, but no land would be made for community college campuses or any other public higher education facility. There is an age-restricted area, Sun City Summerlin that also exists as an enclave with a politically active population but the concerns expressed are for the residents within the community and not the area as whole.*

What Summerlin faces, as does Green Valley and all other neighborhoods with active neighborhood associations, is the task of balancing the need for harmony within each community and the importance of each individual to shape his or her own idea of what it means to dwell in a community. For the most part, the extensive use of Community Codes and Regulations (CC&Rs) means that only a few community members have great power over individuals. As I stated earlier, in successful postmodern mixed communities, respect for persons (human and nonhuman) must be balanced against members taking a collective interest in the well-being of the community. Summerlin would benefit by creating more opportunities for the development of diverse, tolerant, open communities that respect the inherent worth of individuals. I have suggested that there can be degrees in ways communities will reflect attempts to balance individual and community, but as long as community associations focus on short-term economic value over long-term wealth of a community as a whole, Summerlin residents, planners, and community leaders will have to work toward the latter.

## 18.5 My Critique on a Planned Development: Case of Coyote Springs

Sixty miles north of Las Vegas, in the middle of the Coyote Springs Valley, across the line between Clark and Lincoln counties is one of the ambitious attempts to create a large-scale community in the desert (Figure 18.10). Las Vegas valley developer Harvey Whittemore has declared his plans for the size and scope of this development (Figure 18.11):

> ...with as many as 159,000 homes, 16 golf courses and a full complement of stores and service facilities. At nearly 43,000 acres, Coyote Springs covers almost twice as much space as the next-largest development in a state famous for outsized building projects. By comparison, Irvine Co., one of Southern California's largest developers, controls about 44,000 acres in Orange Co.[17]

Though some golf courses are already built, the development's future is on shaky ground (Figures 18.12 and 18.13). Facing a downturn in the economy, the mortgage and housing crisis, coupled with a shortage of water, the development today only barely continues to grow, with a few of the golf courses making up the only sign of growth. Whether Coyote

* Sun City residents were able to block construction of a needed access ramp on to the freeway that encircles the north, west, and southern parts of the valley in spite of numerous requests by other local residents. This ramp was opposed on the grounds that it would decrease property values, increase crime and traffic in the area. It was only after several years that the access ramp given the go-ahead for construction. See http://www.lasvegassun.com/news/2008/nov/21/interchange-lake-mead-boulevard-and-215-beltway-op/ (accessed July 19, 2009).

**FIGURE 18.10**
A created mound with imported desert plants marks an entrance to Coyote Springs.

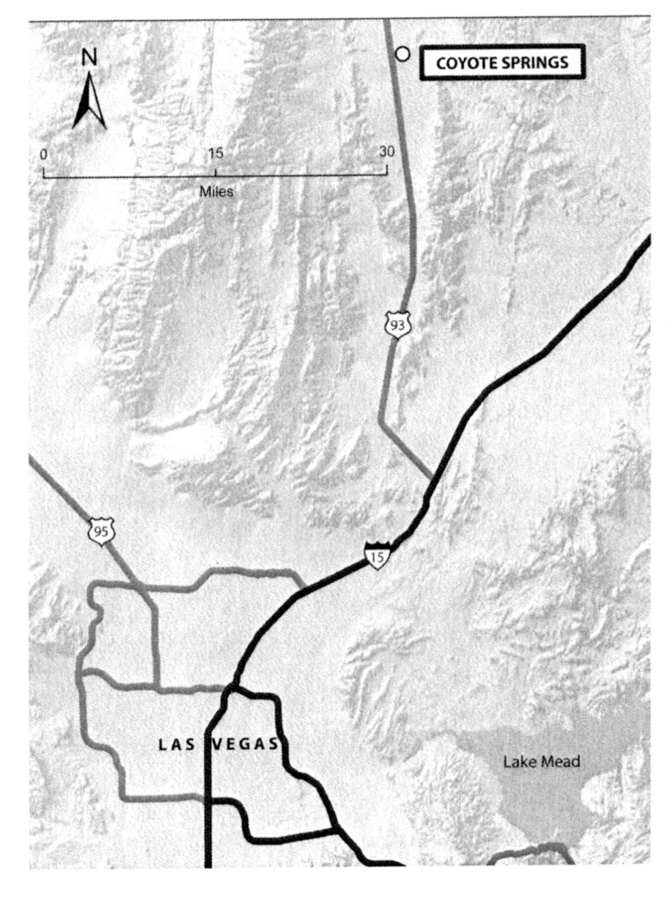


**FIGURE 18.11**
Map of Coyote Springs.

**FIGURE 18.12**
Basic road and infrastructure in Coyote Springs waiting for funding.

**FIGURE 18.13**
Wide-open desert with unpaved road in Coyote Springs.

Springs will live up to its promise of becoming, as the developer's ad campaign claims, *The Next Great Town of Nevada*, or as one website refers to it, "Coyote Springs: Imagining the Next Great Ghost Town" remains to be seen.*

While many environmentalists and others have been hostile to the project from the start, many see Coyote Springs as an opportunity to create communities that were once sought

* http://www.roamingphotos.com/us/nv/coyotesprings/ (accessed July 19, 2009).

in Las Vegas but never found.[18] The website for this development plays on the need for a sense of community today:

> It's the beginning of the next great town of Nevada. Reminiscent of the place you grew up in, where nature and neighborhoods coexist. Your neighbors are your friends. And kids can be kids.*

This strategy touches the deep longing for community on the part of home buyers, to dwell in a place as opposed to merely inhabit. The fragmentation, alienation, and unease that characterizes much of urban life today makes such and campaigns all the more effective. But the upshot of this is that developments themselves do not satisfy this hunger and, in some cases, only make things worse. While what is sought is a sense of belonging, of dwelling in a community, what is delivered is more isolation and dislocation. Gated communities with guard houses may give a false sense of security and a sense that "undesirables" are being kept out, the design of such communities only isolates people further from neighbors in nearby communities behind their own walls. Beautiful scenery and mountains nearby can provide a scenic backdrop, a feature often desired by homeowners, but unless efforts are made to create parks, trails, and other access to the undeveloped desert and mountains, people remain unconnected to the land and often oblivious to what is there and how their own actions will impact it. Native plants and animals are viewed as "pests" to be kept out, even if such species are part of the ecosystem within which the development is a part. But it need not be so. Using the features of true postmodern, mixed communities I presented earlier, it is possible to ask questions about what might be yet done in Coyote Springs in order to make it a community rather than one more development in the desert.

The only major things to be built in Coyote Springs so far are several championship style golf courses (Figure 18.14). The plan as reflected in the website seems to be to make the development a destination community for people wishing to golf extensively.

**FIGURE 18.14**
The one golf course in Coyote Springs with recycled water ponds and terraced pools.

---

* www.villagesofcoyotesprings.com/index_go.php (accessed July 19, 2009).

Around the edges of the golf courses would be expensive homes for enthusiasts of golfing. The approach is not new. Many of the developments in the Las Vegas area have been built up around golf courses. But as was pointed out earlier, there has been a shift in consumer demand away from golf courses which are expensive to create and maintain and require extensive watering and cater only to a small number of residents. Potential homeowners want what Summerlin provided, namely access to many kinds of publicly available trails for casual walking, bicycling, and even more rugged hiking.

So the first question to ask of Coyote Springs is *will the golf courses be central to the development or will there be an extensive system of trails that will enable residents to have easy access to natural areas?* There are several wilderness areas nearby including the Delamar Wilderness Area, the Clover Mountain Wilderness Area, The Arrow Mountains Wilderness Areas, and the Desert National Wildlife Range. The Delamar Wilderness Area will be immediately next to the proposed development, providing an opportunity of quick access of residents to unspoiled natural areas. The close access to the Wildlife Range means that many birds and large mammals will be using the same general locations, offering opportunities for resident interactions with the wildlife.

Another important question to ask of this planned development is *where will people work?* The nearest major urban center, Las Vegas, is more than 60 miles away, meaning that residents would have to commute at least an hour each way. Economic and environmental factors now make living in communities without nearby employment a disincentive to live there. While some employment for the golf courses would be available, and teaching opportunities could occur as the necessary schools for the proposed population get built. But right now there is little to no information about what economic infrastructure is being planned. What is likely is that should the community get built it will initially be a "bedroom community" for those willing to spend time and money on a long commute to Las Vegas. In addition, there is no public transportation available in the entire region, nor are there any plans to create some kind of intra-development system of buses. Some trails have been started, but with development stalled, it is difficult to get any sense of whether pedestrian and bike trials will be as extensive as, say, Summerlin.

Building new homes today provides an opportunity for new, environmentally friendly designs that build with the desert in mind. Pardee Homes, Inc. is listed as the major home developer, but beyond a broad statement of environmental concern in the website* there is little information about what plans there are for taking the opportunity to build desirable homes that incorporate all the latest insights and designed for efficient energy and water use. Nor is there any information about how neighborhood designs might be made so as to encourage a sense of residents being persons in a biotic community and not just homeowners in a neighborhood that cuts them off from each other as well as much of the natural world around them.

One area where Coyote Springs has made progress is in planned water use. The local water resources are limited and with plans for Clark County to utilize more and more water from the rural areas of the state, having enough water for 16 golf courses and up to 159,000 homes remains a challenge. Coyote Springs has incorporated several sustainable and low-impact concepts into the community design. Plans have been made that include "maintaining open space—especially natural open space; sharing water resources with the environment; optimizing available water resources with storm water wetlands and 100% treated wastewater reuse; and providing decentralized water and wastewater

* www.villagesofcoyotesprings.com/index_go.php (accessed July 19, 2009).

infrastructure."[19] So the developers are taking advantage of creating, at least in the planning stages, of the latest technology for water use in the desert. As residents become aware of such basic systems it can help ground them in a sense of place.

While there are forward thinking water use plans, there is no indication yet on other systems such as energy production. The area would be suitable for solar, wind, and even geothermal power productions. So the final question to ask is *where will the energy come from?* If the plan is to provide most of the energy from production sites elsewhere, then the community will lose an opportunity to use local, sustainable power production possibilities, linking residents further with the local area.

Coyote Springs represents an opportunity to show that a truly postmodern, sustainable community can be created. Whether economic conditions and thoughtful community design and planning will enable the community to build or whether the developers will continue to incorporate new, environmentally friendly design features in all aspects of community remains to be seen.

## 18.6 Conclusion

It is possible, and has already been done to a small and limited degree in Las Vegas, to build communities with these postmodern features. Neighborhoods can be built that are not walled enclaves. Homes can be designed for the desert climate with desert landscaping that requires little or no water for maintenance. Parks and common areas can be built that are part of previously existing natural features, such as canyons and waterways. Stores and offices can be built that foster foot and bicycle traffic, instead of demanding greater automobile use. That such features are already present in areas around Las Vegas indicate that it is possible to develop the kind of community with the postmodern features I described.

Furthermore, community is *not* automatically created when the entire community and all of the structures already in place are prefabricated, as it is envisioned in places such as Coyote Springs. The character of a community would best be served by a slow emergence of these values in a participatory process as the specific needs of the community develop. This form of evolution will allow for an organic and flexible structure within the community that responds to the current and future aspirations of and by the residents themselves.

While the goal is to create features that foster community feeling and respect for diverse community members, the planned turnover in home ownership has spawned developments with extensive covenants in the terms of ownership of each home to preserve home and resale values. This is often done in ways that stipulate what may or may not be done to the design of the home, resulting in a bland repetition of landscape and home ornamentation. The legal penalties are enforced, insuring that residents do not attempt to live in a way that might be "different" or disturbing to neighbors, even if this involves retooling homes for solar energy and gray water use. The reliance on neighborhood associations to enforce such covenants underscores the fact that such groups have no common, shared vision of a good life that would render groundless such fears of diminished property values. The pressure to be free, unencumbered individuals able to move on whenever has led to undemocratic pressure to conform to a lifestyle that offends no one and satisfies no one as well.

## References

1. Griffin, D., Introduction: Constructive postmodern philosophy, in *Founders of Constructive Postmodern Philosophy: Peirce, James, Bergson, Whitehead, and Hartshorne* (Albany, NY: State University of New York Press, 1993), p. 1.
2. Giddens, A., *The Consequences of Modernity* (Cambridge, MA: Polity Press, 1990); Giddens, A., *Modernity and Self Identify* (Cambridge, MA: Polity Press, 1991).
3. Venturi, R. and D. Scott Brown, *Learning from Las Vegas: The forgotten Symbolism of Architectural Form* (Cambridge, MA: MIT Press, 1997).
4. Ellin, N., *Postmodern Urbanism* (Cambridge, MA: Blackwell, 1996).
5. Frampton, K., *Labour, Work and Architecture: Collected Essays on Architecture and Design* (London, U.K.: Phaidon Press, 2002).
6. Ellin, N., Shelter from the storm or form follows fear and vice versa, in *Architecture of Fear*, N. Ellin, Ed. (Princeton, NJ: Princeton Architectural Press, 1997), p. 43.
7. Giddens, A., *The Consequences of Modernity* (Cambridge, U.K.: Polity Press, 1990).
8. Kellert, S. and E. O. Wilson, Eds., *The Biophilia Hypothesis* (Washington, DC: Island Press, 1993).
9. Kellert, S., J. H. Heerwagen, and M. Mador, Eds., *Biophilic Design: The Theory, Science, and Practice of Bringing Buildings to Life* (Hoboken, NJ: Wiley & Sons, 2008).
10. Kellert, S., *Building for Life: Designing and Understanding the Human-Nature Connection* (Washington, DC: Island Press, 2005).
11. Knowles, R., *Ritual House: Drawing on Nature's Rhythms for Architecture and Urban Design* (Washington, DC: Island Press, 2006).
12. Leopold, A., The upshot, in *A Sand County Almanac* (New York: Oxford University Press, 1949).
13. Frasz, G., The problem of community, PhD dissertation (Athens, GA: University of Georgia, 1995).
14. Guterson, D., No place like home, *Harpers*, November 1992: 55–64.
15. Bellah, R. N., R. Madsen, W. M. Sullivan, A. Swidler, and S. M. Tipton, *Habits of the Heart* (New York: Harper & Row, 2005), pp. 20–21.
16. Hogan, J., 5 things that make Summerlin unique, *Las Vegas Review Journal*, January 6, 2009, 3AA; See also http://www.summerlin.com (accessed July 19, 2009).
17. Neubauer, C. and R. T. Cooper, Desert connections, *Los Angeles Times*, August 20, 2006. http://articles.latimes.com/2006aug20/business/fi-nevada-20 (accessed July 19, 2009).
18. Schoenmann, J., Builder sees green light in red-flag economy, *Las Vegas Sun*, http://www.lasvegassun.com/staff/joe-schoenmann/ (accessed July 19, 2009).
19. Galuska, C. et al., Palm Springs, Nevada style? Coyote Springs Nevada implements sustainable wastewater treatment and reuse to reduce costs and accelerate the schedule, The Water Environment Foundation, 2006, found at http://acwi.gov/swrr/Rpt_Pubs/wef-pubs.html (accessed July 19, 2009).

# 19

## *Dialogue on Development*

**Richard A. Malloy**

**CONTENTS**



### 19.1 Introduction

We need a sustained and comprehensive dialogue on development in the desert—one that critiques our plans, aspirations, and values on development. Planning without considering the limitations of the natural setting of the desert will set the stage for the loss of time, money, and resources. The ancient cultures of the Southwest had a strong oral tradition where history, traditions, and values were passed to the next generation through stories and myth. Modern society has lost part of the heritage that once served us well. In the recent past, there was a model of planning that involved the delegation of elected or appointed officials to make decisions on behalf of the community for future development. It was assumed that these people would act in a benevolent manner in matters of public welfare, but history has shown that the concept of benevolence has sometimes resulted in *benefit, backroom deals,* and *breaking established rules* by these officials to get a development project completed. The 1960s ushered in a new paradigm in project management at the federal level that provided specific opportunities for the public to have a voice in the planning process. The average citizen now has an opportunity to participate in projects that will affect their lives or property.

## 19.2 Traditional Planning Models

The United States followed several models of city planning and development throughout time. The early settlements on the east coast adopted a colonial development plan that featured a town square as the center of government, church, and business enterprises. The compact colonial city formed a dense grid around the town square to the edge of the urbanized area. Later, industrial cities of the north and midwest created the factory town where large industrial plants were accompanied by multistory tenement housing for the factory workers. After World War II, numerous automobile-dependent communities such as Greenbelt, Maryland, and Levitown, New York, served as models for "suburban" development. At the same time, Sunbelt cities of the Southwest were rapidly developing in a patchwork fashion. Sunbelt city economies were prone to boom and bust business cycles and disperse land development patterns. Sprawling expansion and growth of the urban area were typical; while leaving large tracts of land in the urbanized area with poorly developed transportation and urban infrastructure (see Chapter 14). Along the urban fringe, well-kept developments can be found that house the new elite class that is prospering from the new Sunbelt economy.

We need development in our southwestern cities just as much as anywhere else in the country. The current growth rate from natural population growth and the inflow migration to the region creates a natural demand for new houses, schools, roads, churches, and business development. We need to have an environment where builders and developers can serve the community with its needs to grow and prosper. However, we also need to draw a line when the development interests acts in manners that take away from the continuity of the community, contribute to urban sprawl, or create monetary burdens on the existing residents for the costs of the infrastructure to support this development. Residents of the city have a voice in the development process, but this option is not always used effectively or at the right time to influence the direction of development activities.

The rapid and exponential growth of the Southwest has resulted in a citizenry that grew up in other parts of the country or world that are not faced with the same challenges or limitations of the natural environment. Residents who migrate from northern climates seek the same qualities of their new desert home and community, including extensive green lawn, shrubbery, golf courses, and other amenities that they should have left behind. They often see the native desert landscape as dry, dusty, unappealing in contrast to the well-watered landscape and air-conditioned home. As the number of these people migrating to the Southwest grow, so does the demand for water intensive landscapes, which is unsustainable if we are to conserve the existing water resources available for current and future growth of the desert. These people need to be educated on the natural setting of the desert and how the life choices in homes, recreation, and amenities contribute to the region's viability. We pass these lessons on to our children by example of what we know and what we learn from our new environment. In modern times, children learn through sources outside the family setting such as teachers, coaches, TV, the Internet, social media, and other means that make the dialogue with children on social values challenging to maintain with all of the noise of urban life. We need this community dialogue on development *with* the development process to create a sustainable path to grow responsibly for ourselves, our family, and our future (Figure 19.1).

(a)

(b)

**FIGURE 19.1**
(a) A development in the southeast valley of Phoenix reminiscent of landscapes from other geographic areas with extensive grass, deciduous trees, palms, and white picket fences. (b) Typical ranch style house and landscape in the Phoenix area including a green grass lawn, deciduous trees, and shrubbery pruned as hedges.

## 19.3 Beyond Talk to Solutions

Balancing development and growth to environmental sustainability should be the ultimate goal of the community and development interests. So how can we create an environment for residents and development interests to talk with each other before reaching the point of no return on unsustainable development? The following are some of my thoughts on the nature of development and how all parties can find common ground on community choices.

### 19.3.1 Create a Vision

Effective planning starts with a clear vision of what the goals and objectives are for the proposed action. This vision should reflect the core values and principles of the community at large. Cities are required to develop a general plan at specified time intervals of five, ten or more years. The general plan includes a detailed study of the natural and built environments of the community. The plan starts with an expression of what the community wants to see in the future. The planning guidelines outlined in the plan indicate the values of the community in shaping the course of future development. This vision can be bold in scope or limited to certain areas of development activities, such as schools, roads, or commercial centers. The visioning process for the general plan of a community is an effective place for a concerned resident to enter the dialogue on development of the community.

Development activities are governed by land use planning guidelines set forth in the community's general plan. The Standard State Zoning and Enabling Act of 1922 gave cities the power to enact zoning regulations in accordance with a comprehensive plan for the affected area. In 1926, the Supreme Court handed down a decision in *Village of Euclid v Ambler Realty Co.* that established the constitutional basis for comprehensive

(a)

(b)

**FIGURE 19.2**
(a) Developing the desert. Will it be done responsibly? (b) Native desert slated for future development.

zoning, thus giving cities the power to determine what uses are permitted within the incorporated area, provided it falls within a comprehensive plan for the entire area. In the development of a land use plan, citizens can influence the areas and permitted land uses within their city, which can impact the location and pace of development projects. However, development interests pay close attention to any changes in city land use plans and often have lobbyists and well-paid lawyers to argue on their behalf to minimize any land use restrictions on any parcel of land they have slated for development. To counter development resistance to regulation, the community will need a unified stance on areas of critical importance; otherwise, the system favors those engaged in the process (Figure 19.2).

### 19.3.2 Public Participation

Public participation is essential to the success of a plan or project. The Southwest region has a high percentage of residents from other parts of the country. These new residents have diverse views on the role of government in their lives, making it challenging to arrive at a decision that will please everyone equally. The planner should identify all of the stakeholders of the affected project area and begin a dialogue early in the process to begin to identify the concerns of the local community.

During the 1960s and 1970s, there was a growing number of people who began to question the benevolence of government decision makers in their ability to make independent decisions that adequately address concerns of the community and protect the environment. Government agencies that manage lands or implement projects have public participation requirements where an average citizen can have a voice in the planning process. All federal projects not identified to have categorical exclusion have periods of public comment from the scoping to the development of a draft plan for implementation. This provides the opportunity for dialogue on the course and direction for the proposed project. Some states such as California have similar requirements for state-sponsored projects that impact the environment. On the local level, the town council is a forum for local residents to form a dialogue with the forces of development. Citizens can have a meaningful impact on planning projects if they bring their voices to the planning table at the appropriate time in a respectful manner.

Citizen involvement can provide valuable input to city planners by helping to identify areas of concern for development projects or proposals. Professionals working with the public use several techniques to engage their interests including surveys, public meetings, open houses, workshops, and neighborhood gatherings. Public participation events are designed to garner feedback from the public on proposed actions by the city, state, or federal agency. This is where you the citizen can provide direct and specific comment on a proposed action or project. A well-coordinated effort by concerned citizens can influence the outcome of the proposed project and alter the plan to reflect the stated concerns of the public. However, this input must be provided at the appropriate stage of the planning process. For instance, your opposition to a site development plan will not be taken seriously if you did not provide your feedback when the site selection and plan was under review.

### 19.3.3 Race to the Bottom

Cities are deeply concerned about having a viable economic base. One of the principal means of gaining revenue is through sales tax revenue and jobs through the location of commercial and industrial operations. There is often a fierce competition between cities to lure companies to locate their business within their municipality. The company has the upper hand by having independent power to make their own decision on where to locate their business. Cities jockey for influence with the company by offering incentives from free land, infrastructure, sales tax rebates to exemption from planning and zoning laws. Development incentives for corporations are part of the toolbox that municipalities use to create a sustainable economic base for the community. It would be a step in the right direction if a city did not have to "give away the store" to save the store and instead compete on the merits of its community location and amenities, not its bank account. However, the stakes are too high for cities not to provide these types of incentives to sway development in their direction and too lucrative for the developer not to seek the most favorable deal proposed by several cities at the same time. There is no easy answer to solve this race to the bottom in this quest for cities to obtain business partners that potentially will become the economic base of a community.

### 19.3.4 We Are Not L.A., Really?

Over time, Los Angeles (L.A.), California, has been the poster child for what is undesirable in an urban form. The impacts of a low-density form of urban development, high population growth, and a transportation system geared towards the automobile creates strains on urban life from diminished air quality, traffic congestion and stagnation, and sprawling urban development reduces the quality of life for area residents. Many point to L.A. as what not to do as a model for urban development; however, several Southwest cities are repeating the same pattern, creating a smaller version of the same problems posed by low-density sprawling development. There is a light at the end of this tunnel. As southwestern cities grapple with their own development patterns, there is a growing trend towards efforts to develop or redevelop the neglected urban core. The question is—can the voices of the community speak loud enough to demand a viable alternative to this form of urban development?

### 19.3.5 Who Pays for Development?

New development projects are charged impact fees to cover pact fees to cover the municipal cost of urban infrastructure and city services such as water, sewer, roads, etc. Although the cost to the builder may be significant, the actual cost to the municipality may exceed the money received in impact fees for the development when you consider all of the indirect costs of development. As more and more areas get developed, the city takes on a larger burden to provide police and fire protection, libraries, hospitals, schools, parks, and recreational facilities for its residents. These costs are added to the general operating budget of the city, and all residents end up contributing indirectly to the cost of providing services to new developments. Developers oppose impact fees for new development projects, arguing it hinders growth and puts constraints on development projects. These fees are customarily passed on to the buyer of the new home or property. Can development support itself without a subsidy from other parts of the city budget or should the residents assume that it will take money to grow and develop? (Figure 19.3).

### 19.3.6 Knitting Nature into Development

The old way of developing in the desert was to scrape the land of all of its natural components and replace it with a completely foreign landscape, often reminiscent of other regions of the country. The new way of developing responsibly involves leaving parts of nature that serves the ecology of the place intact, while developing the site to fit the future project. The use of carefully chosen desert-adapted plants in the landscape design can provide beauty, color, and functionality to the property. Concerned residents should consult with a qualified landscape designer during the development of a landscape plan on the selection of plants that are adapted to desert environments. Plants from desert regions require less water, care, and maintenance and can add striking features to the landscape (Figure 19.4).

**FIGURE 19.3**
Will this desert development pay for its impacts to the community?

**FIGURE 19.4**
Incorporating desert-adapted plants with dramatic effects.

### 19.3.7 Promote Urban Infill over Fringe Development

Southwestern cities have been plagued by the contagion of urban sprawl. The rapid expansion of the urban boundary to encompass new development on the fringe is a continuous process that cities are grappling with. The low cost of land in unincorporated areas allows developers to make a larger profit once the land is developed, hence the emphasis to push beyond urban boundaries for new development. This often leaves large swaths of undeveloped land within the urban core and beyond where public infrastructure already exists. Cities should promote and residents should demand that city planners encourage urban infill development. Filling in the urban core with developments that match the existing community matrix will lead to a stronger and more vibrant community (Figure 19.5).

**FIGURE 19.5**
Can cities do a better job at promoting infill development in areas where infrastructure already exists?

(a)

(b)

**FIGURE 19.6**
(a) Walls can separate people from each other. (b) Gates provide security, but they also can keep people from connecting to the community.

### 19.3.8 Walls and Gates

We all believe in property rights, and it is the right of a property owner to define their property in some fashion with walls and gates. Everywhere you go in the Southwest, developments are surrounded by fences, gates, and walls. In the days of the Wild West, fences and walls were required to keep the bad guys out and the livestock in. Now, to some extent, these same features continue this same protection-based design based on fear of "the outside." I would argue the ubiquitous presence of walls and gated communities are contributing to the isolation of segments of the community, rather than helping to relate to each other by developing personal connection and understanding of fellow community members. A postmodern neighborhood design has few walls and low fences that define the property, but still allows for dialogue among neighbors. We should thoughtfully consider whether the number of walls and gates we have really serve the purpose of building community (Figure 19.6).

### 19.3.9 Walkable Cities and Transportation Choices

Most Sunbelt cities are designed for the automobile. A man without a car today is like a man without a horse in the old days—powerless to do anything meaningful in life. The suburban development pattern of predominantly single-family homes fosters the need for an automobile to commute for work and commerce. Some are faced with commutes that can take up to two hours in rush hour traffic to reach the work destination, as the affordable housing stock is increasingly found along the outskirts of the urbanized zone. After living this daily grind for several years, there is a growing number of people that would gladly give up this lifestyle for one that allows for walkable streets that integrate living and working opportunities and promote connection to other members of the community. The notion that a single-family home is the ultimate quest is in question, as more people opt for townhouses, condominiums, and loft apartments in the urban core as alternatives to cheaper houses on the urban fringe. In addition, providing alternative transportation modes such as light rail, bus, and bike lanes provides sensible options for

(a)

(b)

**FIGURE 19.7**
(a) Townhouses provide an alternative to the single-family home. (b) Multistory condominium complexes provide a compact community and efficient use of space. Both developments are urban infill projects with nearby public transportation, bike lanes, and pedestrian pathways.

area residents and makes living in the community possible for those of limited means and physical disabilities (Figure 19.7).

### 19.3.10 Preserving Open Space

Not all space is created equal. Some land holds higher intrinsic and community value than its development potential. The preservation of open space is essential for communities to proactively address. Certain lands have importance for the community to preserve scenic views, serve as recreational corridors, watershed protection, and public parks among others. It behooves a community to identify those areas of open space with critical and intrinsic value to the city before it becomes irreversibly committed to a development project. Phoenix and Scottsdale, Arizona, have been in competition with developers for choice tracts of land for open space preservation. The same tract that holds value for parks and recreation has value for high-end home development in north Phoenix. Fortunately, the current lull in home building due to the housing crisis is allowing municipalities to buy up open space at discounted prices with little competition from commercial developers. Residents concerned with preserving open space in their community should consider participating in public meetings on open space planning, joining a wilderness advocacy group, or volunteer for environmental education projects on preserving the natural environment (Figure 19.8).

**FIGURE 19.8**
Protecting open space is important to preserve scenic views and sensitive lands.

### 19.3.11 Connecting Water with Growth and Development

The rapid pace of development of the Southwest is creating a need to find new ways to deal with the finite amount of water available in the region. Unbridled growth will stretch the water supply and eventually lead to a shortage in times of drought and water uncertainty. There is a compelling need to develop policies that link new development to an assured water supply that will ensure the development has the water resources to sustain the proposed uses of this development. Requirements for development vary by state and region from mandatory to recommended guidelines. Arizona has the most comprehensive regulations for development set forth by the Arizona Groundwater Management Act (AGMA) of 1980. This law requires all new development projects within the active management areas to demonstrate a 100-year assured water supply before the development can proceed. Although not perfect in implementation, other states should seriously consider some form of policy similar to the AGMA that will put some rational planning and accountability for development interests to provide an assured and reliable source of water to support future developments (Figure 19.9).

### 19.3.12 Sustainable Community Design

A vibrant community does not happen by chance. Communities become vibrant when people put forth efforts that allow the residents to live well by putting in place measures that will sustain the community over time. Altogether, effective sustainable community design includes all of the elements presented in this chapter—a comprehensive vision, public involvement, a viable economic base, preservation of community assets, sustainable water supply, transportation choices, and a design that incorporates nature into the fabric of the community. You can experience the effects of sustainable design in communities when you see a community responding by having healthy interactions, active lifestyles, participating in community activities, and contributing to the future of the community on

(a)

(b)

**FIGURE 19.9**
(a) Agricultural lands with existing water rights serve as the source of future assured water supply for home development. (b) Agricultural lands are the site of future home development.

(a)

(b)

**FIGURE 19.10**
(a) Homes under construction. (b) Building homes.

a volunteer basis. Sustainability can be designed into the community, but it is up to the residents to carry the torch for the community if sustainability is to survive as the guiding force for community development (Figure 19.10).

## 19.4 Conclusion

The collapse of the housing market starting in 2008 has exposed one of the weaknesses of the economies of southwestern cities—an overreliance on unsustainable development. I hope we learn the lessons that were so painfully presented to us by the near halt of the regional economy. This is a time that we need this dialogue on development with development.

This is an opportunity to present bold options for development that we may have not considered thus far. We need to have development interests on our side, not as an advisory in this process. The most profound measure of influence on development is to vote with your purse strings by making housing, transportation, or personal choices that reflect sustainability. The development community should provide sustainable development options, but will respond when the market demonstrates that consumers prefer these options.

As a resident of a desert community, you have the power to influence the path of development in your community. There are several opportunities where a concerned resident can provide a voice on development, including project scoping meetings, city council meetings, community visioning workshops, and neighborhood meetings, among others. Some important considerations are to be engaged in the process, act respectfully to all parties in the discussion, understand you can alter the course of a decision when you state your position with facts not emotions, and realize a change in behavior or policy can take a long time to become reality. Provide praise for developers when they take steps in the right direction, but hold them accountable to existing environmental and zoning laws when necessary. Do not give up the effort to make that dialogue with development; you may find you have more power than you think to make a change that works for the community.

# Part IV

# Ecology in Design of Urban Systems

**Margaret Livingston**

As our city ecosystems continue to rapidly evolve, we struggle to better understand and manage the diverse systems and organisms found in these environments and the resources associated with their existence. The following chapters present various discussions about these systems and the design strategies that we employ in our efforts to support them, at the scale of the urban environment.

This part of the book begins with an overview of ecological design by David Orr and its application to our urban ecosystems (Chapter 20). The chapter focuses on design that minimizes destructive impacts of city development through integration with living processes. For example, Orr discusses some of the key elements of ecological design such as improving energy and resource efficiency and the development of closed loop systems that deliver "products of service." The author stresses the success of ecological design through our deeper sense of connection and responsibility to our natural ecosystems, suggesting that without more consensus from us on this need, our efforts will not make much difference.

Chapters 21 through 23 focus on more site-specific strategies for readers that have been gaining popularity in use in urban environments. Heather Kinkade focuses on strategies for increasing water harvesting in cities, stressing the resurgence in the use of this ancient practice by residents in arid, urban environments for the conservation and management of water in their cities (Chapter 21). The author poses several questions for readers to consider to get started with water harvesting, such as whether a system will be a retrofit for an existing building or use of a new integral system, system size, and intended use. She concludes the chapter with a discussion of typical components, the use of a water balance analysis, and an integrated site design, which is intended to match site requirements (e.g., water, energy, food, and aesthetics) with the eventual components of an area (e.g., stormwater runoff, shade from buildings, and vegetation), while improving the function and sustainability of a site.

Chapter 22 by Margaret Livingston offers opportunities for enhancing natural interactions that occur in urban environments, with a focus on the evaluation of habitat preservation and enhancement in urban situations. The chapter begins with a focus on the larger scale of habitats, examining how cities often contribute to the fragmentation of natural systems such as watercourses and large expanses of upland habitats that are critical resources for many species. This overview is intended to inform the reader about how smaller habitats we create in cities can serve as rest stops, or in some cases, new patches of habitat for species displaced by urban growth. The chapter also provides more specific design guidelines for habitat creation of three groups that are relatively well adapted to urban sites: songbirds, hummingbirds, and butterflies. Plant suggestions for attracting each group are also discussed.

In Chapter 23, Dunstan and Livingston discuss practices for retaining some of an original site's composition, but in perhaps a different configuration—the reuse of existing, salvaged plants to new locations in urban environments. This practice stresses the ability to maintain older specimens of arid species, particularly trees and cacti, through effective salvaging. Preserving these "living sculptures," as Dunstan and Livingston refer to them, has become an accepted and valued practice, and the chapter outlines the process for tree salvaging, from site preparation to creating a container ("boxing") around the plant for safe transport and transplanting. This chapter also discusses the practices for salvaging larger columnar cacti such as saguaro for effective transplanting to other locations. In urban areas, this practice has increased dramatically with the implementation of the native plant protection ordinances.

The last two chapters emphasize the future ecological opportunities in urban settings. Chapter 24 focuses on the presentation of an alternative model for sustainable urban living, stressing that a new design "that recovers, recycles, and reuses nutrients lost in the human and animal waste streams is needed." Mark Edwards' discussions on declining freshwater, fossil fuel supplies, soils, and climate change highlight the need for a new model for our food production in the future. Edwards examines green solar energy captured in algae, representing "an agriculture of abundance based on cheap natural resources." He outlines the unique ability of algae to grow sustainable and affordable food and energy (SAFE) production that can assure vitality to urban centers.

Nan Ellin's chapter (Chapter 25) bridges some of the ideas presented in the previous chapters, with a discussion of the concept of integral urbanism. Her chapter emphasizes the creation of linkages and connections in our urban spaces, responding to our rebellion against past sprawl evident in many of our desert cities and the need to effectively connect various land uses and inhabitants, human and animal, for example, in our metropolitan areas. A discussion of some of the changes occurring in downtown Phoenix that reflect concepts of integral urbanism is presented.

It is intended that these chapters represent just some of the concepts and actions that can be used by urban dwellers, aiding in their attempts to support some of the ecological processes that occur in our urban systems. For some readers, this part of the book may also be informative about a few key areas and strategies in ecological design that we have made progress in under urban situations. In contrast, some of the ecological processes that are in desperate need of closer attention and better preservation in our cities for coexistence to occur with us are highlighted.

# 20

# *Ecological Design**

**David Orr**

**CONTENTS**



## 20.1 Introduction

The unfolding problems of human ecology are not solvable by repeating old mistakes in new and more sophisticated ways. We need a deeper change of the kind Albert Einstein had in mind when he said that the same manner of thought that created problems could not solve them. We need what architect Sim van der Ryn and mathematician Steward Cowan define as an ecological design revolution. Ecological design in their words is "any form of design that minimize(s) environmentally destructive impacts by integrating itself with living processes... the effective adaptation to and integration with nature's processes."[1] For landscape architect Carol Franklin, ecological design is a "fundamental revision of thinking and operation."[2] Good design does not begin with what we can do, but rather with questions about what we really want to do.[3] Ecological design, in other words, is the careful meshing of human purposes with the larger patterns and flows of the natural world and the study of those patterns and flows to inform human actions.[4]

Paul Hawken, Amory Lovins, and Hunter Lovins, to this end propose a transformation in energy and resource efficiency that would dramatically increase wealth while using a fraction of the resources we currently use.[5] Transformation would not occur, however, simply as an extrapolation of existing technological trends. They propose, instead, a deeper revolution in our thinking about the uses of technology so that we don't end up with "extremely efficient factories making napalm and throwaway beer cans."[6] In contrast to Ausubel, the authors of *Natural Capitalism* propose a closer calibration between means and ends. Such a world would improve energy and resource efficiency by, perhaps, 10-fold. It would be powered by highly efficient small-scale renewable energy technologies distributed close to the point of end-use. It would protect natural capital in the form of soils, forests, grasslands, oceanic fisheries, and biota while preserving biological diversity. Pollution, in any form, would be curtailed and eventually eliminated by industries

* Adapted with permission from Orr, D., *The Nature of Design Ecology, Culture, and Human Intention*, Chapter 2, Oxford University Press, Oxford, U.K., 2002.

designed to discharge no waste. The economy of that world would be calibrated to fit ecological realities. Taxes would be levied on things we do not want such as pollution and removed from things such as income and employment that we do want. These changes signal a revolution in design that draws on fields as diverse as ecology, systems dynamics, energetics, sustainable agriculture, industrial ecology, architecture, and landscape architecture.*

The challenge of ecological design is more than simply an engineering problem of improving efficiency—reducing the rates at which we poison ourselves and damage the world. The revolution that van der Ryn and Cowan propose must first reduce the rate at which things get worse (coefficients of change) but eventually change the structure of the larger system. As Bill McDonough and Michael Braungart argue, we will need a "second industrial revolution" that eliminates the very concept of waste.[7] This implies, in their words, putting "filters on our minds, not at the end of pipes." In practice, the change McDonough proposes implies, among other things, changing manufacturing systems to eliminate the use of toxic and cancer-causing materials and developing closed loop systems that deliver "products of service," not products that are eventually discarded to air, water, and landfills. The pioneers in ecological design begin with the observation that nature has been developing successful strategies for living on Earth for 3.8 billion years and is, accordingly, a model for

Farms that work like forests and prairies
Buildings that accrue natural capital like trees
Wastewater systems that work like natural wetlands
Materials that mimic the ingenuity of plants and animals
Industries that work more like ecosystems
Products that become part of cycles resembling natural material flow

Wes Jackson, for example, is attempting to redesign agriculture in the Great Plains to mimic the prairie that once existed there.[8] Paul Hawken proposes to remake commerce in the image of natural systems.[9] The new field of industrial ecology is similarly attempting to redesign manufacturing to reflect the way ecosystems work. The new field of "biomimicry" is beginning to transform industrial chemistry, medicine, and communications. Common spiders, for example, make silk that is ounce for ounce five times stronger than steel with no waste byproducts. The inner shell of an abalone is far tougher than our best ceramics.[10] By such standards, human industry is remarkably clumsy, inefficient, and destructive. Running through each of these is the belief that the successful design strategies, tested over the course of evolution, provide the standard to inform the design of commerce and the large systems that supply us with food, energy, water, and materials, and remove our wastes.[11]

The greatest impediment to an ecological design revolution is not, however, technological or scientific, but rather human. If intention is the first signal of design, as Bill McDonough

* The roots of ecological design can be traced back to the work of Scottish biologist, D'Arcy Thompson and his magisterial *On Growth and Form* first published in 1917. In contrast to Darwin's evolutionary biology, Thompson traced the evolution of life forms back to the problems elementary physical forces such as gravity pose for individual species. His legacy is an evolving science of forms evident in evolutionary biology, biomechanics, and architecture. Ecological design is evident in the work of Bill Browning, Herman Daly, Paul Hawken, Wes Jackson, Aldo Leopold, Amory and Hunter Lovins, John Lyle, Bill McDonough, Donella Meadows, Eugene Odum, Sim van der Ryn, and David Wann.

puts it, we must reckon with the fact that human intentions have been warped throughout the twentieth century by excessive violence and the systematic cultivation of greed, self-preoccupation, and mass consumerism. A real design revolution will have to transform human intentions and the larger political, economic, and institutional structure that permitted ecological degradation in the first place. A second impediment to an ecological design revolution is simply the scale of change required in the next few decades. All nations, but starting with the most wealthy, will have to

Improve energy efficiency by a factor of 5–10

Rapidly develop renewable sources of energy

Reduce the amount of materials per unit of output by a factor of 5–10

Preserve biological diversity now being lost everywhere

Restore degraded ecosystems

Redesign transportation systems and urban areas

Institute sustainable practices of agriculture and forestry

Reduce population growth and eventually total population levels

Redistribute resources fairly within and between generations

Develop more accurate indicators of prosperity, well-being, health, and security

We have good reason to think that all of these must be well underway within the next few decades. Given the scale and extent of the changes required, this is a transition for which there is no historical precedent. The century ahead will test, not just our ingenuity, but our foresight, wisdom, and sense of humanity as well.

The success of ecological design will depend on our ability to cultivate a deeper sense of connection and obligation without which few people will be willing to make even obvious and rational changes in time to make much difference. We will have to reckon with the power of denial, both individual and collective, to block change. We must reckon with the fact that we will never be intelligent enough to understand the full consequences of our actions, some of which will be paradoxical and some evil. We must learn how to avoid creating problems for which there is no good solution technological or otherwise[12] such as the creation of long-lived wastes, the loss of species, or toxic waste flowing from tens of thousands of mines. In short, a real design revolution must aim to foster a deeper transformation in human intentions and the political and economic institutions that turn intentions into ecological results. There is no clever shortcut, no end-run around natural constraints, no magic bullet, and no cheap grace.

## 20.2 Intention of Design

Designing a civilization that can be sustained ecologically and one that sustains the best in the human spirit will require us, then, to confront with the wellsprings of intention, which is to say human nature. Our intentions are the product of many things at least four of which have implications for our ecological prospects. First, with the certain awareness of our mortality, we are inescapably religious creatures. The religious impulse in us works like water flowing up from an artesian spring that will come to the surface one way or another.

Our choice is not whether we are religious or not as atheists would have it, but whether the object of our worship is authentic or not. The gravity mass of our nature tugs us to create or discover systems of meaning that places the human condition in some larger framework that explains, consoles, offers grounds for hope, and, sometimes, rationalizes. In our age, nationalism, capitalism, communism, fascism, consumerism, cyberism, and even ecologism have become substitutes for genuine religion. But whatever the ism or the belief, in one way or another we will create or discover systems of thought and behavior which give us a sense of meaning and belonging to some larger scheme of things. Moreover, there is good evidence to support the claim that successful resource management requires, in Anderson's words, "a direct, emotional religiously 'socialized' tie to the resources in question."[13] Paradoxically, however, societies with much less scientific information than we have often make better environmental choices. Myth and religious beliefs, which we regard as erroneous, have sometimes worked better to preserve environments than have decisions based on scientific information administered by presumably "rational" bureaucrats.[14] The implication is that solutions to environmental problems must be designed to resonate at deep emotional levels and be ecologically sound.

Second, despite all of our puffed up self-advertising as *Homo sapiens*, the fact is that we are limited, if clever, creatures. Accordingly, we need a more sober view of our possibilities. Real wisdom is rare and rarer still if measured ecologically. Seldom do we foresee the ecological consequences of our actions. We have great difficulty understanding what Jay Forrester once called the "counterintuitive behavior of social systems."[15] We are prone to overdo what worked in the past, with the result that many of our current problems stem from past success carried to an extreme. Enjoined to "be fruitful and multiply," we did as commanded. But at six billion and counting, it seems that we lack the gene for enough. We are prone to overestimate our abilities to get out of self-generated messes. We are, as someone put it, continually overrunning our headlights. Human history is in large measure a sorry catalog of war and malfeasance of one kind or another. Stupidity is probably as great a factor in human affairs as intelligence. All of which is to say that a more sober reading of human potentials suggests the need for a fail-safe approach to ecological design that does not overtax our collective intelligence, foresight, and goodness.

Third, quite possibly we have certain dispositions toward the environment that have been hardwired in us over the course of our evolution. Wilson, for example, suggests that we possess what he calls "biophilia" meaning an innate "urge to affiliate with other forms of life."[16] Biophilia may be evident in our preference for certain landscapes such as savannas and in the fact that we heal more quickly in the presence of sunlight, trees, and flowers than in biologically sterile, artificially lit, utilitarian settings. Emotionally damaged children, unable to establish close and loving relationships with people, sometimes can be reached by carefully supervised contact with animals. And after several million years of evolution it would be surprising indeed were it otherwise. The affinity for life described by Wilson and others, does not, however, imply nature romanticism, but rather something like a core element in our nature that connects us to the nature in which we evolved and which nurtures and sustains us. Biophilia certainly does not mean that we are all disposed to like nature or that it cannot be corrupted into biophobia. But without intending to do so, we are creating a world in which we do not fit. The growing evidence supporting the biophilia hypothesis suggests that we fit better in environments that have more, not less, nature. We do better with sunlight, contact with animals, and in settings that include trees, flowers, flowing water, birds, and natural processes than in their absence. We are sensuous creatures who develop emotional attachment to particular landscapes. The implication is that we need to create communities and places that resonate with our evolutionary past and for which we have deep affection.

Fourth, for all of our considerable scientific advances, our knowledge of the Earth is still minute relative to what we will need to know. Where are we? The short answer is that despite all of our science, no one knows for certain. We inhabit the third planet out from a fifth-rate star located in a backwater galaxy. We are the center of nothing that is very obvious to the eye of science. We do not know whether the Earth is just dead matter or whether it is, in some respects, alive. Nor do we know how forgiving the ecosphere may be to human insults. Our knowledge of the flora and fauna of the Earth and the ecological processes that link them together is small relative to all that might be known. In some areas, in fact, knowledge is in retreat because it is no longer fashionable or profitable. Our practical knowledge of particular places is often considerably less than that of the native peoples we displaced. As a result, the average college graduate would flunk even a cursory test on their local ecology, and stripped of technology most would quickly founder.

To complicate things further, the advance of human knowledge is inescapably ironic. Since the enlightenment, the goal of our science has been a more rational ordering of human affairs in which cause and effect could be empirically determined and presumably controlled. But something like the opposite has happened. After a century of promiscuous chemistry, for example, who can say how the 100,000 chemicals in common use mix in the ecosphere or how they might be implicated in declining sperm counts, or rising cancer rates, or disappearing amphibians, or behavioral disorders? And having disrupted global biogeochemical cycles, no one can say with assurance what the larger climatic and ecological effects will be. Undaunted by our ignorance, we rush ahead to reengineer the fabric of life on Earth! Maybe science will figure it all out, but I doubt it. We are encountering the outer limits of social-ecological complexity in which cause and effect are widely separated in space and time and in a growing number of cases no one can say with certainty what causes what. Like the sorcerer's apprentice, every answer generated by science gives rise to a dozen more questions, and every technological solution gives rise to a dozen more problems. Rapid technological change intended to rationalize human life tends to expand the domain of irrationality. At the end of the bloodiest century in history, the enlightenment faith in human rationality seems overstated at best. But the design implication is not less rationality but a more complete, humble, and ecologically solvent rationality that works over the long term.

Who are we? Conceived in the image of God? Perhaps. But for the time being the most that can be said with assurance is that, in an evolutionary perspective humans are a precocious and unruly newcomer with a highly uncertain future. Where are we? Wherever it is, it is a world full of irony and paradox, veiled in mystery. And for those purporting to reweave the human presence in the world in a manner that is ecologically sustainable and spiritually sustaining, the ancient idea that God (or the gods) mocks human intelligence should never be far from our minds.

## 20.3 Ecological Design Principles

First, ecological design is not so much about how to make things as it is how to make things that fit gracefully over long periods of time in a particular ecological, social, and cultural context. Industrial societies, in contrast, operate in the conviction that "if brute force doesn't work you're not using enough of it." But when humans have designed with ecology in mind, there is greater harmony between intentions and the particular places in which those intentions are played out that

Preserves diversity both cultural and biological

Utilizes current solar income

Creates little or no waste

Accounts for all costs

Respects larger cultural and social patterns

Second, ecological design is not just a smarter way to do the same old things or a way to rationalize and sustain a rapacious, demoralizing, and unjust consumer culture. The problem is not how to produce ecologically benign products for the consumer economy, but how to make decent communities in which people grow to be responsible citizens and whole people who do not confuse what they have with who they are. The larger design challenge is to transform a society that promotes excess consumption and human incompetence, concentrates power in too few hands, and destroys both people and land. Ecological design ought to foster a revolution in our thinking that changes the kinds of questions we ask from "how can we do the same old things more efficiently" to deeper questions such as

Do we need it?

Is it ethical?

What impact does it have on the community?

Is it safe to make and use?

Is it fair?

Can it be repaired or reused?

What is the full cost over its expected lifetime?

Is there a cheaper and better way to do it?

The quality of design, in other words, is measured by the elegance with which we calibrate means and worthy ends. In Wendell Berry's felicitous phrase, good design "solves for pattern" thereby preserving the larger patterns of place and culture and sometimes this means doing nothing at all.[17] In the words of John Todd, the aim is "elegant solutions predicated on the uniqueness of place."* Ecological design, then, is not simply a more efficient way to accommodate desires as it is the improvement of desire and all of those things that effect what we desire.

Third, ecological design is not apolitical, but is as much about politics and power as it about ecology. We have good reason to question the large-scale plans to remodel the planet that range from genetic engineers to the multinational timber companies. Should a few be permitted to redesign the fabric of life on the Earth? Should others be permitted to design machines smarter than we are that might someday find us to be an annoyance and discard us? Who should decide how much of nature should be remodeled, for whose

* The phrase is John Todd's, see Todd and Todd.[18]

convenience, and by what standards? In an age when everything seems to be possible, who decides on the lines that should not be crossed? Where are the citizens or other members of a biotic community who will be affected by the implementation of grandiose and self-serving intentions? The answer is that they are now excluded. At the heart of the issue of design, then, are procedural questions that have to do with politics, representation, and fairness.

Fourth, it follows that ecological design is not so much an individual art practiced by individual "designers" as it is a community art that involves an ongoing negotiation between the community and the ecology of particular places. Good design results in communities in which feedback between action and subsequent correction is rapid, people are held accountable for their actions, functional redundancy is high, and control is decentralized. In a well-designed community, people would know quickly what's happening and if they don't like it, they know who can be held accountable and can work to change it. Such things are possible only where livelihood, food, fuel, and recreation are, to a great extent, derived locally; when people have control over their own economies; and when the pathologies of large-scale administration are mostly absent. Moreover, being situated in a place for generations provides long memory of the place and hence of its ecological possibilities and limits. There is a kind of long-term learning process that grows from the intimate experience of a place over time.* Ecological design, then, is a large idea but is most applicable at a relatively modest scale. The reason is not that smallness or locality has any necessary virtue, but that human frailties limit what we are able to comprehend, foresee, as well as the scope and consistency of our affections. No amount of smartness or technology can dissolve any of these limits. The modern dilemma is that we find ourselves trapped between the growing cleverness of our science and technology and our seeming incapacity to act wisely.

Fifth, the standard for ecological design is neither efficiency nor productivity, but health beginning with that of the soil and extending upward through plants, animals, and people. It is impossible to impair health at any level without affecting that at other levels. The etymology of the word health reveals its connection to other words such as healing, wholeness, and holy. Ecological design is a healing art by which we aim to restore and maintain the wholeness of the entire fabric of life increasingly fragmented by specialization, scientific reductionism, and bureaucratic division. We now have armies of specialists studying bits and pieces of the whole as if these were, in fact, separable. In reality it is impossible to disconnect the threads that bind us into larger wholes up to that one great community of the ecosphere. The environment outside us is also inside us. We are connected to more things in more ways than we can ever count or comprehend. The act of designing ecologically begins with the awareness that we can never entirely fathom those connections and with the intent to faithfully honor what we cannot fully comprehend and control. This means that ecological design must be done cautiously, humbly, and reverently.

Sixth, ecological design is not reducible to a set of technical skills. It is anchored in the faith that the world is not random but purposeful and stitched together from top to bottom by a common set of rules. It is grounded in the belief that we are part of the larger order of things and that we have an ancient obligation to act harmoniously within those larger patterns. It grows from the awareness that we do not live by bread alone and that the

---

* George Sturt, once described this process in his native land as "The age-long effort of Englishmen to fit themselves close and ever closer into England…" (Sturt, p. 66).

effort to build a sustainable world must begin by designing one that first nourishes the human spirit. Design, at its best, is a sacred art reflecting the faith that, in the end, if we live faithfully and well, the world will not break our hearts.

Finally, the goal of ecological design is not a journey to some utopian destiny, but is rather more like a homecoming. Philosopher, Suzanne Langer, once described the problem in these words:

> Most people have no home that is a symbol of their childhood, not even a definite memory of one place to serve that purpose. Many no longer know the language that was once their mother-tongue. All old symbols are gone... the field of our unconscious symbolic orientation is suddenly plowed up by the tremendous changes in the external world and in the social order.[19]

In other words, we are lost and must now find our way home again. For all of the technological accomplishments, the twentieth century was the most brutal and destructive era in our short history. In the century ahead we must chart a different course that leads to restoration, healing, and wholeness. Ecological design is a kind of navigation aid to help us find our bearings again. And getting home means remaking the human presence in the world in a way that honors ecology, evolution, human dignity, spirit, and the human need for roots and connection.

## 20.4 Conclusion

Ecological design, then, involves far more than the application of instrumental reason and advanced technology applied to the problems of shoehorning billions more of us into an Earth already bulging at the seams with people. Humankind, as Abraham Heschel once wrote, "will not perish for want of information; but only for want of appreciation... what we lack is not a will to believe but a will to wonder."[20] The ultimate object of ecological design is not the things we make but rather the human mind and specifically its capacity for wonder and appreciation.

The capacity of the mind for wonder, however, has been all but obliterated by the very means by which we are passively provisioned with food, energy, materials, shelter, healthcare, entertainment, and by those that remove our voluminous wastes from sight and mind. There is hardly anything in these industrial systems that fosters mindfulness or ecological competence let alone a sense of wonder. To the contrary these systems are designed to generate cash which has itself become an object of wonder and reverence. It is widely supposed that formal education serves as some kind of antidote to this uniquely modern form of barbarism. But conventional education, at its best, merely dilutes the tidal wave of false and distracting information embedded in the infrastructure and processes of technopoly. However well intentioned, it cannot compete with the larger educational effects of highways, shopping malls, supermarkets, urban sprawl, factory farms, agribusiness, huge utilities, multinational corporations, and nonstop advertising. The lessons of these things are human dominance, power, speed, accumulation, and self-indulgent individualism. We may talk about how everything is ecologically connected, but the terrible simplifiers are working overtime to take it all apart.

If it is not to become simply a more efficient way to do the same old things, ecological design must become a kind of public pedagogy built into the structure of daily life. There is little sense in only selling greener products to a consumer whose mind is still pre-ecological. Sooner or later that person will find environmentalism inconvenient, or incomprehensible, or too costly and will opt out. The goal of ecological design is to calibrate human behavior with ecological realities while educating people about ecological possibilities and limits. We must begin to see our houses, buildings, farms, businesses, energy technologies, transportation, landscapes, and communities in much the same way that we regard classrooms. In fact, they instruct in more fundamental ways because they structure what we see, how we move, what we eat, our sense of time and space, how we relate to each other, our sense of security, and how we experience the particular places in which we live. More important, by their scale and power they structure how we think, often limiting our ability to imagine better alternatives.

When we design ecologically we are instructed continually by the fabric of everyday life—pedagogy informs infrastructure which in turn informs us. The growing of food on local farms and gardens, for example, becomes a source of nourishment for the body and instruction in soils, plants, animals, and cycles of growth and decay.[21] Renewable energy technologies become a source of energy as well as insight about the flows of energy in ecosystems. Ecologically designed communities become a way to teach about land use, landscapes, and human connections. Restoration of wildlife corridors and habitats instructs us in the ways of animals. In other words ecological design becomes a way to expand our awareness of nature and our ecological competence.

Most importantly, when we design ecologically we break the addictive quality that permeates modern life. "We have," in the words of Philosopher Bruce Wilshire, "encase(d) ourselves in controlled environments called building and cities. Strapped into machines, we speed from place to place whenever desired, typically knowing any particular place and its regenerative rhythms and prospects only slightly." We have alienated ourselves from "nature that formed our needs over millions of years [which] means alienation within ourselves."[22] Given our inability to satisfy "our primal needs as organisms" we suffer what he calls a deprivation of ecstasy that stemmed from the 99% of our life as a species spent fully engaged with nature. Having cut ourselves off from the cycles of nature, we find ourselves strangers in an alien world of our own making. Our response has been to create distractions and addictive behaviors as junk food substitutes for the totality of body-spirit-mind nourishment we've lost and then to vigorously deny what we've done. Ecstasy deprivation, in other words, results in surrogate behaviors, mechanically repeated over and over again, otherwise known as addiction. This is a plausible, even brilliant, argument with the ring of truth to it.*

Ecological design, finally, is the art that reconnects us as sensuous creatures evolved over millions of years to a sensuous, living, and beautiful world. That world does not need to be remade but rather revealed. To do that we do not need research as much as the rediscovery of old and forgotten things. We do not need more economic growth as much as we need to relearn the ancient lesson of generosity, which is to say that the gifts we have must move, that we can possess nothing. We are only trustees standing for only a moment between those who preceded us and those who will follow. Our greatest needs have nothing to do with possession of things but rather with heart, wisdom, thankfulness, and generosity of spirit. And these things are part of larger ecologies that embrace spirit, body, and mind—the beginning of design.

---

* See also David Abram's remarkable book.[23]

## References


1. Van Der Ryn, S. and S. Cowan, *Ecological Design* (Washington, DC: Island Press, 2006), p. 118.
2. Franklin, C., Fostering living landscapes, in G. Thompson and F. Steiner, Eds., *Ecological Design and Planning* (New York: John Wiley & Sons, 1997), p. 264.
3. Wann, D., *Deep Design* (Washington, DC: Island Press, 1996), p. 22.
4. Orr, D., *Earth in Mind* (Washington, DC: Island Press, 1994), p. 104.
5. Hawken, P., H. Lovins, and A., Lovins, *Natural Capitalism* (Boston, MA: Little Brown, 1999); see also A. Lovins, H. Lovins, and E. von Weizsacker, *Factor Four* (London, U.K.: Earthscan, 1997).
6. Benyus, J., *Biomimicry* (New York: William Morrow, 1997), p. 262.
7. McDonough, W. and M. Braungart, The next industrial revolution, *The Atlantic Monthly* 282(4): 82–92, 1998.
8. Jackson, W., *New Roots for Agriculture* (Lincoln, NE: University of Nebraska Press, 1985).
9. Hawken, P., *The Ecology of Commerce* (New York: Harper Collins, 1993).
10. Benyus, J., *Biomimicry* (New York: William Morrow, 1997), p. 97.
11. Benyus, J., *Biomimicry* (New York: William Morrow, 1997), p. 73; Hunter, J.R., *Simple Things Won't Save the Earth* (Austin, TX: University of Texas Press, 1997).
12. Dobb, E., Pennies from hell, *Harpers* October: 39–54, 1996.
13. Anderson, E.N., *Ecologies of the Heart* (New York: Oxford University Press, 1996), p. 169.
14. Lansing, S., *Priests and Programmers* (Princeton, NJ: Princeton University Press, 1991).
15. Forrester, J., Counter-intuitive behavior of social systems, *Technology Review* January, 73(3): 52–68, 1971.
16. Wilson, E.O., *Biophilia* (Cambridge, MA: Harvard University Press, 1985); Wilson, E.O., *Consilience* (New York: Knopf, 1998).
17. Berry, W., *The Gift of Good Land* (San Francisco, CA: North Point Press, 1981), pp. 134–145.
18. Todd, J. and N. Todd, *From Eco-Cities to Living Machines: Principles of Ecological Design* (Berkeley, CA: North Atlantic Books, 1994).
19. Langer, S., *Philosophy in a New Key: A Study in the Symbolism of Reason, Rite and Art* (Cambridge, MA: Harvard University Press, 1941).
20. Heschel, A.J., *Man is Not Alone: A Philosophy of Religion* (New York: Farrar, Straus, and Giroux, 1951).
21. Donahue, B., *Reclaiming the Commons* (New Haven, CT: Yale University Press, 1999).
22. Wilshire, B., *Wild Hunger: The Primal Roots of Modern Addiction* (Lanham, MD: Rowman & Littlefield, 1998), p. 18.
23. Abram, D., *The Spell of the Sensuous* (New York: Pantheon, 1996).

# 21

## *Rainwater Harvesting and Stormwater Reuse for Arid Environments*

**Heather Kinkade**

**CONTENTS**



## 21.1 Introduction

Collecting and storing rainwater is not a new idea. While the origin of rainwater catchment systems are not known precisely, historical evidence suggests structures for holding runoff water date back to the third millennium BC. Structures have been found in numerous locations including the Negev Desert in Israel, the Mediterranean, India, Greece, Italy, Egypt, Turkey, and Mexico. Historical structures range from saucerlike ground catchments and belowground cisterns to aboveground rooftop runoff storage tanks. Many of the Asian and Middle Eastern countries as well as island communities still use some type of water catchment devices due to low water supply and low water quality.

In the United States, rock cisterns known as the Hueco Tanks in Texas and Tinajas in Arizona trapped rainwater for native dwellers, from the archaic hunters to the Mescalero Apaches, and they later became a stopping point for stagecoach travelers and Jesuit Missionaries.[1] In early communities of Texas, central plazas went beyond social and market places to be collection surfaces for vast underground cisterns, which stored the collected rainwater for use by adjacent shops and homes.

Today, many inhabitants in arid, urban environments are rethinking the use of this ancient strategy for conserving and managing water in their cities. The typical metropolitan area has large commercial and industrial buildings with expanses of impermeable surfaces; more of the land is covered throughout with rooftops, asphalt and concrete, and less rainwater finds its natural intended path to soil/groundwater reserves or surface water lakes and rivers. One plausible solution in arid as well as other environments is to use the freely available natural processes that work for the benefit of watershed maintenance on an individual site basis. Taking advantage of the capacity of plants and the site's soils to aid in absorbing water and filtering pollutants is considered a more sustainable approach to stormwater management (Figure 21.1).

Rainwater harvesting and stormwater reuse guidelines are meant to enhance traditional development practices and techniques to achieve what is known as a low-impact development approach. At the site level, the design approach is focused on passive and active strategies for filtering, detaining, and infiltrating runoff to remove pollutants, reduce runoff contributions to storm sewers, and potentially lessen drainage and erosion problems. The goal of developing a more environmentally sensitive approach is to mitigate the development-generated impacts at the source. Basically, an outcome that achieves a more ecologically and hydrologically responsive development through integration of rainwater harvesting and stormwater reuse techniques on-site.

The reduction of runoff in an urban environment through the use of rainwater harvesting and stormwater reuse structures and techniques will additionally assist with compliance of the Clean Water Act, which regulates the discharge of pollutants through the National

**FIGURE 21.1**
Upside-down umbrellas shade patio below as well as capture rain and direct it to the site's vegetation. (From Kinkade-Levario, H., *Forgotten Rain, Rediscovering Rainwater Harvesting*, Granite Canyon Publications, Forsyth, MO, 2004.)

Pollutant Discharge Elimination Systems (NPDES). The NPDES program, administered nationally by the Environmental Protection Agency (EPA), requires a permit for the discharge of any pollutant from a point source into the waters of the United States. In Arizona, the Arizona Department of Environmental Quality (ADEQ) requires an Arizona Pollutant Discharge Elimination System (AZPDES) general permit authorizing stormwater discharges from development activities into waters of the United States. AZPDES requires best management practices be implemented to minimize pollutant runoff from development activities. Other states may have similar independent pollutant discharge regulations. The implementation of rainwater harvesting and stormwater reuse approaches will assist with meeting the required best management practices as it is highly feasible that the impermeable surfaces proposed for most urban environments will be coated with vehicle oil, pollens, and settled air pollutants or dirt, all of which would be released to the site's retention basins and adjacent watercourses during a major rainstorm. By managing runoff close to its source through site design, low-impact techniques focused on rainwater harvesting and stormwater reuse mitigate these disturbances to the local environment. These water-conserving techniques for nonpotable needs are essential to maintaining growing clean water demands. For example, municipalities and industries combined accounted for the second highest water use—31% in 2006—as reported by the Arizona Department of Water Resources.* Freshwater supplies used for nonpotable requirements can be reduced or even replaced with an alternate nonpotable water source such as harvested rainwater to help limit the overall freshwater/groundwater usage by municipalities and industries. The following discussions focus on the details to consider when planning and designing, and implementing water harvesting techniques.

## 21.2 Determining Reuse Level of Commitment

Before an actual rainwater harvesting or stormwater reuse system can be designed, several basic questions must be answered regarding the following:

- Retrofit existing building or use a new integral system
- System size—large, medium, or small
- Complexity—passive or active collection systems
- Cost—low cost or complex
- Intensity of use and level of commitment or water security required
- Intended use—landscape irrigation, other uses, and/or potable water needs
- Water quality required (mosquito control)

Retrofitting an existing building or landscape generally is more expensive and costly than it is to design a new integral system. Water storage systems can range in size from small to very large:

- Small: <5,000 gal
- Medium: 5,001–25,000 gal
- Large: 25,001–50,000 gal
- Very large: >50,000 gal

* http://www.water.az.gov/adwr (accessed August 18, 2011).

Systems can be complex or simple and active or passive. Complex active systems require pipes, pumps, pressure tanks, and filtration where the simple systems may have all of the same components, but on a much smaller level. The passive systems generally refer to gravity landscape systems. If an active system is selected any type of water storage systems can be designed to fit a site's intended use and supply demands.

With nonpotable rainwater harvesting and stormwater reuse systems, a level of use for the captured water needs to be determined. There are typically four levels of use or levels of commitment to provide a water source for the intended water demand. The four levels of commitment include occasional, intermittent, partial, or full. A full commitment is typically a large storage capacity system that provides all of the water needed by the user (or demand) for an entire year.

## 21.3 Typical System Components

A nonpotable rainwater harvesting or stormwater reuse system has five basic components:

1. Catchment area: the surface upon which the rain falls. It may be a roof, other impervious surfaces and may include landscape areas.
2. Conveyance: transport channels or pipes from catchment area to storage.
3. Roof washing: the systems that filter and remove contaminants and debris. This includes first-flush devices.
4. Storage: cisterns or tanks where collected rainwater is stored.
5. Distribution: the system that delivers the rainwater, either by gravity or pump.

One additional component would be needed to provide a potable rainwater harvesting or stormwater reuse systems; purification. This may include filtering equipment, ultraviolet, chlorination or other methods of disinfection, and additives to settle, help filter, and disinfect the collected rainwater. Depending on catchment surface material and rainfall intensity, a loss of potentially collected rainwater can range from 20% to 70%. This loss is due to runoff material absorption or infiltration, evaporation, and inefficiencies in the collection process.

### 21.3.1 Catchment Area

A catchment area is the defined surface area upon which rainwater falls and is eventually collected. Rainwater harvesting for nonpotable use can be accomplished with any roofing material. Although rooftops are the typical catchment area, patio surfaces, driveways, parking lots, or channeled swales can also serve as catchment areas. Rainwater is slightly acidic, which means it will dissolve and carry minerals into the storage system from any catchment surface.

The total amount of water that is received in the form of rainfall over an area is referred to as the rainwater endowment of that area. The actual amount of rainwater that can be effectively harvested from the rainwater endowment is called the rainwater harvesting potential. Rainwater yields vary with the size and texture of the catchment area. A maximum of 90% of a rainfall can be effectively captured through rooftop rainwater harvesting.

**FIGURE 21.2**
Residential cistern painted to look like a turtle. (From Kinkade-Levario, H., *Forgotten Rain, Rediscovering Rainwater Harvesting*, Granite Canyon Publications, Forsyth, MO, 2004.)

The quality of the captured rainwater depends, in part, upon catchment texture as the best water quality comes from the smoother, more impervious catchment or roofing materials. Captured rainwater quality is also determined by rainfall pattern and frequency. Both the greater the storm event—i.e., the rainfall extent and the quantity of rain that falls—and the shorter the time between storms will affect the catchment surface condition. The larger the quantity of rain and the more often it rains means the catchment area will be cleaner and fewer pollutants will be transported to the first-flush device or to the storage unit during subsequent rainfall events.

If a catchment area is insufficient for quantities required to meet a water demand, supplemental water sources can be added to the rainwater/stormwater quantities. These water sources can include cooling tower blowdown water, air-conditioning condensate, pool water backwash, and gray water. For treatment of any nonpotable water, local codes should be reviewed (Figure 21.2).

### 21.3.2 Conveyance

A commonly used rainwater conveyance system is comprised of gutters with downspouts and/or rainchains. Gutters and downspouts direct rain from rooftop catchment surfaces to cisterns or storage tanks. Gutters and downspouts can be easily obtained as a standard construction material, or they can be specifically designed to enhance a building facade and maximize the amount of harvested rainfall.

### 21.3.3 Roof Washing

Roof washing is the initial process in reducing the debris and soluble pollutants that may enter a rainwater harvesting system. Roof washing systems may use one or several components to filter or collect debris and soluble pollutants, including gutter leaf guards, rain heads, screens, and/or first-flush devices.

Roofs, like other large, exposed areas, continuously receive deposits of debris, leaves, silt, and pollutants on their surfaces. All rainwater dislodges and carries away some of these deposits, but, during any given rainfall, the stormwater that falls first carries the highest concentration of debris and soluble pollutants. First-flush devices collect and dispose of this initial rain before it contaminates previously harvested and stored rainwater.

Capacities of first-flush devices may vary depending on the catchment size and ultimate use of the rainwater. Rainwater collected from a rooftop will typically be cleaner than rainwater collected from a ground level surface or pavement area. Thus, the storage capacity of the rooftop first-flush device does not need to be as large as a ground level surface catchment area first-flush. Rainwater collected from surface or pavement areas where dirt and debris are more prevalent may require longer settling periods for suspended solids and an absorbent material to remove oil and grease. Therefore, a more sophisticated and larger capacity first-flush device is typically required.

Use of a first-flush device is especially important when rain events follow a long dry period; during dry spells, debris and other pollutants build up on catchment surfaces. In this case, a large volume of water may be required to remove the catchment surface contaminants surpassing the volume allowed in a specified first-flush device. This means some contaminants will not be diverted and will enter the rainwater storage system. When a second rain event closely follows one that was strong enough to sufficiently "wash" the catchment area, use of the first-flush device during the second rain event may not be required. However, if the first rain event was not strong enough to move the catchment area contaminants, diversion of the second rain event's initial rainwater runoff to the first-flush device may be warranted. Multiple first-flush devices may be required on large surfaces as the time needed for the dirty water starting farthest from the first-flush is greater than dirty water closer to the device. In this case, the dirty water located some distance away will mix with cleaner water close by the devise. This can be avoided by having several first-flush devices evenly spaced apart in the catchment area.

### 21.3.4 Storage

Most of the components of a rainwater harvesting system are assumed costs in a building project. For example, all buildings have a roof and some have gutters and downspouts. Most homes and businesses also have irrigation systems and landscape materials placed around the structures. The cisterns or storage tanks represent the largest investment in a rainwater harvesting system because most homes and businesses are not initially fitted with a storage system.

Most cisterns and tanks have three distinct components all of which need to be waterproofed: base, sides, and a cover. They also contain several minor components including water inlet, water outlet, access hatch, overflow pipe, and means of draining. A typical storage cistern is covered and made of stone, steel, concrete, ferro-cement, plastic, or fiberglass. A storage system should be durable, attractive, able to withstand the forces of standing water, watertight, clean, smooth inside, sealed with a nontoxic joint sealant, and easy to operate. A tight cover is essential to prevent evaporation and mosquito breeding, and to keep insects, birds, lizards, frogs, and rodents from entering the tank. Cisterns and tanks should not allow sunlight to enter or algae will grow inside the container and the water will not age correctly.

Some storage tanks contain settling compartments to encourage any roof or pavement runoff contaminants to settle rather than remain suspended. Storage tanks can have an inlet from a sand filter or directly from the gutters through a leaf and debris filter. They must also have an overflow equal in size to the inlet flow rate, and an outlet or drain. The overflow should daylight to a landscape basin or an adjacent drainage system.

**FIGURE 21.3**
Commercial office building with a butterfly roof hidden behind the parapet directs rainwater to a rainchain and retention basin. (From Kinkade-Levario, H., *Forgotten Rain, Rediscovering Rainwater Harvesting*, Granite Canyon Publications, Forsyth, MO, 2004.)

The outlet leads to the distribution system. Some systems—especially if they are a sole source of water or if landscape irrigation requires supplemental water—may have an inlet pipe from an alternate, makeup water (water supplied to compensate for losses) source such as a municipal water supply. Whenever an alternate, makeup water source is used with a storage system, an air gap (a typical size is 14 in.) must be maintained between the high water line or highest flood line in the storage cistern/tank and the inlet of the alternate water. The overflow line should be placed to maintain the maximum high water line in the cistern/tank. An additional security to avoid contamination of the alternate water supply is a reduced pressure double backflow device which should be installed in the alternate water line prior to it reaching the storage container. A reduced pressure double backflow device also may be required on the service line from mains/municipality (Figures 21.3 and 21.4).

### 21.3.5 Distribution

Stored rainwater may be conveyed or distributed by gravity or by pumping. If a tank is located uphill or above the area proposed for irrigation, gravity may be sufficient for the system. Most plumbing fixtures and appliances including drip irrigation systems require at least 20 lb per square inch (psi) for proper operation, while standard municipal water supply pressures are typically in the 40–80 psi range. Pumps, rather than elevated tanks, are typically used to extract both below-grade and at-grade cistern- or tank-stored water. Submersible or at-grade pumps may be used in any rainwater storage system. Self-priming pumps with floating filter intakes and automatic shutoffs—for times when water levels are insufficient—are optimal equipment. The bottom few inches of stored water will typically contain very fine sediment and should be avoided if possible. A storage system may

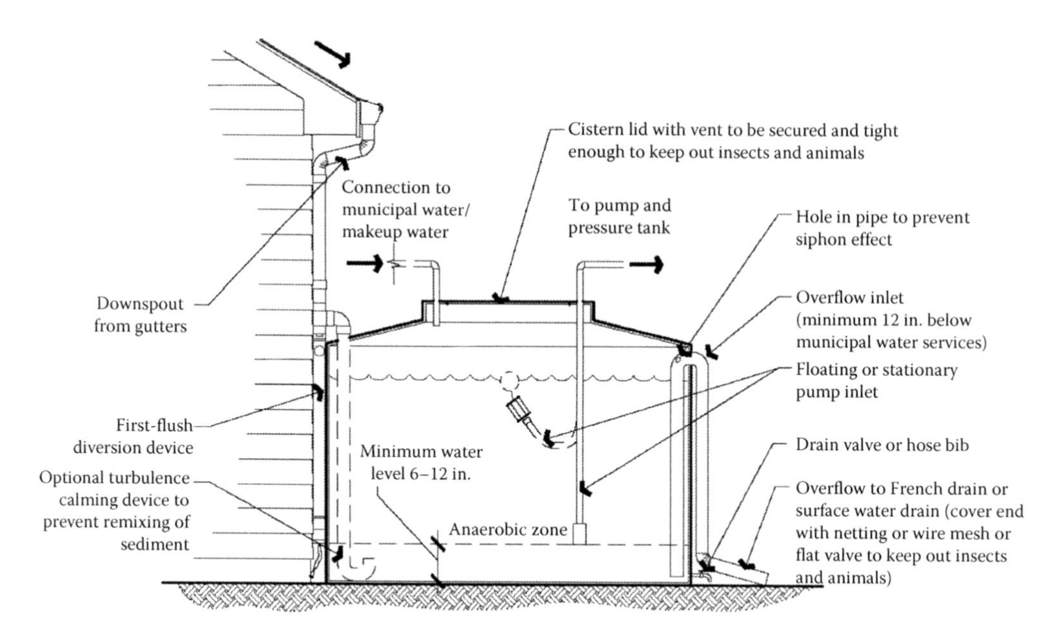


**FIGURE 21.4**
Typical residential rainwater harvesting cistern.

require an inflow smoothing filter or turbulence dissipater depending on the proximity of the rainwater inlet and pump intake or the amount of time rainwater is left to settle before the pumping is initiated.

The storage system overflow may act as a distribution system that delivers excess water to an adjacent landscape. All overflows exposed aboveground should have some means of stopping pests from entering the storage system. Fine screens may be placed over the pipe ends and, in areas of high rain quantities, water traps—similar to sinks and toilets—may be used.

For landscape irrigation, stored rainwater may go through additional filters before it is directed into an irrigation pump and distribution lines. This may be necessary to avoid clogging the irrigation system. For a potable water system, the water must go through a purification process (Figure 21.5).

## 21.4 Water Balance Analysis

A water balance analysis or water budget allows a designer to determine how much rainwater can be collected by the project catchment area, including rooftop and ground level areas. A water budget provides a supply and demand analysis on a monthly basis and provides water quantities for sizing cisterns. It will also determine how much, if any, supplemental water is needed to augment a system. The budget will allow a designer to redesign a project to increase or reduce the catchment area to meet the water demands of a landscape or the amount of potable water desired. An alternate water source may be needed for a few years in addition to rainwater to supplement water needs until plants are established. Eventually, harvested and natural rainfall may be adequate for plant needs.

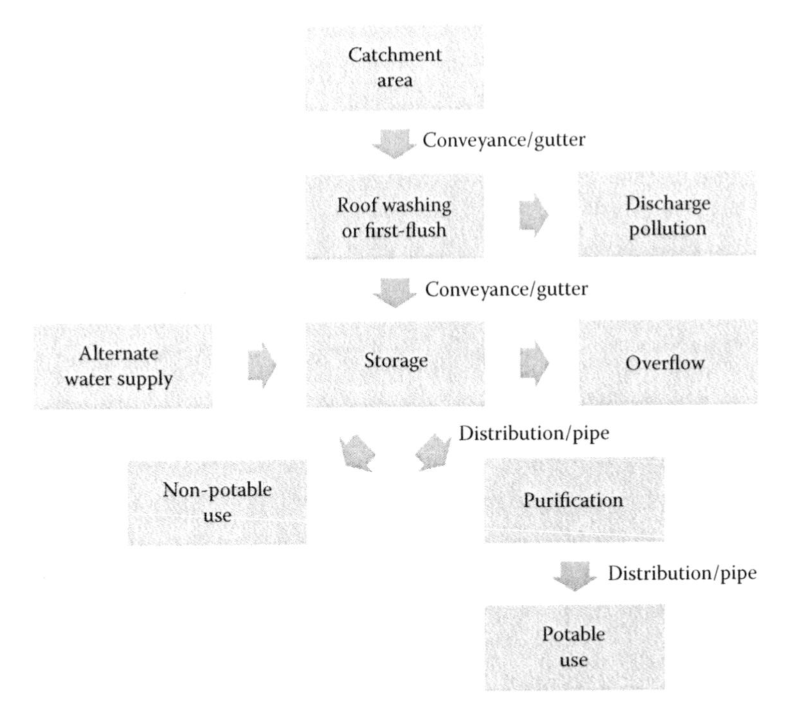


**FIGURE 21.5**
Typical rainwater harvesting system components.

Several items are required to prepare a water budget: average annual rainfall data, a site plan, and a water demand quantity. For a potable system, a residence's monthly water usage table would be required. In some states a landscape water budget, a rainwater harvesting plan, and an implementation plan are required to be submitted with development plans. In October 2008, Tucson, Arizona, was the first in the nation to require commercial properties to submit rainwater harvesting plans. Their City Ordinance, Chapter 6, Article VIII Rainwater Collection and Distribution Requirements, outlines the requirements for commercial properties to provide 50% of the site's estimated yearly landscape water budget from rainwater harvested on-site. They allow a 3 year plant establishment period before the 50% is enforced. This Tucson City Ordinance has set the stage for others to follow* (Table 21.1).

## 21.5 Integrated Site Design

Integrated site design is intended to match site requirements (e.g., water, energy, food, and aesthetics) with the eventual components of an area (e.g., stormwater runoff, shade from buildings, and vegetation) while improving the function and sustainability of a site. This process requires an integrated design process based on a multidisciplinary approach. Every step of the design phase should be evaluated to pursue opportunities to achieve

* http://library.municode.com/html/11294/level3/PII_C6_AVIII.html (accessed August 19, 2011).

**TABLE 21.1**

Nonpotable Water Budget Worksheet

| Month | Total Landscape Irrigation Requirement for Established Plants | Available Runoff Supply | Runoff Minus Landscape Irrigation Requirement | Excess Runoff to Storage (Not Used Each Month) | Accumulative Storage | Irrigation Requirement from Storage (Required to Supplement Irrigation Requirement) | Irrigation Requirement from Municipal Supply (No Rainwater in Storage Tank) | Excess Runoff to Overflow |
|---|---|---|---|---|---|---|---|---|
| January | | | | | | | | |
| February | | | | | | | | |
| March | | | | | | | | |
| April | | | | | | | | |
| May | | | | | | | | |
| June | | | | | | | | |
| July | | | | | | | | |
| August | | | | | | | | |
| September | | | | | | | | |
| October | | | | | | | | |
| November | | | | | | | | |
| December | | | | | | | | |
| Total monthly gallons | | | | | | | | |

*Source:* Kinkade-Levario, H., *Forgotten Rain, Rediscovering Rainwater Harvesting*, Granite Canyon Publications, Forsyth, MO, 2004.

multiple benefits. A designer must determine how each site element can serve multiple functions (e.g., aboveground water tanks could provide stored water, shade, privacy screening, and noise abatement).

A design of a water saving, integrated site should focus on three general components: (1) site grading, and opportunities for structures, materials, and strategies for (2) passive systems, and (3) active systems. These are briefly discussed in the following sections.

### 21.5.1 Site Grading

There are two types of rainwater harvesting techniques, passive and active. Where active systems deal with capturing rainwater and storing it in cisterns or tanks, passive systems work with land contours and gravity to collect, detain, and slow down stormwater while routing it through a site. Rainwater is low in salt and contains some nitrogen, both beneficial to plant growth. Implementation of passive rainwater harvesting techniques typically does add costs to a traditional site design. Although, incorporating passive concepts into a site design early in the process aids in reducing costs and maximizing long-term benefits.

The first step in incorporating harvesting structures and techniques into a design is to analyze the site's watershed, which includes off-site land that contributes stormwater runoff to or receives stormwater runoff from the site. The overall goal is to manage small amounts of water at higher elevations and as it passes through a site to reduce stormwater volume at lower elevations of a site. Organizing the site into multiple small watersheds where the stormwater is slowed and allowed to disperse is the method most used to attain this goal. This reduces the stormwater's erosive nature and sediments transport off-site. The stormwater should be exposed to as much soil surface as possible, using passive rainwater harvesting structures to increase infiltration and natural soil storage of the site's stormwater.

The following table identifies several low-impact development objectives and their corresponding development technique, which support rainwater harvesting and stormwater reuse concepts (Table 21.2).

### 21.5.2 Structures, Materials, and Strategies for Passive Systems

Relatively simple structures can be incorporated on a site to enhance its harvesting potential, including microbasins, swales, French drains, and rain gardens. These are discussed below, followed by descriptions of some of the common materials and strategies used in passive systems: porous pavement, mulches, and use of Xeriscape guidelines.

#### *21.5.2.1 Microbasins*

Microbasins are small catchment areas that are best for low flow stormwater volumes, slowing stormwater and allowing infiltration rates to increase. These basins can be located parallel to one another or in an alternating pattern allowing overflows to be slowed and allowing additional infiltration. The microbasins can be tree wells, planter islands—with curb cuts allowing stormwater to enter—or just small depressions next to a path or drive. Microbasins can be designed with several variations including multiple sizes and shapes for supporting planting groups with similar water needs.

#### *21.5.2.2 Swales*

Swales are best for stormwater occurring in low to medium flow volumes. These are not large channels for moving water, but are small depressions meant to slow sheet flow and

**TABLE 21.2**
Low-Impact Objectives and Techniques

| Low-Impact Development Objectives | Low-Impact Development Techniques | On-Lot Bioretention | Wider and Flatter Swales | Maintain Sheet Flow | Clusters of Trees and Shrubs in Flow Path | Provide Tree Conservation/Transition Zone | Minimize Storm Drain Pipes | Disconnect Impervious Areas | Save Trees | Preserve Existing Topography | LID Drainage and Infiltration Zones |
|---|---|---|---|---|---|---|---|---|---|---|---|
| Minimize disturbance | | • | | • | • | • | • | • | • | • | |
| Flatten grades | | | • | • | | | • | | | • | • |
| Reduce height of slopes | | | | | | | • | | | • | • |
| Increase flow path (divert and redirect) | | | • | • | • | | • | • | • | | |
| Increase roughness | | • | | • | • | • | • | • | • | | • |

*Source:* Kinkade-Levario, H., *Forgotten Rain, Rediscovering Rainwater Harvesting*, Granite Canyon Publications, Forsyth, MO, 2004.

to allow longer standing/infiltration periods. Swales can be located next to sidewalks, paths, and driveways and typically direct stormwater toward vegetation and away from buildings. They can be designed to follow or parallel the contours as well as be designed to be at a slight angle from the contour. Swales designed at a slight angle from the contour are more appropriate for larger areas such as parks, but the smaller swales can be designed as small pocket swales similar.

#### *21.5.2.3 French Drains*

French drains and rain gardens are meant to rapidly infiltrate stormwater and to remove standing water from surface view. French drains are typically lined transport channels leading to subsurface storage areas or overflow infiltration areas, or providing a faster transport to plantings such as a rain garden.

#### *21.5.2.4 Rain Gardens*

Rain gardens use the concept of bioretention, a water quality practice where plants and soils remove pollutants from stormwater naturally. Proposed retention/detention basins can be placed at the lowest elevation on a site to assist with infiltrating excess stormwater. Large basins are perfect locations for rain gardens, which are created with layers of soil, sand, and organic mulch. These layers naturally filter the stormwater as it flows into a basin/rain garden and as the stormwater infiltrates through the layers. Excess water is stored in the soil voids or infiltrates through the soil. Rain gardens can also be designed with man-made perforated crate-like boxes that quickly remove standing surface water by creating larger subsurface water storage through the presence of subsurface voids, which will eventually empty as the infiltration process continues. An alternative to this system would be the use of an underground natural rock or sand bed.

A second location for rain gardens can be relatively small scale areas located in landscape strips such as parking lot islands or along drives which can be used to naturally infiltrate stormwater. In any rain garden without supplemental irrigation, plants must be drought tolerant to sustain dry periods between harvesting events.

#### *21.5.2.5 Porous/Pervious Pavements*

Traditional asphalt and concrete surfaces are designed to be virtually impermeable, with stormwater runoff often mitigated through the use of storm sewers and detention/retention systems. Porous paving has proven to be durable enough for parking areas, pedestrian uses, and some road surfaces. These materials are designed specifically to reduce, or in some cases eliminate this stormwater runoff and direct the replenishing benefits of rainwater naturally into the ground below.

Porous paving surfaces such as porous asphalt, porous concrete, or porous pavers will permit water to infiltrate rapidly. Use of porous asphalt and concrete should be restricted to parking lots and local roads since they support lighter loads than standard asphalt and concrete. Porous concrete can typically support 1800–2400 psi and interlocking pavers are designed to meet a minimum of 8000 psi.[2] Underlying soil strength also contributes to the pavement strength and its porosity. This underlying soil layer and geotextile should generally be effective in detaining pollutants from the infiltrated water. If pollution is of special concern a collection pipe could be used to transport the filtered water to a specified area for further treatment. This collection pipe is also effective for low permeability soils. Porous concrete and pavers are cement-based and will not release harmful chemicals into the environment whereas an oil-based asphalt might. With any porous surface, the greater the slope the less time water has to infiltrate making the porous surface ineffective, therefore flatter areas are more appropriate for porous materials.

Segmental concrete unit pavers offer unique advantages such as

- Resistance to severe loads
- Flexibility of repair
- Low maintenance
- Exceptional durability
- Consistently high quality

Concrete pavers have openings in the pavement surface that facilitate rainwater infiltration, thereby reducing or eliminating stormwater runoff and maximizing groundwater recharge and/or storage. These pavers also provide the following benefits:

- Lessen or eliminate downstream flooding, streambed, and bank erosion
- Decrease project costs by reducing, or eliminating drainage, and retention systems required by impervious pavements and reduce cost of compliance with many stormwater regulatory requirements
- Provide a highly durable yet permeable pavement surface capable of supporting vehicular loads

- Eliminate curing time and stress cracking or degradation of the surface that is typically found in traditional asphalt due to the numerous joints in the pavers
- Eliminate visible cuts and resurface scars seen in traditional asphalt or concrete when underground repairs are made
- Eliminate buckling or building due to heat and weight of vehicles as can be seen in traditional asphalt
- Reduce surface heat load due to reduction in surface area compared to impervious materials

The pavers can be installed with a different colored paver designating a parking space stripe or lane stripe thereby eliminating repainting for the life of the pavers—approximately 20 years. Pavers can also be sealed, which reduces any stains that may occur due to oil spills or other spilled substance, but this can affect the porosity. As with any porous pavement, some reduction in porosity may also occur due to organic growth and fine accumulation, but these pavers generally maintain porosity for a greater time frame than porous asphalt, which requires cleaning yearly. The pavers should be cleaned every 3–5 years—a 4 year cycle is recommended—with a commercial street sweeping or vacuuming equipment. Additional aggregate fill material can be added to the openings at the same cleaning time, if necessary.

The recommended concrete pavers can provide a low runoff coefficient of 0.3–0.5 as compared to a higher traditional asphalt or concrete runoff coefficient of 0.95. Studies have proven that these pavers can, depending on the subbase soil and surface slope, infiltrate up to 5 in. of rainfall per hour prior to becoming saturated. Use of these pavers provides a surface that is 70%–100% pervious and not a surface that is 70%–100% impervious like traditional asphalt and concrete. Studies have shown that permeable pavers are more expensive to install than the traditional asphalt, but are less costly than asphalt to maintain over time. Following are some initial typical cost guidelines in a per square feet installed price format:

Asphalt: $0.50–$1.00

Porous concrete: $2.00–$6.50

Grass/gravel pavers: $1.50–$5.75

Concrete pavers: $5.00–$10.00

A more accurate price comparison would involve the cost of a full stormwater management paving system where the full system with asphalt and storm drains, catch basins, and ungrounded pipes could be closer to $9.50–$11.50/ft$^2$ installed and the concrete paver costs would remain closer to $5.00–$10.00/ft$^2$ installed. Use of the concrete pavers would eliminate the need of traditional stormwater infrastructure. All of these prices are estimated per manufacture guidelines and would need to be verified for any project to understand a precise cost comparison. The cost of eliminating retention basins and increased buildable area or smaller site requirements could also be added into the cost to give an even larger long-term price difference.

#### 21.5.2.6 Mulches

All passive rainwater harvesting structures should include a top mulch layer to assist in reducing evaporation of the stormwater runoff captured in the structures. Mulches can be organic or inorganic as long as they do not inhibit infiltration.

#### *21.5.2.7 Using Xeriscape Principles and Native Plants*

Most urban landscapes are irrigated with municipal supplies, treated to potable/drinking water standards unnecessary for most landscapes. More sustainable landscapes can be maintained on what is naturally provided, in an efficient manner using water harvesting techniques. Xeriscape, a water-conserving landscape design approach, promotes seven basic principles:

1. Planning and design
2. Soil analysis and improvement
3. Practical areas of turf
4. Appropriate plants
5. Efficient irrigation
6. Mulching
7. Proper maintenance

This approach groups plants with similar water requirements. These planting zones typically include a *native* zone where once plants are established they can survive on natural rainfall.

In some cases, landscape designs are created entirely with plants that can survive on rainfall once established, often termed *natural landscaping*. The purpose of natural landscaping is to preserve and reintroduce indigenous plants from the site, a practice that may eliminate the need for supplemental watering after plants are established. This movement is especially appropriate in arid locations where water is scarce. The water-saving benefits of a natural landscape supplemented with harvested rainwater irrigation include

- Reduced peak water demand
- Reduced groundwater overdraft and contamination
- Reduced water costs
- Improved long-term water utility revenue stability
- Reduced runoff, soil erosion, and costs for stormwater management
- Creation of distinctive landscapes that represent the natural biotic characteristics where the site is located
- Reduced energy costs for landscape maintenance
- Reduced plant disease, rot, and mortality caused by over watering

### 21.5.3 Active Systems

Buildings with expansive rooftops can be considered a watershed within a site and typically provide a fairly clean water source for supplemental irrigation water and possibly other nonpotable water uses. A few of these uses are discussed to illustrate how active systems can be designed (Figures 21.6 and 21.7).

#### *21.5.3.1 Vehicle Washing and Automated Carwashes*

According to the International Carwash Association, water used to wash a vehicle can range from 15 gal per vehicle for self-serve washes to 50–60 gal per vehicle for in-bay (stationary)

**FIGURE 21.6**
Cistern at a school recreational center is integral to the design. (From Kinkade-Levario, H., *Forgotten Rain, Rediscovering Rainwater Harvesting*, Granite Canyon Publications, Forsyth, MO, 2004.)

**FIGURE 21.7**
Interior view of school recreational center cistern. (From Kinkade-Levario, H., *Forgotten Rain, Rediscovering Rainwater Harvesting*, Granite Canyon Publications, Forsyth, MO, 2004.)

automatic washes to 66–85 gal per vehicle for conveyor type washes.* Carwashes that contain reclaimed systems—systems that separate grit, oil, and grease from wash and rinse water and then filter it for reuse—can reduce water use by more than half. Although, even with a reuse system 10–20 gal of makeup water is needed per vehicle washed. According to a recent study on Phoenix carwashes, water losses can additionally occur through evaporation and fine mists, especially in warmer areas or during hot periods in Arizona, and those losses have been estimated to range from three to seven and one-half gallons per vehicle.

Water-efficient measures should be required of all individual carwash facilities proposed for any site along with the need for all facilities to reuse as much water as possible. Harvested rainwater from adjacent individual buildings and associated pavement could be used to supplement water for a carwash. If a permeable pavement is chosen, the water filtered by the pavement could be harvested and reused for a carwash makeup water system. This would allow the pavement to capture most solids instead of requiring a separate first-flush system. Carwash reuse systems require the used, silt-laden water to pass through three settling tanks before reuse; therefore, if a little silt is in the harvested rainwater (makeup water), it will be filtered out in the typical carwash water reuse process. The final rinse will typically be a reverse osmosis process initiated with clean municipal water, which is required for reducing water spotting, but the basic wash cycle can include harvested rainwater. The use of rainwater has the added benefit that it is naturally soft water and will potentially allow a carwash to use less soap. Use of harvested rainwater will assist in reducing the demands placed on the municipal water supply.

#### 21.5.3.2 Water Losses and Reuse by Mechanical Equipment

Cooling towers use a significant amount of water to maintain the air-conditioning/cooling process. Cooling towers have been identified as often being the largest single user of water in commercial and industrial buildings. The basic function of a cooling tower is to use evaporation to lower the temperature of water that has been heated for some building operating process. Cooling towers typically lose water in three ways: evaporation, bleed-off (or blowdown), and drift. All of these losses are replaced with makeup water. The evaporation process involves the main cooling component while the bleed-off water flushes the high concentrations of total dissolved solids (TDS). The drift is uncontrolled water loss in the form of mist or droplets carried away by airflow or winds. Some additional loss may occur through valve leaks or draw downs for various miscellaneous uses. Makeup water replaces all these possible water losses in a cooling tower.

In most cooling towers, the primary opportunity for conserving water is to reduce the amount of makeup water required to replace bleed-off water. In order to use harvested rooftop rainwater as makeup water for the cooling tower, the rainwater constituents would need to be determined. The TDS and individual dissolved constituent would affect the chemistry in the cooling tower operating water and adjustments might need to be made. Most solids could be settled or filtered out of harvested rooftop rainwater leaving the dissolved content as the issue for reuse. Rainwater for makeup water should be disinfected with ultraviolet prior to use. In some areas the large quantity of makeup water cannot be met by the collected rainwater and the opposite situation could be conducted; i.e., instead of using rainwater for a cooling tower, the blowdown water from the cooling tower could be added to the rainwater supply to increase alternate water for building water needs.

* http://www.hanna-sherman.com/water-chem/reclaimII.html (accessed April 30, 2003).

**FIGURE 21.8**
Rooftop gutters extended to direct water to a permeable channel for rainwater infiltration. (From Kinkade-Levario, H., *Forgotten Rain, Rediscovering Rainwater Harvesting*, Granite Canyon Publications, Forsyth, MO, 2004.)

## 21.6 Conclusion

Meeting water shortages is a challenge, but challenges can be opportunities. Rainwater harvesting and stormwater reuse structures and strategies are the opportunities presented for any landowner and municipality to use in implementing a low-impact design and development process. Implementation of these water-saving opportunities will reduce the percentage of municipal water used for nonpotable purposes. In turn, harvesting structures and strategies provide support for on-site vegetation which aid in reducing temperature extremes, pollutants, and stormwater runoff. It has been shown that passive rainwater harvesting can be incorporated into designs with little additional costs and that the potential to create an active rainwater harvesting system can be implemented with minor effort (Figure 21.8).

It is recommended that the long-term benefits of rainwater harvesting and stormwater reuse opportunities be evaluated against the initial costs of alternatives to assure that long-term benefits outweigh the short-term costs. Long-term maintenance and performance targets should be set for a project to guarantee the elimination of contaminants that may get into the groundwater or kill adjacent plants—and to guarantee continued water-saving strategies. Survival in arid environments depends on the existence of water; we need to capture it effectively when it rains as the opportunities for harvesting this valuable resource can leave as quickly as it comes.

## References

1. Kinkade-Levario, H., *Forgotten Rain, Rediscovering Rainwater Harvesting* (Forsyth, MO: Granite Canyon Publications, 2004).
2. Interlocking Concrete Pavement Institute, Structural design of interlocking concrete pavement for roads and parking lots, *Tech Spec* 4: 1–8, 2002.

# 22

## *Designing Habitats in Urban Environments**

**Margaret Livingston**

**CONTENTS**



### 22.1 Introduction

Maintaining biological linkages among urban and natural areas has long been considered an important concept among conservationists, land managers, wildlife biologists, and landscape ecologists involved in human and wildlife interactions.[1] These linkages, for example, support crucial ecosystem functions including facilitation of genetic diversity, population maintenance, and seasonal movements of species. Linkages typically consist of vegetative patches within a developed matrix that are connected by corridors to larger natural areas. Studies have suggested that particular types of wildlife utilize these corridors; however, the direct benefits with regard to wildlife population dynamics are still questioned.[2] Furthermore, there has been concern about the possible negative effects of enhancing the connectedness of corridors and patches, such as transmission of contagious disease, increased fire potential due to additional fuel from larger massing of plants, and increased transport of invasive plant species.[3] Adopting a restorative approach to the design of urban patches that incorporates the spatial patterns and species found in existing patches of natural vegetation associated with wildlife habitat may minimize these negative impacts. This chapter discusses a design framework for developing such areas, in addition to the significance of selection and arrangement of plants for attracting wildlife in urban spaces.

* This chapter is based on the article, Livingston, M., "Landscape design for attracting wildlife in southwestern urban environments" for the *Urban Wildlife Conservation 4th International Symposium*, 1999.

## 22.2 Design Approach

### 22.2.1 Site Analysis

A site analysis serves as an evaluation of the opportunities and constraints of an area in relation to the objectives of a project. It is critical that the initial scope of an analysis for attracting wildlife is performed at a larger scale than intended for most created wildlife areas. This is due to potential flows and interactions between transient wildlife and the regional resources surrounding the proposed wildlife habitat site. Using a landscape ecology approach to the initial site analysis is appropriate. In this approach, the focus starts with the spatial patterns of the land uses in the urban matrix and the expected interactions associated with existing land uses. For example, developed land, created and natural open spaces, wetlands, and watercourses are evaluated and prioritized relative to their significance to the purpose (species attempting to attract), size, and location of the proposed wildlife area. These land uses should then be further categorized and evaluated for quality of wildlife habitats based on an assessment of resources such as plant communities and natural water sources.

Initial analysis at this larger scale allows recommendations to be developed relating to connectedness of the habitat site with existing patches of habitat, appropriateness of habitat location, and compatibility with the existing plant communities of a large metropolitan area. For example, an analysis of Tucson at this larger scale would indicate various plant communities associated with the Sonoran Desert scrub such as Arizona Upland and Lower Colorado River Valley subdivisions.[4] These plant communities provide different habitat benefits for a variety of wildlife. In urban areas, where such habitats are limited, it is important to seek out opportunities where these plant communities can be preserved, or perhaps recreated in a series of adjacent wildlife-friendly backyards. The habitat patches that result from this type of connectedness can provide a significant respite for migratory birds and support populations of native birds.[5]

Further site analysis is also done at a local scale where evaluation of potential edge effects between the wildlife areas and adjacent land uses are considered. Various studies have shown the significance of edge effects on species diversity and population size.[6] At this scale, the designer evaluates and prioritizes various locations in relation to these edge effects such as natural watercourse and created park buffers adjacent to other land uses. Other site-specific elements such as nearby water, roads, buildings, and existing vegetation are considered for their microclimatic effects. For example, water from a natural source such as a seep may provide water for wildlife and nearby plants whereas a swimming pool would ameliorate site temperatures and provide a milder microclimate for frost-tender plants.

### 22.2.2 Vegetation Analysis

A more detailed analysis of vegetation is done following general plant community assessment as described in the previous section (Table 22.1).

Ideally, species identification, structure (height), and density should be evaluated prior to choosing a site for creating a wildlife habitat. This evaluation of existing vegetation indicates its appropriateness for the wildlife species being targeted. For example, large, existing, nonnative trees in Tucson such as eucalyptus (*Eucalyptus* spp.) are considered useful perches for raptors such as hawks and owls whereas Gambel's quail prefer the cover provided by low-branching shrubs.[7] Therefore, depending on your goal, it might

**TABLE 22.1**

Vegetation Analysis of Existing Plants for a Proposed Urban Habitat for Birds in Tucson, Arizona

| Species | Common Name | Nest | Cover | Food | Type[a] |
|---|---|---|---|---|---|
| *Acacia constricta* | Whitethorn acacia | X | X | X | S |
| *Acacia farnesiana* | Sweet acacia | | X | X | T |
| *Acacia greggii* | Catclaw acacia | X | X | X | S |
| *Atriplex canescens* | Four-wing saltbush | X | X | X | S |
| *Berberis* sp. | Red barberry or agarita | X | X | X | S |
| *Bouteloua* sp. | Grama grasses | | | X | S |
| *Calliandra eriophylla* | Fairy duster | | | X | S |
| *Carnegiea gigantea* | Saguaro | X | | X | C |
| *Celtis pallida* | Desert hackberry | X | X | X | S |
| *Celtis reticulata* | Netleaf hackberry | X | X | X | T |
| *Chilopsis linearis* | Desert willow | X | X | X | T |
| *Condalia* sp. | Greythorn | X | X | X | S |
| *Cupressus glabra* | Arizona cypress | | X | X | T |
| *Dodonaea viscosa* | Hopbush | | X | X | S |
| *Encelia farinosa* | Brittlebush | | | X | S |
| *Ephedra trifurca* | Desert jointfir | | X | X | S |
| *Eriogonum* sp. | Buckwheat | | X | X | S |
| *Fallugia paradoxa* | Apache plume | | X | X | S |
| *Ferocactus* sp. | Barrel cactus | | | X | C |
| *Forestiera neomexicana* | New Mexico olive | X | X | X | T |
| *Fouquieria splendens* | Ocotillo | | | X | S |
| *Justicia californica* | Chuparosa | | | X | S |
| *Larrea tridentata* | Creosote | X | X | | S |
| *Lycium* sp. | Wolfberry | X | X | X | S |
| *Olneya tesota* | Ironwood | X | X | X | T |
| *Opunita* sp. | Prickly pear | X | X | X | C |
| *Opuntia* sp. | Cholla | X | X | X | C |
| *Parkinsonia floridum* | Blue paloverde | X | X | X | T |
| *Parkinsonia microphyllum* | Littleleaf paloverde | X | X | X | T |
| *Parkinsonia praecox* | Palo brea | X | X | | T |
| *Penstemon* sp. | Penstemons | | | X | S |
| *Pithecellobium flexicaule* | Texas ebony | | X | X | T |
| *Poropis pubescens* | Screwbean mesquite | X | X | | T |
| *Prosopis glandulosa* | Honey mesquite | X | X | X | T |
| *Prosopis velutina* | Velvet mesquite | X | X | | T |
| *Quercus turbinella* | Scrub oak | | X | X | T |
| *Rhus microphllya* | Desert sumac | | X | X | S |
| *Robinia neomexicana* | New Mexico locust | | X | | T |
| *Sambucus mexicana* | Mexican elder | | X | X | T |
| *Sapindus drummondii* | Western soapberry | X | X | | T |
| *Simmondsia chinensis* | Jojoba | | X | X | S |
| *Stachys coccinea* | Scarlet betony | | | X | S |
| *Washingtonia fiilfera* | California fan palm | X | | X | T |

[a] T, tree; S, shrub; C, cactus.

be appropriate to remove species such as a eucalyptus and provide a native tree, such as a mesquite, which can provide cover, food, and nesting resources for native birds. From a design standpoint, existing vegetation at the site level also indicates a potential starting point for selection of additional plant species and placement.

### 22.2.3 Analysis of Other Resources

Climate, topography, and soils also influence site appropriateness and should be considered at each level of analysis. For example, topography can be associated with significant variation in native vegetation communities (i.e., a watershed basin) at a larger scale and enhance water catchment or create a warm microsite for an area. In addition, knowledge about variation in rainfall patterns and amounts for a region, erosion potential, and general soil conditions provide additional information relating to the site's capacity and hence likely success of a design for a particular site. At this point the designer is prepared to initiate design development of the wildlife habitat.

### 22.2.4 Development of the Design

Following completion of the site analysis, the design process begins with development of a program and conceptual plan for the selected site. The program includes the constraints and opportunities of the site, as indicated by the site analysis, and takes into account the needs of wildlife viewers (if they are involved in the process). For example, questions about whether viewing areas are desired and what type of wildlife the user wishes to attract are addressed in this phase. The conceptual plan involves the general arrangement of design elements such as plants, viewing areas, and the circulation of users through the proposed habitat. The next phase of design development focuses on the specific requirements for the various design elements.

## 22.3 Design Guidelines

Designs for wildlife should emphasize the selection of habitat elements, primarily plants that focus on providing shelter, food, nesting spots, and water. However, these requirements can easily accommodate an aesthetic focus in addition to the creation of wildlife habitat. Following are some simple guidelines for creating an effective and attractive wildlife habitat:

1. *Arrange plants in large groupings*: Large groupings of plants will provide the opportunity to create layering of plant species from tall, peripheral, overstory trees and large shrubs to the smaller shrubs, accents, and ground covers (Figure 22.1).

   This arrangement is pleasing to the eye due to its natural sequential appearance from tall to low-growing species, and such layering provides a variety of shelter and nesting spots for wildlife.
2. *Create vertical, horizontal, and seasonal variety*: Using a diverse selection of plants will provide food for a longer period of time throughout the year, attract more species, and will create visual interest in the habitat.

**FIGURE 22.1**
Large groupings of plants provide continuity to the design and effective cover for wildlife.

**FIGURE 22.2**
Massing of particular plant species provides a unifying effect to designs with high species diversity.

3. *Use plant massing*: In contrast to variety, it is important to incorporate some masses of a single plant species that unify the design (Figure 22.2).

   Too many individual species can add disjointedness to the design—adding masses of certain plants creates unity and can also increase pollination and fruit production for those species.
4. *Create open spaces*: Providing open spaces in wildlife areas is just as important as creating plant masses. These areas serve as an aesthetic contrast to the variety of plant groupings and also create areas for wildlife interactions (such as sunning or bathing) and viewing.

The following discussions cover more specific design guidelines, elements, and plants for areas designed for native birds, hummingbirds, and butterflies. However, some of these guidelines will coincide with needs for a range of wildlife types.

### 22.3.1 Bird Habitats

It is important to acknowledge the potential for backyard wildlife areas to become nodes within a series of associated patches that serve as stopovers and habitat for native birds.[5] These areas are particularly critical considering the current status of natural bird habitats. Unfortunately, while we protect wild birds with state and federal laws that apply to parks and refuges, the vast majority of land remains in private ownership with no formal protection.[5] Organizations such as the National Audubon Society and Nature Conservancy are actively identifying specific locations that are critical to birds, thereby increasing the opportunities for potential linkages between created and natural bird habitats.[5] Depending on the location, design guidelines will vary according to site resources and the bird species desired to be attracted. However, there are some general guidelines that are important for the design of any bird habitat:

1. *Mimic nature in species selection and arrangement*: Native plants particularly play an important role in bird habitats due to the interactions between indigenous birds and associated plants over many years. Native plants that have coevolved with native birds are more likely to provide appropriate nutrition for the birds when it is needed, produce fruits that provide supplemental water, and provide suitable cover for nesting and shelter.[5,7] It is also important to recognize the wide range of plants necessary for creating a habitat that can provide food and shelter. For example, selecting plants that have dense foliage and thorns provide nesting opportunities and cover for many native birds (Figure 22.3).

   Arrangement of plants should mimic patterns of existing native plant communities. These layers of plant growth are used for a multitude of purposes. Birds may build their nests in the layer of tall shrubs and find food below by scratching through leaf litter. Again, it is typical to arrange tall species along the periphery of the habitat, followed by large understory shrubs or small trees closer to the view point (in many cases, the home), and layers of small shrubs, groundcover, and annuals such as wildflowers and grasses in the foreground.

**FIGURE 22.3**
Providing thorny shrubs to intertwine with other native species provides effective cover and food for wildlife.

2. *Create both variety and same-species clumps in the habitat*: Different birds require different food in different seasons. Therefore, it is important to have enough variety in the habitat to sustain birds year-round, if possible. On the other hand, it is also important to maintain some masses of high fruit-producing species to provide very visible, massed displays for recognition.
3. *Leave nature alone*: This is probably the most difficult goal to achieve in a created wildlife habitat. With any type of planned landscape, it can be quite difficult for managers and homeowners to leave litter and old branches or tree snags in the habitat. However, these provide perches, nesting cavities, and insects for food. Minimal raking of areas contributes to litter accumulation that can harbor food for ground-feeding birds such as thrashers. On the other hand, maintaining some open areas is important for dust baths and sunbathing.
4. *Avoid use of exotic plants*: Exotic trees and shrubs should be avoided when possible due to their potential to spread into native habitats, and their reduced ability to provide food and shelter for native birds. Minimize turf areas since they provide little habitat or food for birds and may be associated with factors that have a negative effects on birds such as fertilizer and pesticide use.[5]
5. *Provide water*: A circulating water source is preferred because traditional birdbaths can encourage mosquitoes and cause the spread of disease (Figure 22.4).

   If a birdbath is used, it should be cleaned with a stiff brush every day and should be no deeper than 3 in.[8] A simple system can be installed using a separate zone on a drip irrigation system that provides flow each day.

**FIGURE 22.4**
Small "wildlife waterer" provides a water source with a reservoir (basin) that is flushed out with each cycle from a drip emitter system.

**TABLE 22.2**

Native[a] Plants for Bird Habitats

| Species | Common Name | Plant Type | Use |
|---|---|---|---|
| *Acacia greggii* | Catclaw acacia | Tree/shrub | Food, cover |
| *Acacia constricta* | Whitethorn acacia | Tree/shrub | Food, cover |
| *Atriplex lentiformis* | Quailbush | Shrub | Food, cover |
| *Baileya multiradiata* | Desert marigold | Forb | Food |
| *Bouteloua curtipendula* | Sideoats grama | Grass | Food |
| *Carnegiea gigantea* | Saguaro | Cactus | Food, shelter |
| *Celtis pallida* | Desert hackberry | Shrub | Food, cover |
| *Celtis reticulata* | Western hackberry | Tree | Food, shelter |
| *Ceratoides lanata* | Winterfat | Shrub | Food, shelter |
| *Chilopsis linearis* | Desert willow | Tree | Food |
| *Encelia farinosa* | Brittlebush | Shrub | Food |
| *Larrea tridentata* | Creosote | Shrub | Cover |
| *Lycium* species | Wolfberry | Shrub | Food, cover |
| *Mimosa dysocarpa* | Spiny mimosa | Shrub | Food |
| *Muhlenbergia rigens* | Deer grass | Grass | Food |
| *Olneya tesota* | Ironwood | Tree | Food, shelter |
| *Opuntia spinosior* | Cane cholla | Cactus | Shelter |
| *Parkinsonia florida* | Blue paloverde | Tree | Food, shelter |
| *Prosopis velutina* | Velvet mesquite | Tree | Food, shelter |

[a] Sonoran, Chihuahuan, Great Basin, and/or Mohave Deserts.

Birds consume fruits, buds, flowers, seeds, and the nectar of plants, as well as the insects that are found on them. For example, quail, doves, and finches consume large quantities of seeds whereas mockingbirds and thrashers prefer fruits and berries when they are available.[7] The insects that birds consume are typically associated with a healthy, functioning habitat. On the other hand, many plants rely on birds for pollination and dispersion of seeds over the landscape. This relationship is indicated by the numerous plant characteristics that have evolved to ensure the appeal of their fruits to specific birds. For example, many fruits ripen when bird migration reaches its peak, have fruit sizes appropriate for bird consumption, and are brightly colored.[5] Table 22.2 provides a list of plants that provide the variety in growth type and attractants for a southwestern bird habitat.

### 22.3.2 Hummingbird Habitats

Many of the design guidelines for landscapes that attract birds also apply to hummingbirds. In addition, it is important to remember that hummingbirds have relatively high food requirements compared to other birds because of their small size and high metabolic rate.[9,10] Because of this, more of hummingbirds' activities tend to be tied to their food sources (nectar and insects) than for other birds. For example, Anna's hummingbirds will attempt to stay year-round in a location where food persists,[10] and migrating hummingbirds look for resting stops or nesting sites where they initially find adequate food.[9]

**FIGURE 22.5**
Variety of plants (fairy duster, ocotillo, and penstemon) that produce nectar at various times of the year for hummingbirds.

Because of this heavy reliance on constant and plentiful food sources, habitats should emphasize a wide variety of flowering species to maintain an extended flowering season (Figure 22.5), and a healthy plant community to promote the presence of insects.

The resourceful Anna's hummingbirds will seek out remnants of pollen or nectar on half-frozen blossoms if fresh blooms are not available at the end of the warm season.[9] In some cases, noninvasive exotics that flower during late winter, such as aloe vera (*Aloe vera*) and cape honeysuckle (*Tecomaria capensis*), are used to extend the flowering season of these areas. Species choices should be guided by hummingbird preference of pink, red, purple, or orange color and single flowers of a trumpet or tubular shape.[9]

Nectar in plants that attract hummingbirds has a sugar content that matches a ratio of about three or four parts water to one part white cane sugar.[9] If sufficient food is not provided by plant nectar, syrup feeders can be used as supplement. If this is the case, feeders should be properly cleaned every 3–4 days to prevent spread of disease and spoiled syrup (due to high temperatures). Table 22.3 provides a list of plants that would provide the variety in growth type and nectar for a southwestern hummingbird habitat.

### 22.3.3 Butterfly Habitats

There have been dramatic reductions in the ranges of rarer butterflies due to the modification and devastation of their habitats.[11] Increases in agriculture and other land uses have led certain butterfly species to be restricted to isolated areas. Due to the highly varied plant communities in this area, southeastern Arizona is considered a critical region in terms of butterfly habitat, providing an environment for more than 240 species of butterflies in 6 counties.[12] Butterfly habitats in urban communities can supplement adjacent natural environments and provide extended habitats during the colder months. For example, it has been noted that some species are able to expand their populations and thrive in a human-dominated landscape.[11]

Habitats should accommodate various life cycles and activities of the butterflies, typically associated with seasonal changes.[12] Butterfly activity increases in the spring due

**TABLE 22.3**

Native[a] Plants for Hummingbird Habitats

| Species | Common Name | Type | Notes |
|---|---|---|---|
| *Acacia willardiana* | Palo blanco | Tree | Cover |
| *Anisacanthus thurberi* | Desert honeysuckle | Shrub | Food |
| *Calliandra californica* | Baja fairy duster | Shrub | Food |
| *Fouquieria splendens* | Ocotillo | Accent | Food |
| *Hesperaloe parviflora* | Red yucca | Accent | Food |
| *Justicia californica* | Chuparosa | Shrub | Food |
| *Lobelia cardinalis* | Cardinal flower | Small shrub | Food |
| *Mimosa dysocarpa* | Spiny mimosa | Shrub | Food |
| *Penstemon parryi* | Parry's penstemon | Small perennial | Food |
| *Parkinsonia microphylla* | Foothills paloverde | Large shrub/small tree | Cover |
| *Salvia dorrii* | Desert purple sage | Small shrub | Food |
| *Salvia greggii* | Autumn sage | Small shrub | Food |
| *Zauschneria californica* | Hummingbird trumpet | Small shrub | Food |

[a] Sonoran, Chihuahuan, Great Basin, and/or Mohave Deserts.

to greater availability of nectar and larval food plants. During this time, males use hilltops for mate location and frequently perch on the tallest shrubs or trees in natural communities.[12] In the hottest months, butterflies cluster in lower elevation locations, and gatherings of butterflies in moist pockets on the ground occur during the rainy season in July and August, when the adults take in moisture and salt.[12] The fall season is a productive time for egg production, until freezing nights cease adult activity. Created habitats may include some nonnative species that extend the period of availability of nectar-rich plants.[13,14] For example, exotic species such as lantana (*Lantana* spp.) and rosemary (*Rosmarinus officinalis*) offer nutritional support during the late fall and early spring when limited nectar is available from native plants. However, use of bush lantana (*Lantana camara*) should be avoided to prevent potential spread into natural areas.

Design of a butterfly habitat follows the general design guidelines relating to the arrangement of plants and variety. However, there are slight variations and additional guidelines that should be used that take into account some of the seasonal needs of butterflies:

Provide a variety of plants that feed larval and adult butterflies.

Food plants need to provide for two different stages of the butterfly's life: larval and adult. Therefore, it is important to have an adequate mix of plants that will support larvae and other plants that will be available for the adults:

1. *Emphasize massing of plants*: Massing should be relatively greater in butterfly habitats compared to bird habitats. This is due to the greater recognition factor of plant masses rather than a singular plant by adult butterflies. Use of accent plants (unique shape such as agaves or yuccas) within the many masses can create effective contrast and interest for the butterflies and habitat visitors.
2. *Provide sunny, wind-protected locations*: Butterflies are cold-blooded and need sunlight to warm the muscles they use to fly, and protection from wind when feeding (Figure 22.6).
3. *Provide a puddle*: Butterflies require a shallow puddle or moist soil for water (see Figure 22.4). A slow dripping emitter near a water-loving plant can fulfill this need.

**FIGURE 22.6**
Sunny locations and plants such as butterfly mist provide a small effective habitat for butterflies.

Table 22.4 provides a selection of plants that would provide the variety in growth type and food for a southwestern butterfly habitat.

## 22.4 Summary

Successful wildlife habitat areas can be created through an emphasis on thorough site analysis, careful arrangement of plantings and other design elements, and attention to plant species selection. Site analysis should provide the designer with knowledge of the existing natural communities that support wildlife and indicate the possibilities and limitations that exist at both the community and site-specific levels.

While this chapter focuses on creating habitats for a particular type of wildlife, many plants such as velvet mesquite (*Prosopis velutina*), catclaw acacia (*Acacia greggii*), or Baja fairy duster (*Calliandra californica*), will support more than one type of wildlife. For example, common urban lizards, such as the ornate tree and desert spiny lizards, are supported by increased diversity of plants for food sources (insects), thorny vegetation (escape cover), and structural diversity that creates habitat for terrestrial and arboreal lizards.[15] Therefore, if desired, it is possible to create a single habitat for a variety of wildlife types. In this case, attention should be given to needs of the diverse wildlife and the overall design of the project. For example, an area with concentrated massing and full sun locations for butterflies can transition into an area with more overstory species and variety in structure for birds, hummingbirds, and lizards (Figure 22.7). A few species of plants can be repeated throughout the design to provide unity and hence effectiveness of the overall composition. With these few requirements and a large number of plant species to choose from (Tables 22.2 through 22.4), a landscape design for wildlife habitat can be created to meet the spatial confines of any urban enthusiast.

**TABLE 22.4**

Native[a] Plants for Butterfly Habitats

| Species | Common Name | Type | Notes |
|---|---|---|---|
| *Acacia angustissima* | Fern acacia | Shrub | Larval food |
| *Acacia greggii* | Catclaw acacia | Tree | Adult food |
| *Ageratum corymbosum* | Butterfly mist | Shrub | Adult food |
| *Aloysia gratissima* | Bee bush | Shrub | Adult food |
| *Asclepia linaria* | Pineleaf milkweed | Shrub | Larval, adult food |
| *Bebbia juncea* | Sweet bush | Shrub | Adult food |
| *Calliandra californica* | Baja fairy duster | Shrub | Larval food |
| *Celtis pallida* | Desert hackberry | Large shrub | Larval food |
| *Chrysothamnus nauseosus* | Rabbitbrush | Shrub | Adult food |
| *Dalea frutescens* | Black dalea | Shrub | Larval food |
| *Dyssodia pentachaeta* | Golden dyssodia | Small perennial | Larval/adult food |
| *Eriogonum fasciculatum* | Flattop buckwheat | Small shrub | Adult food |
| *Eupatorium greggii* | Eupatorium | Small shrub | Adult food |
| *Eysenhardtia orthocarpa* | Kidneywood | Large shrub/small tree | Adult food |
| *Justicia californica* | Chuparosa | Small shrub | Adult food |
| *Lysiloma thornberi* | Featherbush | Large shrub/small tree | Adult food |

[a] Sonoran, Chihuahuan, Great Basin, and/or Mohave Deserts.

**FIGURE 22.7**
A mix of plant species and large massings of particular ones, along with some open space, creates a diversity of habitats for wildlife. Addition of neighboring trees and mid-story shrubs would further increase habitat opportunities.

## References

1. Mann, C.C. and M.L. Plummer, Are wildlife corridors the right path?, *Science* 270: 1428–1430, 1995.
2. Simberloff, D. and J. Cox, Consequences and costs of conservation corridors, *Conservation Biology* 6: 493–504, 1987.

3. Westbrooks, R.G., *Invasive Plants: Changing the Landscape of America* (Washington, DC: Federal Interagency Committee for the Management of Noxious Exotic Weeds, 107 pp., 1998).
4. Brown, D.E., C.H. Lowe, and C.P. Pase, A digitized classification system for the biotic communities of North America, with community (series) and association examples for the Southwest, *Journal of Arizona Nevada Academy of Science* 14 (suppl. 1): 1–16, 1979.
5. Kress, S.W., *Bird Gardens* (New York: Brooklyn Botanic Garden, 1998).
6. Paton, P.W.C., The effect of edge on avian nest success: How strong is the evidence?, *Conservation Biology* 8: 17–26, 1994.
7. Arizona Native Plant Society and Tucson Audubon Society, *Desert Bird Gardening* (Tucson, AZ: Arizona Native Plant Society and Tucson Audubon Society, 1997).
8. Terres, J.K., *Songbirds in Your Yard* (New York: Cromwell Co., 1968).
9. Holmgren, V.C., *The Way of the Hummingbird* (Santa Barbara, CA: Capra Press, 1986).
10. Stiles, F.G., *Food Supply and the Annual Cycle of the Anna Hummingbird* (Berkeley, CA: University of California Press, 1973).
11. Pollard, E. and T.J. Yates, *Monitoring Butterflies for Ecology and Conservation* (London, U.K.: Chapman & Hall Press, 1993).
12. Bailowitz, R.A. and J.P. Brock, *Butterflies of Southeastern Arizona* (Tucson, AZ: Sonoran Arthropod Studies, Inc., 1991).
13. Arizona Native Plant Society and Sonoran Arthropod Studies Institute, *Desert Butterfly Gardening* (Tucson, AZ: Arizona Native Plant Society and Sonoran Arthropod Studies Institute, 1996).
14. Xerces Society, *Butterfly Gardening* (San Francisco, CA: Sierra Club Books, 1990).
15. Seely, J.A., G.P. Zegers, and A. Asquith, Use of digger bee burrows by the tree lizard (*Urosaurus ornatus*) for winter retreats, *Herpetological Review* 20: 6–7, 1989.

# 23

# *Native Plant Salvaging and Reuse in Southwestern Deserts*

**Allan Dunstan and Margaret Livingston**

CONTENTS



## 23.1 Introduction

Salvaging native desert plants prior to developing a site has been a recognized practice in the Southwest since the late 1970s. Before this time, only a relatively small fraction of the population in the Southwest consistently used native plants in their landscape designs. For the most part the surrounding desert environments of urban areas were perceived as hostile places—interesting but not representing the kind of landscape composition desired in our designs.[1,2] Residents moving to desert regions often favored plants that reminded them of the areas from which they came. Species such as Aleppo pine (*Pinus halepensis*) and mulberry (*Morus alba*) were planted in great numbers and generally required higher water and maintenance than our native species. In addition, some of these plants had greater susceptibility to disease and grew to heights that exceeded the desert canopy, introducing a new set of ecological variables to the region.

These trends in rapidly growing urban areas in the Southwest have had a heavy impact on the native plant communities. To accommodate growth, large expanses of desert scrub were converted to farmland first along the edge of the urbanized areas and then replaced by urban housing and industrial development. Land uses replacing the natural desert setting during this period had no connection with the native plant communities, often appearing like landscapes typical of the Midwest or Northeast. Furthermore, migration of people from other areas often encouraged the creation of a more exotic landscape from the professionals involved in developing new subdivisions and commercial centers.

During this time, large expanses of colored rock over black plastic with a few cacti and boulders were often considered a "desert landscape." However, a growing movement of people who understood or appreciated the desert environment began to rethink the exotic landscape as one that best represents the vision for the replacement of the natural environment. Over time, through education and outreach, the concern for incorporating some of the native desert plants back into the landscape through design and regulation has taken hold throughout the region. This article discusses some of the developments in striking a balance between respecting the nature of the desert and developing with land restoration practices.

The availability of native plants for desert landscapes was relatively scarce until the foundation of native plant nurseries, such as Mountain States Nursery in Glendale, AZ, Sierra Valley Farms in Beckwourth, CA, and Bernardo Beach Native Plant Farm, Albuquerque, NM. The effort to generate desert plants and educate the public had begun through these efforts, but there was a parallel need to develop horticultural techniques to remove, protect, and restore emblematic desert trees and cacti prior to site development. Such efforts have elevated the landscape aesthetics of these species by providing larger, unique specimens of several native plants of the Southwest.

### 23.1.1 Early Days of Tree Salvaging Practices in Arizona

In 1979, a mechanical design engineer named Phil Hebets, quit his job at Garrett Corporation (now Honeywell), and started Sonoran Desert Designs, a design-build landscape company. Phil's innovative concepts included the liberal use of native trees and shrubs, not just cacti. At the time mature trees were unavailable in nurseries and Phil challenged a friend, Don Fedock, to salvage a mature blue palo verde (*Parkinsonia florida*). Fedock developed a specialized boxing process and the tree salvage industry was born. The technique was taught to Maurice Bosc of Sonoran Desert Designs, who made great strides in large tree boxing and moving technology. Subsequently, Phil and Al Dunstan founded Desierto Verde as a native plant nursery to complement and oversee the salvage operation.

In 1983, Don Fedock collaborated with Steven Carothers of SWCA environmental consultants on salvaging efforts for the Ventana Canyon Resort in Tucson. About 400 trees were salvaged by Fedock to complement the 300 saguaros and hundreds of smaller cacti salvaged by SWCA. The result was a successful revegetation effort, replacing the mix of plants that existed prior to construction. The same boxing process used for the lower desert plants also proved to be successful for oak trees and other higher elevation species. In 1985, the Arizona-Sonora Desert Museum near Tucson contracted with Desierto Verde to move a variety of mature plants for the Mountain Habitat exhibit being constructed. The museum decided they wanted the exhibit to look like a finished product the day it opened. Over 100 trees were moved including two 25-ft-wide Arizona white oaks (*Quercus arizonica*); other species salvaged included emory and silverleaf oaks (*Quercus emoryi* and *Quercus hypoleucoides*), Chihuahua and Apache pines (*Pinus leiophylla* and *Pinus engelmannii*), and Arizona walnut (*Juglans major*).

The end product of the salvage process is a tree that exhibits much more character than a typical nursery-grown plant. For example, salvaged ironwoods (*Olneya tesota*) reflect years of overcoming the adversities of lightning, dust storms, frost, intense summer heat, and drought with gnarly trunk formations, twisting branches, and dead wood. These "living sculptures" became desirable features that were placed at key focal points in upscale developments. Using native plants soon became an accepted practice for virtually every type of development.

As a result of this process, increased use of salvaged native trees in high-end projects created a perception of high value which is consistent with the scarcity of the trees of such stature as a natural resource. The trend toward using a greater number of desert trees in place of imported nonnative species gained momentum in the late 1980s with home buyers and landscape architects using native plants in their landscape designs.

## 23.2 Salvaging Guidelines

Over the last few decades, experimentation in field techniques involving a variety of desert species has led to some general salvaging guidelines for the Southwest. In general, minimizing the stress created when attempting to change the location (that is, the environmental factors) to which a native plant has been adapting to since germination is the critical goal. Severing the roots, thereby cutting off a significant portion of the plant's ability to take up water and nutrients, is the most obvious stress factor. But other factors may have just as much impact on the potential survival of a salvaged plant. Orientation to the sun, soil type, drainage conditions, neighboring plants, microclimatic conditions (temperature, wind, humidity), and even geologic features will all change to some degree as a result of the transplant. The following sections focus on some factors considered when salvaging the two most common plant types involved in this practice, cacti and trees.

### 23.2.1 Cacti Salvaging

Salvaging saguaros and other cacti has become fairly common in the last few decades.* Transplanting older cacti (generally individuals greater than 6 ft) is considered rather risky, with some experts questioning the practice and considering it to be an ineffective practice.[3,4] Before starting the process, plants are tagged, typically on the south side, to maintain the plant's orientation during transplanting, thereby preventing sunburn to the epidermis. Garden hose wrapped around saguaros or sections of carpet wrapped with rope are often used as handles during the move, considering that their enormous weight can be quite unwieldy during the process.

Cacti are dug out approximately 12–18 in. from the base of the plant and the larger roots are usually trimmed at approximately 2 ft deep. Wrapping very strong nylon rope around the roots and affixing the rope ends to the lifting apparatus ensures the spines are not damaged during the lifting. In some cases, plants may be placed in holding areas until a site is ready for positioning salvaged individuals. These holding areas typically contain a soil mix consisting of approximately 50% sand and 50% compost or planting

* http://cals.arizona.edu/pubs/garden/az1376.pdf (accessed December 30, 2009).

mix (complete with soil sulfur), with native soil included in some cases. Protection from sunburn is a significant concern during this transition time; depending on the time of year, shade cloth may be needed to temporarily protect the plant from sunburn following transplanting. Cacti are replanted at the same depth as the original site and the planting hole is filled with a sandy soil and tamped in to stabilize the plants. Irrigation is generally applied after a couple of weeks and continued every few weeks, depending on the growing conditions. A watering moat surrounding the cacti is the common irrigation technique created with the soil in the planting area. An armless saguaro, well planted, should not require staking.

### 23.2.2 Tree Salvaging

In many respects, moving a tree and accompanying root ball, weighing hundreds of pounds to many tons, is an engineering problem. The goal is to disturb the root ball and the many feeder roots as little as possible. In addition, damage to the structure of the tree is a concern. The process of removing a native tree from its original habitat may be divided into four phases: presalvaging preparation, sideboxing, bottoming, and maintenance. A brief description of each phase follows.

#### *23.2.2.1 Presalvage Preparation*

In an ideal situation, the tree would be root pruned up to a year in advance and a supplemental watering and nutritional program implemented. As a practical matter, some situations allow the opportunity to prewater the plant 2 weeks before digging begins, and in many cases, even this is not economically feasible.

When the salvage process begins, a specialty trimming crew arrives and first removes the underbrush and any debris around the base of the tree. In general, it is desirable to trim only minimally to maximize photosynthesis, usually less than 30%. Some of the lower branches had to be removed to allow crews access for the boxing process. Other pruning is done for horticultural reasons. For example, crossing branches which will eventually cause injury to one another are removed, along with any dead wood that impedes the aesthetic value. Sometimes, however, a seasoned dead branch may actually enhance the appeal of the tree such as ironwood.

#### *23.2.2.2 Sideboxing*

Crews arrive and begin by marking a previously determined box size on the ground. Nearby plants and obstacles as well as machine access must be considered when orienting the box. A trench is excavated along the marked boundary line around the tree. To the extent possible the trench is tapered inward to match the shape of the box to be placed around the root ball. Roots are cut by hand as they are encountered; with clean cuts desirable. Once the root ball is exposed, the four box sides are placed and secured with steel banding and nails (Figure 23.1).

Any spaces between the box and root ball are backfilled with dirt and packed to eliminate air pockets. Additional beams are placed across the top of the box and secured against the trunk to assure that the root ball does not shift during moving. The sideboxed tree is watered and left to recover from the digging process. Another crew is responsible for watering the tree regularly until the bottom of the box can be attached (bottoming). Use of a soil probe will determine watering efficiency.

**FIGURE 23.1**
Completion of sideboxing and attachment of cables for moving the boxed specimen involve machine and hand labor.

### *23.2.2.3 Bottoming*

After a period of time, the tree is prepared for the "bottoming" process. Within reason, the interim between sideboxing and bottoming should be maximized. Generally, a minimum of 3 weeks for plant stabilization is required to provide a guarantee for the salvage work. Four to eight weeks is probably ideal depending on the time of year. The practicality and effectiveness of continuing the watering in the field must be considered. This watering is usually done by a crew with a water truck under difficult conditions. The amount and duration of each watering with this scenario is not nearly as suitable for the plant as what can be applied using drip irrigation in a nursery setting. The cost of truck watering is significantly higher.

Placing the bottom on the box without causing the root ball to break apart is a very challenging part of the process especially in sandy soils. The root ball is gradually undercut and pulled over with a winching device and chain. Roots are cut as they are encountered just as with sideboxing. The big difference is gravity; the winching process creates uneven forces that tend to crack the root ball, causing soil to fall away and expose the feeder roots vital to survival of the tree. Therefore, the winching of root balls and cutting of roots become critical steps, with the soil type playing a major role in the result. Once the box has been tipped over to approximately a 45° angle, the bottom is placed, nailed, and further secured by bands proceeding vertically around the box. The bottom of very large box sizes are carefully secured one board at a time using a tunneling method.

In most cases, a hole is cut near the bottom of the box side to allow any soil that fell away during bottoming to be replaced and packed into the completed box.

### *23.2.2.4 Maintenance*

The boxed tree is left for a few days to allow the soil to settle and moved to a holding yard for ongoing maintenance. Proximity to the salvage site, machine access, and availability of pressurized water are key factors in selecting the on-site nursery location. However, in some circumstances, plants must be transported many miles to an available nursery site. In most cases, the trees are lifted from the dug area by a wheel loader, placed upright on a

FIGURE 23.2
Boxed tree being moved to a nursery for recovery prior to transplanting.

semitrailer, and driven to the nursery (Figure 23.2). As desert tree salvage normally occurs before roads are in place, creating passable routes is one of the real-world problems faced by the contractor. A rough road, and possibly a long ride, inevitably disrupt a root ball and becomes one more stress factor for the recovering tree to overcome.

Assuming a water source is available, the assembled trees are irrigated with a drip system (Figure 23.3). While automatic controllers are an essential tool, crews must check the operation of the system every time a watering is scheduled. Each tree is in a precarious state and missed irrigations can cause many deaths. Checking irrigation lines, adjusting emitter placement, repacking soil around root balls when necessary, and changing the duration of watering are all essential functions of the maintenance crew. The goal is to achieve a uniform moisture level throughout the root ball. It also important to assure that

FIGURE 23.3
Salvaged specimens receiving irrigation and awaiting transplanting.

the moisture content remains within an acceptable range until the next watering and a soil probe is helpful for this task. Weather patterns, even short-term spikes in temperature, wind velocity and humidity, are critical variables in the success of salvage tree survival.

### 23.2.3 Tree Spading

If a large tree spade is used for transplanting, much of the manual labor needed has been eliminated. Generally, the spade captures over 90% of the root ball. The trees are then immediately transported to the new site for transplanting. However, in many cases, tree boxing is required due to the need to retain plants until a new site is ready for salvaged species.

### 23.2.4 Other Factors Affecting Tree Salvaging Success

Site assessments to determine the box size to be placed on each tree are a major component of the tree salvage process. General guidelines are based on trunk caliper for each species. However, soil type and drainage patterns also come into play. A larger box size captures more feeder roots but presents greater handling problems. In the end, the ideal box size is the smallest which will support a healthy transplant. Boxes are sized in 6 in. increments in conformance with nursery industry standards. Typical tree boxes range from 36 to 72 in. Shrubs like creosote (*Larrea tridentata*) and jojoba (*Simmondsia chinensis*) are typically salvaged in 24–48 in. boxes.

A related issue is the use of machines versus manual digging. The best argument for the manual method is the original premise that less disturbance to the feeder roots typically increases the success rate. Excavation by hand enables workers to carefully expose and cut roots without shaking the root ball. Most salvage operators use a backhoe to dig trenches around the tree. The problem with this method is the digging bucket tends to tear roots rather than making a clean cut. With larger roots, the action of the teeth catching the roots can cause significant disturbance to the root ball. However, if the root zone is hand excavated, machinery may be used to complete the digging process as long as the roots are not touched by the machine. A qualified operator is essential for this method. On average, salvage companies have found that a given tree can be successfully salvaged using one size smaller box with the manual digging method. For larger sizes (greater than a 48 in. box), a combination of machine and manual digging can be very effective. A backhoe is used to remove the larger quantities of soil near the surface but the manual method is used in the zone where most of the roots occur.

The intent of the foregoing is to give the reader some appreciation of the intricacies and practical issues related to saving native trees and cacti on a scale that can positively impact the environment of our metropolitan areas. It is not intended to be a step by step "how to" manual. The current results reported barely scratched the surface of all the considerations relevant to successfully salvaging naturally occurring desert species. In addition to dealing with climatic conditions, the salvage industry will most likely contend with several levels of regulatory constraint and the driving forces of economic expansion. Each of these factors provides both opportunities and challenges.

## 23.3 Challenges

Some laws and the regulatory agencies responsible for protecting native species may seem at odds with the goal of saving native plants from destruction and maximizing their use in landscapes. For example, laws designed to protect saguaro (*Carnegiea gigantea*) cacti from theft may discourage the salvage of smaller cacti such as hedgehog (*Echinocactus* spp.) and pincushion (*Mammillaria* spp.). More specifically, the permitting cost which makes economic sense for a plant worth hundreds of dollars (saguaro) is prohibitive for those with a value of $10 or so. Likewise, the bureaucratic process of obtaining wide load permits may work well for mobile home transporters but create major obstacles for companies trying to move a 25 ft-wide paloverde down the road. Some of the common regulatory issues related to native plant salvage are briefly discussed, using the state of Arizona as an example, as well as other challenges faced in the salvaging process.

### 23.3.1 State Native Plant Laws

Until 1989, Arizona operated under a native plant law, promulgated in 1929, which focused on preventing cactus theft and the cutting of desert trees for firewood. The enforcement division of the Arizona Commission of Agriculture and Horticulture were affectionately known as "Cactus Cops." Their relationship with the fledgling native plant salvage industry in 1980s was primarily adversarial. It was a very real possibility that a legitimate salvage operator could be put out of business for an inadvertent error in not attaching a tag to a tree. No one anticipated that thousands of native trees, although not theft targets, would be moved around on semitrailers routinely.

The salvage industry helped organize a coalition that came to be known as the Science and Industry Group to affect meaningful changes to the native plant law. The group included representatives from the Arizona Nursery Association, individual nursery businesses, the Native Plant Society, cactus movers, the Desert Botanical Garden, and many more. The result was Senate Bill 1086, which became effective in 1990 with the following major provisions:

- Created five categories of protected native plants each with appropriate rules based on need. The common native trees were put in a "salvage assessed" group with streamlined procedures to encourage saving them.
- Required land owners to notify the State of their intent to clear land to allow the possibility of salvaging trees, since it was not required by the State of Arizona.

The new law also enabled a comprehensive revision of the list of protected species including a "highly safeguarded" (essentially endangered species) category.

A final challenge for the Science and Industry Group was to keep agency staff members from nullifying some of the positive changes made by the law with restrictive regulations. One major issue was the proposal to make tagging of propagated native plants required even though the law specifically excluded them from the definition of "protected." This procedure would have placed a huge administrative burden on nurseries growing native plants discouraging their use. Fortunately, the agency avoided this item.

At the time of this writing there is a similar proposal related to protected species growing in developed areas (as opposed to growing in the wild). At present, if such plants are to be moved, the transporter may use a "blue tag" obtained from the Department of Agriculture to identify the plant and avoid the possibility of being questioned as to the legality of moving the plant. The department is considering making the blue tag a requirement due to the problem of determining whether a plant that may be moved came from a natural desert area or from someone's front yard. As propagated native trees are now grown to large sizes on tree farms, the question could be raised as to permitting and tagging these trees if an enforcement official cannot distinguish a farm-grown individual from a

**FIGURE 23.4**
Salvage crew marking location of saguaros using GPS.

naturally occurring one. Nevada has a similar regulation, requiring permits and transport tags for native cacti and yuccas. Harvests on private land for commercial purposes require a native flora harvest registration permit, if removing six or more individuals.

Local ordinances have been implemented in many Southwest cities. A key requirement is the submittal of an inventory of the protected native plants on site identifying each plant with a tag number and sometimes a GPS point (Figure 23.4). The inventory consists of a listing of the plants. City personnel review the inventories with particular interest in trees determined to be unsalvageable. An explanation must accompany any such determination; the most common are poor health of the tree and poor soil conditions.

An ordinance enacted by the City of Phoenix requires a two-step approval process. The first is a Landscape Inventory Plan which includes the basic site inventory. The second is a Landscape Conservation Salvage Plan which (involves) grading and drainage issues and details of the salvage effort.

Pima County also requires a site inventory but takes a different approach to mitigation. Based on assessments of viability (health) and transplantability there is a formula which determines a number of smaller plants that must be added to the landscape to replace any plants for which salvage is not attempted or is unsuccessful.

To summarize, tremendous improvements in regulatory actions have been enacted. However, it seems we will always be walking a thin line between the prevention of abuses and encouraging desired uses of certain native plants.

### 23.3.2 Economic Challenges

From a developer's perspective, getting a project approved and permitted through the local jurisdiction is a long and arduous process even without any requirements related to native plants. In the early days of salvaging, owners and builders were not too enthralled with another layer of regulation. However, as it became evident that tree salvage was an economically viable process, attitudes became more positive. Salvaging plants on-site saves about 50% compared to purchasing a similar one at a nursery. And as more cities and counties pass ordinances, it is becoming more and more difficult to find replacements.

From the perspective of saving the plants, it is not the cost of the process that presents a challenge. Timing is generally the problem. Construction schedules are driven by very weighty financial considerations such as minimizing interest costs, propitious grand opening dates, and overall economic cycles. Layering the governmental approval maze over the financial realities makes it very difficult to plan ahead in a way that would benefit the native plants.

The time window between project approval and the construction start is usually very limited. Rushing the tree salvage effort inevitably decreases the survival rate whereas having more time allows the plants to better adapt to each step of the process. With the luxury of time, more steps could be taken to prepare the trees before boxing starts. Well-intentioned efforts to start the tree salvage early can result in severe financial penalties. In one instance recently, a builder was fined about $25,000 for allowing the native plant salvage to start before a building permit was finalized. The risk is that if the project hits a snag and is not approved, the trees may be dug unnecessarily.

Another facet of this challenge is the time of year at which development starts. Depending on species, there is a significantly lower success rate for tree salvage done during the months October through March. Survival can drop from more than 90% down to 60% or worse. While it may be unrealistic to expect development to stop for a 5 month period, there may be some solutions if we took a global approach to saving native plants. As the trend is toward more long-range regional planning, perhaps areas designated for development could be addressed, relative to salvaging strategies, during the summer even if the permit process has not been completed. If the anticipated project is delayed or not approved, the worst-case scenario would involve using the trees on another current project. When the delayed project needs trees, they could be supplied from another project.

### 23.3.3 Horticultural Practices

While individual historic specimen trees have been moved in other areas of the United States for many years, the concept of saving all salvageable materials in the path of development is in its infancy. The needs of desert trees, particularly large, salvaged desert trees are unique. Our body of knowledge related to desert tree salvage is relatively small; much is to be learned about how the stress factors related to transplanting a mature tree affect its physiology. Of particular interest is how the effects vary based on seasonal changes in daylight hours, angle of the sun, and temperature extremes. Most importantly, what can be done to ameliorate the stress and improve survival rates?

Part of the challenge to improve horticultural practices is economic. Native plant salvage has become an extremely competitive business with the majority of participants being landscape contractors. Setting aside research and development time and money is difficult because margins are narrow even in good economic times. The narrow time window for salvaging on a site previously discussed makes it even tougher to conduct meaningful and controlled experiments.

When a successfully boxed tree is replanted, follow up and maintenance procedures are more critical than with a nursery grown plant. The average customer has the misconception that if native trees can survive without supplemental water in their native habitat they should be able to survive in a landscape. The challenge is to get just the right amount of water to a root system that has been significantly cut back without overwatering it. We recommend that salvaged trees receive supplemental water for at least 2–3 years if not indefinitely. But too much water can be detrimental, especially in heavy soil.

Planting depth is important for any plant. With a salvaged tree, maintaining the original soil level appears even more critical because an older, stressed tree is much more susceptible to fungus and bacterial infections caused by continuous moisture against the trunk.

## 23.4 Impacts

The net result of effective salvaging practices and related regulatory ordinances is that the typical planned community can potentially increase the density of native plant species on site upon development. Furthermore, salvaging native plants ultimately adds value to any given project due to lower clearing expenses and costs to replace trees of similar size. Beyond the tangible cost savings, the availability of salvaged native trees enables landscape architects and others to relatively quickly create a mature desert context in designs. Massed plantings of mature, native species can give our parks, civic centers, office complex atriums, and other public gathering places just such a feeling. Deciduous desert trees strategically planted provide significant energy savings by providing shade in the summer and letting sunlight through in the winter. On a larger scale, the use of our native desert trees has begun to create a canopy of shade that helps to mitigate the urban heat island effect and establish valued microsites for understory plantings and inhabitants.

The recent drought years have focused much attention on water conservation. While salvaged native plants do require water to supplement rainfall, there is potential for significant water savings compared to a nondesert landscape with exotic plants and lawns. There are a few keys to realizing the full potential savings:

- A drip system operated by a controller with enough programming flexibility to allow appropriate adjustments for the time of year and weather changes.
- A water harvesting system which could be as simple as land contouring to direct rainfall, including roof runoff, to the root zones.

Landscapes featuring mature native trees also impact the local ecosystems and habitats for animals, birds, and insects. It is generally accepted that providing more mature, native trees rather than small, transplanted plants will support a greater amount and number of native wildlife (see Chapter 22). Furthermore the network of thorny branches and spines associated with many desert species provide effective cover for escape and possible nesting sites.

## 23.5 Future Outlook

Native plant salvage has enabled us to change the face of development in the desert Southwest in the last 20 years. However, there is still great potential for advancement in this field. Improving survival rates, particularly in the winter months will make even more large native plants available. To do this we must devote significant resources to research and development with a focus on science. An attitudinal shift is part of the process. What is the value of a 100 year-old ironwood? Beyond economics, does such a tree have intrinsic value? Does it have a right to live that supersedes our desire to place a building on its footprint? We do not all have to become "tree huggers" but we can begin to allocate more resources to their preservation. As we have seen, our efforts will pay economic dividends.

## References

1. Martin, C., K. Peterson, and L. Stabler, Residential landscaping in Phoenix, Arizona, U.S.: Practices and preferences relative to covenants, codes, and restrictions, *Journal of Arboriculture* 29: 9–17, 2002.
2. Spinti, J., R. St. Hilaire, and D. VanLeeuwen, Balancing landscape preferences and water conservation in a desert community, *HortTechnology* 14: 72–77, 2004.
3. Emming, J., Transplanting large saguaros the right way, *Cactus and Succulent Journal* 79(4): 159–163, 2007.
4. Harris, L., C. Funicelli, and E. Pierson, Survival evaluation of transplanted saguaros in an urban housing development and golf course development, *ESA Conference Proceedings*, Tucson, AZ, p. 82, 2002.

# 24

# *Sustainable Urban Living: Green Solar Energy for Food and Biofuels Production*

**Mark Edwards**

**CONTENTS**



## 24.1 Introduction

The current urban living model is not sustainable in the U.S. West and Southwest because key natural resource flows will be insufficient to sustain lives and lifestyles. The natural resources supplying food, energy, and transportation will be severely impacted by net zero water allocation for irrigating crops, net zero oil exports from OPEC, and significant climate changes. A new sustainable urban living design that recovers, recycles, and reuses nutrients lost in the human and animal waste streams is needed. Conservation and reuse must include freshwater as well as nutrients because both limit the available food and biofuel supply.

Additional populations will add to urban living costs and create more air, soil, and water pollution. The only feasible solution requires a transformation to an ecological positive source of food and fuel that uses no or minimal fertile soil, freshwater, fossil fuels, or pesticides. This ecological design must either conserve and reuse nonrenewable, fossil inputs or find new inputs for food and energy production that are cheap and will not run out. A food and energy supply that produces with sunshine, surplus $CO_2$ and wastewater would be sustainable and conserve fossil resources for conventional food crops and urban populations.

Limits to key natural resources will force substantial design changes in food and energy production in order to support sustainable urban living in the U.S. West and Southwest. Natural resource flows that support the food supply such as freshwater, fossil fuels, fertile soils, and fertilizers are diminishing while the demands of urban dwellers for food, energy, and transportation continue to expand. Additionally, the air, water, and soil pollution caused by fossil resource consumption lowers the quality of life for people living in both rural and urban areas.

Rural populations pushed out of farming have migrated to urban centers where they became dependent on others for food. In 1800, only 3% of the world's population lived in cities but over half live in cities today. In 1950, only 83 cities had populations exceeding 1 million but today 468 cities have populations of more than a million.* The U.N. forecasts that the current urban population of 3.2 billion will rise to nearly 5 billion by 2030, when three out of five people will live in cities.[1] Rural to urban migration results not only in fewer food producers but more energy consumption because urban dwellers consume about four times as much energy as their rural relatives.[2]

One of the main challenges facing urban designers will be to provide adequate quantities of nutritious and affordable food for urban inhabitants. People expect food to be readily available and cheap, as it is today. Unless we change the ecological design of our food supplies and conserve fossil resource inputs, some cities will perish due to insufficient food and freshwater, while others will find their air and water too polluted for healthy living. The following discussions about our freshwater fossil fuel supplies, soils, and climate change emphasizes the need for a new model for our food production in the future.

## 24.2 Freshwater Limits

The West holds far too little freshwater to support food production for expanding populations even before global warming melted 50% of the snowpack and evaporated lakes and reservoirs.† City faucets and fountains are going dry because farmers are putting far too many straws in the ground and removing fossil water reserves. The aquifers on which food production depends are being depleted rapidly and are likely to expire within a generation.[3]

Water is the primary limitation to food production from crop plants. Without sufficient freshwater delivered on time, crops fail and the land reverts to its natural state—which in much of the West is prairie or desert—and human populations must migrate to more productive areas.

Therefore, production and yield are directly related to water use. A decrease in applied water stresses crops and decreases yield. More irrigation has doubled food production over the last 30 years but at the unsustainable expense of tripling the freshwater consumed.‡ About 80% of all freshwater use in the West goes to irrigation and much of that water is lost to evaporation and plant transpiration. Much of the 300% increase in water consumption occurred because new croplands were expanded into deserts. Irrigation creates higher yields but consumes more water, especially in hot sunny regions with high evaporation rates. Irrigation often loses 50% of the applied water before it reaches the crops from leaks and evaporation.[3] Over half of the irrigation water comes from groundwater aquifers which are being depleted at 3–300 times nature's replacement rates.§

* http://www.citypopulation.de/world/Agglomerations.html (accessed August 10, 2011).
† http://www.wcc.nrcs.usda.gov/cgibin/westsnow.pl (accessed August 10, 2011).
‡ http://www.fao.org/nr/water/topics.html (accessed August 10, 2011).
§ http://www.fao.org/nr/water/news/clim-change.html (accessed August 10, 2011).

In normal years, California produces over half of the vegetables and fruits consumed in the United States. The state irrigates 9.6 million acres using roughly 34 million ac ft of water either from lakes, reservoirs, and rivers or pumped from groundwater. Lester Snow, director of the California Department of Water Resources, said in 2009 that California may be in its worst drought in history.* The Central Valley Authority that distributes irrigation water announced a zero allocation to many crop regions. The Bureau of Reclamation estimated that 1 million ac would be put out of production and another 2 million ac would grow less food than normal. He called the situation grim.[4] Agriculture consumes 81% of the water in the state while providing only 3% of the state's revenue. Agriculture in California competes constantly with cities for access to water that is diverted mostly from the northeastern part of the state to the fertile deserts of the Central Valley and southern California.

Drought also plagues the West, South, and Eastern United States. In the fall of 2008, farmers in Texas received no rainfall and lost their winter wheat crop and cannot plant spring crops because there is no soil moisture. Texas farmers rediscovered the "WW" problem that farmers have known for eons; "If there is insufficient Water to germinate Weeds, food crops do not grow either."

Biofuel crops consume huge amounts of water that is neither sustainable nor practical. Land crops consume far too many fossil resources that will become unaffordable or unavailable such as freshwater, fertile soil, fossil fuels, pesticides, herbicides, and agricultural chemicals. In addition, land crops create severe erosion and pollution that poisons the air, soils, and surface and groundwater.

Land crops grown west of the Mississippi consume irrigation water, much of which comes from fossil aquifers that are not replenished by annual rains. An acre foot of water, 326,000 gal, covers an acre 1 ft high. The USGA Water Use Report reported that several arid Western states such as Colorado, New Mexico, and Arizona applied an average of 5.5 ac ft to irrigate crops while the High Plains averaged about 2.5 ac ft of water.† The High Plains get about a third of their water in rain—in wet years.

A single acre of irrigated corn consumes 3 ac ft—about 1 million gallons of water.‡ An acre of irrigated corn produces about 140 bu of corn, which yields 350 gal of ethanol. Therefore,

- Production of 1 gal of ethanol consumes 3000 gal of water.
- Each gallon of ethanol made using irrigated corn wastes 12 tons of consumptive use water.
- The West cannot support biofuel production using irrigated crops and still have enough water for urban centers.

Water rights are legal commitments by citizens, cities, and other entities for control of precious surface and groundwater. Many cities are buying water rights from farmers in order to provide urban water.§ Cities are assured water but decimate local food supplies which means more fossil fuels are needed to transport food. Even in normal rainfall years, many urban areas have over 100% of the local water promised under contract. In dry years when water becomes insufficient, cities simply extract more groundwater. The pumping strategy works until the wells go dry.

* http://www.owue.water.ca.gov/agdev/ (accessed August 10, 2011).

† http://pubs.usgs.gov/circ/2004/circ1268/htdocs/text-ir.html (accessed August 10, 2011).

‡ Personal interview, Gerry Sanders, Salt River Project, June 2007. Three acre feet is standard but some farmers in the Southwest get 6.5 ac ft for their crops. The USDA Water Use Report, 2002, indicates and average of 5½ ac ft were delivered for the western state, which approaches 2 million gal of freshwater per acre.

§ http://www.ewg.org/node/18330 (accessed August 10, 2011).

A solution to the challenge of water, a critical issue throughout the ages, remains possibly the most vital global issue today, especially for the West. Solutions to the problem of water present only two alternatives:

- Find, harvest, and transport more water.
- Develop food and energy sources that require no or minimal fresh water.

Alternative one replicates the unsustainable actions of the last 50 years—using wider pipes, larger pumps, and deeper holes to mine more nonrenewable water faster to grow crops. Even if energy for pumping were free, this approach depletes aquifers within a few decades and makes the land unsustainable for both crops and people. Powerful modern pumps draw water from deeper and deeper wells—at an unsustainable cost of both power and water. Unsustainable use of fossil water foretells severe hunger and starvation for future generations.

Alternative two leaves the aquifers in place to support people, businesses, and food crops now and for future generations. The world desperately needs an efficient agriculture system that produces nutritional food and reliable clean energy and does not rely on freshwater.

## 24.3 Consumption of Fossil Fuels

Industrialized agriculture is a costly business and fossil resources are the currency used to grow food. After cars, food production consumes more fossil fuel than any other sector of the economy—about 20%.[5] Industrialized agriculture depends on fossil fuels for farm machinery, food processing, packaging, transportation, fertilizers, herbicides, and pesticides (Figure 24.1). Farmers may cross a field six to nine times on tractors, trucks, or harvesters to produce each crop.* Tractors pulling plows, disks, cultivators, planters, spray equipment, and harvesters consume huge amounts of fuel (Figure 24.2).

David Pimentel and Ted Patzek analyzed the fossil energy inputs to U.S. corn production and concluded that machinery and fuel, used to reduce human and animal labor, total about 25% of the fossil energy input and the remaining 75% is invested in agricultural chemicals to increase crop productivity.[6] Failing access to fossil resources, modern agriculture could produce only a fraction of current food production because the majority of fossil energy goes to enhance crop productivity.

Globally, farmers put millions of tons of fossil fuel-based herbicides, pesticides, and fungicides on crops to control undesired weeds and pests. These agricultural chemicals are produced using extensive fossil fuels and chemicals. The unintended consequence of expanded use of agricultural chemicals, besides pollution and human health problems, can be seen in resistance figures. In 1950, there were about 10 species of insects resistant to pesticides. Today, there are over 600. Similarly, the number of weeds with herbicide resistance was near zero in 1950 but there are more than 400 today.[7] Even though insecticide use has increased 10-fold, crop loss from insects is double the level it was in the 1940s—about 13%.† Pest resistance forces farmers to continually add more chemicals which consume more fossil fuel.

* Personal interview, Marvin Morrison, Arizona farmer, July 2003.

† In brief, *Environment*, September 2001, p. 8.

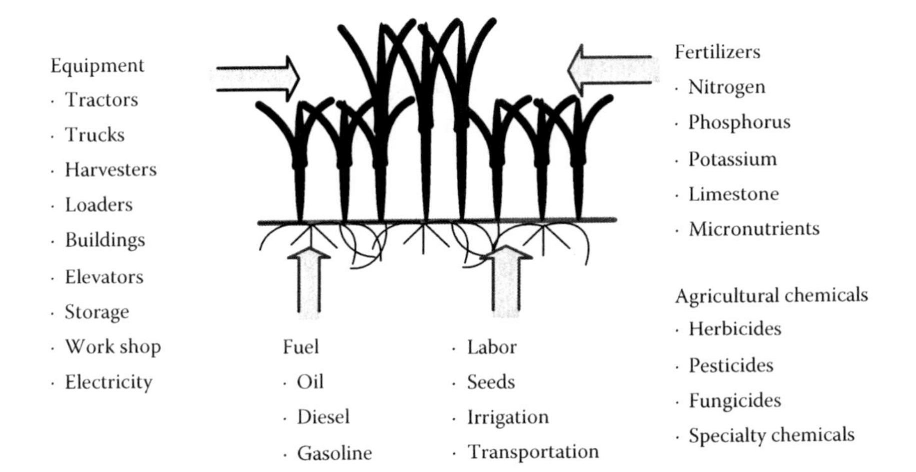


**FIGURE 24.1**
Fossil inputs necessary for modern agriculture.

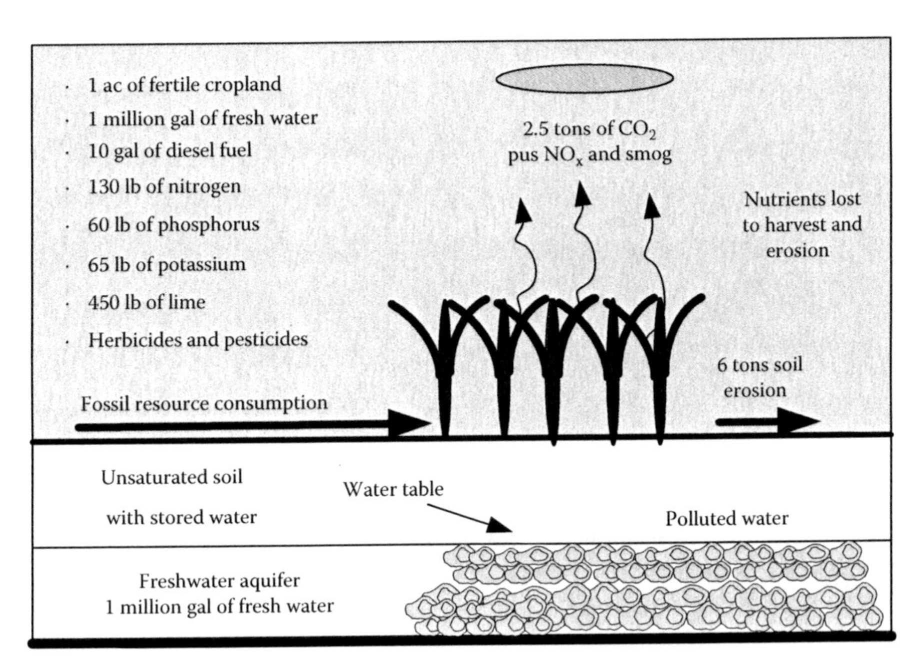


**FIGURE 24.2**
Environmental impacts from modern agricultural practices.

These agricultural chemicals erode into the groundwater and enter human foods and municipal water. The City of Des Moines, Iowa, ground zero for ethanol production, will spend $455 million on water treatment to reduce agricultural pollutants from city water.[8] Urban centers throughout the West will have to follow suit because EPA studies show that 37% of U.S. lakes are unfit for swimming due to runoff pollutants.[9]

Farmers exhibit fossil fuel exuberance because they pay only about 20% of the true cost of their fossil inputs thanks to government shelters and subsidies.* Farmers are sheltered from paying for externalities (social and environmental costs) caused by agricultural pollution of air, soils, and groundwater, the loss of nonreplaceable fossil water and agricultural chemicals, the cost of military protection of oil assets, and the health costs associated with agricultural pollution. Farmers pay nothing for their contribution to climate change and they do not reimburse fishermen for the billions in lost revenue due to the off shore dead zones caused by agricultural runoff.

The U.S. government subsidizes fossil agriculture on the order of $12 per gal of diesel fuel.[10] Farmers may pay $3 per gal but the price reflects only the direct cost of the fossil fuel supply chain to the farm. Farmers benefit from huge tax subsidies to the oil industry such as the oil depletion allowance. Terry Tamminen calculated in *Lives per Gallon*, that the true cost of fuel was $15 per gal.[11] These subsidies are built into all agricultural inputs including water, power, equipment, and agricultural chemicals.

Subsidies and cost calculations reflect neither future costs nor resource loss. As fossil resource supplies such as water and agricultural chemicals diminish, they will increase in cost. Farmers pumping water from fossil aquifers lower the aquifer's water level which means it will cost more for all farmers 10 years from now because they will have to use more energy and larger pipes to pump the water. The children of today's farmers will face the end of resource reserves and have to give up farming. Their family farm will lose most its value. If those young people were allowed to put a price on the loss of fossil resources, it would be far higher than the current subsidized price of fossil fuels, water, and agricultural chemicals.

After cars, agriculture consumes the most energy in the United States, about 20%. The West and Southwest are the heaviest users.[12] Agriculture is responsible for about 37% of America's air pollution and the majority of soil and water pollution.[13] Reasonable arguments may be made for subsidizing the domestic food supply but not for corn ethanol.[14] Famers will produce the target 9 billion gal of ethanol in 2009. Production will consume 40 million prime cropland acres, 2 trillion gal of freshwater, 5 billion gal of fossil diesel fuel, and millions pounds of agricultural chemicals. The 100 million tons of corn used for ethanol production will consume nonrenewable resources, pollute air, soils, and groundwater and create health problems for people and animals while replacing less than 3% of U.S. oil imports.[15] Removing food from the market drives up food prices in the United States and globally while also pushing up food prices for meat producers.

Those who debate peak oil miss a far more perilous concern, net zero oil exports. The problem for fossil fuel consumers occurs because the cash infusion from oil exports in Venezuela, for example, stimulates domestic consumption of government subsidized $0.19 a gallon gasoline. As reserves fall, oil prices rise, bringing in more cash and further increasing domestic consumption to the detriment of exports.

Geologists Jeffrey Brown and Samuel Foucher have modeled the net export problem and show that once oil production in an exporting country peaks and begins to decline, exports drop sharply.† Due to increased domestic demand, only about 10% of post-peak oil production is exported. Their most likely case scenario predicts that the top five oil producers will approach net zero exports around 2031. Net zero oil exports means there is no oil to buy. Another petroleum engineer, Jean Laherrère, assumed greater Saudi oil reserves and projected net zero exports by 2050.[16] Unfortunately, a decade before net zero

* http://farm.ewg.org/sites/farmbill2007/region1614.php?fips=00000 (accessed August 10, 2011).

† Brown, J. and Foucher, S. Peak oil versus peak exports, *Association for the Study of Peak Oil and Gas Conference*, 2010. http://aspousa.org/2010/10/peak-oil-versus-peak-exports/ (accessed September 12, 2012).

exports occur from Saudi Arabia, fossil fuels will be too expensive for industrial food production in many areas, especially high energy use sectors like the West and Southwest. The net zero export problem makes finding a food supply that produces no or minimal fossil fuels both mission-critical and urgent for the survival of human societies.

Fossil fuel challenges range far beyond concerns about supply and include subsidies, consumption, pollution, health impacts, prices, and supply chain. Each must be significantly reduced to avoid catastrophic outcomes. A solution to aim for is the development of a suite of green energy sources that are sustainable and nonpolluting while designing ways to significantly reduce energy consumption.

## 24.4 Soils and Soil Nutrient Losses

Globally, one-third of the prime cropland has been so degraded it had to be abandoned over the last 30 years.[17] An additional third is so degraded that farmers must use significantly more fertilizers to achieve normal yields. Farmers in the United States abandon cropland due to soil wear out, soil erosion, and salt invasion from irrigation and tidal surges. The West is especially vulnerable to irrigation salt build up because most croplands are irrigated and the heat evaporates the water, leaving salt.

Soil nutrients create a serious problem because they are ravenously consumed rather than conserved with industrial agriculture. Growing a food crop for one season removes about 50% of the applied soil nutrients which are lost to the field when the crop is harvested.[18] The field loses another 30% of the applied fertilizer to erosion from wind, rain, and irrigation. Therefore industrial agriculture forces framers to apply about 80% of the needed crop nutrients fresh to the field each year. Without nutrient replacement with fertilizer, the next crop lacks critical nutrients and production diminishes or fails. Farmers found that adding more fertilizer and irrigation, significantly increased production. However, mined (inorganic) fertilizers are inefficiently absorbed by plants so farmers must add substantially more fertilizer than the crop actually needs.

The Green Agricultural Revolution allowed short-term food productivity gains by mining trillions of gallons of fossil water and substituting inorganic fertilizers for organic nutrients. However, this fossil food strategy is sustainable only until about a decade before the first of the Magic 21 fossil resources needed for industrial agriculture runs out (Table 24.1).[19] Unfortunately, several fossil resources will run out or become unaffordable before our children reach midlife.[20]

Phosphorus is often a limiting nutrient in natural ecosystem because the supply of available phosphorus constrains the ecosystem size. Food and biofuel production depend on substantial phosphorus fertilizer which must be mined, packaged, transported long distances, and stored before application on fields. World governments are beginning to recognize the strategic value of phosphorus which is also used in the food, munitions, and chemical industries. India is running low on matches and fireworks as factories run short of phosphorus. The Brazilian government is debating whether to nationalize privately held mines that supply the fertilizer industry. In 2009, Beijing imposed a 170% tariff on phosphate rock exports to try to secure enough for its own farmers.* The U.S. phosphorus

* http://www.icis.com/Articles/2009/02/01/9097380/china-delays-phosphate-export-duty-hike.html (accessed August 10, 2011).

**TABLE 24.1**

The Key 21 Fossil Natural Resources for Food Production

| Primary Inputs | Macronutrients | Micronutrients |
|---|---|---|
| Freshwater | Nitrogen—N | Magnesium—Mg |
| Fertile soil | Phosphorus—P | Boron—B |
| Fossil fuels | Potassium—K | Copper—Cu |
| Fine seeds | Calcium—Ca | Chorine—Cl |
| | Carbon—C | Iron—Fe |
| | Oxygen—O | Molybdenum—Mo |
| | Hydrogen—H | Manganese—Mn |
| | Sulfur—S | Nickel—Ni |
| | | Zinc—Zn |

production has dropped 20% over the last 3 years forcing phosphorus imports from Morocco.

Peak phosphorus use will occur before 2040, which will drive up prices dramatically on remaining stocks.[21] Economically recoverable reserves for several other plant nutrients are seriously constrained. Armin Reller has been investigating world supplies of metals and estimates that the world will run out of copper in 25 years and zinc in 20–30 years.[22] Plants need zinc to propagate which is critical for people and animals that depend on the seeds of plants (the fruit of the vine), for food.

Inorganic soil nutrients dissolve quickly in water which makes them ideal for plant absorption through the roots. The downside of high solubility is that these nutrients are easily rinsed out of topsoil by rain or irrigation water. Therefore, farmers must add substantial additional fertilizer every year which drives up costs. The high solubility also means these agricultural chemicals find their way into well water, wetlands, and estuaries.

Soil degradation and nutrient loss reduces yields for years before the land must be abandoned. Global grain production continues to increase but yield increases are decreasing each decade (Figure 24.3), largely due to soil degradation.*

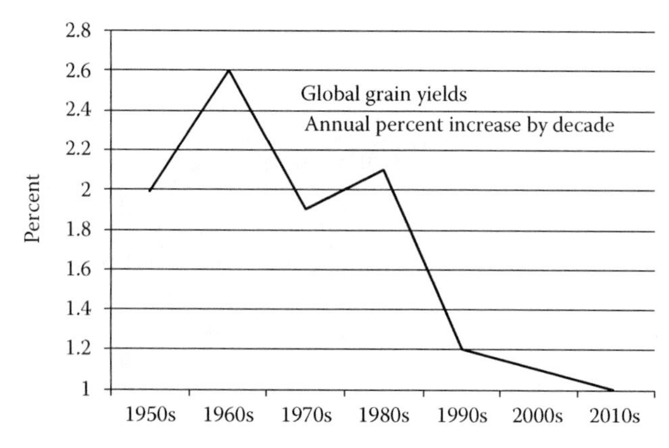


**FIGURE 24.3**
Increase by decade global grain yields—annual percent.

* http://www.who.int/whr/2004/annex/topic/en.annex_2_en.pdf (accessed October 1, 2004).

Farmers typically cultivate fields to remove weeds and apply fertilizer with planted seeds in order to maximize germination and early growth. Cultivation and planting occurs in the spring when strong winds are common. Soil organics near the surface that have absorbed the applied agricultural chemicals are carried from rural to urban areas where pollution impacts large populations.

A new ecological design for food crops is needed, conserved, and reused rather than consuming fossil natural resources. The ecology of this new food source would provide substantial benefit if it did not require fertile soils, freshwater, and fossil fuels and did not generate agricultural pollution.

## 24.5 Climate Change

Many scientists, including James Hansen at NASA, believe that global warming is accelerating and may be approaching a tipping point where climate change acquires a momentum that makes it irreversible. The consensus is that we may have a decade to turn the situation around before this threshold is crossed.[23]

Global warming that included drought, wildfires, and fierce storms was largely responsible for food price spikes in 2008 which caused food riots in 40 countries. These insurrections disrupted national economies, spurred food theft, and resulted in hundreds of deaths. Several countries created policies that prohibited food hoarding, waste, and even exports.

Secretary of Energy and Nobel-Prize-winning physicist Steven Chu said in his first interview as secretary that California's farms and vineyards could vanish by the end of the century and its major cities could be in jeopardy if Americans do not act to slow the advance of global warming.[24] He also predicted that 90% of the Sierra Nevada snow pack on which California cities and agriculture depends would be gone by the end of the century.

More heat will compound food insecurity caused by variable rainfall and will increase the incidence of agricultural droughts caused by elevated evaporation from soils, transpiration from plants, low soil moisture, and high rates of water runoff from hard pan soils when it rains. Excess heat causes virga—rain that evaporates before it hits the ground. The challenges to food and biofuel production from global warming are summarized in Table 24.2.

A new ecological design is needed for production of food and biofuels that is effective in spite of climate chaos.

## 24.6 Addressing the Goal of Abundant Agriculture

Green solar energy captured in algae represent an agriculture of abundance based on cheap natural resources that will not run out, sunshine, wastewater, and $CO_2$ (Figure 24.4). Green solar employs photosynthesis to store solar energy in carbon molecules in algal plant chemical bonds. Depending on the species, algae may be 60% by weight food energy for people, nutrient energy for plants, or fuel energy for vehicles. Green solar has no growing season since it grows so fast that about half of the biomass may be harvested daily. Most species have a preferred production period which typically corresponds to land plants but

**TABLE 24.2**

Climate Chaos Impacts on Food and Biofuel Production

| Factor | Description |
|---|---|
| Heat | Increased temperatures cause heat stress in food crops which can significantly diminish their productivity and lead to plant death and crop failure. |
| Hot winds | Increased temperatures and dry winds evaporate soil moisture and increase the need for freshwater irrigation. |
| Water scarcity | Water, the critical resource for sustainable food production, has passed its tipping point as global warming causes food crops to need more water but water in many growing areas' water sources have been degraded, depleted, or diverted. |
| Rising sea levels | Oceans will consume millions of acres of prime cropland on coasts and river deltas and tidal and storm surges will destroy millions of acres of cropland from sea salt invasion. |
| Ocean acidity | Dissolved $CO_2$ in the oceans diminish fisheries, destroy shellfish, and dissolve coral reefs that protect coasts and estuaries. |
| Higher ocean surface temperatures | Heat creates the energy that intensifies storms, hurricanes, and typhoons. Heat also changes the rainfall patterns and leads to drought and severe forest fires experienced in the western United States. |
| Extended spring and fall | Spring is starting a week earlier and fall lasts an extra week, enabling pest vectors—bugs, fungi, molds, mildews, viruses, and weeds—to multiply earlier and sometimes survive the winter. |
| Rain patterns | Shifts in rain patterns will cause huge losses of cropland that lack the infrastructure for irrigation. |
| Wildfires | Range lands and forests are especially vulnerable to heat and drought, and winds drive catastrophic wildfires such as those in California in 2008 and 2009. |
| Loss of snow pack and glaciers | Snow packs are down 50% which means faster runoff and heavy flooding in the spring. Reservoirs, creeks, and rivers may be only half full when irrigation is needed later in the growing season. Melting snow packs and glaciers mean less river water for irrigation and human use. California announced a net zero irrigation water allocation for many farmers in the San Joaquin Valley. |
| Blowing dust | While the U.S. Midwest experienced severe flooding in 2008, Texas and Oklahoma lost millions of acres of crops to drought and blowing dust. Dust decimates crops, amplifies drought by removing soil moisture, and erodes thin topsoil. |

they can be cultivated year-round. Algaculture produces pure $O_2$ while sequestering $CO_2$, so it provides a positive atmospheric footprint while conserving scarce fossil resources.

Green solar provides a portable energy source and grows biomass with solar energy stored in forms that may be used for a variety of purposes:

*People*—organic protein in food

*Animals*—organic protein in fodder

*Fowl*—natural protein for birds

*Fish*—natural protein in fish feed

*Land plants*—organic nitrogen fertilizer

*Fire*—high energy algal oil for cooking and heating

*Cars*—carbohydrates refined to gasoline/ethanol for transportation

*Trucks and tractors*—high energy clean, green diesel

*Trains, boats, and ships*—high energy clean diesel

*Planes*—high energy, clean aviation gas and jet fuel

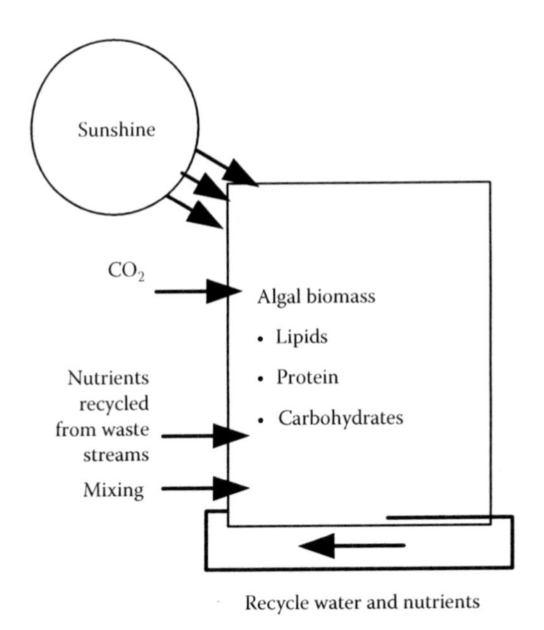


**FIGURE 24.4**
Abundant agriculture—algae cultivation.

Algae evolved in millions of independent moist environments which created over 75,000 known species and possibly 10 million total species.[25] The plant grows in nearly every ecosystem and offers a wide variety of nutrient profiles for its many hungry consumers. Algae are at the base of the food chain and serve as food for 100 times more organisms than any other plant on Earth. Multiple algal species have adapted to every known ecosystem on Earth but many prefer heat, sunshine, and brackish water.

Algae are far more productive than other biofuels sources because algae do not put energy into producing roots, stems, trunks, and leaves because they grow in water. Algae are energy positive because the energy cost of algaculture and downstream processing is less than the energy yield of the algae oil, which is produced using solar energy in the process of photosynthesis. The energy produced, clean, green diesel or jet fuel, provides about 30% more energy than gasoline and about 50% more than ethanol.[26]

Algae's energy potential is 30–100 times higher than corn ethanol production per acre and algae's productivity advantage for protein is similar. Other parameters such as coproducts, growing requirements, and ecological footprint may be even more critical to the choice of changing to sustainable algaculture than oil productivity differences. Algae's potential remains theoretical because scaled production has not yet been achieved. However, significant production breakthroughs are occurring now.[27]

## 24.7 Green Solar Geography

The ecology of the West and Southwest are ideal for algal production (Table 24.3). These regions offer the unique combination of sunshine, warm weather with few frosts, and low-cost flat, noncropland. The Southwest has numerous brine aquifers with water that cannot support agriculture but are ideal for algae production.

**TABLE 24.3**

Southwest Environmental Qualities

| Need | Description | West and SW |
|---|---|---|
| Sunny days | The more the better because algae grow slower on cloudy days | 360 days |
| Temperature | 60°F–110°F | 350 days |
| Few frost days | Altitude <1000 ft to minimize frost | <1000 ft<br>5 days of frost |
| Flat, cheap land | Noncropland, undeveloped | Hundreds of square miles of desert |
| Waste or brine water | Algae get nutrients from waste, brine, lake, or ocean water | Oceans of brine water |

Algae are sustainable because growth requires only a tiny fraction of the fossil inputs—energy, water, land, fertilizers, herbicides, and pesticides—required for land-based plants like corn, citrus, cotton, or cattle. Algae can remediate the nitrogen, phosphorus, and other pollution from agriculture in groundwater and wastewater, conserve those polluting nutrients, and reuse them to produced food and energy. Algae production is ecologically positive because it has minimal input needs and, in closed biofactories where no or few waste products are produced to leach into the soil, float on the wind or fill waste dumps.

## 24.8 Green Solar Value Chain

When the challenges of algal production are resolved, hungry, thirsty, and cold people may share in algae's green promise for sustainable and affordable food and energy (SAFE) production. Algae are uniquely positioned to provide a value chain of products and solutions for critical human needs (Figure 24.5). The value chain includes sustainable and affordable foods for people, fish, fowl, and animals. Algae can provide liquid transportation fuels such as green diesel and jet fuels that displace fossil fuels and enable communities and countries to become oil independent.[28]

Algae oil burns cleanly with little black smoke particulates because it is a vegetable oil. Algae grown locally for clean burning oil for cooking and heating fires can end smoke death and disability for the millions of mothers and children who inhale black smoke particulates as they cook over wood, coal, or dung.[29]

Algae's ability to clean waste and brine water offer a high-value solution for the water-starved West and Southwest. Israel, which also has arid lands, currently recycles 87% of its municipal water through algal ponds to clean the water.* Coal fire power plants and industrial manufacturers can flue their chimney smoke through algal ponds to sequester the $CO_2$ and heavy metals, which will improve the quality of living for urban dwellers downwind from smoke plumes.

Algal biomass production can provide not only valuable freshwater but fodder for grazing animals. Algal fodder can reduce overgrazing, which can save grasslands and forests from being denuded. Algal carbohydrates can be used to produce paper which can save forests or made into biodegradable plastic that does not fill waste dumps.[29]

* Personal interview with Israeli algae expert, Professor Amos Richmond, April 2008.

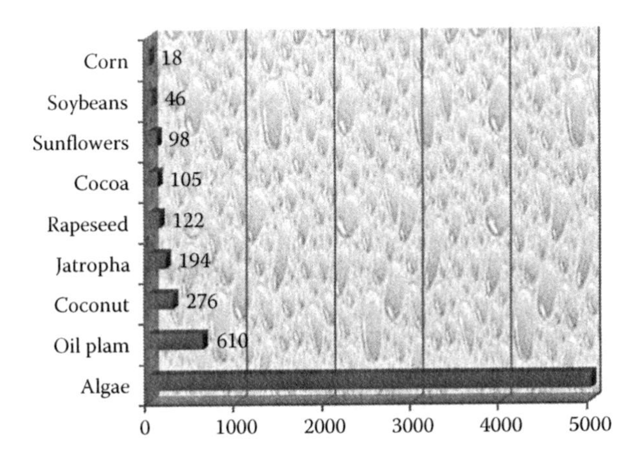


**FIGURE 24.5**
Oil production potential—gallons per acre per year.

Many farmers cannot afford expensive fossil fertilizers. Dairy and livestock farmers can use green solar to recover and recycle 60% of the energy of the original plants that lies in the animal manure. The same process can recover 80% of the original plant nutrients, and can provide rich fertilizer without the use of fossil fuels that farmers can use directly on their land. Algal fertilizer provides rich organic matter that builds up humus, aids in water retention, and protects soil from erosion.

Algal biomass is not a full solution to hunger because the plant is low in calories, at about 2%. However, saving forests from grazing animals will allow communities to plant legumes and nut trees that can provide needed calories.

Knowledge and capability transfer for SAFE production will enable communities to end their dependence on long distance food transportation. Teaching people to grow their own food and energy near urban centers makes a lot more sense than creating dependence on food from distant sources. The new urban design may also enable people to practice SAFE production in urban centers, on rooftops, balconies, and vacant lots. As more green solar producers gain experience, many new innovative products and solutions will appear (Figure 24.6).

Nature's first food production system, algaculture, offers a wide range of potential benefits. The revenue generated from an algae industry could exceed the other bioscience niches. It is too early to predict which algae products will produce the most revenue but several appear very promising, including

*Liquefied energy*—biodiesel, jet fuel, ethanol, or methanol

*Foods*—high protein replacement for grains such as wheat, corn, and soybeans

*Health foods*—Spirulina, vitamins, special nutrients, and minerals

*Animal foods*—high protein food grains that match the nutrient needs of beef, dairy, poultry, and aquaculture

*Medicines*—nutraceuticals, vaccines, pharmaceuticals, and high-value medicines

The algae industry business models are very attractive because with relatively modest investments, high value products can be produced that can be sold for substantial profits. However, the West has seen failed attempts before at building industries

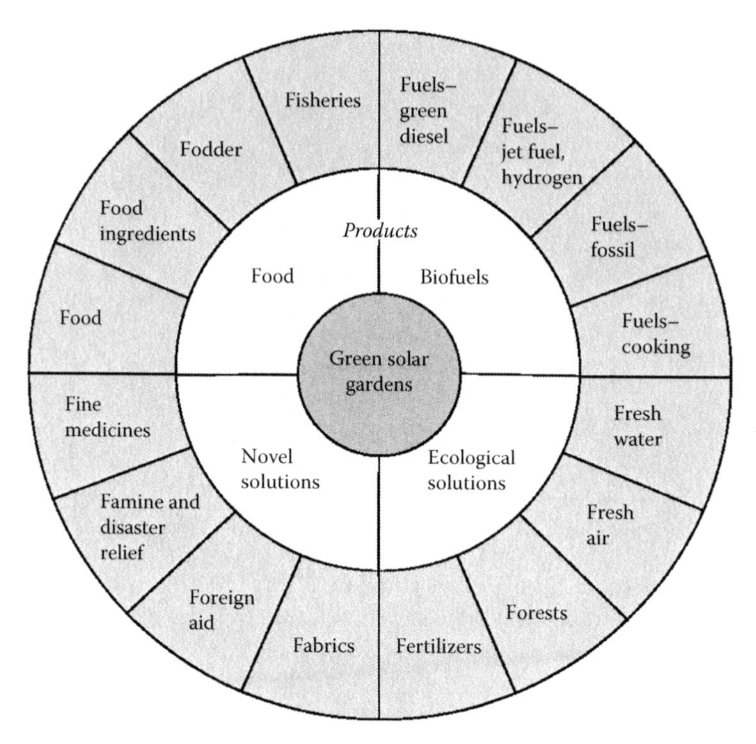


**FIGURE 24.6**
Green solar products and solutions.

around new crops such as guayule, a plant that produces natural rubber, and jojoba, with a seed than produces oil. Similarly, early attempts at new growing systems such as hydroponics failed to live up to their hype. However, none of these new products offered a strong competitive advantage or an expansive product mix. Algae offer an intriguing ecological design that meets and aligns with the conservation needs of the West and Southwest.

## 24.9 Algae Production Aligning with Ecological Design

David Orr, in his article in this book on ecological design, suggests mimicking nature on six dimensions in order to craft successful ecological models. Table 24.4 illustrates how algae could provide an effective model.

Algae's adaptability make in sufficiently malleable to fit numerous ecological design variations based on situational need such as access to waste $CO_2$, wastewater, or other needs. In the near future, every wastewater treatment facility, dairy, power plant, and other air, soil, and water polluter may build algal production systems to turn their waste streams into revenue-producing products. Transforming waste streams to food and energy near urban centers will reduce transportation costs significantly while assuring a reliable and affordable source of food.

**TABLE 24.4**

Algae and Ecological Design

| Characteristic | How Green Solar Energy Aligns with Ecological Design |
|---|---|
| Farms that work like forests and prairies | Algae sequester carbon and produces $O_2$. Each pound of algae captures two pounds of $CO_2$. Algae produce about 60% of the Earth's $O_2$ daily, more than all the forests and fields combined. |
| Buildings that accrue natural capital like trees | Algal production on rooftops, balconies, and the sides of buildings produces food and energy plus valuable coproducts that create natural capital. |
| Wastewater systems that work like natural wetlands | Algae can remediate industrial, human, animal, and other wastewater faster, more completely and more reliably than natural wetlands. Harvesting the algae enables the recycle and reuse of valuable energy, nutrients, and chemicals that were lost in the waste stream. Algae bioaccumulate the diluted elements and chemicals into a solid form which makes possible recovery that is not economically feasible in the liquid waste stream. |
| Materials that mimic the ingenuity of plants and animals | Algae have been adapting for over 3.5 billion years and have all of the construction elements necessary to build any plant material. Land plants evolved from green algae about 400 million years ago. |
| Industries that work more like ecosystems | The algal industry can provide SAFE while creating positive impacts on the environment, including cleaning air, soils, and water. |
| Products to become part of cycles resembling natural materials flows | Algal production requires only plentiful and cheap natural resources that will not run out including sunshine, $CO_2$, and wastewater (or brine water to supply nutrients). Algae can transform the natural carbon cycle and sequester carbon from industrial sources for 5000 years. |

## 24.10 Conclusion

Urban living design must adapt to the nontrivial challenges of insufficient freshwater, fossil fuels, and climate chaos. Scarcity of resource flows impacts urban dwellers in many ways and puts the affordability and availability of food, energy, and transportation in jeopardy. Algae have the unique ability to grow SAFE production that can assure vitality to urban centers.

Our collaborative task is simple: design, develop, demonstrate, and diffuse SAFE production systems.

## References

1. Lewis, M., 21st century cities: Megacities of the future, *Forbes*, June 11, 2007.
2. Pimentel, D. and M. Pimentel, *Food, Energy and Society*, 3rd edn. (New York: CRC Press, 2007).
3. USGS, Ground-water depletion across the nation (U.S. Geological Survey Fact Sheet 103-03, November 3, 2003).
4. Gorman, S., California farms lose main water source to drought, *Reuters*, February 20, 2009, p. 1.
5. Brown, E. and N. Elliot, *On-Farm Energy Use Characterizations* (Washington, DC: American Council for an Energy-Efficient Economy, 2005), p. 4.

6. Pimentel, D. and T. Patzek, Ethanol production using corn, switchgrass, and wood: Biodiesel production using soybean and sunflower, *Natural Resources Research*, 14: 65–76, 2005.
7. Halweil, B., Pesticide-resistance species flourish, in L.R. Brown, ed., *Vital Signs 1999* (New York: W.W. Norton, 1999), p. 24.
8. Beeman, P., State will spend millions to improve water quality, *Des Moines Register*, May 10, 2009, A1.
9. Gilliom, R.J., Barbash, J.E., Crawford, C.G., Hamilton, P.A., Martin, J.D., Nakagaki, N., Nowell, L.H., Scott, J.C., Stackelberg, P.E., Thelin, G.P., and Wolock, D.M., 2006, *Pesticides in the Nation's Streams and Ground Water*, 1992–2001 (U.S. Geological Survey Circular 1291), p. 27; *Quality of Our Nation's Water* (Washington, DC: Environmental Protection Agency, 1994).
10. Tamminen, T., *Lives per Gallon: The True Cost of Our Oil Addiction* (Washington, DC: Island Press, 2006), p. 60.
11. Tamminen, T., *Lives per Gallon: The True Cost of Our Oil Addiction* (Washington, DC: Island Press, 2006), p. 62.
12. USDA, *U.S. Agriculture and Forestry Greenhouse Gas Inventory* 1990–2005, p. 81.
13. USDA, *U.S. Agriculture and Forestry Greenhouse Gas Inventory* 1990–2005, p. 86.
14. Edwards, M., *BioWar I: Why Battles over Food and Fuel Lead to World Hunger* (Tempe, AZ: CreateSpace, 2007), p. 43.
15. Edwards, M., *Green Algae Strategy: End Oil Imports and Engineer Sustainable Food and Fuel* (Tempe, AZ: CreateSpace, 2008), p. 93.
16. Nedler, C., Oil exports may soon dry up, *The Futurist*, 43:2, March–April 6, 2009, 23–26.
17. Pimentel, D. and M. Pimentel, *Food, Energy and Society*, 3rd edn. (New York: CRC Press, 2008), p. 27.
18. Natural Resource Conservation Center, *Model Simulation of Soil Loss, Nutrient Loss and Soil Organic Carbon Associated with Crop Production* (Washington, DC: U.S. Department of Agriculture, June 2006), p. 104.
19. Edwards, M., *Crash! The Demise of Fossil Foods and the Rise of Abundance* (Tempe, AZ: CreateSpace, 2009), p. 2.
20. Edwards, M., *Crash! The Demise of Fossil Foods and the Rise of Abundance* (Tempe, AZ: CreateSpace, 2009), p. 16.
21. Cordell, D., J. O. Drangert, and S. White, The story of phosphorus: Global food security and food for thought, *Global Environmental Change*, 19(2): 292–305, 2009.
22. Cohen, D., Earth's natural wealth: An audit, *New Scientist* 23: 4–41, 2007.
23. Hansen, J., M. Sato, P. Kharech, G. Russell, D.W. Lea, and M. Siddall, Climate change and trace gases, *Philosophical Transactions of the Royal Society A* 365: 1925–1954, 2007.
24. Tankersley, J., California farms, vineyards in peril from warming, U.S. energy secretary warns, *Los Angeles Times*, February 4, 2009, p. C1.
25. Graham, L. and L. Wilcox, *Algae: Biology and Biotechnology* (Upper Saddle River, NJ: Prentice Hall, 2000), p. 8.
26. Kurki, A., A. Hill, and M. Morris, Biodiesel: The sustainability dimensions, National Sustainable Agricultural Service, IP281, 2006.
27. Edwards, M.R., Who is producing Algae? *Green Algae Strategy: End Oil Imports and Engineer Sustainable Food and Fuels* (Tempe, AZ: Independent Publisher Book Awards, IPPY, 2008), pp. 163–187.
28. Edwards, M.R., *Green Algae Strategy: End Oil Imports and Engineer Sustainable Food and Fuels* (Tempe, AZ: Independent Publisher Book Awards, IPPY, 2008), p. 172.
29. Edwards, M., *Green Solar Gardens: Algae's Promise to End Hunger* (Tempe, AZ: CreateSpace, 2009), p. 54.

# 25

## *Desert Urbanism**

**Nan Ellin**

**CONTENTS**



### 25.1 Introduction

A house where I once lived in Cincinnati came with a small grape ivy plant in the hall bathroom. I adopted the plant and watered it regularly. But oddly, it never grew. It did not die, but during the 2 years I lived in the house, it never sprouted a leaf. While residing in Phoenix, Arizona, I have been reminded often of that grape ivy. Leaving it behind for the next inhabitants, it became emblematic for me of so many North American cities that, although may be surviving, are clearly not thriving.

Just as we are a part of nature, so are our habitats, including our cities. Many of our expressions implicitly acknowledge this organic quality of places. For instance, we typically describe a dull place as "lacking character" in contrast to a "lively" place. The French describe the dull place as lacking soul (*Il n'a pas d'âme*) and the lively one as *"animé"* (animated, spirited, or soulful). Over the last half century, however, urban development has treated the city as a machine for efficiently sheltering and protecting, as well as moving people, money, and goods. However, these well-intentioned efforts to cleanse the city of illness and to render it more efficient have gone too far. Globalization and attendant standardization have been endangering the soul and character of our landscapes and our selves, as manifest in sprawl, the growing perception of fear, a declining sense of community, and environmental degradation.

While this downward spiral continues, it has been countered in recent years by a marked upward spiral. Indeed, a quiet revolution has been underway in urban design and planning, born of a frustration with reactive and escapist trends, aiming to heal the wounds inflicted upon the landscape over the last century. This revolution is quiet because its practitioners are not united under a single banner and because its sensitivity to people and the environment translates into interventions which may not call attention

* Adapted in part from Ellin, N., *Intregal Urbanism* (New York: Routledge, 2006).

to themselves. Nonetheless, numerous stones have been thrown around the globe over the last decade, and their small-but-growing ripples are beginning to dramatically reshape our physical environment while enhancing life quality.

## 25.2 Integral Urbanism

I have described this body of proactive urban design and planning as "integral urbanism."[1] Integral urbanism seeks to redress dispersal and fragmentation by recovering earlier city-building wisdom while also accommodating contemporary technologies and lifestyles. Rather than free-standing single-use buildings connected by freeways along with rampant (sub)urban sprawl which separates, isolates, alienates, and retreats, integral urbanism emphasizes connection, communication, and celebration.

Integral urbanism is characterized by five qualities: *hybridity, connectivity, porosity, authenticity,* and *vulnerability.* While modern urbanism espoused the separation of functions, integral urbanism reaffirms their symbiotic nature by bringing activities and people together at all scales. These various integrations can be accomplished through cross-programming buildings and regional plans—spatially (plan and section) as well as temporally. Examples of cross-programming include the office building with basketball court and daycare center, the community center and library, the intergenerational community building (combining day care, teenage community center, adult education, and seniors center), the public school/community center, the integrated parking structure (into office, residential, and office buildings), the movie theatre/restaurant, and the urban plaza by day/movie theatre at night. When successful, the efficiencies allowed by these integrations conserve energy and other resources while decreasing social isolation, thereby empowering people to envision alternatives and implement change most responsively and creatively (Figure 25.1).

In contrast to the modern attempt to eliminate boundaries and the postmodern tendency to ignore or alternatively fortify them, integral urbanism seeks to generate porous membranes. These membranes might also be described as thresholds, or places of intensity. By allowing for diversity (of people and activities) to flourish, this approach seeks to reintegrate (or integrate anew) without obliterating differences, in fact, preserving and celebrating them. As such, it might be regarded as a form of "urban acupuncture" that clears blockages and liberates *chi,* or the life force, supporting urban and economic revitalization. Applied to existing built environments as well as new development, these interventions may have a tentacular or domino effect by catalyzing other transformations (Figure 25.2).

In sum, integral urbanism emphasizes networks, relationships, connections, interdependence, communities, translucency or transparency, permeability, flux, flow, mobility, catalysts, frameworks, and process rather than boundaries, independent objects, individuals, walls, permanence, final products, master plans, and utopias. Practicing integral urbanism entails: (1) integrating parts of the city that fragmented over the last century: live, work, create, and recreate; (2) identifying what is integral to a place, its DNA, and building upon these assets, rather than focusing on deficits and problems; and (3) engaging in urban acupuncture by removing blockages along "urban meridians," just as acupuncture eliminates blockages along the energy meridians of a body. This practice emphasizes the basic theories of landscape ecology, with a larger focus on patterns of

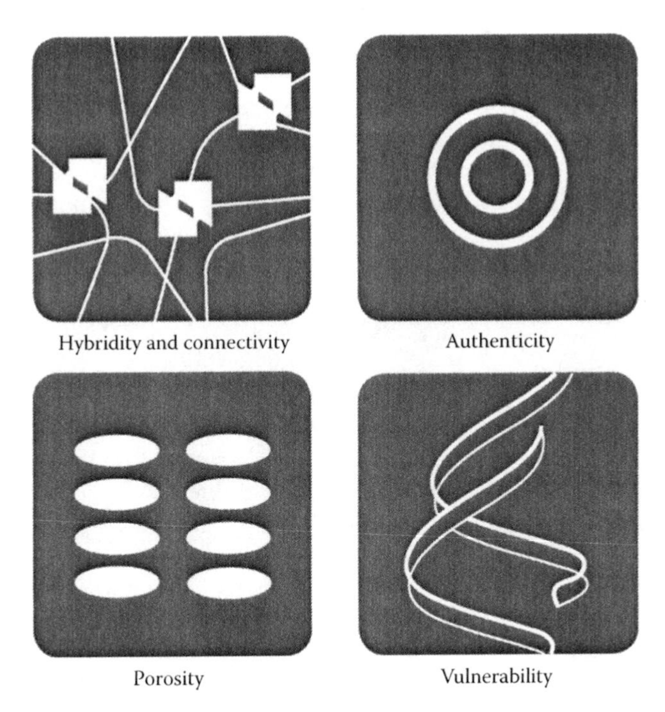


**FIGURE 25.1**
Integral urbanism qualities. (Coutesy of Barbara Ambach.)

**FIGURE 25.2**
Vitality in Phoenix.

corridors and patches of a region, and their critical linkages in the underlying matrix of a system, such as an urban environment.[2]

Political and economic trends supporting the practice of integral urbanism include widespread opposition to urban sprawl, interest in conserving the environment and preserving historic urban fabrics, the rise of regional governments, the renaissance of central cities, the exponential growth of neighborhood associations and community gardens, the establishment of community land trusts, and transformations wrought by the new economy (e-commerce, partnering, and technological convergences). Social trends reflect and reinforce the political and economic ones, expressing a frustration with the fragmented landscapes produced by conventional urban development and a craving for the excitement, spontaneity, and sense of flow characteristic of truly urban places.

## 25.3 Learning from Mistakes

In Western society, generally, we have been witnessing a gradual reorientation toward valuing slowness, simplicity, sincerity, spirituality, and sustainability in an attempt to restore the connections that have been severed over the last century between body and soul, people and nature, and among people. If the 1960s witnessed the "*We* generation" calling for peace and love; the 1970s the "*Me* generation" with a focus on self-awareness and self-actualization; the 1980s the "*Whee* generation" characterized by materialism and escapism; the 1990s the "*Whoa* generation," placing a self-imposed brake upon the rapid changes that were wreaking havoc upon our landscapes and our well-being; then perhaps the new millennium has been spawning a re-generation, with a clear-eyed vision and the courage to rebuild our towns and cities, revitalize our communities, restore what has been taken from the earth, and realign design with the goal of supporting humanity.

For architects and planners, this has been apparent in the shift away from using the machine as a model for buildings and cities to seeking models simultaneously in ecology and new information technologies (e.g., thresholds, ecotones, tentacles, rhizomes, webs, networks, the World Wide Web, the Internet). In contrast to earlier models that bespoke aspirations for control and perfection, these current models suggest the importance of connectedness and dynamism as well as the principle of complementarity. Rather than seeking the norm, average, or center, there is clearly a fascination among urban designers with what happens along the border, the edge, and the in-between. There is interest, for instance, in the ecological threshold where two ecosystems meet, where competition and conflict coexist with synergy and excitement, if it is thriving. At the same time, the attitude among urban designers toward rapid change has been shifting. From attempting to deny or control change, an attitude characterizing most of the twentieth century, we are now witnessing an acceptance or even embracing of change.

The earnest but ultimately misguided modernist dictum that form follows function was largely supplanted by the deeply cynical late twentieth-century tendency for form to follow fiction, finesse, finance, and foremost fear.[3] In the promising approaches we are currently seeing, form is once again following function, but function is redefined. Rather than primarily mechanistic and instrumental, function is understood more holistically to include emotional, symbolic, and spiritual functions. From "less is more" (modernism) to "more is more" (postmodernism), the byword has become "more from less." Some implications of these trends for urban design include greater emphasis on mass transit along with transit-oriented developments, quality public spaces, urban infill, and mixed-use development.

One byproduct of the early twentieth-century quest for efficiency was zoning. Introduced one century ago, when the car was first mass-produced and -consumed, zoning segregated functions that had been integrated from time immemorial. As people are mutually interdependent, however, so are our activities as expressed in city form. And cities only thrive

**FIGURE 25.3**
Sprawl in Phoenix. (Photo courtesy of Tomoko Yoneda.)

(are only sustainable) when these interdependencies are allowed to flourish. We are now belatedly recognizing the problems wrought by zoning and the need for reintegrating fragmented urban areas.

As a number of recent studies demonstrate, sprawl takes an enormous toll on our physical and mental health contributing to automobile fatalities, obesity, asthma, workplace woes (and decreased productivity), and suburban sorrows (boredom and isolation). Research also highlights the importance of urban vitality for economic vitality since corporate headquarters and young talented people choose to locate in the most vibrant cities (Figure 25.3).[4]

## 25.4 Retrofitting Metro Phoenix

Phoenix, Arizona epitomizes urban sprawl. Currently the sixth largest city in the United States, Phoenix is growing the fastest of the 15 largest cities and expected to be the third largest city by 2020. The urban form of the Phoenix metropolitan area rarely betrays its desert setting because virtually 90% of it was constructed since the World War II, a period when local topography, climate, culture, and history played little role in shaping cities. This period additionally privileged moving cars over moving people, making the Phoenix metropolitan area larger than the Los Angeles metropolitan area as well as seven states, including Massachusetts. As sprawl replaced natural desert and agricultural lands with highways, suburban tract housing, and shopping malls, the nascent central cities of this polycentric metropolis were largely abandoned.

Phoenix is paying for its youth, then, by lacking the built-in strategies of older and wiser desert cities that protect pedestrians from the blazing sun such as compact urban cores with a range of passive cooling devices. Instead, the settlement pattern virtually prohibits walking while contributing to social isolation and environmental degradation. The highly publicized "heat-island effect," the result of substituting asphalt for biotic landscapes, intensifies the magnitude of this problem, as well as the need for remedying it.

The pendulum of city-building logic has been swinging back over the last decade to privilege human comfort and well-being along with the creation of pedestrian-friendly environments. But Phoenix inherits an urban legacy of misdirected decades prior. How can this urban region retrofit itself to offer a truly urban experience? How can this desert city protect and honor its distinctive landscape? How can this sprawling city become pedestrian-friendly *as well as* auto-friendly? How can this young city be authentic, true to its roots, have a sense of place, or a "there" there? And how can it gracefully accommodate people of different ethnicities, races, social classes, age groups, and abilities?

These questions are both logistical and existential. The ways they are resolved will determine whether and how Phoenix rises again. Phoenix will not accomplish these goals by continuing to build tract housing, freestanding office towers, and shopping malls surrounded by seas of asphalt parking. That is certain. But nor should it attempt to start over *tabula rasa* or emulate pre-automobile nineteenth-century cities.

A certain density is necessary to achieve an urban experience, but Phoenix need not emulate the high-density model of cities such as New York or Chicago. Although building and population density will need to intensify in certain pockets throughout the Phoenix metropolitan region, most important is enhancing "programmatic density," or the juxtaposition of diverse activities. Rather than separate living from working from recreating and so forth, the predominant pattern presently, these activities should occur in close proximity, albeit not necessarily in the same fashion they coexisted in the pre-automobile era. The outcome would be new hybrids that pool human and natural resources to the benefit of all, conserving energy, time, money, water, fuel, building materials, paper, and more. Such an integration of activities would reduce commuting, enhance convenience, preserve the natural environment, and greatly increase the amount of quality public space along with the opportunities for social interaction, spontaneous as well as planned.

Without reverting to a preindustrial model or continuing along the same bleak path, Phoenix could truly become a "Metroasis," where swaths of pristine Sonoran desert wind through low-lying neighborhoods surrounding a network of urban cores and corridors comprised of shops, restaurants, cafes, workplaces, cultural institutions, and urban housing. Many of these vital hubs of activity would appear where canals meet major streets, leveraging the region's 181 miles of canals, initially constructed by early inhabitants of the region 2 millennia ago and rebuilt during modern times (Figures 25.4 and 25.5).

A "canalscape" initiative has been working toward the creation of Metroasis through:

- Private investment: Creating vital hubs of urban activity where canals meet major streets.
- Public investment: Enhancing the common area alongside the canals to encourage alternative non-motorized transportation and to provide alternative energy sources (wind, hydro, solar, geothermal).[5]

The idea of canalscape in Phoenix supports enhancement of green infrastructure and greenways, a growing trend in many urban areas, and ultimately reinforces linkages, natural and created, in the larger urban matrix.

By introducing mixed-use urban infill along recreational and commuting corridors, canalscape would furnish a distinctive quality of life that combines urbanity with nature. Canalscape offers an alternative to sprawl, quality places to gather, beautiful and comfortable recreational opportunities, alternative transportation routes throughout the region (walking or biking instead of driving), and homegrown non-polluting energy for local use. Distributed

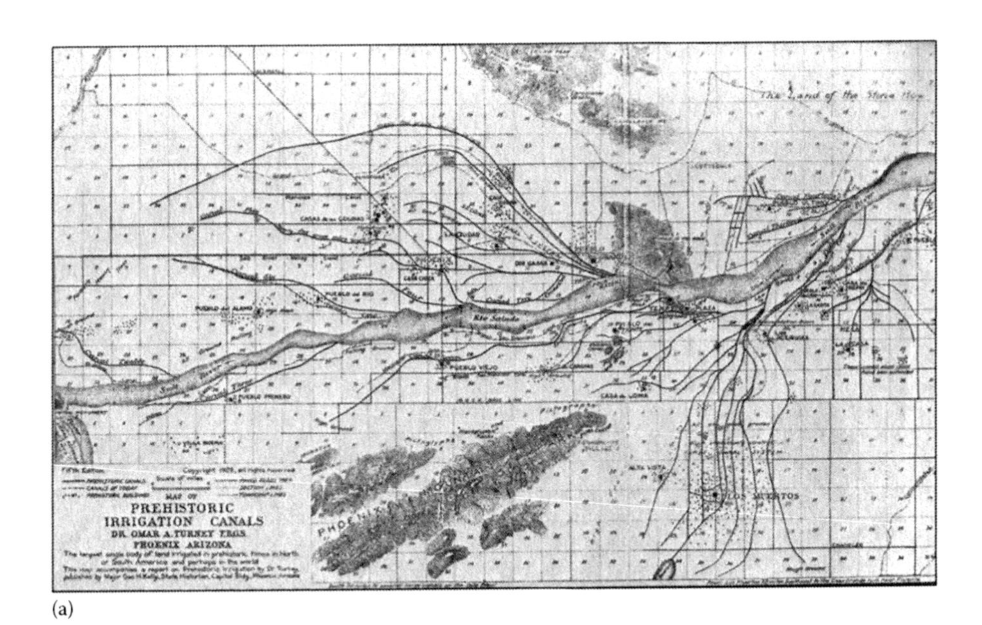


(a)

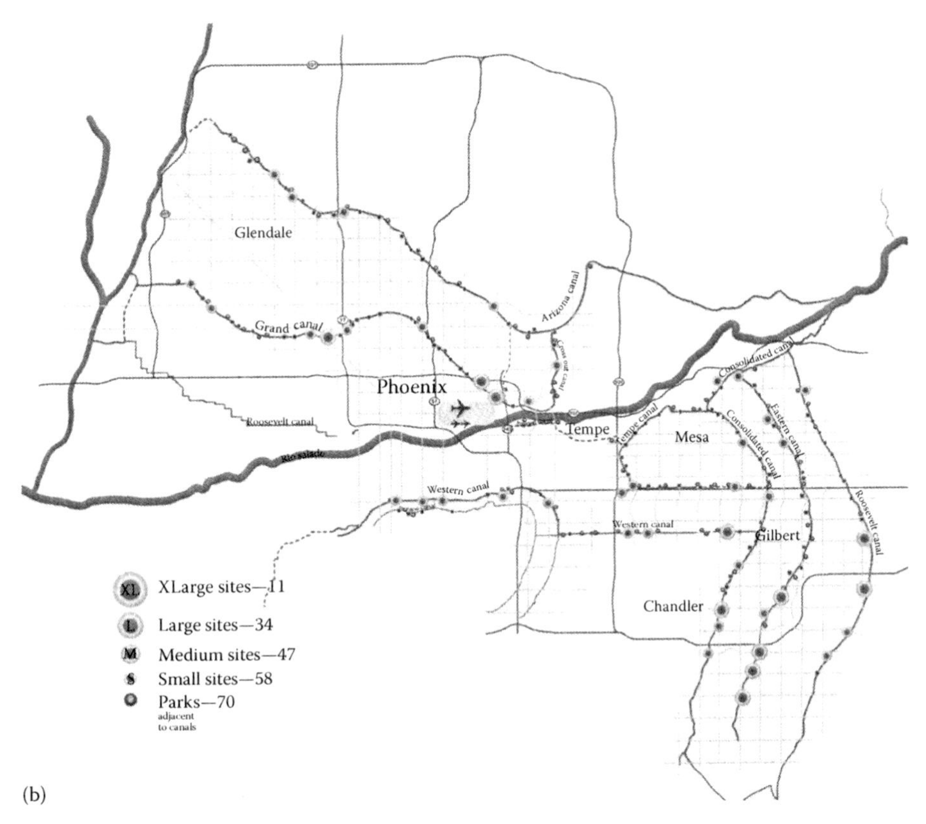


(b)

**FIGURE 25.4**
(a) Prehistoric canal system map. (Map courtesy of Omar A. Turney.) (b) Potential sites for canalscape. (Map courtesy of Francisco Cardona.)

canalscape

A Sustainable Desert Urbanism for Metro Phoenix

(a)

(b)

(c)

**FIGURE 25.5**
Canalscape—A sustainable and authentic urbanism. (a) Existing canalscape site. (b) Proposed canalscape design. (Courtesy of Jens Kolb.) (c) Conceptual image of the proposed Canalscape design. (Courtesy of Jens Kolb.)

along the urban energy meridians of the Phoenix region, canalscape would perform urban acupuncture, bringing health and well-being to the city and its inhabitants.

While the canals offer distinctive settings for vital urban hubs, these are also emerging at street intersections and along certain arterials, particularly along the new light-rail corridor. Throughout the metropolitan area, creative entrepreneurs with keen intuitions about what is right for here and now—not architects or planners—have been making indelible marks on the urban and cultural landscapes of the Phoenix area.

In a disinvested area of downtown Phoenix, for instance, former investment banker David Lacy purchased a two-storey building with the goal of opening a bakery on the ground floor and living above. Willo Bread opened its doors in 1999 and was embraced by the community for which it became a hearth writ large. After numerous requests from customers for a place to linger, Lacy purchased the space next door and opened an urbane café/restaurant My Florist in July 2001 which immediately became extremely popular both as a neighborhood restaurant and a popular gathering spot for the downtown business crowd at lunch and the opera, symphony, and theatre crowd in the evenings. Its landmark neon sign, dating from 1947 when it was indeed a floral shop, has been described as "a tower of flower power."[6] In need of more parking and wishing to improve the view across the street, Lacy leased a former Dolly Madison Bakery Building and arranged for an art gallery to locate there. He subsequently purchased adjacent properties to open a market, pastry shop, and more.

Soon after the opening of Willo Bread, photographer/developer Wayne Rainey purchased a 12,000 ft$^2$ warehouse and converted it into a diverse and thriving creative community, including two large shooting areas, an art gallery, a graphic design firm, a film/video company, an architectural firm, offices of an arts and culture magazine, and more. All share a conference room with a concrete tabletop salvaged from the bathroom floor, dressing rooms and makeup areas, high-speed Internet connections and workstations, and a full kitchen. Rainey selected the name monOrchid for this cooperative to suggest "many petals to make one flower." In deciding exactly which petals may join this collective enterprise, Rainey explains, "Latent potential is the descriptor we most look for in a project." Reflecting on the day-to-day workings of the monOrchid community, Rainey remarks, "It makes for an interesting life. There are creative projects conceived every day and the energy levels sustained are nothing less than phenomenal." We look at things from so many perspectives. Having gone through the process of growing such a business changed the way we look at our work too." In addition to monOrchid, Rainey also owns the nearby Holga's, a two-storey apartment building that he sandblasted and converted into live/work spaces for artists as well as a gallery. The Chinese characters prominently displayed on the building façade symbolize love, interdependence, and balance (Figure 25.6).

In a mid-twentieth-century suburban district northeast of downtown Phoenix, a bustling hub of activity has been sprouting over the last 6 years, thanks largely to the dedication, diligence, and vision of Craig and Chris De Marco. Straddling a small parking lot are the wine bar/restaurant Postino's, located in an old post office, and the market/bakery/café/pizzeria/flower shop La Grande Orange, both extremely popular among neighbors as well as people who travel long distances to enjoy the quality and character of these enterprises. La Grande Orange is designed by Chris de Marco along with architect Cathy Hayes (whose office is a few doors down) with custom display cases by Hayes as well as fixtures salvaged from a 1940s Los Angeles high school gymnasium. The bathrooms are outfitted with Philippe Starck appliances. Each entrepreneur owns her/his own business. "The whole thing," Craig explains, is "synergistic, based on creating a certain energy." De Marco describes this corner of Phoenix as a "constantly moving and breathing space" with the recently added restaurant Radio Milano, an anticipated taco stand along its east side, and more.

Numerous other creative entrepreneurs have been sprinkling the Phoenix region with pockets of soul and character providing unique combinations of coffee shops, restaurants, bookstores, bike stores, yoga studios, art supply stores, wine bars, boutiques, and more.

**FIGURE 25.6**
Phoenix art scene.

## 25.5 Slash City

I have described the attitude as well as outcomes embraced by canalscape as well as these creative entrepreneurs as *Slash City* (*/city*) because of the hybrid functions (this-slash-that) and the emphasis on the slash itself, on what happens where these functions meet.[1] The city celebrates the boundary, edge, or threshold as the place where people, things, and ideas converge. It does this by not only retaining the integrity of each activity and group of people (e.g., bakery, restaurant, residence, gallery), but also allowing for easy movement between them through the creation of porous "membranes": separators that are physically, visually, and/or symbolically permeable. This is an urbanism that refuses to stay within the lines, that seeps through (seeping of views, people, and activities, inside and outside).

Conventional big box stores, schools closed off from their surrounding communities, and gated communities are not porous. The */city* is. This approach and the landscape it generates reflect the complementary human urges to merge (connect) and to separate (for distinction, individuation), with the ongoing tension and dynamism these generate. In the process, we are also integrating—or slashing—the professions that divided and subdivided: architecture/planning/landscape architecture/engineering/interior design/industrial design/graphic design/fashion design/sculpture/painting/performance art/etc. We are perhaps "slashing" (in the sense of deconstructing) some existing structures of thinking, acting, and building for the sake of reconstructing. This reconstruction features permeable boundaries (the slash itself) that become thresholds of diversity: biodiversity, social and cultural diversity, artistic diversity, and commercial diversity.

The pop culture trend to *slash* television programs bears some interesting parallels with slash urbanism. Published in "slash zines" and on numerous websites, *slash* is a subgenre of *fanfic* which involves rewriting television programs through recontextualization (filling in the gaps between episodes), expanding a series timeline to past or future, refocusing (shifting attention from main characters to secondary ones), moral realignment (for instance, transforming villains into protagonists), genre shifting (for instance, converting a drama into a comedy), crossing over (combining programs), character dislocation (moving characters to another time and/or place), personalization (injecting oneself), emotional intensification, and eroticization. Applied to urbanism, we "slash" the city when we become actively involved with it, when we challenge convention by recombining elements in new ways, and when we activate places that have laid dormant, or manifest intensities in places where they have only been latent.

Neither idealizing the past nor escaping the present, the incipient canalscape and already impactful contributions of numerous creative entrepreneurs are generating a distinctive desert urbanism. They are making slits in the urban fabric, slashing Phoenix, in a fashion described by Ellen Dunham-Jones and June Williamson as "incremental metropolitanism," thereby placing a brake on the economic, cultural, and environmental devastation of sprawl. To the extent this upward spiral succeeds in countering the downward one, this could be a region whose majestic landscape joins an ancient civilization with contemporary urban sophistication to produce an unparalleled quality of life.[7]

## 25.6 The Beauty of the Child

Prometheus was bound to a rock for stealing fire from the gods and giving it to people. Adam and Eve were banished from the Garden of Eden for eating fruit from the Tree of Knowledge. And the Babylonians were forced to speak mutually unintelligible languages and scattered across the earth for attempting to build a tower to heaven and achieve notoriety. As allegories about our desire for knowledge, power, and control, these cautionary tales advocate against hubris and for humility. And they admonish against excessive rationality, invoking instead wonder, awe, mystery, and sanctity. They serve as reminders to acknowledge and celebrate our human qualities in contrast to the dual temptation to become god-like or machine-like, a temptation particularly endemic to architects and planners.

Listening to the minute differences noticed by my then 6 year old daughter Theodora (now 16) between her Star War Legos and the characters from the movie, I remarked that she remembered many more details from the movie than I did. Theodora responded, "Children who found the beauty of being a child when they were a child remember these things. I guess you didn't find the beauty of being a child when you were a child. Or you lost it." Struck by the painful recognition of my innocence lost, I asked, "Is there any way I could get it back?" She thought for a moment and replied, "If you lose the beauty of the child, it's very hard to get it back. You get it by playing a lot. Maybe if you have lots of fun as a grown-up, you can get it back." After a moment, she added, "And once you've got your child back, there's nothing that can stop you from doing anything" (Figure 25.7).

Just as the "beauty of the child" can get buried beneath the responsibilities of adulthood, so the vitality of a city—its soul and character—can disappear if squeezed into a rational and overly-prescribed master plan. Rather than throw any discipline or planning to the wind, perhaps we might rethink how and when to apply them. Keeping a place's "child" alive, or bringing it back to life, would not mean zero intervention but instead a gentle guidance that is responsive, flexible, playful, and nurturing, permitting self-realization.

(a)

(b)

**FIGURE 25.7**
(a) Downtown Phoenix 2002. (b) Downtown Phoenix 2007.

Practicing integral urbanism in Phoenix and other cities may offer the soul food necessary for revitalization, allowing them and us to blossom and truly thrive. Not merely survive.

## References

1. Ellin, N., *Integral Urbanism* (New York: Routledge, 2006).
2. Forman, R. and M. Godron, *Landscape Ecology* (New York: John Wiley & Sons, Inc., 1986).
3. Ellin, N., *Architecture of Fear* (Princeton, NJ: Princeton Architectural Press, 1997); Ellin, N., *Postmodern Urbanism* (Princeton, NJ: Princeton Architectural Press, 1999).
4. Florida, R., *The Rise of the Creative Class* (New York: Basic Books, 2002).
5. Ellin, N., *Canalscape*, 2009, Salt River Project (SRP).
6. Best of Phoenix, *New Times*, 1996.
7. Dunham-Jones, E. and J. Williamson, *Retrofitting Suburbia: Urban Design Solutions for Retrofitting Suburbia* (New York: John Wiley & Sons, 2008).

# Part V

# Urban Sustainability

Anthony C. Floyd

Urban desert settlements have a long and rich history. The very cradle of civilization was in the desert regions of the Middle East. The ancient urban centers of Mesopotamia, Egypt, the Roman Empire, and numerous Islamic empires provide time-tested examples of desert-adapted urban living. Narrow streets, courtyards, cross ventilation, daylight/glare control, and passive cooling systems all contributed to a pedestrian-scale, mixed-used, medium-density urban environment. Today, we know more about our planet, its regional ecosystems, and the environmental impacts of human activity. We are by far more technologically advanced, but recent history shows urban development is wreaking havoc on the natural world.

How can we partner with the desert in co-creating architecture and communities that are sensitive to the character of the place? Richard Malloy starts off this part with a chapter entitled "Settlement, Growth, and Water Security for Southwest Cities" (Chapter 26). This chapter provides a historical perspective on how the major southwest cities developed into the modern urban metropolitan centers they now have become. In addition, he discusses developments in establishing a secure water supply to provide sustainable futures for the growth and development of these cities. In Chapter 27, "Creating Tomorrow," Vernon D. Swaback identifies seven challenges toward ecologically based urban designs in the deserts of the southwest: (1) desert culture; (2) indigenous design; (3) regional open space; (4) complexity and integration; (5) effective transportation; (6) technology, awareness, and behavior; and (7) heroic design and commitment. With the lessons of the great southwest Native American settlements of the past, Swaback envisions future desert cities of the southwest that combine new technologies, diverse densities, and mixed-use developments with ecologically sensitive planning.

The City of Scottsdale has proven that it takes an informed citizenry coupled with long-term thinking to protect and enhance the desert in the midst of urban development. In Chapter 28, "Desert Vernacular: Green Building and Ecological Design in Scottsdale, Arizona," Anthony C. Floyd describes the role of Scottsdale's planning and building policies in creating environmental-sensitive urban development in the Sonoran Desert of central Arizona. From its early stormwater management practices and hillside ordinance to its environmentally sensitive land ordinance and green building program, Scottsdale has consistently created planning policies, ordinances, and design guidelines that protect and enhance the character of the Sonoran Desert environment. With the city's recent adoption of the International Green Construction Code, Scottsdale is continuing to move to ever-higher levels of ecological understanding for sustainable urban desert communities in the southwest.

In Chapter 29, "Sustainable Energy Alternatives for the Southwest," David Berry describes the emerging transformation of the electric supply and demand system in the desert of the Southwest from one dominated by central station fossil-fueled power plants to a cleaner energy future that relies much more on renewable energy and energy efficiency. The current power generation infrastructure will be difficult to sustain and replicate in the future, from both an environmental and economic perspective. Berry discusses energy efficiency as the first resource alternative followed by renewable energy technologies that can help utilities manage their risks. Urban areas have great potential to integrate distributed renewable energy facilities with energy-efficient site and building design. The abundant supply of solar resources can not only increase the use of clean energy resources to meet regional retail demand but is also a valuable economic resource for exporting in the form of electricity to other regions.

It is often said that suburbia is the antithesis to true sustainable urban communities. In Chapter 30, "Search for a Lean Alternative," urban visionary Paolo Soleri discusses his arcology concept as the lean alternative to suburbia and hyperconsumption. The construction of Arcosanti began in 1970 and continues today as the first arcology and urban laboratory in the high desert of Arizona. Arcosanti represents a viable and positive solution to population growth, urban living, resource efficiency, transportation, net energy utilization, food production, preservation of natural habitats, affordable housing, global warming, and ultimate recycling. In the Arcosanti arcology, many systems work together, with efficient circulation of people and resources, multi-use buildings, and passive solar orientation for lighting, heating, and cooling. Soleri presents his latest development, the lean linear arterial city, as an elongation of the arcology principle. This lean linear arcology is a dense and continuous urban ribbon consisting of interlinked city modules designed to take advantage of regional wind patterns and solar radiation that is so prevalent in the deserts of the southwest.

In Chapter 31, "Creating Sustainable Futures for Southwestern Cities: The ProtoCity™ Approach in the Ciudad Juarez, Mexico/El Paso, Texas Metroplex," Pliny Fisk III introduces a model for evolving into the "green city of the future," where decisions related to the built environment are informed by a full life cycle of resources in which waste and by-products are utilized as resources. Key to understanding and regenerating an "ecology of place" is to establish a framework in which to understand the conditions under which a city/region is evolving. Using lessons from the early development of Austin's Green Building Program, the Center for Maximum Potential Building Systems has addressed the more complex conditions confronted by the border metroplex of El Paso, Texas, and Juarez, Mexico. The authors identify the Development Ladder (conceptual tool) to address a city/region's state of development and the ProtoScope (operational tool) as a systemic representation of how a city/region could potentially function and evolve from experiences with other cultures in the world. Information, currency, energy, and material flows determine a city/region's position along the Development Ladder while a ProtoMetric identifies the biophysical characteristics of a place, involving ecology, hydrology, climate, and geology/soils.

# 26

# *Settlement, Growth, and Water Security for Southwest Cities*

**Richard A. Malloy**

**CONTENTS**



## 26.1 Introduction

The southwestern cities of Las Vegas, Phoenix, Tucson, Albuquerque, and El Paso are the largest centers of urban population in the desert Southwest and have been among the fastest growing areas in the country in recent decades. The recent economic downturn in the United States starting in 2008 hit the Southwest region extremely hard, forcing state and local agencies to make deep and painful cuts due to the dramatic fall in revenues to fund government operations. Today, these cities are currently grappling with some of the highest home foreclosure and unemployment rates in the country with limited signs of any immediate change in this condition. This pause in the pace of urban development of these cities exposed the fault in the economies of southwestern cities—an overreliance on growth and development activities for sustaining the economy. Overall, this may be a good thing if community leaders take the time to examine options to restore balance to the urban environment. The lessons learned from this and past cycles of boom and bust can provide lessons to put forth a new vision that will sustain the region with solutions that better serve the residents of southwestern desert cities.

This chapter will describe the process of urbanization in the southwestern cities of El Paso, Albuquerque, Las Vegas, Phoenix, and Tucson from their natural environments to their current metropolitan settings. In particular, the need for securing water for these growing cities is a fundamental concern to provide a sustainable future for growth and development. Much like a living organism, without adequate water, cities will not grow to their full potential. In summary, this chapter will provide a snapshot of these urban centers in their founding, development, and efforts to balance growth, water security, and conservation.

## 26.2 Background

While the European settlers were busy building up the eastern and midwestern urban communities of the United States, the desert Southwest was largely an uninhabited region. This rugged desert landscape has been molded by the natural hydrologic and climatic cycles, which always seems to restore the land to a state created by the thousands of years of navigating a delicate balance of hydrologic processes and climate. Over the last century, this fragile landscape has turned into some of the most intensely developed areas in the United States. The Southwest urban center of El Paso, Albuquerque, Las Vegas, Phoenix, and Tucson has evolved over time to balance the struggles between life and death of the biotic communities. This rapid pace of development brings to mind a concern about the process of urbanization, which often irreversibly alters the complexity of the natural setting.

The Southwest desert region was once a highly desirable place to live for people with health problems or those seeking a place with a closer connection to nature. By the turn of the twentieth century, the quality of life of southwestern cities was once the highest marketing aspect of the region for health seekers. Today, people move to the Southwest for jobs, reasonable land values and affordable housing, among other factors. Urban sprawl has altered an environment that once featured clean, dry desert air to one that has deteriorated to the point that the quality of the air can be detrimental for people with poor health to live without discomfort in the urban centers.

Growth and development are expanding the land area of urban cities from the conversion of native desert or agricultural land to urban developments. Urbanization is an irreversible process of transformation from the natural to the built environment. In his book entitled *Urban Society: An Ecological Approach*, Hawkley states that "movement from the simple, highly localized unit to the complex and territorially extended system is a growth process. We use the term urbanization to refer to that process."[1] Southwest cities are now rapidly expanding into their urbanization process faster than most people could have previously imagined. While some older, industrialized cities in the East and Midwest are experiencing decline in population and vitality, Sunbelt cities are blessed with newfound prosperity and development. Growth is not always achieved from the most noble and ecologically sound reasons. Speculation and profit have been drivers of Western land development for many decades. Some cities have actively taken on the community's role in balancing environmental protection and economic development; other areas are still at the mercy of speculative interests for sustaining the community. Unfortunately, once the natural landscape of the desert has been adversely transformed through human intervention, the result of this intervention can have a long-term impact on future development potential.

## 26.3 Early History

The earliest settlers of the region were the *Hohokum* Indians that settled along the Salt and Gila Rivers in modern day Arizona and developed extensive irrigation canals and aqueducts. The Hohokum Indians proved to be remarkable engineers to manage the land for crop production. These sedentary Indians farmed the river valleys with crops of corn and squash and lived in simple mud brick houses. To the north in high desert country, the *Anazasi* Indian tribes inhabited the dry desert mesas with their cliff dwellings and pit houses. The Anasasi were largely hunting and gathering tribes that made unique pottery and jewelry designs. The ancient tribes appeared to vanish from the region about AD 1150 leaving no trace for the tribes that followed into the region. To this day modern anthropologists can only theorize as to the demise of the ancient cultures with ideas such as ecological disasters or disease, but without conclusive evidence (Figure 26.1).

A few early accounts were documented by explorers. Juan de Oñate made famous the *Camino Real* or Royal Road, a highway that linked the area to the interior of Mexico. Oñate was in search of resources from the newly founded territories to enrich the Spanish Crown. His relationship with the native cultures was tolerant, but sometimes barbaric. To the west, Father Eusebio Francisco Kino laboriously established the Jesuit missionaries of Tubac and San Xavier del Bac in southern Arizona and Mexico. Kino's missions were largely successful in attracting the natives to support the agrarian, Christian-based community, and some missions are still in use today (Figure 26.2).

Until the Mexican–American War in 1846, the desert Southwest was under the control of New Spain ruled from the European continent. The region was sparsely settled with few small towns of any size. The population was largely Hispanic farmers and ranchers that homesteaded pastoral lands or farms. Small settlements were established periodically along the major rivers of the region where a more permanent source of water could support a growing community. A large portion of the Southwest was added to the United States after the controversial Mexican–American War through the Treaty of Guadalupe Hidalgo in 1848. Additional lands were acquired from Mexico through the Gadsden Purchase in 1853, which added additional lands south of the Gila River to the border with Mexico and is the current international boundary. To the east, the State of Texas was formed by cessation from Mexico and annexation in 1845. The current state boundaries include some areas claimed by Texas in the years after joining the Union (Figure 26.3).

## 26.4 Original Settlements

The regional settlements of the major southwestern cities were strategically located near major rivers in relatively flat, defensible landscapes. By the time European settlers were

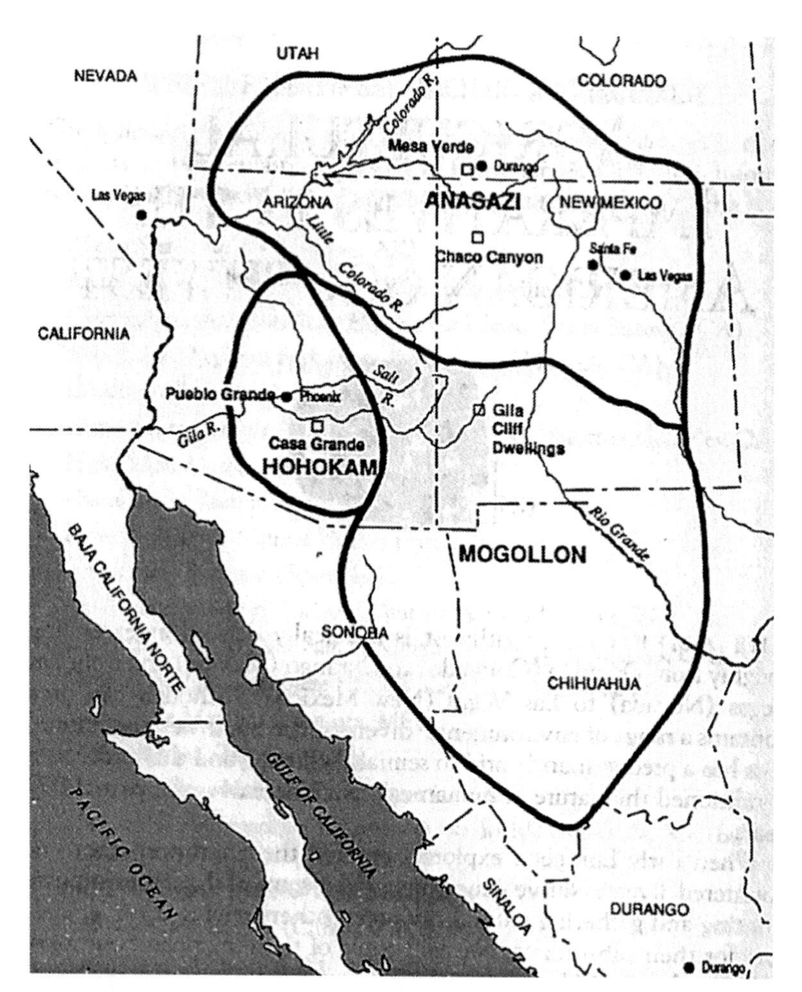


FIGURE 26.1
Geographic ranges of ancient southwestern Indian cultures.

passing through the region, the major concern was protecting the community from periodic Indian attacks. Apache and Comanche tribes carried out regular raids on anyone who passed through the region. Not until the surrender of Geronimo, the infamous Apache warrior, was the threat of hostilities considered close to manageable for the local population. As a result most of the settlements remained close in proximity and established outposts for protection against the Indian threat.

### 26.4.1 El Paso

In 1598, Juan de Onate declared El Paso del Norte in the name of New Spain. The pass was a stop along the *Camino Real* (Royal Road) that linked with the interior of Mexico. Just south of El Paso del Norte, Father Garcia de San Francisco established a mission in what is now the City of Juarez, Mexico. This rapidly developing city was situated right on the U.S.–Mexican border in the Chihuahuan Desert. The town became a major stop on the Butterfield Overland mail coach route. El Paso quickly developed a reputation as a lawless

**FIGURE 26.2**
Camino Real map. (Courtesy of the National Parks Service.)

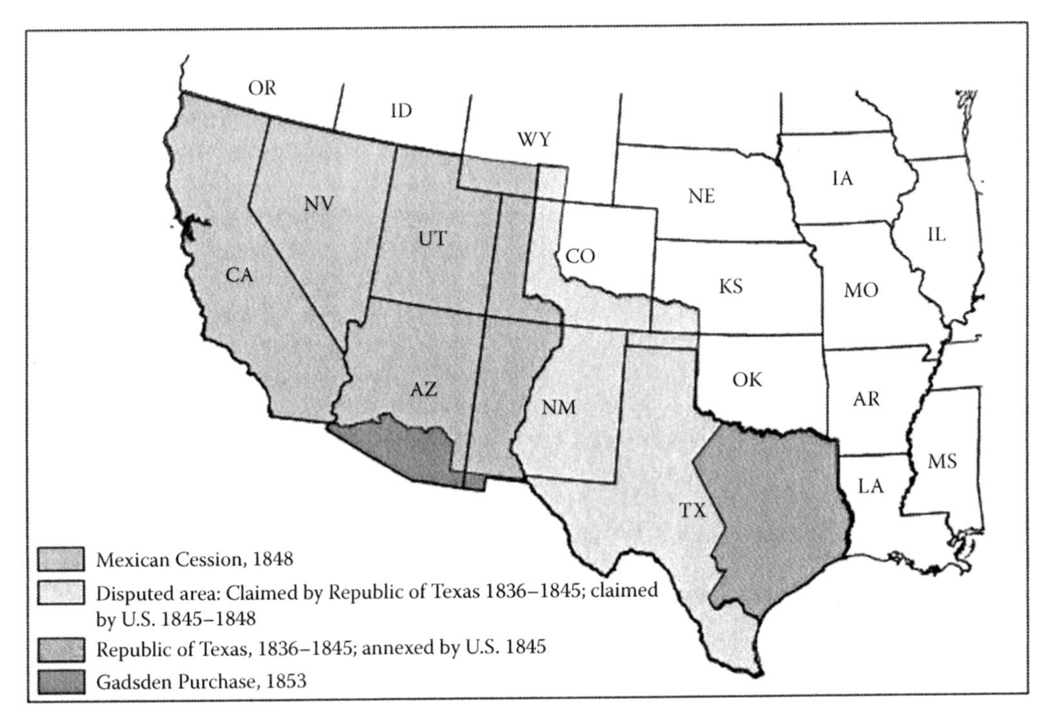


**FIGURE 26.3**
Map of southwestern land acquisitions. (Courtesy of U.S. Geological Survey, Reston, VA.)

center where gunslingers and rustlers had their way about town. This fertile valley was the home of pastures, farms, and vineyards along the Rio Grande River.

During the period from 1852 to 1868, the Rio Grande River experienced severe and violent floods that shifted the course of the river to the south. This diversion of the river added several hundred acres of land to the United States due south of the urban center of El Paso, now considered valuable developable land. Mexico made repeated claims to the Chamizal lands in dispute, citing that previous treaties with the United States define the center of the Rio Grande River as the international boundary at the time of the treaty. Both countries agreed to arbitration in 1910 that resulted in a proposal to return the lands to the historic boundaries and transfer of lands back to Mexico. The United States refused to accept this settlement and continued to ignore Mexico's pleas to settle the dispute. The issue was finally resolved when John Kennedy agreed to settle the dispute in accord with the 1911 arbitration proposal and was eventually signed by President Johnson in 1967. Both countries shared the cost of channelization of the Rio Grande and the exchange of the lands to Mexico.[2] The *Chamizal Dispute* is an important milestone in southwestern history in urbanized settings where the issue of land was an emotional, valuable and an object of national pride, more on the part of the Mexican perspective, but smoothed the path of diplomatic relations with Mexico who was still angry about the U.S. land grab of northern Mexico territories after the Mexican–American War.

### 26.4.2 Albuquerque

In 1540, the Spanish explorer Francisco Vasquez de Coronado traveled through the New Mexico region in search of the Seven Cities of Cibola. After spending the winter camped along the Rio Grande River, Coronado proceeded north to establish the town of Santa Fe in 1610. The Spanish sought to control the Pueblo Indians and found a strong rebellion ensued by the 1580s and drove the Spanish to the south for over 10 years until it was recaptured. The villa of Albuquerque, named after the Duke of Spain, was founded in 1706 and was the next major westward stop along the Camino Real from El Paso del Norte. By 1880, the railroad had entered the territory, and, in 1885, the town of Albuquerque was founded. In 1889, the University of New Mexico opened its doors as the new higher-education center of the region. The town remained part of the New Mexico territory until 1912 when New Mexico became the 47th State in the Union.

### 26.4.3 Las Vegas

The dusty town of Las Vegas was named after the Spanish term that means "the meadows." The site was one of the few artesian springs that emerged from the parched landscape. The Mormon Church had grand plans for building a string of settlements from Salt Lake City to the Pacific Ocean. In 1855, a fort was constructed by the Mormons to build farming and mining communities, but was left abandoned by 1857. Las Vegas had been part of territorial New Mexico when the Mormons built their fort in 1855, but the western segment of the territory became part of Arizona in 1863. When the new state of Nevada was created, Las Vegas was part of Mojave County, Arizona.[3]

Las Vegas had a bad reputation from some illicit activities, such as whiskey running, prostitution, and cattle rustling. The region was promoted and profited by the flamboyant and controversial Senator William Clark, a developer who had a questionable reputation. The city was founded in 1905 after a land auction of 110 ac creating the Las Vegas town site and was governed as part of Lincoln County until 1909 when it served as the seat of the newly created Clark County. The city benefited greatly from state legislation that legalized gambling in 1931 and the signing of the Boulder Canyon Project Act in the same year by Calvin Coolidge. This act began the largest, most ambitious government construction project; the Hoover Dam.

### 26.4.4 Phoenix

While the Spanish explorers paid attention to other southwestern regions, the Salt River Valley was largely undisturbed throughout the colonial period. In 1865, the U.S. Army established Fort McDowell about 20 miles north of the Salt River to defend the area from hostile Indian attacks. About the same time, a former soldier, John William (Jack) Swilling, saw great potential in rekindling the old Hohokum canal system to irrigate the valley. Swilling, aware of the potential for revitalizing the land, called the settlement Phoenix, after the legendary bird that rose from the ashes with new life. In April of 1870, a 320 ac parcel was issued to the town site of Phoenix. In 1881, the town site was incorporated into a city and John T. Alsap served as the first mayor of Phoenix. The completion of the Roosevelt Dam and the Arizona and Grand canals fueled the interest in settlement of the Salt River valley, which allowed settlers to irrigate large tracts of land once limited by the availability of reliable sources of water for agriculture and development.

### 26.4.5 Tucson

Tucson was founded by Hugh O'Connor in 1775 under the direction of the Spanish Crown to locate and establish an outpost along the Santa Cruz River. O'Connor selected a site on the east side of the Santa Cruz River to establish Presidio San Augustin. Tucson was a name taken from the Indian designation of a local landmark spring located at the base of the modern day Sentinel Peak. Tucson was added to the United States in the territory that was acquired in the Gadsden Purchase in 1854. Tucson was the territorial capital from 1867–1877 when it was relocated to Prescott. Over the next decade, the Tucson delegation worked feverishly to regain the territorial capital each time the legislature met. In 1885, Tucson was awarded a $25,000 appropriation to found the University of Arizona. However, the local residents were shocked that they were awarded the university rather than the territorial capital as a prize and almost forfeited the award until a benefactor stepped forward with the required land contribution for the new university (Table 26.1).[4]

**TABLE 26.1**
Southwest City Characteristics

| | | Desert City Characteristics | | | |
|---|---|---|---|---|---|
| **Feature** | **El Paso** | **Albuquerque** | **Las Vegas** | **Phoenix** | **Tucson** |
| Year city founded | 1873 | 1885 | 1905 | 1881 | 1885 |
| County | El Paso | Bernalillo | Clark | Maricopa | Pima |
| Elevation | 3710 | 5000 | 2174 | 1117 | 2389 |
| River | Rio Grande | Rio Grande | Virgin | Salt | Santa Cruz |
| Annual rainfall | 8.81 in. | 8.88 in. | 4.13 in. | 7.6 in. | 14.10 in. |
| Average temperatures | 77/49 | 71/42 | 81/54 | 85/57 | 84/57 |
| Nearby mountain range | Franklin | Sandia | Spring | South Mountain | Santa Catalinas |
| Desert community | Chihuahuan | Chihuahuan | Mojave | Sonoran | Sonoran |

## 26.5 Securing Water

The desert Southwest is an area defined by the scarcity and unpredictability of water supply. The settlements of southwestern cities were established on what was then flowing rivers; although maybe not large in size, they provided an adequate supply of water for the small population of these areas at that time. At first, these young cities were able to provide water by primitive means, by windmill pumps, aqueducts or even water delivery services by a water wagon. All of these desert cities were sited near known sources of perennial water in streams, rivers, and artesian wells. As the population began to grow, the cities began to recognize the need for a more sustained long-term water supply. The future of these desert communities was tied to securing a permanent, reliable source of water.

In the first decade of the twentieth century, the federal government began serious consideration of water reclamation projects to address the problems presented by the periods of drought and flooding that plagued the major rivers of the region. Major flooding of the Salt River in the 1890s and later the diversion of the Colorado River began flooding the Salton Basin in 1905 to form an inland sea before the river course was corrected. At the same time, frequent and unpredictable changes in the other major rivers such as the Salt, Verde, Santa Cruz, Gila, and Rio Grande presented a concern for the development and public safety for area leaders to address. After careful evaluation, the Bureau of Reclamation chose the Salt River Dam project to be the first major project to be undertaken by the federal government to address concerns about western water issues.

The Salt River Dam was the first water reclamation project initiated by the federal government in 1902 after successful attempts by Benjamin Fowler and George Maxwell to secure funding in Washington for the passage of the Newlands Act to fund the construction of the Roosevelt Dam. In 1904, the Salt River Valley Water Users Association was formed to negotiate a contract to repay the federal government for the construction of the dam. It was dedicated by Theodore Roosevelt on March 18, 1911 with much fanfare. The Roosevelt Dam had an immediate economic and social impact on the Phoenix area. An extensive canal system was soon constructed that carried water to areas that were previously desolate, and farms and ranches along the path of the canals were allowed to prosper. The success of the Roosevelt Dam fueled ideas for bolder and far-reaching water projects in the West. The idea that modern engineering could harness the West created great excitement for investment and growth possibilities, particularly with the eastern and midwestern establishments (Figure 26.4).

In 1928, Congress passed the Boulder Canyon Project Act that authorized the construction of the Boulder Canyon (later renamed to Hoover Dam) project. Before construction, the federal government reached an agreement with the states for the division of river allotments. The river was separated into upper and lower divisions. Each division would share the water proportionately with considerations for future settlements with Mexico. The Boulder Canyon project was the largest public works project ever undertaken by the federal government. This project provided thousands of jobs for willing workers, many of whom migrated from areas deeply affected by the depression-era economy, and served as a catalyst for growth of the otherwise desolate area around Las Vegas, Nevada (Figure 26.5).

New Mexico and West Texas had a similar dilemma with the Rio Grande River. After contentious debate among delegates from Texas, New Mexico, and Mexico, and the benefactors of the dam project, it was decided that the site at Elephant Butte Lake was the best

**FIGURE 26.4**
Roosevelt Dam under construction on the Salt River. (Courtesy of the Salt River Project.)

**FIGURE 26.5**
Backside of the Hoover Dam. (Courtesy of the U.S. Bureau of Reclamation.)

location to build this reclamation project. In 1906, the United States signed a treaty with Mexico for the equitable distribution of Rio Grande water through the delivery of water to the *Acequia madre* at Juarez of 60,000 ac ft a year. As with the Roosevelt Dam project, a water users group was formed to reimburse the federal government with the dam construction costs over time. The dam was complete in 1916 for a cost of $5.2 million. A hydroelectric plant was added later in 1937 to provide electric power to the region (Figure 26.6).

**FIGURE 26.6**
Elephant Butte Dam. (Courtesy of the U.S. Bureau of Reclamation.)

**TABLE 26.2**
Dams of the Southwest

| Southwest Dam Data | | | |
|---|---|---|---|
| **Fact** | **Elephant Butte Dam** | **Hoover Dam** | **Roosevelt Dam** |
| Constructed | 1912–1916 | 1931–1936 | 1903–1911 |
| River system | Rio Grande | Colorado | Salt |
| Storage capacity | 2,109,423 ac ft | 28,537,000 ac ft | 2,910,200 ac ft |
| Height | 301 ft | 726.4 ft | 356 ft |
| Crest length | 1,674 ft | 1,244 ft | 723 ft |
| Crest elevation | 4,414 ft | 1,232 ft | 2,218 ft |
| Concrete used | 629,500 $yd^3$ | 3,250,000 $yd^3$ | 606,000 $yd^3$ |
| Drainage area | 28,900 $mile^2$ | 167,800 $mile^2$ | 5,830 $mile^2$ |

*Source:* U.S. Bureau of Reclamation. http://www.usbr.gov/dataweb/dams.

In spite of the successes of the Roosevelt, Elephant Butte, and Hoover Dams, central Arizona was left without a permanent water solution (Table 26.2). A dam project was proposed in Tucson at Sabino Canyon in 1936 and received widespread support until the Army Corps of Engineers put the local contribution for the dam construction at $500,000. The political leaders at this time were not able to muster the financial backing within the community, and the idea of banking water in southern Arizona was tabled. Inaction on dealing with the larger water problem in Tucson left the region with no choice other than to rely on groundwater pumping for the foreseeable future. Some relief was on the way, however, with a long-awaited construction of the Central Arizona Project (CAP), a multibillion dollar federal project that would divert Colorado River water into central and southern Arizona. The CAP took many decades to become a completed project (Figure 26.7).

The CAP was originally proposed in 1947 but was not authorized until 1968 due to disagreements on the merits of this project with other regional stakeholders. Political wrangling over appropriation of funds ensued in the years following the authorization of the CAP. The debate over the CAP forced the enactment of the broad-sweeping Arizona

(a)

(b)

**FIGURE 26.7**
(a,b) Central Arizona project canal. (Courtesy of the U.S. Bureau of Reclamation.)

**TABLE 26.3**
Central Arizona Project Facilities

| **Central Arizona Project** | |
|---|---|
| Construction | 1973–1993 |
| Service area | Lake Havasu—Tucson, Arizona |
| Storage facilities | 1 |
| Diversion facilities | 3 |
| Aqueducts | 325 miles |
| Tunnels | 15 miles |
| Pumping plants | 15 |

*Source:* U.S. Bureau of Reclamation. http://www.usbr.gov/dataweb/html/crbpcap.html.

Groundwater Act in 1980. This act established active management areas in Arizona and provided guidelines for water conservation and recharged targets for replenishing groundwater depletion within the management areas (Table 26.3).

## 26.6 Water Sources

Sources of water for each of the growing southwestern cities have prompted a need to find increasingly larger water reserves as population growth sharply increased the demand for water delivery to the new developments in the urbanized areas. Without an assured water supply, the future of any of these cities will be in doubt. Local area decisions on water management and planning have played a significant role in the effectiveness of securing these water resources for the growth of the cities. However, regional hydrologic patterns, geography, and climatic factors govern the sustainability of the locality.

### 26.6.1 El Paso Area

El Paso has relied on water from the Rio Grande and groundwater pumping as the primary water sources. El Paso shares water withdrawals from the Hueco and Mesilla *bolsons* (aquifers) with the City of Juarez, Mexico. The longevity of the underground aquifer is tied to the hydrologic cycle of the Rio Grande River. With the current rate of water mining from the watershed, a serious problem is on the horizon for the region. El Paso city water managers are attempting to address the problem through the development of new water resources, recovery, and reuse of treated wastewater and aggressive conservation programs. By the 1990s, the Hueco Bolson aquifer was losing 3 ft annually, which created an urgent need for the El Paso Water Utilities (EPWU) to find new sources of water. EPWU responded by purchasing 202 ac of land near the Franklin Mountains to collect and manage storm runoff. In addition, EPWU entered into an agreement with the Department of Defense to construct the largest inland water desalination plants capable of producing 27.5 million gal of water a day (Figure 26.8).[5]

### 26.6.2 Albuquerque

The city of Albuquerque relies primarily on ground and surface water reserves for the municipal water supply from the Middle Rio Grande watershed. This aquifer is derived from deep basin fill deposits about 14,000 ft in depth; only 2,000 ft of this constitutes the aquifer system. The aquifer is an enclosed basin region surrounded by the Sandia, Manzanita, Los Pinos, and Jemez Mountains. Currently, the Middle Rio Grande aquifer is in danger of depletion due to excessive groundwater pumping. The Rio Grande River is now experiencing periods of low or no flow during the dry season.

**FIGURE 26.8**
El Paso water utilities desalination plant. (Courtesy of El Paso Water Utilities.)

**FIGURE 26.9**
San Juan Chama Azotea tunnel outlet. (Courtesy of the U.S. Bureau of Reclamation.)

In the 1960s, the Bureau of Reclamation established the San Juan–Chama water diversion project to bring much needed water to the growing New Mexico region. The City of Albuquerque was awarded about 48,000 ac ft per year from the diversion project. The water was channeled through 26 miles of tunnels, across the Continental Divide, and into the El Vado and Abiquiu reservoirs. The traditional water plan called for the use of groundwater pumping with the release of San Juan–Chama water from the reservoirs to enhance the groundwater uptake (Figure 26.9).

### 26.6.3 Las Vegas

Las Vegas relies primarily on withdrawal of water from Lake Mead and pumping of groundwater to support the city. Early settlements in the allocation of Colorado River water left Nevada with a paltry sum of water in relation to the already developed neighboring states of California and Arizona. At the time of the Colorado River Compact, Nevada was a desolate state with little hope of supporting a significant urban population and was only allocated an annual withdrawal allowance of 300,000 ac ft, compared to Arizona's 2.8 million and California's 4.4 million ac ft (MAF). Today, the demographics of the Southwest have shifted to a point where Nevada is now the fastest growing urban population in the region. In addition, the natural rainfall of the region is the lowest of any of the desert cities, thus limiting the ability of sustained reliance on groundwater resources. Nevada has put forth a decade-old plan to allow the state to withdraw more water from Lake Mead. Nevada argues that it has the surface water rights to 128,000 ac ft from the Muddy and Virgin Rivers. The state feels that it has the right to recapture the water from the reservoir instead of building expensive pipelines to transport the water within the state. California and Arizona strongly object to this plan and the decision will have to be made in Washington. In a recent agreement, Arizona will allow water-starved Nevada to take as much as 1.25 MAF of Arizona's Colorado River allotment in exchange for about $330 million.[6]

On a large scale, Nevada has an uphill battle to find additional water resources to add to its water portfolio to support current and future needs for development. Solutions to augment Nevada's water supply have involved the funding of water conservation projects in exchange for water supply and controversial proposals to construct pipelines from outside the state and region to deliver water.* In addition, Nevada and other Colorado River Basin states has provided funding for the construction of the Drop 2 reservoir along the lower Colorado River to retain water that might otherwise be lost and sent downstream without beneficial use.† Nevada has been aggressive in curtailing water uses with restrictions and a program to pay residents to remove turf grass and install desert landscaping.

### 26.6.4 Phoenix

Phoenix has been blessed with having the ability to draw on several sources of water to accommodate growth. The Salt River Project that was established in 1911 has provided a continuous supply of water to the cities by the canal system. In addition, many independent water providers in the region use or augment their water supply through groundwater pumping. The construction of the CAP has added an additional source of water from the Colorado River. As a result, Phoenix has benefited from being at a unique geographical position in the hydrologic regime. Even with the more abundant water resources available to the city, Phoenix is still facing obstacles to unbridled growth through the state water regulations set forth in the Arizona Groundwater Act of 1980. The onset of drought conditions in recent years caused the Salt River Project to reduce their 2003 allotments to cities for only the third time in the 50-year history of the utility. The reservoirs were at 27% of normal capacity by the end of 2002.[7] Today, Phoenix is aggressively working on water conservation and public education on water conservation.

### 26.6.5 Tucson

Early efforts to develop water banking and storage projects in the Tucson area suffered a major defeat from public support in paying for these large-scale projects. As a result, Tucson has been forced to rely on groundwater pumping as the primary source of water for the city. Tucson Water Company established wells along Valencia Road south of the city as the main supply source. The growing need for urban water motivated the city to begin purchasing land in the Avra Valley region to serve as water farms to be piped into the Tucson Valley. In addition, Tucson began purchasing water companies outside the city limits so that it could begin a basin-wide management strategy for water resources.

In 1990, the CAP canal to Tucson was completed, thus providing long anticipated relief from the water deficit problems of the region. Problems soon arose from the use of CAP water for residential use. Colorado River water is harder and contains a larger share of total dissolved solids than local groundwater. More importantly, CAP water began causing corrosion problems in the residential water infrastructure. After angry protests from area residents, Tucson Water backed off the direct use of CAP water in the potable water system. In 1995, Tucson voters approved the Water Consumer Protection Act that restricted the use of CAP for residential use, unless it conformed to the local water quality standards. The water supply situation in Tucson is still a critical issue. Long-term groundwater pumping is causing areas of land subsidence, and area wells are going dry

* http://www.8newsnow.com/story/6963917/big-water-battle-brewing-with-snake-valley-utah?redirected=true (accessed April 7, 2012).

† http://ag.arizona.edu/azwater/awr/mayjune08/feature1.html (accessed April 6, 2012).

quickly. As Tucson grapples with the challenge of finding ways to best manage its water resources, growth and development will continue to deplete the existing water resources of the city. Efforts to conserve water through a tiered water rate structure, public education and awareness on water conservation, and regulations requiring the harvesting of rainwater for new developments, among others, are strategies adopted to help reduce water use by its residents.

## 26.7 Water Policy

The development of the West by the turn of the twentieth century raised several concerns from the western states on how to ensure fair and equitable use of the rivers. The growth in California in particular had several of the region's leaders concerned that this growth could lead to disproportionate use by some states. The proposed construction of a dam near Boulder Canyon prompted the need to establish guidelines for how the Colorado River water would be shared amongst the states. In 1922, the six Colorado Basin states met to sign the Colorado River Compact, which divided the river into two water management groups: the upper, including Wyoming, Utah, Colorado, and New Mexico, and the Lower Basin that included Arizona, Nevada, and California. Arizona refused to sign the compact, citing fears of California's overreach in the collective management of the water portfolio. Each basin was allotted 7.5 MAF of water. The division of the Lower Basin allocated 4.4 MAF to California, 2.8 MAF to Arizona, and 0.3 MAF to Nevada. Arizona eventually signed onto the compact in 1944, but remained grossly unhappy about this arrangement.

Arizona took its case to the Supreme Court, resulting in the 1963 decision in *Arizona v California*. This case clarified the division of future surpluses of water on the river, as well as some disputes over water rights. In addition, the outcome of this case propelled the CAP to be approved by Congress in 1968. CAP was one of the largest aqueduct projects in the United States. The aqueduct was built to divert water from the Colorado River through central Arizona with potential extensions to New Mexico. In response to a federal ultimatum to reduce its groundwater use, Arizona adopted stringent water management policies in 1980 entitled the Arizona Groundwater Management Act (AGWA) to be administered by the Arizona Department of Water Resources (ADWR). The act created five "active management" areas that included all of the urbanized areas in the state including Phoenix, Tucson, Pinal, Prescott, and Santa Cruz. All development projects within the AMAs are required to demonstrate that a 100-year supply of water is available to support this development. Developers protested the requirements of AGMA's ability to limit development outside municipal water providers' service area. ADWR backpedaled on this regulation by establishing the Central Arizona Groundwater Replenishment District (CAGRD). This new entity was set up to draft legal documents to ensure that an equal amount of water that was withdrawn by the development is replenished in the AMA. There is no requirement that the water be returned to the same aquifer or within close proximity to the place it was withdrawn. Currently, CAGRD serves as the water bank on paper for disconnected development outside the urbanized areas. The CAGRD is, at the same time, a step in the right direction for preventing overdraft of groundwater and a measure that has ecological flaws in the manner of providing sustainable "place-based" water management (Figure 26.10).

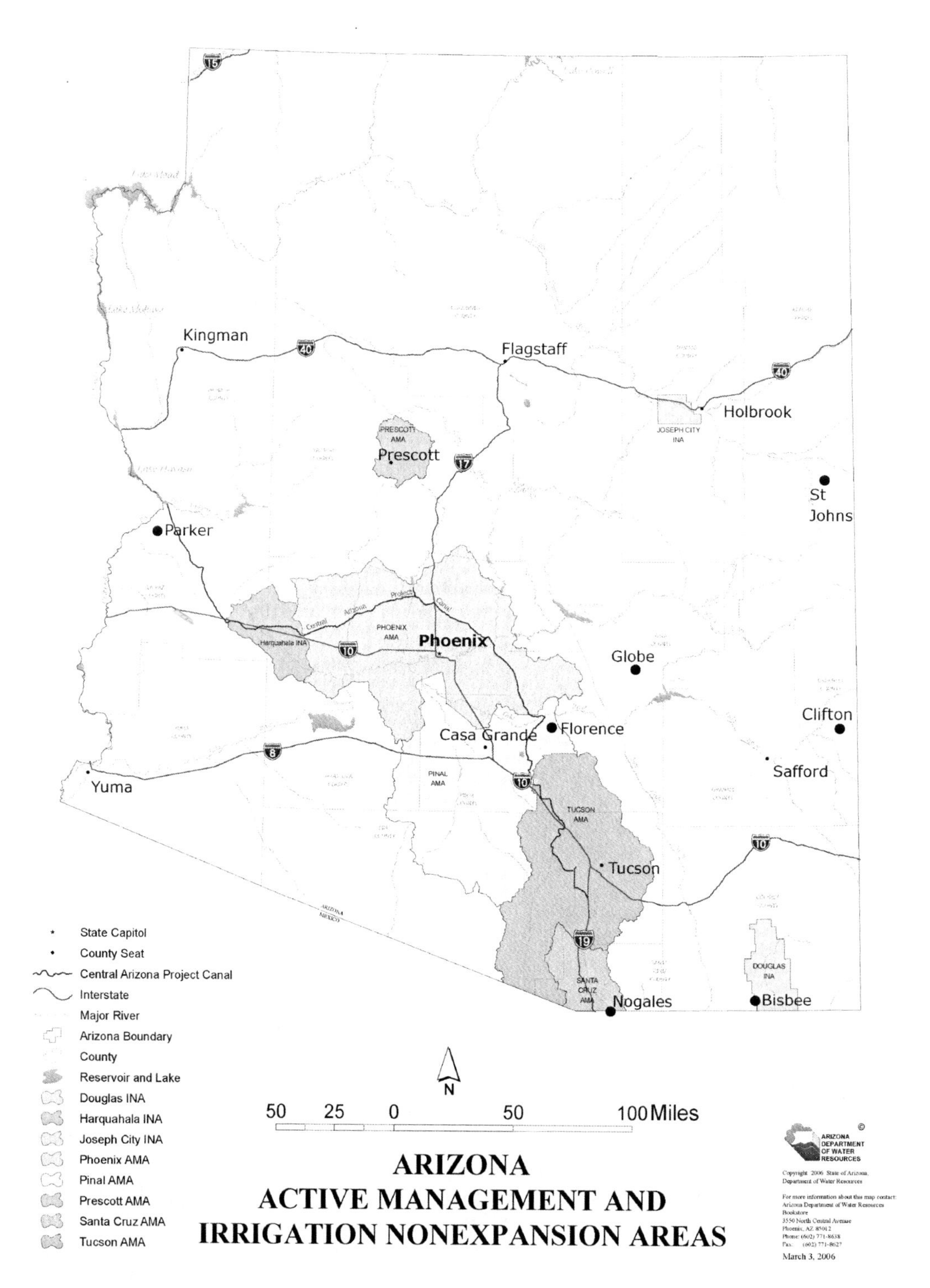


**FIGURE 26.10**
Arizona Department of Water Resources' active management areas. (Courtesy of the Arizona Department of Water Resources.)

With continuing problems from the overallocation of water resources of the Colorado River, the U.S. Secretary of the Interior issued water shortage guidelines in 2007 for how all the water rights of the river in a time of shortage will be managed.* A trigger point was established on Lake Mead and water levels below this level trigger proportional reductions by all of the Colorado River Basin states. This is the result of a river that has been overallocated and has reached what Peter Glick terms "peak water" limit of the river capacity to balance the river flow and the use of this resource.[8]

## 26.8 Water Problems

Even with the massive public investment in dam construction in the Southwest, water is still a problem for the hyperconsumption rate of growth experienced by the urban centers. In some circles, the mere existence of dams on the rivers is seen as detrimental to the hydrologic cycle. Environmentalists argue that the natural flood cycle carried sediments downstream to enrich the valleys with alluvial silt, which settles out in the area behind the dam. The resulting water released downstream is clear and regular, not conducive to any sedimentation or ecosystem health. The issue has been very controversial. In 1998, then Secretary Bruce Babbit sanctioned a well-publicized release of water at Lake Mead in an experiment on stream health. The debate continues between environmental activists concerned with the effects on the ecosystem health and biotic communities and those seeking to manage the dam for utilitarian purposes of power and water delivery. For the time being, the merits of dams are generally regarded as helpful to the human inhabitants of the desert regions.

Tucson has struggled with water issues and public support for infrastructure development. While the city of Tucson was able to add numerous wells from new water sources along the Old Nogales Highway in 1954 and 1968, the result of this water mining caused a massive die off of the extensive Mesquite bosques by the lowering of the groundwater table in the area around the San Xavier del Bac mission.[9]

With the construction of the San Juan Chama in New Mexico, a concern arose with the impact of this project on a native fish called the silvery minnow. The habitat of this small but ecologically significant fish was put in jeopardy by the completion of this project. There are several ongoing legal and administrative actions pending over the division of water by the San Juan Chama under the Endangered Species Act. The outcome of these claims by the project may be years away from being resolved and may not ever satisfy either of the concerned stakeholders of the project (Figure 26.11).

The Colorado River Compact, which was formed among western states in the 1920s, has been under scrutiny as a shift in demographics has dramatically changed over the last 80 years. The state of Nevada, at the time of the compact, was a desolate place with little hope of sustained development. In token, a paltry allotment of 300,000 ac ft a year was awarded to the state, a small amount in comparison to Arizona's and California's share. Growth of southwestern cities has added increased burdens on the infrastructure to support the growing urban populations. Despite major water projects such as dams and canals, these cities began to experience increased demands for water, causing the rapidly growing municipalities to search for new solutions to managing water resources in the urban areas.

* http://www.usbr.gov/lc/region/programs/strategies.html (accessed January 20, 2012).

**FIGURE 26.11**
Endangered silvery minnow. (Courtesy of the U.S. Fish and Wildlife Service.)

One manner that has been used to alleviate the water demand was in the conversion of agricultural water right to urban use, which has has resulted in a net gain of water for the city, as residential water use is less intensive than that of agricultural crop production.

Conserving water is an issue that is of critical concern for southwestern cities. Each area has met this concern with policies and programs that have had varying degrees of success. In 2002, Western Water Advocates published a study of urban water use of western cities.[10] Although this study is now over 10 years old, this study provided a useful comparative analysis of water usage between southwestern cities in regard to water planning, usage, and conservation measures. Figure 26.12 shows a comparison of southwestern cities and

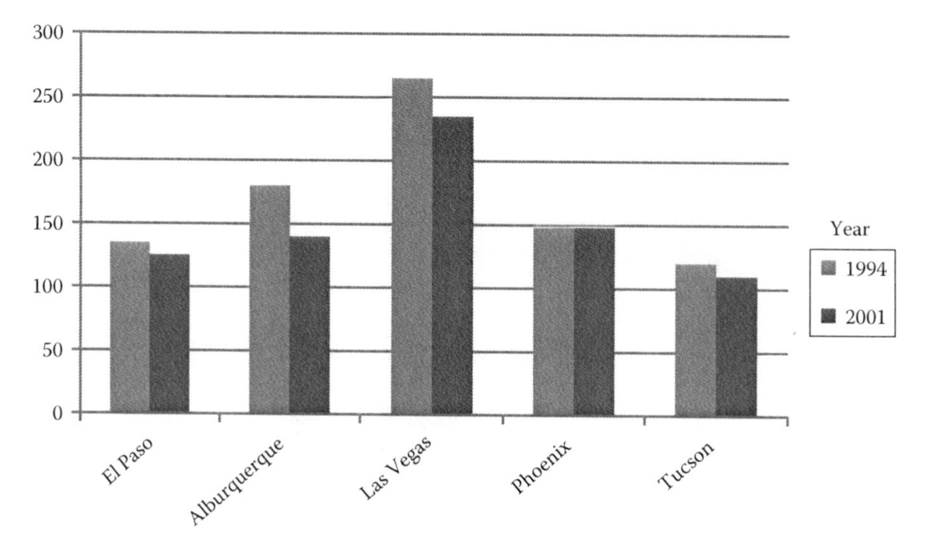


**FIGURE 26.12**
Changes in single family residential per capita water consumption 1994–2001.

the water-usage rates of urban water users. The surprising observation is that Las Vegas is in the most water-deprived climatic zone, and yet it has the highest per capita consumption rate of all of the cities. Tucson and El Paso have the lowest consumption rate. Water use decreased in most cities, but the increase in population is increasing the overall water demand on the water providers; however, the conservation lifestyle has not yet become a way of life for most of these cities.

## 26.9 Economic Development and Water

Southwestern cities early on attracted people because of the warm climate and wide open spaces. The completion of the Roosevelt Dam in 1911 opened a floodgate of interest from people interested in farming and ranching along the newly constructed canal system. Phoenix, Tucson, Albuquerque, and El Paso were actively competing with each other for eastern clients seeking a cure for tuberculosis. The dry desert climate was promoted by many to help those people afflicted by this devastating illness. Many, in fact, were cured through their convalescence in the desert. In El Paso, Drs. Charles Hendricks and Albert Baldwin established large successful sanatoriums to treat tuberculosis patients. In Phoenix, Hotel Adams was considered one of the finest hotels in the region, located at the corner of Central Avenue and Adams Street in the downtown district. Other stores began to prosper in the newly thriving urban centers, such as the M. Goldwater and Brothers, a successful Jewish merchant relative of the now infamous Barry Goldwater.

Military bases played a large role in the development of southwestern cities. At first, the bases served as a means of protection against the savage Indian raids. As the area matured, the military bases had become a valuable part of the economy by providing employment and steady income for area suppliers and were often responsible for helping to improve local infrastructure that provided service to the base. Ft. Bliss in El Paso was established in 1849 and continues to function as an active military base today. Ft. Lowell in Tucson was an active fort, but later was abandoned after the Civil War ended. Ft. McDowell in Phoenix played a key role in providing protection for the settlements along the Salt River Valley. During WWII, Davis–Monthan Air Force Base in Tucson, Nellis Air Force Base in Las Vegas, Luke Air Force Base in Phoenix, and the Sandia Complex added to the local economies by adding jobs and large financial impact to the locality. A 2002 report estimated that the payroll and expenditures from military operations in Arizona contribute over $1.5 billion to the state economy and employs over 41,000 people.[11]

Copper mining was a large economic interest from southern Arizona to El Paso. The Florilla mining company opened in El Paso in 1899, followed by the El Paso Tin and Smelting Company in 1909. Other mining operations developed as new rail lines were able to haul the heavy ore to smelting facilities in El Paso; the Farrah clothing manufacturing company opened a plant that employed more than 5000 garment workers. Cotton was a crop of choice for southwestern farmers in dry land or irrigated croplands. Desert lands were used to graze cattle where no other crop would survive.

In other parts of the Southwest, low land costs and relative cheap labor costs were drawing businesses from across the country. By the 1950s, Phoenix had become the most aggressive seeker of corporate firms by creating a favorable business climate for large defense and aerospace firms to relocate. Phoenix's geographic location about midpoint between the major manufacturing centers of Chicago and southern California made the city attractive to cost conscious firms. The first major player in the Phoenix market was Motorola. By 1960, the corporation had three major plants in the valley with over 5000 employees.[12] Phoenix boosters had a well-coordinated team of leaders that actively sought out clean, high technology companies over all other southwestern cities. The result of the booster efforts created a diverse industry and technology manufacturing that formed a diverse economy that was unparalleled in the region. Tables 26.4 and 26.5 demonstrate the difference in manufacturing and wholesale sales in region. It is clear to see that Phoenix outpaces all other cities in scale of goods produced.

## 26.10 Settlement and Growth

After WWII, the growth rate of southwestern cities accelerated rapidly. The low cost of land, favorable business climate, and warm weather were factors that drew people to settle in these Sunbelt cities. One of the catalysts for the suburban postwar boom was a Phoenix developer named Del Webb. His visionary low-cost developments provided affordable houses to people in search of good clean housing stock in the desert. A Webb legendary development in Sun City to the west of Phoenix was a monumental success in marketing his active retirement community. His developments sold as quickly as they could be built.[13] Low-density suburban development was the driving force of most southwestern cities. New development began to spring up on the fringes of the cities. The race to develop new land on the edge of the urban center continued to expand the urban limits of cities, sometimes with adverse consequences for the existing residents.

Annexation is the primary means that cities acquire lands to expand the urban boundaries. Annexation in Arizona only requires 51% of the landowners to be annexed to agree to the measures. Phoenix adopted aggressive annexation planning strategies starting in the 1960s, which led to a rapid expansion of the urbanized boundaries in all directions.[14] In New Mexico, annexation can be carried out by petition and arbitration through the Roswell Law, which allows any city to annex an area by resolution if the area borders the city on two sides. Albuquerque's aggressive use of annexation to acquire land prompted the state legislature to pass a law to halt any future annexation by the city unless it had 100% approval by the landowners.[15] Figures 26.13 and 26.14 show the rapid and exponential growth in land area and population of southwestern cities, particularly after the 1960s. Alternatives to low-density sprawl have been proposed by various groups, including growth boundaries and targeted incentives for development and transportation alternatives, but the free market forces won out in the Southwest to comprehensive policies on growth (see Chapter 14).

Phoenix continued an aggressive annexation policy to expand the boundaries of the city. Development continued at such a rapid pace it neglected essential areas deep within the urban core of the city. Even by the 1950s, a pattern of decline in the Central Business District was observed. As urban development was pushed to the fringe, parcels of urban land remained vacant in the city center. Recent efforts by the City of Phoenix to work with Arizona State University to develop a Capital Center university in the downtown area

**TABLE 26.4**

Manufacturing Outputs of Southwestern Cities from 1967 to 1997

| | Manufacturing (NAICS 31–33) | | | | | | | | | |
|---|---|---|---|---|---|---|---|---|---|---|
| | 1997 | | 1987 | | 1977 | | 1967 | | 2007 | |
| City | Estmts. | Value Added (Millions) | Estmnts. | Value Added (Millions) | Estbmts. | Value Added (Millions) | Estbmts. | Value Added (Millions) | Value Added (Millions) | Estbmts. |
| El Paso | 599 | 3,115.9 | 507 | 1,377.8 | 386 | 689.9 | 244 | 184.9 | 579 | 2,897.9 |
| Albuquerque | 592 | D | 499 | 673.6 | 483 | 288.7 | 212 | 69.8 | 562 | 1,507.2 |
| Las Vegas | 274 | 273.0 | 159 | 138.5 | 150 | D | 81 | 19.6 | 225 | 275.6 |
| Phoenix | 1,706 | 10,283.8 | 1,559 | 4,665.3 | 1,284 | 1,923.4 | 790 | 517.3 | 1,637 | 6007.5 |
| Tucson | 520 | 1,767.5 | 497 | 1,745.3 | 303 | 287.1 | 204 | 47.8 | 484 | 1,234.0 |

*Source:* U.S. Census Bureau. http://www.census.gov/.
D, as listed in the census document.

**TABLE 26.5**

Wholesale Trade for Southwestern Cities from 1967 to 1997

| | Wholesale Trade (NAICS 42) | | | | | | | | | |
|---|---|---|---|---|---|---|---|---|---|---|
| | 1997 | | 1987 | | 1977 | | 1967 | | 2007 | |
| City | Estmts. | Sales (Millions) | Estmnts. | Sales (Millions) | Estbmts. | Sales (Millions) | Estbmts. | Sales (Millions) | Estbmts. | Sales (Millions) |
| El Paso | 950 | 5,954.5 | 876 | 2,629.3 | 650 | 1,271 | 489 | 532 | 2,961 | 4,462.8 |
| Albuquerque | 919 | 3,630.1 | 1,031 | 2,717 | 714 | 1,325 | 484 | 435 | 866 | 4,844.1 |
| Las Vegas | 500 | 2,208.8 | 349 | 1,172.3 | 285 | 425 | 177 | 138 | 449 | 1,709.5 |
| Phoenix | 2,447 | 21,996.9 | 2,412 | 12,583.2 | 1,658 | 4,830 | 1,144 | 1,420 | 2,332 | 29,631 |
| Tucson | 615 | 1,801.0 | 682 | 1,270 | 464 | 674 | 324 | 199 | 641 | 9,596.5 |

*Source:* U.S. Census Bureau. http://www.census.gov/.

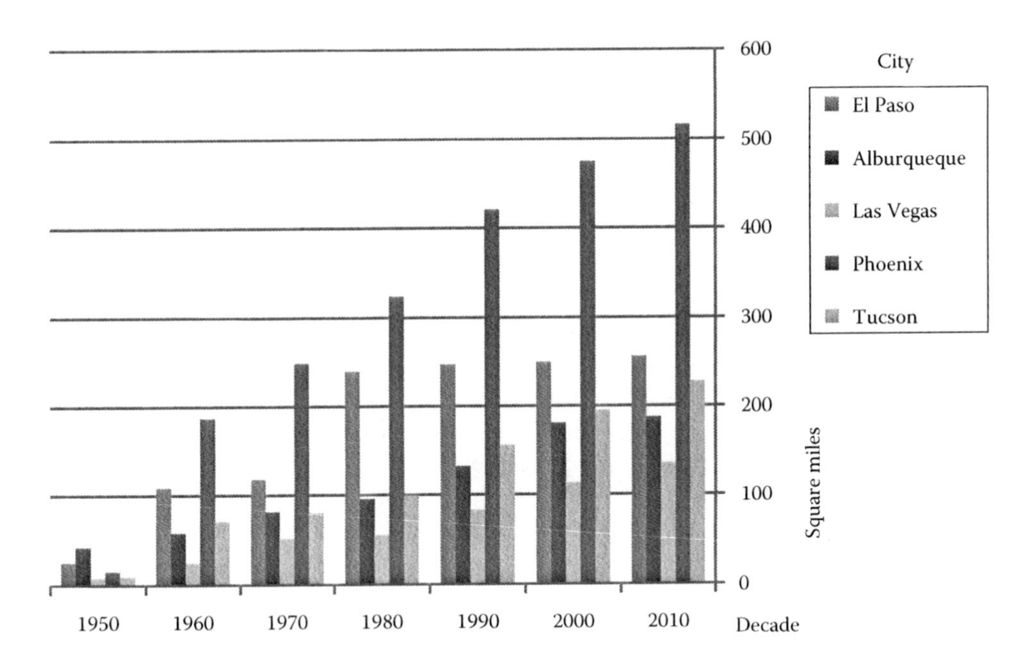


**FIGURE 26.13**
Southwestern city land area.

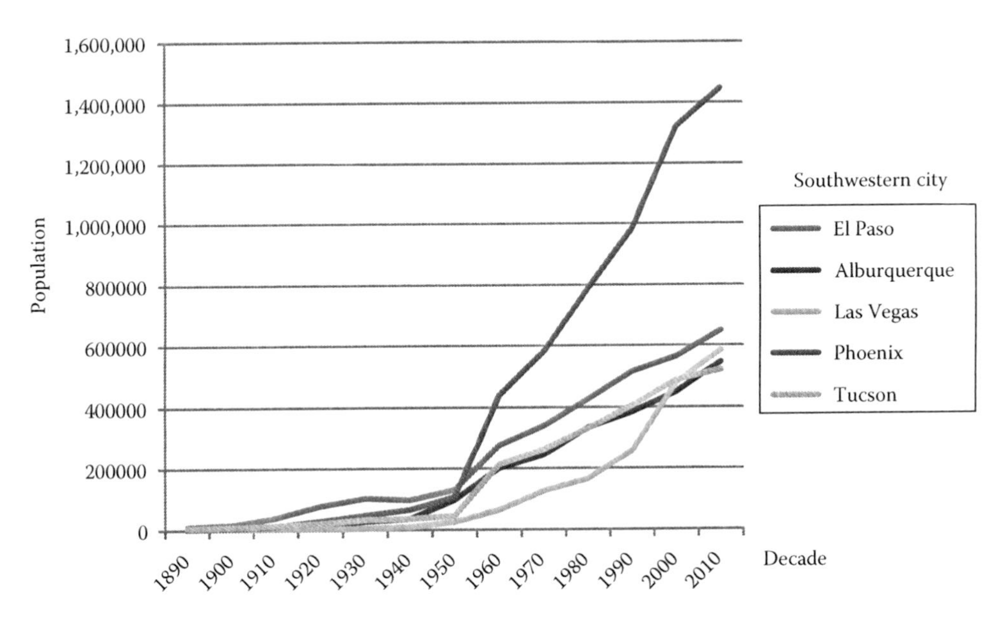


**FIGURE 26.14**
Southwestern city population.

and the presence of T-gen, a biotechnology consortium, the construction of a light rail system, and other development projects have significantly revitalized the urban core and will help ensure this city will be able to sustain itself for decades to come.

## 26.11 Conclusion

The Southwest has an appeal of character and uniqueness that sets it apart from other regions of the country. The history of these towns from the days of cowboys and Indians left a mark on the establishment of these emerging Sunbelt cities. The rapidly urbanizing desert cities are attracting new residents seeking prosperity and enjoyment from these urban settings. The fragile balance of the region could easily be lost without an understanding of how these great cities developed. The scarcity of water will define the region and dictate the places where development can occur. Las Vegas and Tucson are in the most critical position to contemplate the future of the urban area with imminently diminishing water reserves for development. We hope that the people of today will carry forth responsible development that will continue to serve the existing community and maintain a viable city to live, work, and play.

In the end, the southwestern cities will continue to grow and develop in the United States. In summary, we will close with the following reflections about this dynamic region including

*...the good*

Southwestern cities have an abundance of sunshine and seasonable weather that has been a force of attraction for people from other areas to consider relocation or investments. Area leaders have shown leadership in attempting to attract relatively clean industries that would provide high-paying jobs and community development projects. The quality of life through economic diversity and stability is an attractive element of the southwestern urban centers.

*...the bad*

The increase in population and diversity has come at the cost of creating more urban pollution and congestion from industrial and transportation activities. The quality of air that was once an attraction to these desert cities is now an area of health concern. The current pace of development will continue to add to the mounting problems in the region. In addition, areas that have refrained from comprehensive solutions to critical community problems will suffer the consequences by limiting the development of the urban center. Tucson, for example, is now experiencing critical concerns for urban development due to the lack of availability of long-term water resources to support development activities.

*...the ugly*

Urban sprawl is creating an aesthetically unappealing landscape of disjunct houses and developments and human activities. The effect of private development that favors low-density urban planning has created stains on municipal infrastructure and urban transportation systems. Also, the loss of the natural desert in some areas around the rapidly urbanizing cities that are being replaced by urban wastelands has left a landscape that holds little attraction for the urban or natural community in the desert (Figure 26.15).

**FIGURE 26.15**
Aerial photo of Phoenix in 1970, showing low-density urban sprawl development pattern. (Courtesy of the U.S. Environmental Protection Agency.)

## References

1. Hawley, A.H., *Urban Society: An Ecological Approach* (New York: John Wiley & Sons, 1981).
2. Galdys, G. and Liss, S., *The Chamizal Dispute, the Handbook of Texas* (Austin, TX: The Texas State Historical Society, 2002).
3. Land, B. and Land, M., *A Short History of Las Vegas* (Reno, NV: University of Nevada Press, 2002).
4. Luckingham, B., *The Urban Southwest: A Profile History of Albuquerque, El Paso, Phoenix, Tucson* (El Paso, TX: Texas Western Press, 1982).
5. Scott, D., Water shortage tests El Paso's conservation efforts, Governing the states and environment, *Energy and the Environment*, February 2012.
6. McKinnon, S., State may deal water to Nevada, *Arizona Republic*, December 3, 2004, http://www.klamathbasincrisis.org/other%20places/nev_azwtrplan120304.htm (accessed September 28, 2012).

7. McKinnon, S., City faces SRP water limits, *Arizona Republic*, August 4, 2002.
8. Gleick, P., Peak water, *Circle of Blue: Reporting the Global Water Crisis*, Saturday, January 22, 2011, http://www.circleofblue.org/waternews/2011/world/peter-gleick-peak-water/ (accessed September 28, 2012).
9. Gelt, J., Henderson, J., Seasholes, K., Tellman, B., and Woodard, G. (Eds.), Looking to the past to understand the present, in *Water in the Tucson Area: Seeking Sustainability* (Tucson, AZ: Water resources Research Center, University of Arizona, 1998).
10. Smart water report: A comparative study of urban water use across the southwest (Denver, CO: Western Resource Advocates, 2002).
11. Economic impact of Arizona's principal military operations (Phoenix, AZ: The Maguire Company, 2002).
12. Luckingham, B., *Phoenix: The History of a Southwest Metropolis* (Tucson, AZ: University of Arizona Press, 1989).
13. Finnerty, M., *Del Webb: A Man, a Company* (Flagstaff, AZ: Heritage Publishers, 1991).
14. Vander Meer, P., *Desert Visions and the Making of Phoenix 1860–2009* (Albuquerque, NM: University of New Mexico Press, 2010).
15. Logan, M.F., *Fighting Sprawl and City Hall: Resistance to Urban Growth in the Southwest* (Tucson, AZ: University of Arizona Press, 1995).

# 27

## *Creating Tomorrow*

**Vernon D. Swaback**

**CONTENTS**



### 27.1 Introduction

The spirit of architecture and planning cries out for authentic commitment and design. For the latter part of the nineteenth and all of the twentieth century, this meant breaking beyond imitating the past into expressions of our own time. Some saw this as an abusive forgetting of our heritage. Others like Frank Lloyd Wright advocated for our youthful democracy to have an architecture of its own. In the early 1970s, what the architectural profession celebrated as "Post Modernism"[1] was more a literary invention than anything to do with the significance of architecture. The result was too-often nothing but an awkward grafting of iconic remnants of the past onto box-like buildings of the present, with neither adding meaning to the other. The architectural profession's celebration was short-lived. Within two decades after Mario Botta had become one of the most highly regarded postmodern practitioners, he repudiated the movement, saying that he found it to be a disgusting paving-the-way for anything goes. He went on to describe postmodernist ideas as being the products of the barbarian architecture, justifying buildings that resulted in a colossal waste of energy.[2] To this, I would add that the movement was also a literary justification for the imitation of the worst kind.

By the end of the twentieth century, the search for architectural freedom pushed the limits of whatever engineering would permit, resulting in buildings of complex and unusual shapes. At its best, this exploration was accompanied by a dialogue that focused not only on form but also on space and place. This led to a merging between urban design, architecture, and landscape architecture, with the emphasis remaining on inventive forms and favoring whatever lent itself to dramatic imagery. In the more extreme cases, building shapes were contorted with little or no relationship to purpose, simply because technology made it possible to do so. At the turn of the century, this search for the dramatic had gone beyond the meaningful to produce the merely curious.

In the wake of this obsession with imagery, we are beginning to pursue a deeper commitment to purpose. After centuries of gradual change, starting in the 1950s, important metrics including population, water use, the increasing number of motor vehicles, the concentration of atmospheric $CO_2$, ozone depletion, and the loss of tropical rain forest began to spike upward.[3] This is now fostering a more integrated need to understand ecologically sustainable objectives that will revolutionize and broaden the reach of architecture, planning, and related disciplines. It is nothing less than the world of design reaching out to embrace the design of the world. While development patterns of the past will not take us to this more comprehensive engagement, the design professions have the greatest potential to move us toward ever-higher levels of ecological understanding and performance.[4]

James Rouse, one of the twentieth century's most philosophic developers, insisted that we are not coming up with the right answers because we are not asking the right questions. As an exploratory start in asking the right questions and with a focus on the Southwest, consider how we might answer the following:

1. Do you believe that our now dominant patterns of growth can lead to a better future than our recent past or will continuing in this direction lead to a deteriorating quality of life?
2. Of that which you have observed or experienced, what planning and development activities provide hope for a future you would like to see happen? Which cause you the most concern?
3. Do you believe that the Southwest is blessed with a compelling sense of urgency that could inspire a youthful, creative spirit?
4. If the advantages of climate, natural beauty, open government, human vitality, and imagination could form the basis of a heightened level of commitment, could you foresee a future brighter than our past?
5. In the spirit of creating a ground-swell urgency for good design, what guiding principles would you regard to be among the most significant?
6. What qualities do you want most for your own life that money *alone* cannot buy and you cannot have unless a great many others have them as well?
7. If you were given the opportunity to lay the groundwork for a high-performance desert community of whatever size and location you choose, what key differences would you advocate from what now exists in our dominant patterns of development?

The purpose of these questions is to get beyond the far easier and by now tiresome recitations about what we do not like and rather thoughtlessly call "sprawl." If we are serious about beneficial change, we must know that for the future to go beyond replicating the past and we must decide to do something different in the present. Among the beliefs that are not helpful are those in this editorial statement from an Arizona newspaper. "The rugged truth is that most of our local desert is doomed and has been since the Europeans first brought their concept of property ownership here."[5] This view assumes the presence of people, you and me, to be ruinous to the desert. What if the reverse were true? What if instead of exploiting the desert character, we became involved in a cocreative understanding as to how to live in partnership with what makes it so special?

For more than three decades, my firm has been engaged in community programming and planning for both government and private sector developers. The following seven challenges summarize a wide variety of concerns.

### 27.1.1 The Desert Culture

Cities in the Southwest have long topped the list of the nation's dynamic growth areas, replacing untouched desert with homes and businesses at an unprecedented rate. When confronted with statistics concerning the speed at which open land is becoming suburbia, even the most astute observers can be stunned. At one extreme are those that exploit the advantage of being in the right place at the right time. At the other extreme are those who see a paradise being destroyed and vow to stop the insanity.

Between these two extremes, there is a range of less obvious voices that can be critical to our future success as both occupants and stewards of the land. For those of us who represent this third view, it would be helpful to hold in mind two questions. First, if we agree to set our sights on creating a sustainable community, how might we raise our level of dialogue and what long-range commitments should we be considering in the present to achieve that goal in the future?

The second question concerns the visual character and quality of the environment. If what we build were to draw its inspiration from the Sonoran desert, what would that look like? These two questions are meant to address the full spectrum of community development, from high-rise urban centers to that which occurs in the vastness of the rural landscape.

To effectively design for the future, we must be willing to reconsider anything that now stands in our way, including how we talk to each other and how we translate our personal desires into a more collective sense of understanding and action. Years of low-level dialogue in public debate have resulted in polarized standoffs and single-issue arguments that have been both costly and self-defeating. Our new dialogue must include a deeper understanding about what works and what does not—for all of us and in the long term.[6]

A heightened awareness, including all that occurs under the banners of smart growth, green architecture, and sustainable design, is all about building and living in tune with the nature and character of the land (see Chapters 28, 30, and 31). To understand our desert heritage while creating appropriate desert settlements are two sides of an inseparable quest. If we do not excel in both, we will end up with neither.

What if instead of identifying with the plains of the Midwest or the character of our coastal regions or all the other nondesert places we have visited or once called home, we could begin to feel a kinship to the builders of Pueblo Bonita and Canyon Chelley? What if instead of using generated power to overcome the heat of the desert, we designed our activities and structures around adjustments to the climate of these special places?

And what if we added to our collective memory a kind of imagined past in which we felt some bond with the Anasazi and the Spaniards or with the ranchers who so recently worked the land where our houses now sit? Could we not design for a heritage that adds richness to our daily lives in the here and now? Would not this be more appropriate than the tendency to see everything through the preconceptions we carry with us from nondesert environments?

Those who long ago inhabited the Southwest found meaning in the mythic vastness and qualities of the land (Figure 27.1). Instead of shutting themselves up in little boxes, they found a way to have an ongoing dialogue with the environment. They and we live with a

(a)

(b)

(c)

**FIGURE 27.1**
(a,b) Univision Studio in Phoenix. Rammed earth, tensile fabric, and weathering metals. (c) Univision Studio at night.

contradiction that has been experienced by all desert people. It is the need to be sheltered from the scorching heat of the sun while celebrating the gift of living openly under the sky.

If we were living in the eastern states, we might take pride in the number of buildings our community had on the National Historic Register. That is not who we are. Instead, we have more national parks, national monuments, and historic sites than any other region. It is all part of our reason to feel our own special connection with the land.

To the extent that we value authenticity, our future success hinges on a kind of personal rebirth—a sense of belonging to the desert in a way that commits us to a stewardship beyond the rights of ownership. We may have the constitutional right to do as we please, but we need to respond to a higher purpose than anything allowed by codes or ordinances. Few places on earth are as blessed by nature as the Southwestern deserts. We need to get to the point where a felt sense of that heritage enriches everything we are and do.

### 27.1.2 Indigenous Design

Growing out of the first, the second challenge is to nurture an indigenous approach to design until it becomes the dominant character of our structures and landscape. Just a few

**FIGURE 27.2**
Entry garden at Swaback Partners Studio Offices in Scottsdale, Arizona. No imported palette can compare with the beauty of native trees, shrubs, and grasses of the desert Southwest.

decades ago, desert plants were not to be found in our nurseries. It was common for those who moved to the desert to treat the land as a blank slate, waiting to be made over in the image of wherever they came from. Pioneering nurserymen who began nurturing indigenous plants were thought to be growing weeds (Figure 27.2).

We have come a long way in understanding the value of desert trees and shrubs. Desert plants are now protected by ordinance, and restoration procedures have been developed to a high art. Palo verde (*Cercidium florida*), palo brea (*Cercidium praecox*), and mesquite (*Prosopis* spp.) trees surround single-family dwellings and have become a welcomed feature in highly urbanized areas.

The use of desert plants has expanded beyond the prickly varieties to those that provide year-round color, all attracting the sights and sounds of birds. No imported palette can compare with the beauty of native trees, shrubs, and grasses (Figure 27.3). Mexican poppies (*Eschscholzia californica*) and honeysuckle (*Anisacanthus thurberi*), globe mallow (*Sphaeralcea ambigua*), gallardia daisies (*Gaillardia aristata*), and penstemons (*Penstemmon* spp.) provide richly varied forms and colors. Add ocotillo (*Fouquieria*), saguaros (*Carnegia*), and prickly pear (*Opuntia*) and the indigenous variety produces a far richer environment than any less appropriate landscape could hope to achieve (see Livingston, this volume). Plants that have been adapted to the Mojave (Las Vegas) and Chihuahuan (Albuquerque) deserts include native mesquite (*Prosopis*), apache plume (*Fallugia paradoxa*), deergrass (*Muhlenbergia rigens*), bear grass (*Xerophyllum tenax*), candellia (*Euphorbia antisyphilitica*), and claret cup hedgehog (*Echinocereus triglochidiatus* var. *melanacanthus*).

**FIGURE 27.3**
Streetscape scene outside the Swaback Partners Studio in Scottsdale, Arizona.

(a)

(b)

**FIGURE 27.4**
Skyfire Residence. (a) Nightscene. (b) From the outside looking in.

The Southwest already excels in its exploration of a regional architecture that belongs here and no place else. What is needed is to have the understanding represented by our best work become more widespread. From the earliest Hohokam and Anasazi cultures to the finest work of the present, the desert has produced an architecture that radiates a space-loving sense of mystery and creativity (Figure 27.4).

No amount of worldly sophistication can ever equal the unique character of the desert. We may be exhilarated by high-rise buildings, stadiums, and other symbols of urban triumphs, but unless they reflect an indigenous sense of belonging, all such additions can ever hope to accomplish is to diminish our desert settlements by making them more and more like every place else. We may need to be reminded that those of us who call home, the Mojave, Sonoran, Chihuahuan, and Great Basin Deserts, are all living in the same ecosystem. Whether we live downtown in a mid-rise apartment or on a ranch at the end of a dusty road, we are all inhabitants of the desert.

**FIGURE 27.5**
Frank Lloyd Wright's Taliesin West, Scottsdale, Arizona.

Frank Lloyd Wright considered the Sonoran desert to be among the most creative places on earth. How we live with what we have been given is up to us, including the danger that we could lose what makes our environmental experience so special. The threat of that loss is worthy of on-going evaluation (Figure 27.5). The next time we get excited about something to be added to the land, we should ask a simple question: will the addition, however appealing in itself, make the community more like every place else or will it reinforce that which makes the environment such a special place?

The Southwest character is a great teacher. Its textures, colors, and geometry are varied, bold, and dramatic. The flavor of the Southwest has many moods. Watch the cloud shadows pass over the land and surrounding mountains. Something new is always being revealed. And so it must be with any human habitation worthy of such magnificent settings. Everything we build, from the most rural to the most urban environments, should express a connection to the timeless spirit of the land.

### 27.1.3 Regional Open Space

The third challenge is to create a regional network of open space as "large chunks of healthy natural landscape" after the manner of Frederick Law Olmsted and Calvert Vaux.[7] Once established, this open space system would serve as the organizing framework for the connecting circulation system and all other land uses including valuable vegetating, drainage, and regional connectivity for wildlife. No matter how many details can easily get in the way of thinking at a regional scale, this network should go well beyond present needs to address the ultimate build—out for as far as we can imagine. Anything less will simply continue the frontier notions of piecemeal decisions that no longer serve (Figure 27.6).

Frontier values are those that assume there are always new landmasses and untold riches yet to be discovered. While it is true that the Southwest has many thousands of undeveloped acres that may seem to be out there in the great beyond, most everything in private ownership is being analyzed as a candidate for future growth with little to guide

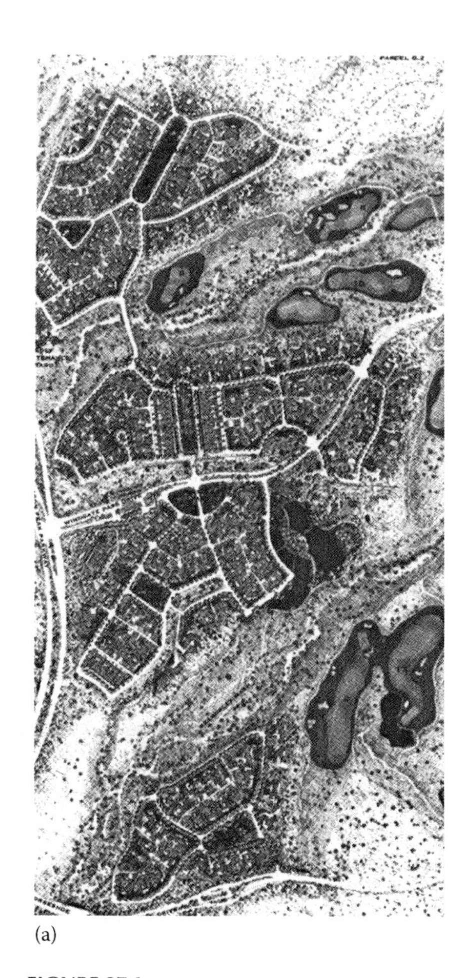

(a)

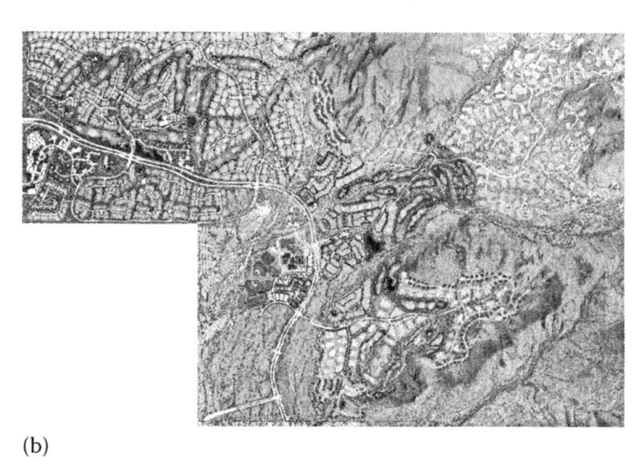

(b)

**FIGURE 27.6**
(a) Wash Preservation at DC Ranch, Scottsdale, Arizona. (b) DC Ranch Master Plan.

such thinking beyond the ownership boundaries of each separate parcel. No one has ever referred to Manhattan as "sprawl" because its past and future growth has always been informed and limited by natural boundaries. In the great Southwest, few such natural or obvious edges appear to exist.

There are many arguments against establishing a regional or statewide system of open space. The first is that we do not need it because, for example, in Arizona less than 17% of the entire state is in private ownership. This is like saying, because we have the oceans, we do not need lakes, rivers, and streams.

What is open land today will not remain so for the future. As an example, we have long become accustomed to viewing as open space the lands held by the Southwest Indian communities whose economic base remained agricultural while the rest of the Southwest was rapidly developing otherwise. This growth around the Indian lands now make them commercially appealing settings for hotels, lakes, golf courses, casinos, shopping centers, and all manner of industrial uses. Add to this that in Arizona, the State Trust lands exist for the benefit of the public schools. By constitutional mandate, the state must seek to lease or sell these lands at full market value, which usually means whatever form of development can most profitably exploit the resource. So in addition to the privately owned land that can be developed, we can add 27.5% for the Indian land and another 14% for the land held in trust by the State.[8]

Another argument used to oppose a regional or statewide network of open space is its triple cost. First, the land has to be purchased, generally at fair market value. Once in the public domain, it is removed from the tax rolls, and lastly, there is the cost of ongoing maintenance. For much of the open-space system, the only answer to such concerns is that the long-term value is worth the public investment. In other cases, in addition to land requiring outright purchase, the open space network can be expanded wherever continued private use can be maintained with benefit to the land. The traditional uses of forestry, farmlands, citrus groves, ranching, and recreation are all candidates for combining private use with public benefit. The devastation of our recent desert and forest fires has created an urgency for more intelligent land management. Conservation groups are beginning to work with private landowners and municipalities on everything from agriculture and ranching to a more ecological approach to the treatment of golf courses and the design of wetland gardens as natural systems for reclamation.

An exciting example of a public/private approach is described in *Beyond the Rangeland Conflict* by Dan Dagget. This book chronicles the story of 11 ranches where the land is being restored in the process of being fully utilized. In his review, Gary Paul Nabhan, the award-winning author and director of the Center for Sustainable Environments at Northern Arizona University, states that the book explodes, "the false dichotomies of user versus conservator and rancher versus environmentalist."[9] Dagget's book portrays an integrated, highly complex view of open space as something that can be used while being effectively preserved wherever public/private entities are able to choose cooperation over confrontation.

The regional open-space network I advocate is not like Portland's urban limit line that exists as a temporary barrier to be moved as needed. Nor is it a greenbelt surrounding any one community in the manner of a jurisdictional buffer. As much as possible, regional open space systems should be established by way of ecologically defined edges, irregular, inviolate, and determined by the nature of the land, and beyond the reach of political decisions to change over time.

### 27.1.4 Complexity and Integration

The fourth challenge is to correct the wrong turn we took when suburban development began to associate *separation* with the creation of *value*, including stratified housing projects, isolated office parks, and big-box shopping centers (see Chapter 25). Everything was put neatly in its place, easy to build, see, and sell. The creation of sustainable value requires a far more complex integration of uses. The urban fabric will become not only more efficient, but more alive, including proximity to agriculture and community gardens. We will come to treat the all-out, thoughtful integration of uses as a design-based alchemy for doing more with less (Figure 27.7).

This approach will differ greatly from suburban developments that are based on the assumption of mutual intolerance. Because community interaction has been so intentionally designed out of suburbia, it would be possible for every household to dislike the others, and all would still go reasonably well. We have so accepted this as the norm that we do not realize what a price we pay for our lack of community. An assumed inability to cooperate requires more paving, while offering less natural open space. The typical suburban development offers more pseudo privacy with less genuine individuality and more autonomy with less awareness, caring, and security. Just as no individual space can mean

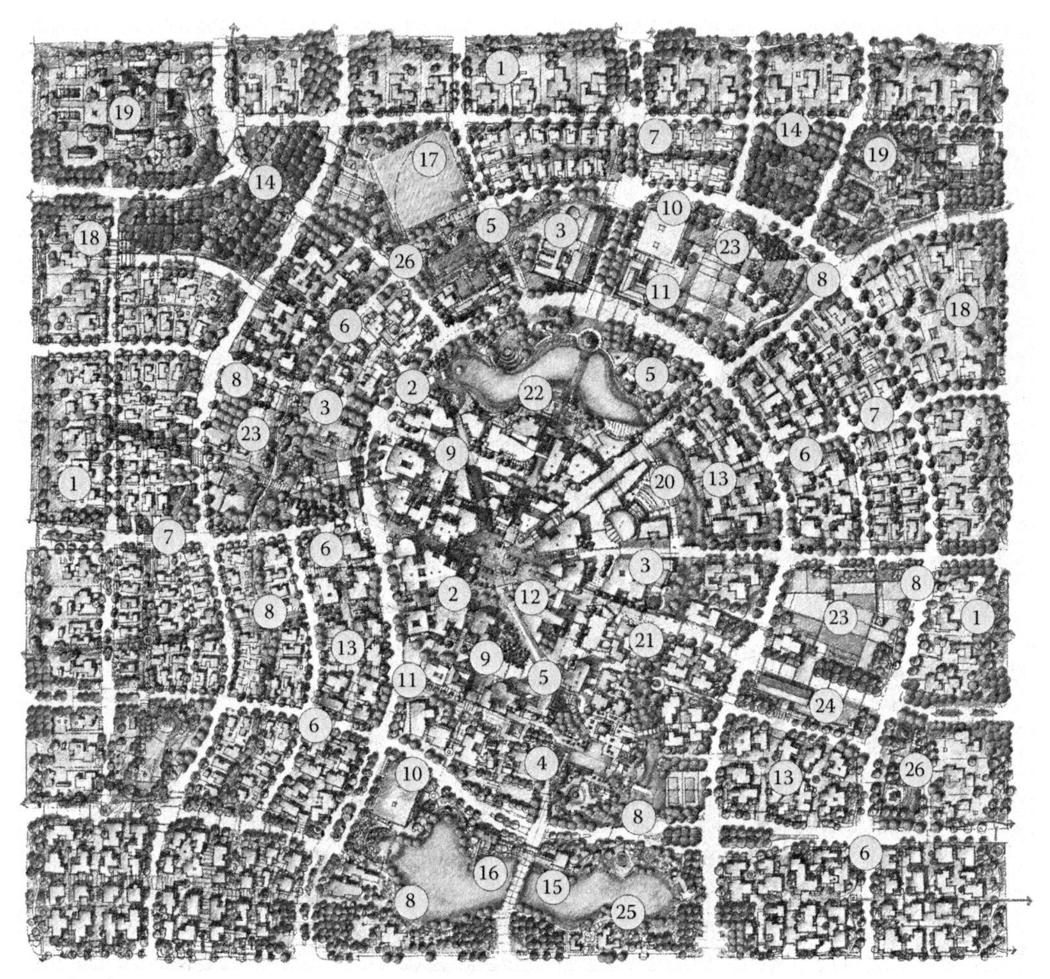

Land uses

1. Half acre lots and paths
2. Offices and commercial
3. Schools
4. Boutique hotel
5. Organic gardens
6. Attached residential
7. Detached residential
8. Community trails
9. Live / work districts
10. “Stable” on-call vehicles
11. “Commissary” good and services
12. Auto free mixed use
13. Auto free residential
14. Orchards and citrus groves
15. Ecology center
16. Metered house
17. Sports fields
18. Custom residences
19. Family compounds
20. Art school and recital center
21. Design center
22. Community commons
23. Fields and farmers market
24. Health and wellness spa
25. Biological filtration gardens
26. Community parks

**FIGURE 27.7**
200 ac site integrating 26 land uses.

very much unless it has a well-proportioned relationship to its surroundings, no house can achieve its full potential unless it exists in supportive balance and harmony with its overall setting. Until our sense of home can include more than whatever we can lock and call our own, we will be short-changing our experiences and paying more for less.

Economic and market forces cannot be ignored but they can be considered within a broader perspective. In the new paradigm, developers, citizens, and government officials will speak of the spirit of community with as much conviction as we now speak of codes and ordinances. Words like interdependent, holistic, integrated, indigenous, even spiritual, will be used as easily and as frequently as we now use words like coverage, setbacks, and density.

We will know that the future is in good hands when we put our trust in the language of human values, with only occasional references to the transactional measures of codes and contracts. In far too many instances, it is still the other way around, but that is all about to change.

Whether it is the villages of Europe or the American towns of the nineteenth century or the newly master-planned communities of the Southwest, the places we find most appealing have one thing in common. They could not be as they are without some degree of authority for establishing agreed-upon rules.[10] This inevitably involves carefully placed limitations on personal freedom. No significant planning or community of merit can occur without cooperation, and cooperation is only possible when there is an allegiance to a framework of limits. Without the agreed-upon limitations of a musical score, the finest symphony orchestras would be incapable of producing anything but aggravating noise. The analogy is most appropriate. Like the musicians, we are all individuals who are free to be our best, but our highest achievements will always depend on our ability to nurture artful, interdependent relationships with others.

### 27.1.5 Effective Transportation

With the possible exception of density, no issue has inspired more outrage with less clarity than that of how we transport ourselves from one place to another. To add to the confusion, we tend to treat dissimilar problems as though they are the same. Congestion, gridlock, air pollution, mobility for the nondriver, the loss of human scale, and energy conservation may be related, but they are not the same (see Chapter 14).

Consider the following as a starting point for a more shared understanding. It is a mistake to try to get rid of congestion. All the great cities of the world experience varying degrees of congestion. It does not even require automobiles. The tightly packed pedestrian lines at Disneyland are a testimony to the pervasiveness of congestion. It occurs any time a great many people all want to go to the same place at the same time. Where human interaction is desirable, as in a vibrant retail scene, a parade or a sporting event, congestion is not only part of the experience, without it, the result would be considered a failure.

It would be an even greater mistake to plan on a future in which we give up our automobiles. Tomorrow's cars will become more varied in size and performance, more fuel efficient, and possibly capable of connecting to community-wide guidance devices that increase the capacity of our street systems while decreasing accidents. What will remain constant is our need and desire for personal mobility.

My views differ from those who see the automobile as destructive to the urban fabric and public transit as its savior. By now, we should all know that this matter deserves more open dialogue than simply choosing up sides.

Here is why I do not foresee us all giving up our cars for public transit. Reduced to its fundamentals, successful public transit must satisfy only two requirements: (1) It must go everywhere; (2) All the time. A common-sense third requirement is obviously cost. Those using public transit are more likely to be price-sensitive than those who can afford to drive everywhere. These requirements explain why personal transit in the form of cars is overwhelmingly popular throughout the United States. They also explain why public transit remains a marginal transportation mode in most parts of the country. Not only do cars go wherever roads are provided, whenever the driver wishes, but automobile transportation costs are paid largely by users, who buy, maintain, and insure their own cars and who fund a good part of highway construction and repair costs through gas taxes.[11]

By contrast, public transit routes do not go everywhere and service usually emphasizes weekday commutes, with abbreviated schedules and longer headways for nights and weekends. Moreover, the costs of equipment, operations, maintenance, insurance, and employee wages and benefits are publicly funded, with fares reimbursing only a fraction of total cost.

This sets up a chicken-and-egg situation. On the one hand, public transit cannot be considered a serious alternative unless it is at least reasonably convenient. On the other hand, building transit routes to a reasonable degree of convenience that attracts patronage is prohibitively expensive in many regional situations. Rail transit requires large amounts of energy, so a partly filled train—which is unavoidable until a reasonably complete system is in place—is the mass transit equivalent of a single-occupant SUV. Many of these downsides to public transit are overcome by buses. Buses are the unappreciated workhorses of regional transit.

Considering the multitude of operating examples that exist around the world and, in particular, the demonstrated relationships between urban form, density, and transportation hardware, we should have long ago stopped looking for any simple either–or answers.

In summary, seven ways to increase the efficiency of urban mobility and individual vehicular trips include (1) car and van-pooling, (2) strategic scheduling to lessen traffic during peak demand, (3) delivery systems that distribute goods and services more efficiently than individual back-and-forth trips, (4) teleconferencing, telemedicine, and telecommuting, in which one "travels" with zero hulk at the speed of light, (5) a redesign of the present automobile, including its size, guidance systems, and fuel source, (6) more compact, interrelated land-use patterns with provisions that make walking or riding a bicycle a viable option, and (7) going beyond ownership to shared use in the form of taxis or the subscription use of shared vehicles and, where appropriate, public transit.

Although rail and bus-ways potentially offer superiority in speed over cars, for urban commuting, each transit stop along a rail or bus line unavoidably reduces possible time advantages compared with driving a private automobile or taking a taxi. Ideally then, the most effective, time-competitive public transit would accommodate the largest number of passengers with the smallest number of stops. In the United States, this ideal situation occurs in only a handful of places like the Manhattan subways and the BART corridors in the San Francisco Bay area.[12]

### 27.1.6 Technology, Awareness, and Behavior

The sixth challenge concerns our ecological understanding and response. The first decade of the twenty-first century combines unprecedented advances in technology with greater than ever human awareness and comparatively little change in human behavior. The problem is that everything to do with creating ecologically sustainable environments is at least 75% dependent on changes in human behavior with the contribution of innovative technology, contributing 25% or less.

High strength steel and the invention of the elevator created the possibility for centralized, high density, cities. Moving in the other direction, the technology of vehicles, ships, aircraft, first wire then wireless transmission, the computer, and the Internet, all greatly increased the possibility for decentralization. Human behavior has opted in favor of the latter, all in the direction of what Frank Lloyd Wright called Broadacre City, the city that would be "everywhere and nowhere." Three-quarters of a century after Wright published

*The Disappearing City*, Richard Ingersoll wrote in *Sprawltown*, "Almost without notice the city has disappeared." Ingersoll goes on to say, "Though people continue to live in places with names like Rome, Paris, New York, and Beijing, the majority of the inhabitants of the developed world live in urban conditions somewhere outside the center city."[13]

With respect to the built environment, in so many ways, our understanding of technology and behavior has never been more focused. Governments, developers, designers of all kinds, engineers, manufacturers, builders, educators, professional societies, and journalists have all joined in the chorus of advocacy for planning and design that is smart, green, and sustainable. But we must still design for the future on a case-by-case basis. Many urban areas are land-locked and, having been built out, can only build up. Timing sets another decisive difference. So much of what we associate with great European cities was already formed or transformed a century and a half ago. Within the United States, New York was shaped in the 1910s and Los Angeles largely in the 1950s. These and all other cities will continue to change. Many of today's most dynamically growing cities—for example, those in Asia, instead of creating something new are simply playing catch-up along the lines of nineteenth century patterns.

Against this background, the desert Southwest stands out as having one of the greatest potentials for a new direction that combines the behavioral desire for decentralization made possible by technology, with the nature-inspired integration that ecology demands.

While the future is unknown and unknowable, there are always insights from which to make long-range decisions. From the local to the global conflict is now the normal dynamic, given the world's growing need for fuel, food, and freshwater. Where possible, the desired pattern of growth will be in the direction of decentralized, high-performance settlements designed around sources for alternative fuels, fresh produce, and recycled water. In keeping with this direction, the Southwest enjoys an abundance of solar energy, vast stretches of fertile soil, and highly developed systems for water management, microfiltration, and the biological treatment of wastewater.

This leads us back to human behavior. Will our municipalities be willing to engage in regional planning? On the developer's side, if the more simplistic approach that produced suburban sprawl remains an option, to what extent will developers be inclined to pursue the greater rewards that require greater complexity?

In like manner, will potential buyers be willing to embrace the complexity and rewards of community as being more desirable and more valued than the notion of independence associated with conventional low-density subdivisions?

Will individuals be ready to take advantage of the benefits of live/work communities and shared-use vehicles instead of requiring the conventional separation between home and office and maintaining our emotional attachment to personal vehicles?

Will financing be available for what conventional wisdom will view as plans organized around complex systems of spatial connections? And will citizen activists herald the advent of demonstrably "green" communities or will they oppose anything that differs from what already exists?[14]

My observations and experience suggest that the answer to all such questions will be in whatever direction can be shown to provide for a genuinely richer, more healthful way of living for less cost. Codes will be changed to allow for multiple generations to inhabit the same house as well as the design and development of custom family compounds that replace the present line-up of look-alike houses.

A live/work version of the cottage industry of 200 years ago is now possible, with the added benefit of real-time audio and visual communication, both local and global. Instead of finding solutions to some of our more pressing suburban ills, we will simply eliminate the problems. The automobile reduced our reliance on the old form of centralized city. The digital revolution has the potential of reducing our reliance on the automobile.[15]

Employment, education, and health care all face the need to do more with less. The old paradigm of centralization was reflected in large factories, large schools, and large hospitals. Economies of scale required that the brightest individuals and the best hardware be given the greatest utilization and that meant bringing as many people as possible to centrally located places. Digital communication now adds an explosive thrust in the other direction. Access to work, access to education, and access to health care no longer mandate physical proximity.

The latest master-planned communities include communication infrastructures that connect houses and neighborhoods by way of computerized networks. The digital revolution is making new systems and opportunities as normal to our lifestyle as yesterday's use of the telephone. Telephony, telemetry, and other forms of telecommunications-based services are decreasing passenger miles for those who do not want to spend all day in their cars, while providing needed care to those who do not have that choice. Medicine can be dispensed electronically, with follow-up calls to confirm that it has been taken. If the calls are not answered, help is sent with greater dispatch than the family doctor of old could have hoped to provide (Figure 27.8).

### 27.1.7 Heroic Design and Commitment

We have suggested the need to extend the world of design to the *design of the world*. This could be seen as arrogant if it were not such an obvious necessity. Furthermore, it is a commitment

**FIGURE 27.8**
Bicycle and pedestrian path system at DC Ranch. Creating compact, interrelated land-use patterns makes bicycling a viable transportation option.

to extend ones caring and commitment beyond the boundaries of any one project. For all of human history, the metrics of population growth, consumption of natural resources, the rate of species extinction, and climate change have all been gradual, until now.[16]

These and other critical changes to the underlying dynamics that affect us all are destined to produce future environments very different from what is so familiar today. The coming changes in how we relate to each other and to the earthly home we all share are inherent in multiple trajectories already in play. While the degree of change to be experienced is not a choice, the character and quality of what such changes represent offer unprecedented opportunities for sustainable success. Both the opportunity and threat were long ago characterized by Buckminster Fuller as "Utopia or Oblivion."[17]

Having started with seven questions and worked our way through the prior six challenges, what remains is to put it all together into a positive prophesy for human life on earth. The seventh and final challenge is not only to preserve the natural environment but also to create heroic structures and settlements that respect and enhance their surroundings. An essential ingredient will be the creation of artful places that exert an emotional, gravitational pull upon the soul. Such creative contributions of humanity will become critical components of nature itself. We will understand our well being as something inseparable from the health of the natural systems that surround and sustain us and we will need new tools to get us there.

Many southwestern municipalities continue to struggle with Euclidean zoning and other mechanisms that are beginning to be seen as outdated and detrimental to the pursuit of smart development and green design. Others are taking steps forward to revamp their approach (see Chapter 13). A case in point is the City of Mesa and its Gateway Center, which is destined to become a new urban hub. The key to this vision is based on a public/private partnership between the City of Mesa and DMB, one of the southwest's leading community developers.

In 2006, DMB purchased the 3200 ac General Motors Proving Grounds with the goal of developing a new kind of urban desert environment. Swaback Partners is working closely with the Developer and City on both the physical planning and new tools for guiding the development. An analysis of existing zoning standards resulted in all parties agreeing that many of the City's past tools were not appropriate for the creation of twenty-first century desert urbanism. Mesa's zoning standards, like those enacted by many cities following World War II, had focused on separating land uses, emphasizing vehicular circulation, and inadvertently, embracing sprawl. For the Mesa Gateway Center, the new zoning, approved by the City Council at the end of 2008, is a hybrid system of codes and ordinances that include components of traditional zoning, performance-based zoning, and form-based code. Elements of the form-based code look to guide a seamless palette of (1) Open Space, (2) Civic Space, (3) Neighborhoods, (4) Villages, (5) Districts, (6) Regional Campus, (7) General Urban Center, and (8) An Urban Core.

Among our most significant twenty-first century goals, must be the creation of communities that become our greatest works of art, for which most structures will be experienced as background texture. Just as most trees in a landscape, or brush strokes in a painting, or notes in a symphony do not call undo attention to themselves, so it must be with the structures of a well-considered community. But background texture is only part of the story. Think of the timeless beauty of symbolic places that continue to inspire the imagination. King Ludwig II's Neuschwanstein Castle and the hillside villages of Italy and France all occupy sensitive settings in which they add something special. Closer to home, Taliesin West, and the ruins of Mesa Verde, Keet Seel, and Batatakin are enduring attractions. Their presence becomes part of the mystery and magic of nature itself.

**FIGURE 27.9**
Frank Lloyd Wright's Fallingwater in Bear Run, Pennsylvania.

Fallingwater in Bear Run, Pennsylvania, is very likely the most photographed house in the world (Figure 27.9). In David Pearson's, *Earth to Spirit: In Search of Natural Architecture*, he writes: "Frank Lloyd Wright … always attempted to build in harmony with the land. Completed in 1936, Fallingwater is the archetypal expression of Wright's spirit of the land and sense of place. Perched over the Bear Run waterfall, the building's various cantilevered levels glide over the water like the branches of a tree."[18]

The reason for bringing up this nondesert example is that such sentiments frequently accorded to Fallingwater are worthy of deeper analysis. Its design consists of huge slabs of concrete cantilevered over a waterfall with concrete footings anchored directly into a living stream. If the house did not already exist, and Frank Lloyd Wright was not so globally honored, how many of today's environmentalists would give their enthusiastic endorsement to the concept? I would guess the most accurate answer would be very few, if any. Calling attention to this example is simply to point out that too much of what goes under the banner of environmentalism has its own lack of understanding. However much we respect and defend the untouched land, it is also our home. In addition to the preservation of open space, what we build requires, not only sensitivity, but also boldness.

## 27.2 Summary

Under the title of Creating Tomorrow, we have offered a glimpse of the future, starting with an overview of the past. We have advocated the need for a regional connectivity of open space with all design elements expressing an indigenous sense of belonging. We observed that settlement patterns will become more decentralized, varied, and artful, and their multimodel transportation systems will be as personal as possible. Beyond the reach of technology will be the contribution of behavioral adjustments all in the direction of both stewardship and community.

In the context of human history, the earliest settlements were individual. A move toward centralization was fostered by many needs with the three most dominant being the need for defense, trade, and the desire to congregate for religious purposes. Unprecedented technology permitted decentralized, mainly agrarian settlements and the exploration of previously unknown territories. The centralizing force of the industrial revolution attracted workers off the land and into the factories, resulting in the separation of home and work. Advances in technology reversed the flow, allowing for commerce to decentralize into the countryside.

With the ability to communicate around the world faster than driving to work, time replaced distance. A first for humanity is that it has become possible to be both local and global at the same time and in the same place. Add to this that we are rediscovering the connection between our personal health and well being with that of nature and the community. The result is that we are well on the way to designing the new form of city that Frank Lloyd Wright long ago described as being "everywhere and nowhere."[19]

The past separation of housing, life, and work will be replaced with a far richer mix between community and the sustaining essence of nature. The now dominant repetition of same size, detached houses on uniformly sized lots, occurring in lock-step patterns along gridded streets and alleys have had their day. They will be replaced with mixed-use compounds for life and work as artfully integrated and widely varied as the musical performances that range from solos and chamber ensembles to orchestras and operas. Individual transportation devices, some shared, and others owned, will become more efficiently accommodated while nature and the pedestrian will be more richly celebrated. Technology will extend the economies of scale and high performance will become more personal. The result will be a kind of full-time version of the way-of-the-life patterns that are now more associated with the provisions of a resort (Figure 27.10).

Other, now dominant urban elements to be replaced are the isolated, hermetically sealed block buildings, and high rises. Their replacements will be richly terraced multilevel structures in which natural light, fresh air, and community gardens will be the norm. To visualize this new building form, think of a twenty-first century merger between the planted terraces of Machu Picchu and the Arcologies of Paolo Soleri, where indoor/outdoor living is the norm.

The traditional single-use suburbs will be replaced by what Witold Rybczynski characterizes as a blend between the Greenwich Village ideal described by Jane Jacobs and the Broadacre City ideals illustrated by Frank Lloyd Wright.[20] Wright's own Taliesin West provided an engaging laboratory for understanding how we might envision a live/work, doing more with less, cultural community appropriate to the desert environments of the Southwest.

We inhabit the desert regions against a background of great human endeavor. We stand on the shoulders of everything from the ancient settlements of the Hohokam and Anasazi to all that subsequent generations have created at their best. We are at a pivotal moment in the history of the Southwest and ready to move with unprecedented sensitivity and effectiveness. We have an enlightened citizenry and the capability of our builders and developers is second to none. Our architects and landscape architects are among the finest in the world for creating arid-region environments. All that remains is to raise our level of dialogue until the seven challenges outlined in this chapter become the seven pillars of our new reality.

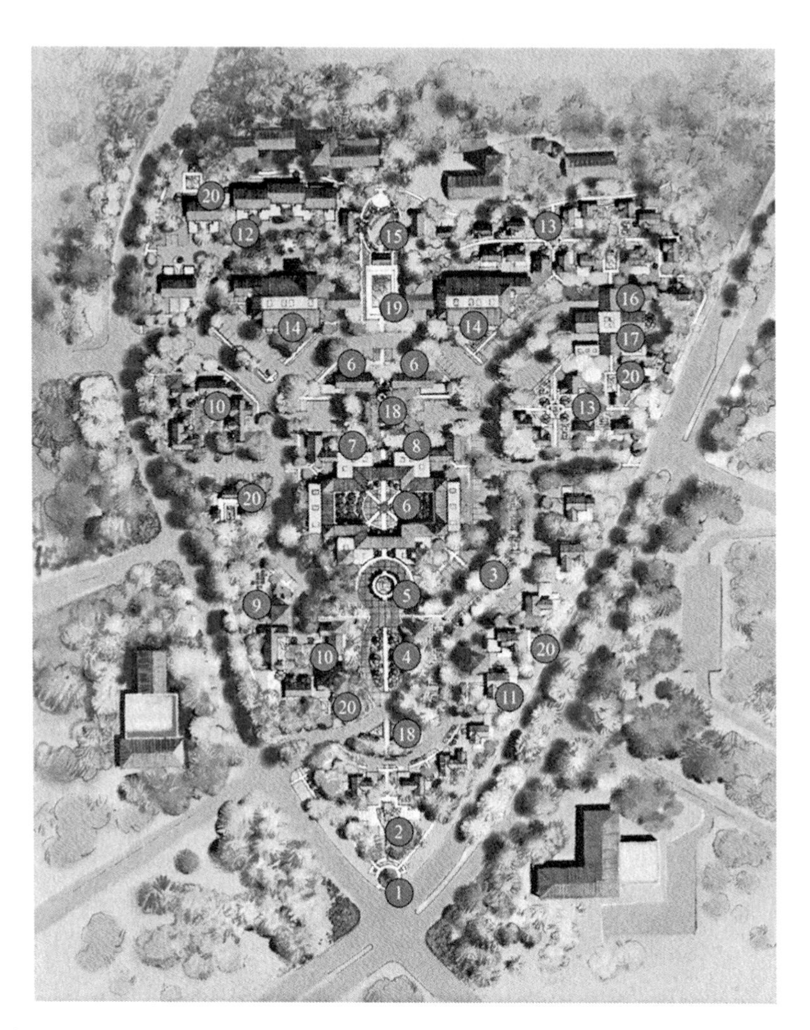


**FIGURE 27.10**
Mixed-use orchestration.

## References

1. Jencks, C., *Language of Post-Modern Architecture* (London, U.K.: Academy Editions, 1977).
2. negenter@worldcom.ch (Nold Egenter), Botta declares defeat of post modernism, April 9, 1997. Available online at http://home.worldcom.ch/negenter/486aDL_BottaEndPoMo_01.html (accessed August 20, 2011).
3. Steffen, W., A. Sanderson, P. D. Tyson, J. Jäger, P. A. Matson, B. Moore III, F. Oldfield, K. Richardson, H. J. Schellnhuber, B. L. Turner II, and R. J. Wasson, Global change and the earth system, 2005. Available online at http://www.igbp.kva.se/documents/IGBP_ExecSummary.pdf (accessed August 18, 2011).
4. Farson, R., *The Power of Design, a Force for Transforming Everything* (Atlanta, GA: Ostberg Library of Design Management, Greenway Communication, 2008), p. 1.
5. *Scottsdale Progress Tribune*, editorial section, June 15, 1996.

6. Swaback, V. D., *Designing the Future* (Tempe, AZ: The Herberger Center for Design Excellence, Arizona State University, 1997), p. 1.
7. Rybczynski, W., *City Life, Urban Expectations in a New World* (New York: Scribner, 1995), p. 125.
8. Arizona State Land Department, Land Ownership, http://www.land.state.az.us/acris/htmis/ownership.html (accessed August 20, 2011), 2009.
9. Dagget, D., *Beyond the Rangeland Conflict—Toward a West That Works* (Layton, UT: The Grand Canyon Trust and Gibbs-Smith, 1995), Back Cover.
10. Ehrenhalt, A., *The Lost City, Discovering the Forgotten Virtues of Community in the Chicago of the 1950's* (New York: Basic Books, A Division of HarperCollins, 1995), p. 21.
11. Johnson, H. B., *Order upon the Land* (Oxford, England: Oxford University Press, 1976), p. 39.
12. Swaback, V. D., *Creating Value, Smart Growth and Green Design* (Washington, DC: Urban Land Institute, 2007), p. 21.
13. Ingersoll, R., *Sprawltown, Looking for the City on Its Edges* (Princeton, NJ: Princeton Architectural Press, 2006), p. 3.
14. Swaback, *Creating Value, Smart Growth and Green Design* (Washington, DC: Urban Land Institute, 2007), p. 29.
15. Kotkin, J., *The City: A Global History* (New York: Modern Library, 2006), p. 149.
16. Brown, L. R., *Plan B 3.0, Mobilizing to Save Civilization* (New York: W.W. Norton & Company, 2008). Available online at http://www.earth-policy.org/books/pb3/pb3_table_of_contents (accessed August 12, 2011).
17. Mumford, L., *The City in History, Its Origins, Its Transformations, and Its Prospects* (New York: Harcourt, Brace & World, 1961), p. 9.
18. Pearson, D., *Earth to Spirit: In Search of Natural Architecture* (San Francisco, CA: Chronicle Books, 1994), p. 77.
19. Wright, F. L., *An Autobiography* (New York: Duell, Sloan, and Pearce, 1943), p. 320.
20. Rybczynski, W., *City Life, Urban Expectations in a New World* (New York: Scribner, 1995), p. 233.

# 28

# *Desert Vernacular: Green Building and Ecological Design in Scottsdale, Arizona*

**Anthony C. Floyd**

CONTENTS



## 28.1 Introduction

To what extent can the environmental criteria of green buildings shape the architectural vernacular in the upper Sonoran Desert of southwestern United States? How can regional design guidelines contribute to a smaller ecological footprint? This chapter will outline Scottsdale, Arizona's continuing efforts to develop effective design guidelines for a growing urban community concerned with preserving the natural desert with its flora, fauna, and geological features. One of the fundamental concerns for Scottsdale: to integrate development and the natural desert setting in an environmentally responsible manner. This represents the application of an architectural vernacular that relates the buildings to their environment. Vernacular buildings use locally suited design solutions for locally derived social and environmental conditions. True Vernacular architecture evolves over time to reflect the changing cultural and environmental context in which it exists. It is sometimes misrepresented as culturally dated and antiquated but is an important reference point in the design of climate responsive and environmentally responsible buildings. By using current building technologies combined with the passive design techniques that respond to site conditions, climate and regional resources, an architectural vernacular unfolds (see Chapter 31).[1]

Environmentally responsible buildings are designed to be less impactful and in turn harmonious with local bioclimatic and environmental conditions. This is one of the core objectives of most green building initiatives including the Leadership in Energy and Environmental Design (LEED) building certification program. Green building as defined by the Scottsdale Green Building Program is

> a whole-systems approach through design and building techniques to minimize environmental impact and reduce the energy consumption of buildings while contributing to the health of its occupants.[2]

The design of green buildings encompasses a range of design features from passive building design principles (e.g., orientation, massing, and shading) to active building systems (e.g., mechanical ventilation, automated lighting controls, dynamic glazing). The degree to which the environmental features are revealed in a building's esthetic is determined by the building envelope and its interface with site, climate and the context of place, both ecologically and culturally.

Architectural history depicts a dichotomy ranging from the romantic "primitive hut" to the autonomous living machine. There are numerous examples and approaches that demonstrate integration, harmony, and the dichotomy of buildings in their "natural" setting (Figure 28.1). Over the ages, desert regions have historically demonstrated a certain vernacular approach. Unique building characteristics have evolved as a result of regional climatic and geographic conditions.

Over the past 30 years, the city of Scottsdale has strived to develop a unique Sonoran Desert urban vernacular character, which reflects the lifestyle and values of the community. During the 1970s a Development Review Board was established to address the design of buildings reflecting the context of the local site and place. However, the review and approval of projects often led to inconsistent design outcomes. With the advent of the Scottsdale's Sensitive Design Program in the mid-1990s and the establishment of the Green Building Program in 1998, the city has made significant strides in developing environmentally sensitive design criteria that respond to the bioclimatic conditions of the Sonoran Desert region (see Chapter 27).

**FIGURE 28.1**
This courtyard house in San Miguel de Allende, Mexico, is based on a historical tradition of interior outdoor living spaces that allows for cross ventilation, daylighting of interiors, and shading and cooling from the evaporative cooling effect of water fountains and the evapo-transpiration of vegetation.

## 28.2 Context of Place

> It is in the nature of any organic building to grow from its site, to come out of the ground into the light, the ground itself always held as a basic component part of the building. The land is the simplest form of architecture. Buildings, too are creatures of the earth and sun.[3]
>
> **Frank Lloyd Wright**

Any geographic region is the product of a long period of co-evolutionary forces involving the interactions of geological, climatic, and biological interrelationships. If enough time passes before a major ecological upheaval, complex networks of symbiotic relationships will develop involving synergistic energy and material flows. Such an evolved ecosystem defines the particular character and dynamics of a place.

An environmental approach to building concerns the ways that design can grow, respond to, engage in, and benefit from the life forces of a specific region. The track of the sun, the conditions of the sky, the climate, flora and fauna, and the nature of the site are significant environmental forces that influence design (for more detailed information on these subjects see Parts I and II). The effects and experiences of each force are made to be place-specific through the interactions of each with the particular geology, geography, latitude, and longitude of the place.[4] Each of the unique desert regions of the southwest have geographic characteristics that allow for specific design approaches specific to local site conditions.

Throughout history, architecture and building technologies have responded to environmental forces in resourceful ways. The sun, the climate, and the site have all shaped architecture as much as have the materials from which buildings have been built. Survival and comfort have depended on responses to the cycles of day and night, the changing seasons, and to shifting climatic patterns.[5]

In today's urban desert communities, we have not made the most appropriate use of our most abundant resource—the sun. The sun regulates and guides our daily lives. Environmental and esthetic benefits are manifested when buildings are designed to reveal the solar cycles of day and night, the shifts of the seasons, and the climate of place.[6]

## 28.3 Toward a Desert Vernacular

Over thousands of years, humans have developed shelter based on local materials and renewable sources of energy. Building form, materials, and appurtenances such as overhangs and window treatments make up the vocabulary of architectural vernacular. They provide for function and comfort as well as an esthetic character. Besides cultural factors, one of the main determinates of vernacular architecture is climate and access to material resources.

In the desert regions of the world, people constructed houses with thick walls and small openings to keep out the heat and glare of the sun during the day. In Egypt, Iraq, India, and Pakistan, deep loggias, projecting balconies, and overhangs cast long shadows on exterior walls. Wooden and marble lattices fill exterior openings to screen the glare of the sun while permitting breezes to pass through.[7] Massive walls are used for their time-lag effect.

FIGURE 28.2
Semi-enclosed courtyards in hot–dry desert climates allow for cross ventilation and the venting of hot air.

The exterior surface colors are usually very light to minimize the absorption of solar radiation. Since there is usually little rain, roofs are flat and consequently available for additional living and sleeping space during the summer nights.[1] Outdoor areas cool quickly after the sun sets due to the rapid radiation of heat to the clear night sky. Buildings are often clustered together to shade one another and the public spaces between them. Semi-enclosed integrated courtyards provide for a tempered outdoor living space with shade and evaporative cooling from plants and fountains (Figure 28.2). Historically, many distinctive changes in architectural form occur to address the challenges of excessive desert heat.

## 28.4 Scottsdale, Arizona

### 28.4.1 Background and History of Scottsdale, Arizona

Founded by Army Chaplain Winfield Scott in 1888, Scottsdale has matured from a tiny desert farming community into one of the premier examples of the new West—urbane, sophisticated, and cultured."[8] This growth has been a challenge for respecting and maintaining the integrity of the Sonoran Desert ecosystem. While the Mojave and Chihuahuan Deserts extend into parts of Arizona, none have the diversity found in the Sonoran. Within the Sonoran region, there is considerable variation in vegetation and wildlife due to differences in temperature, elevation, and bi-modal rainfall pattern (see Chapters 3 and 4). Scottsdale averages 328 days of sunshine and 9.41 in. of rainfall per year.[9] The geographic area that makes up Scottsdale is home to an abundant and diverse collection of desert plant species, many of which are unique to the region.

The modern City of Scottsdale, Arizona is located immediately east of Phoenix, Arizona along the northern boundaries of the Sonoran Desert (Figure 28.3). The northern and most significant area of the city is situated within pristine native desert that is bordered by the McDowell Mountains to the northeast. Scottsdale has an estimated population of 230,000 and is the sixth largest city in Arizona. With a land area of 185 mile$^2$, the city stretches 31 miles from north to the south.[10]

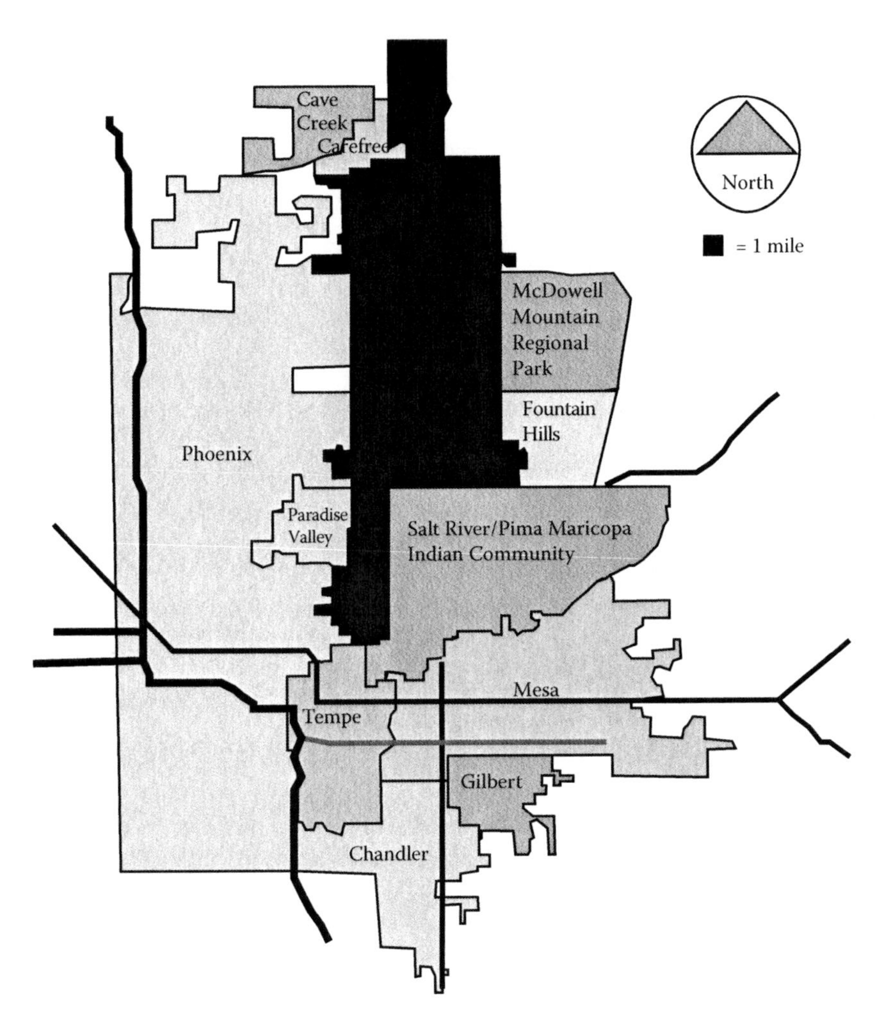


**FIGURE 28.3**
Map of Scottsdale and surrounding communities.

Unlike typical suburban communities, Scottsdale is a net importer of employment and serves as a major tourist, retail, entertainment, and cultural arts destination with a walkable downtown. Although not all local major resorts are located in the city, Scottsdale contains the core of specialty shopping, art galleries, and recreational facilities, and many of the cultural and sporting events that attract and sustain the local tourism industry. The high quality of the city's visual environment is an important component of maintaining this industry.[11]

Despite Scottsdale's resort and tourism destination, it is still considered a highly desirable livable community. The things citizens like least about living in Scottsdale are growth and traffic.[12] Maintaining the integrity and beauty of Scottsdale's unique environment is an ongoing priority for the community. History shows a commitment by the city's residents to support the strongly held community values of protecting, preserving, and sustaining Scottsdale's unique desert environment.[10] Hence, the way buildings are designed and integrated into the desert environment are central to these key community values.

Scottsdale's economic and environmental well-being depends a great deal upon the community's distinctive character and its natural amenities. Some of the many reasons people state for living in Scottsdale include desert beauty, open space, and a quiet, clean and safe environment.[13] These attributes have been nurtured by city ordinances, programs and guidelines intended to protect the community's esthetic and environmental qualities. To this end, the city has adopted the following planning and environmental instruments: (1) General Plan and Character Area Guidelines; (2) Sensitive Design Principles; (3) Green Building Program; (4) Environmentally Sensitive Lands Ordinance (ESLO); and (5) Native Plant Protection. Each of these instruments will be discussed in the following sections.

### 28.4.2 Scottsdale's General Plan and Character Area Guidelines

The General Plan is a statement of goals and policies that work as the primary tool for guiding the future development of the city. It is an expression of our collective vision and direction for the future of Scottsdale and how we want future growth and the character of the community to occur over the next 10–20 years.[14]

On a daily basis the city is faced with tough choices about growth, housing, transportation, neighborhood improvement, and service delivery. A General Plan provides a guide for making these choices by describing long-term goals for the city's future as well as policies to guide day-to-day decisions. As such, Scottsdale's vision is to be a community that[15]

- Demonstrates its commitment to environmental, economic, and social sustainability and measures both the short- and long-term impacts of our decisions
- Creates, revitalizes, and preserves neighborhoods that have long-term viability, unique attributes and character, livability, and connectivity to other neighborhoods in the community and that fit together to form an exceptional citywide quality of life (i.e., the whole is greater than the sum of its parts)
- Facilitates human connection by anticipating and locating facilities and infrastructure that enable human communication and interaction and by promoting policies that have a clear human orientation, value, and benefit
- Respects the environmental character of the city with preservation of desert and mountain lands and with innovative ways of protecting natural resources, clean air, water resources, natural habitat and wildlife migration routes, archaeological resources, vistas, and view and scenic corridors
- Builds on its cultural heritage, promotes historical and archaeological preservation areas, and identifies and promotes the arts and tourism in a way that recognizes the unique desert environment in which we live
- Coordinates transportation options with appropriate land uses to enable a decreased reliance on the automobile and more mobility choices
- Maintains or improves its high standards of appearance, aesthetics, public amenities, and levels of service
- Recognizes and embraces change: from being predominantly undeveloped to mostly built out, from a young town to a maturing city, from a bedroom community to a net importer of employees, and from a focus on a single economic engine to a diverse, balanced economy
- Simultaneously acknowledges our past (preservation of historically significant sites and buildings will be important) and prepares for our future

- Promotes growth that serves community needs, quality of life, and community character
- Recognizes and embraces the diversity of the community by creating an environment that respects the human dignity of all without regard to race, religion, national origin, age, gender, sexual orientation, or physical attributes

Scottsdale's General Plan is divided into six chapters that are based on the Six Guiding Principles of the CityShape 2020 citizen participation process[16]: character and lifestyle, economic vitality, neighborhoods, open space, sustainability, and transportation.

Twelve "Elements" or sections of the General Plan contain the city's policies on character and design, land use, open spaces and the natural environment, business and economics, community services, neighborhood vitality, housing, transportation, growth issues, human services, protection of desert and mountain lands, economic vitality, and the character of neighborhoods.

Neighborhoods hold a unique identity that when grouped together complete Scottsdale's identity as single community. Urban, suburban, rural, and native Sonoran Desert characters provide a broad pallet of expressions. It is the city's responsibility to oversee connections, transitions and blending of these neighborhood characters to ensure the community comes together to create a unified composition.

These policies in the General Plan are implemented and detailed through ordinances and ongoing procedures of the city, including the Zoning Ordinance, Subdivision Ordinance, and Design Guidelines. The General Plan is further reinforced through recommendations from city Boards and Commissions and the decisions made by the City Council.

### 28.4.3 Sensitive Design Principles

The Scottsdale Sensitive Design Program is a comprehensive compilation of policies and guidelines related to desirable aesthetic qualities and unique character attributed to the Sonoran Desert. The Sensitive Design Principles were developed with guidance from the City Council, Planning Commission, Development Review Board, and citizen groups to serve as an overlay to existing planning guidelines and regulations (Figure 28.4). These principles outline the city's design expectations and are based on the overall belief that development should respect and enhance the unique climate, topography, vegetation, and historical context of Scottsdale's Sonoran Desert environment, all of which are considered amenities that help sustain the community and its quality of life. One of the major goals of the program is to address design and sustainability in the context of regional architectural character. The principles serve as a planning tool in the review and evaluation of proposed developments with respect to environmental responsive design.

Scottsdale's Sensitive Design Principles are listed in the following order[17]:

1. The design character of any area should be enhanced and strengthened by new development: Building design should consider the distinctive qualities and character of the surrounding context and, as appropriate, incorporate those qualities in its design. The project should take into account the evolving context of an area over time.
2. Development should recognize and preserve established vistas through siting, orientation of buildings, and the protection of natural features including desert washes, boulders, and archaeological and historical resources.

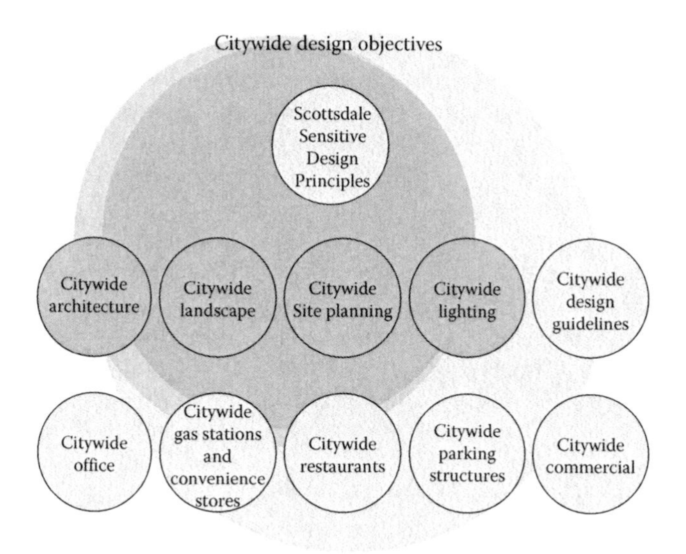


**FIGURE 28.4**
Scottsdale Sensitive Design Principles provide uniform design objectives from the city wide planning level down to the building site level and applies to all building project types.

3. Development should be sensitive to existing topography and landscaping. The design should respond to the unique terrain of the site by blending with the natural shape and texture of the land while minimizing disturbances to the natural features of the site.
4. Development should protect the character of the Sonoran Desert by preserving and restoring natural habitats and ecological processes.
5. The design of the public realm, including streetscapes, parks, plazas, and civic amenities, is an opportunity to provide regional identity and character to the community. Streetscapes should provide design continuity among adjacent uses through landscaping, textured paving, street furniture, public art, and integrated infrastructure.
6. Development should integrate alternative modes of transportation, including bicycles and transit service access, within the pedestrian network that encourage social contact and interaction within the community.
7. Development should show consideration for the pedestrian by providing landscaping and shading elements including pedestrian connections to adjacent development. Design elements should account for human scale and the daily/seasonal angles of the sun with a sensitivity to building configuration and massing.
8. Buildings should be designed with a logical hierarchy of masses to control the visual impact of height and size. The building design should also highlight important building volumes and features, such as the building entry and courtyards.
9. The design of the built environment should respond to the desert environment. Interior spaces should be extended into the outdoors both physically and visually. Incorporate regional materials with natural integral colors and coarse textures that are associated with the Sonoran Desert region. The materials should be used to

provide visual interest and richness, particularly at the pedestrian level. Materials should be used honestly and reflect their inherent qualities. Make use of deep roof overhangs and recessed windows.

10. Buildings should strive to incorporate resource efficient healthy building practices. Utilize design strategies and building techniques that minimize environmental impact, reduce energy consumption, and conserve water.
11. Landscape design should respond to the desert environment by utilizing a variety of mature landscape materials that are indigenous to the arid region. The character of the project should be expressed through the selection of planting materials in terms of scale, density, and arrangement. The landscaping should complement the built environment while relating to the various uses, such as shading and buffering.
12. Site design should incorporate techniques for minimizing water irrigation needs by providing desert-adapted landscaping and preserving native plants. Water, as a landscape element, should be used judiciously. Water features should be placed in semi-enclosed or at least partially shaded locations where pedestrian activity occurs such as entry courtyards or patio areas.
13. The extent and quality of lighting should be integrally designed as part of the built environment. A balance should occur between the ambient light level and the designated lighting needs. Lighting should be designed to minimize glare and invasive overflow to conserve energy while reflecting the character of the area. Strike a balance between using natural light as a part of the building function and regional architectural expression.
14. Signage should consider the distinctive qualities and character of the surrounding context in terms of size, color, location, and illumination. Signage should be designed to be complementary to the architecture, landscaping, and design theme for the site, with due consideration for visibility and legibility.

### 28.4.4 Green Building Program

Scottsdale has long held a leadership position in developing environmental planning initiatives. From the recreational multiuse Indian Bend Wash Greenbelt 7.5 mile stretch (Figure 28.5) and the 36,400 acre McDowell Sonoran Preserve (Figure 28.6) to the Sensitive Design Principles described earlier, Scottsdale's has achieved national recognition in its efforts to harmonize the built environment with the Sonoran Desert (see Chapter 13). These programs have served as a foundation for the development of the city's Green Building Program.

The underlying principle of "green building" as defined in Scottsdale is *to minimize environmental impact of buildings and associated site development*. Green building broadens the regulatory perspective of the built environment with the recognition that "nothing we do happens in isolation." By connecting the building project to the regional environment, the project design parameters are broadened in the regional context of energy, resource conservation, and environmental impacts.

The City of Austin, Texas established the first green building program in the early nineties and there are now over 50 municipal residential and commercial green building programs in the United States (see Chapter 31).* In 1998, the City of Scottsdale developed

* https://my.austinenergy.com/wps/portal/aegb/aegb/home (accessed August 20, 2011).

**FIGURE 28.5**
Gabion tiered walls in a section of the Indian Bend Wash Greenbelt. The seven and a half miles of lush parkland provide lakes, golf courses, swimming pools, many recreational facilities, and an extensive multiuse path system for skating, biking, walking, and jogging.

Arizona's first green building program through a collaborative effort of a citizen advisory committee, staff and the support of city's executive leadership. Originally focusing on new single family residential construction, the program added criteria for commercial projects in 2001 and multifamily projects in 2005. Scottsdale developed its own voluntary green building rating criteria based on the geographic and climatic conditions of the northern Sonoran Desert region. The rating criteria are structured as a flexible, point-based system, containing both mandatory and optional green building measures organized by categories.

The city provides technical assistance, expedited plan review, educational programs and promotional incentives to residential and commercial builders, architects, developers, and project owners to encourage them to participate in the program. Public acceptance has continually matured to the point that by 2006, one of every three new home building permits was approved under city's green building program. The City has also developed residential remodeling guidelines for retrofitting existing housing in a more sustainable and environmentally responsible manner.

In March 2005, Scottsdale became the first city in the country to require all new City buildings, and major renovations to be designed, constructed and certified at the LEED Gold level of certification under the U.S. Green Building Council.[18] This groundbreaking policy represents a major commitment in city leadership toward achieving a healthy, sustainable, and desert appropriate developed community. Scottsdale's Granite Reef Senior Center (LEED Gold) along with a downtown fire station (LEED Platinum), trailhead facility (LEED Platinum), and library (LEED Gold) are among the first city buildings constructed under this policy (Figure 28.7). At the time of this publication, there were over nine LEED certified city facilities.

In July 2011, Scottsdale adopted the International Green Building Construction Code (IgCC) as a voluntary code to replace the city's existing commercial green building rating program. The IgCC is a new overlay code designed to work in unison within the existing framework of building codes. It was developed by the International Code Council (ICC),

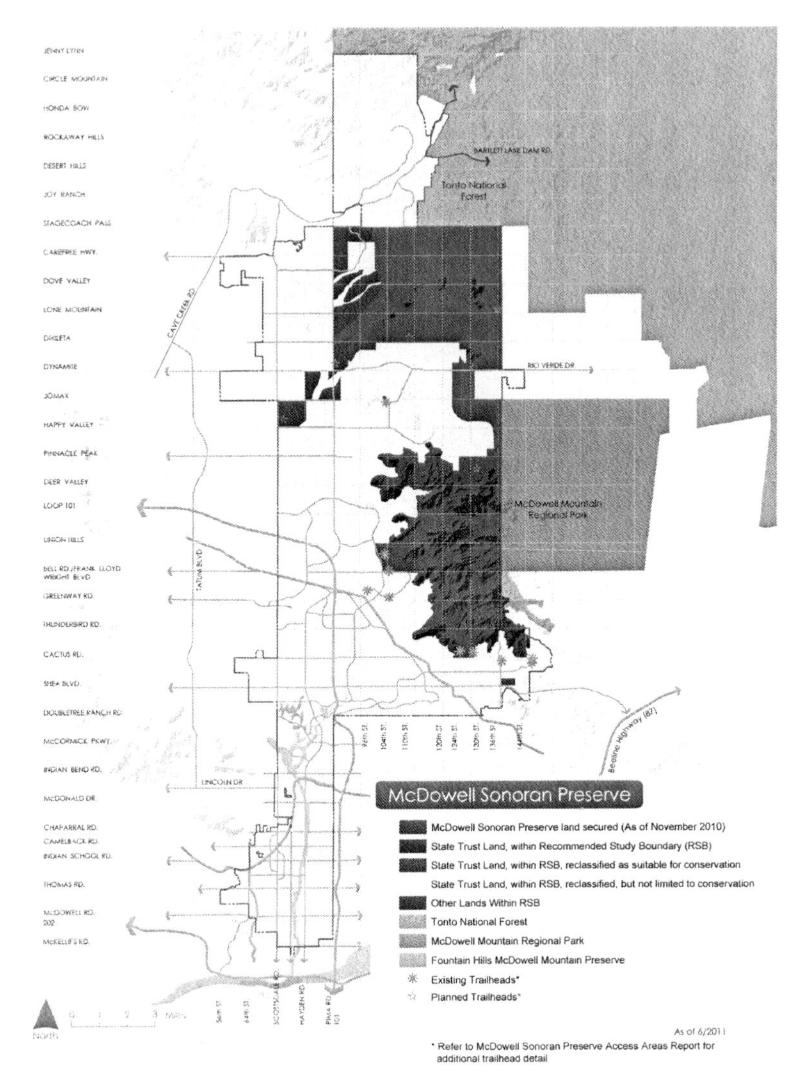


**FIGURE 28.6**
Scottsdale has acquired or protected over 14,000 ac of desert and mountain land through preservation efforts and the implementation of the McDowell Sonoran Preserve. The total area proposed for preservation is 36,400 ac or 56 mile$^2$. This represents 30% of the city's land area. (Courtesy of the City of Scottsdale, Arizona.)

the predominant U.S. code development organization which publishes model construction codes for jurisdictional adoption including building, fire, mechanical, plumbing, and energy codes. The IgCC is not a rating system with points, nor does it have rating levels like the LEED rating system. It is a baseline code for minimum measures in the areas of site development, materials, energy, water, and indoor environmental quality in the context of the climate and regional resources. Scottsdale has amended the IgCC to be compatible with the city's environmental policies and design guidelines including shading requirements on the east and west glazing of commercial buildings. The IgCC has been further amended to integrate it into the city's established green building plan review and inspection process involving the issuance of "green building permits" and "green certificates of occupancy" based on compliance with the IgCC. The IgCC is a national model code that will in time

**FIGURE 28.7**
City of Scottsdale Fire Station 2 received a LEED Platinum certification in 2009 for its exemplary integrated design features. The building design includes rainwater harvesting, passive cool tower, regional building materials, and renewable energy systems (LEA Architects).

standardize and unify baseline green building criteria for all communities subscribing to ICC building codes.

### 28.4.5 Environmentally Sensitive Lands Ordinance

The ESLO is an overlay set of zoning regulations and guidelines designed to achieve environmental sensitive development throughout 134 square miles of native desert and mountain areas of northern Scottsdale.[19] By protecting environmentally sensitive lands, the ESLO also provides for public health and safety by controlling erosion and maintaining the natural hydrology of the area. The ordinance requires that a percentage of each property be permanently preserved as natural area open space (NAOS) and those specific environmental features including vegetation, washes, high slopes, mountain ridges, and peaks be protected from development (Figure 28.8).

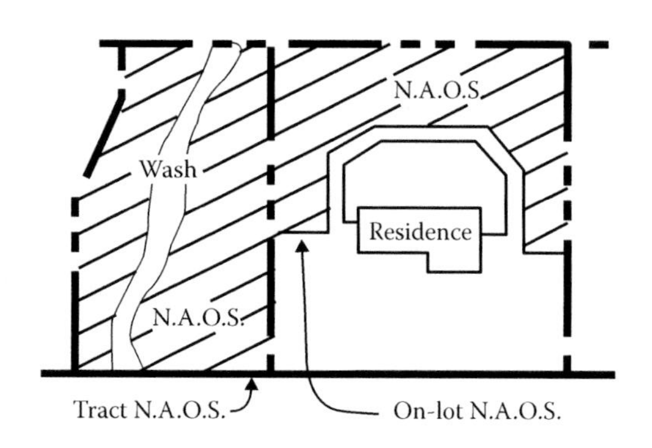


**FIGURE 28.8**
Natural area open space.

The ESLO has a direct impact on the citizens of Scottsdale as its key provisions determine the location and design of residential, commercial, industrial, and institutional development in two-thirds of the City. Application of the ESLO has resulted in the preservation of over 9000 acres of Sonoran Desert open space while protecting citizens from potential flooding, erosion, and visual blight (see also Chapters 10 and 13).

ESLO is not intended to deny the reasonable use of the land, but rather guide its use in ways that are sustainable and recognize the unique features this setting provides. ESLO encourages development that blends into and respects the character of the natural desert setting as follows:[20]

1. Streets should be kept to a minimum on steeper slopes and should be designed to avoid unnecessary exposed cuts and fills.
2. Grading and construction should be kept within clearly identified building envelopes so that NAOS areas are not damaged.
3. Development should not intrude on or damage boulder features or major boulders.
4. Washes should be left in their natural state wherever this is feasible. If it is necessary and appropriate to modify a natural watercourse, the modifications should be minimal and the watercourse should be restored to a natural condition.
5. Allowable building height is measured from the natural grade, which encourages buildings to follow the form of the natural topography.
6. Discourage the use of subdivision perimeter walls.
7. Rear and side walls on larger lots are required to be set back from the property lines to allow for wildlife movement and stormwater flow.
8. Walls are not allowed to cross watercourses.
9. Use restrained site lighting, which do not spill glare onto adjacent properties.
10. Limit use of nonnative plants to enclosed yard areas (enclosed by a solid wall); nonnative plants with the potential of reaching over 20 ft in height are not allowed.

The NAOS required by ESLO can be either natural desert that has been undisturbed by development activity or previous developed areas that has been restored to the desert terrain and vegetation of its natural condition. The amount of NAOS required to be set aside with each development is based upon two factors—the landform area and land slopes. The NAOS requirement increases from the lower desert to the hillside landform areas and from land slopes under 2% to those over 25% (Table 28.1). The NAOS requirement ranges from a low of 15%–20% to a high of 80% of the total property area.

### 28.4.6 Native Plant Protection

Native vegetation plays a vital role in the dynamic system of the Sonoran Desert. Its presence helps to prevent erosion, provides food and shelter for desert wildlife, and acts to shade the desert floor and minimize the urban heat island effect. In addition, native vegetation requires less water and maintenance than nonindigenous plant materials. In most cases, salvaging existing plant material is more economical and achieves a natural desert appearance in a shorter amount of time.

Scottsdale adopted the Native Plant Ordinance as a way to preserve the unique native character of the Sonoran Desert under a system of responsible community development.[21] Many desert trees and cacti are slow growing and can take decades to reach maturity.

**TABLE 28.1**

NAOS Requirements

| Land Slope | Lower Desert (%) | Upper Desert (%) | Hillside (%) |
|---|---|---|---|
| 0%–2% | 20 | 25 | 50 |
| Over 2% up to 5% | 25 | 25 | 50 |
| Over 5% up to 10% | 30 | 35 | 50 |
| Over 10% up to 15% | 30 | 45 | 50 |
| Over 15% up to 25% | 30 | 45 | 65 |
| Over 25% | 30 | 45 | 80 |
| Minimum NAOS after reductions | 15 | 20 | 40 |

The amount of NAOS required to be set aside with each development is based upon two factors—landform and land slopes.

Factors such as the size, form, or location of certain mature specimen plants, such as the saguaro cactus (*Carnegiea gigantea*) or ironwood tree (*Olneya tesota*), make finding a comparable nursery-grown tree for replacement extremely difficult if not impossible (Figure 28.9). Therefore, leaving such plants in place or salvaging them for incorporation into landscaping is both environmentally and economically beneficial (see Chapter 23).

Any development project which will affect designated native plants is required to submit a native plant site plan detailing the existing location and proposed treatment of each native plant. Protected plants should optimally remain in place. Those plants that must be moved are required to be salvaged and replanted within the project site. Native vegetation enhances the projects' aesthetic appeal while conserving the desert habitat.

## 28.5 Summary

Scottsdale is home to Frank Lloyd Wright's Taliesin West School of Architecture. We are continually inspired by this exemplary desert complex located at the base of the McDowell Mountains in North Scottsdale (Figure 28.10). Over 50 years ago, Mr. Wright wrote:

> I don't see how we can consider ourselves as civilized, cultured people if we live ignorant of the nature of our environment; if we do not understand what we do to make it. Where the buildings that we live in are false, where they do not represent truth and beauty in any sense, where they are merely stupid or merely copying something that's not understood. Because believe me, when you understand a thing you will not copy it. A copycat is a copycat because he does not understand.[22]

For thousands of years, humans have devised vernacular solutions based on an informed nature of the environment. Since the industrial era, much of the sustainable approach for living in the desert has been ignored. Any building that neglects its setting in a desert environment unnecessarily relies on excessive energy in the form of nonrenewable resources to support mechanical cooling and ventilation systems. Such systems enables a building to "work" in a climate it essentially ignores. The sole reliance on mechanical solutions does not recognize environmental constraints that can be turned into unique esthetic

**FIGURE 28.9**
Protecting native plants through Scottsdale Ordinance. (Courtesy of the City of Scottsdale, Arizona.)

**FIGURE 28.10**
Taliesin West in Scottsdale—when Wright set out to design and build Taliesin West, his goal was to integrate the structures with the "nature" of the desert, its soul, and with the desert's nature, its physical characteristics. (From Lucas, S.A., *Taliesin West: In the Realm of Ideas*, The Frank Lloyd Wright Foundation, Scottsdale, AZ, 1993.)

opportunities.[23] Do we choose to ignore or celebrate environmental challenges? I would say Scottsdale has chosen to continue the effort to integrate the innate characteristics of the desert environment. By being responsive to place, we are responsive to the environment. By being responsive to the environment, we become responsive to our use of energy and natural resources.

As a statement of goals and policies, a general plan is only as good as the choices made and actions taken on a daily basis regarding development proposals in the context of land use, natural open space, infrastructure, density, heat island effect, transportation, community services, pedestrian access, and connectivity. A community *must be actively engaged with long term community vision if the intention is to build a sustainable environment*, viable economy, and balanced community (see Chapters 18, 25, 27, and 31). This is the challenge of all city planning agencies and regulatory authorities.

Maintaining the integrity of Scottsdale's unique environment is an ongoing priority for the community. The city's efforts on desert preservation and biodiversity have been internationally recognized. Scottsdale's planning and development policies, ordinances and guidelines have evolved over the past 50 years with an early recognition of the unique characteristics of the Sonoran Desert including its climate, landforms, native plants and animals, and historical/cultural attributes. Implementing desert preservation, environmentally sensitive development and green building, Scottsdale has learned to work with the Sonoran Desert environment. Scottsdale's greatest challenge with development is low density urban sprawl and the dominance of the automobile. We are embracing new tools such as the LEED for Neighborhood Development (ND) green rating program and the city's adoption of the International Green Construction Code (IgCC). The LEED ND program can help guide city policy toward smart project location and linkages, neighborhood patterns, green infrastructure, heat island mitigation, and transportation impact alternatives. Scottsdale adopted the IgCC in July 2011 as a voluntary to replace the city's existing commercial green building rating program. Although the IgCC is designed as a mandatory code, Scottsdale will use the code as an integrated part of the plan review and inspection process culminating in the issuance of a "green certificate of occupancy." Scottsdale will consider subsequent IgCC editions for possible adoption as a mandatory code. This will be based on a number of factors including the number of successfully completed IgCC projects, the state of the local economy and the will of city council.

By minimizing the impact of development and utilizing the natural resources of site and region, Scottsdale will strive to embody the desert's bountiful riches as a part of a vernacular Sonoran Desert architecture. This outlook will create a fundamental shift in the way buildings look and perform in the desert as opposed to buildings that are out of place and time. By generating greater awareness in the development community and implementing the broader community vision, the desert communities of the southwest can become more environmentally responsive and move toward a true vernacular expression of place.

## References

1. Lechner, N., *Heating, Cooling, Lighting: Design Methods for Architects* (New York: John Wiley & Sons, 2009).
2. City of Scottsdale, Scottsdale green building program brochure (Scottsdale, AZ: City of Scottsdale, April 2004).

3. Rattenbury, J., *A Living Architecture* (Rohnert Park, CA: Pomegranate, 2000), p. 42.
4. Guzowski, M., *Daylighting for Sustainable Design* (New York: McGraw-Hill, 2000).
5. Guzowski, M., *Daylighting for Sustainable Design* (New York: McGraw-Hill, 2000), p. 3.
6. Guzowski, M., *Daylighting for Sustainable Design* (New York: McGraw-Hill, 2000), p. 4.
7. Fathy, H., *Architecture for the Poor* (Chicago, IL: The University of Chicago Press, 1973).
8. Arizona Department of Commerce, Community profiles: Scottsdale, Arizona (Phoenix, AZ: Arizona Department of Commerce, 2009), http://www.azcommerce.com/doclib/commune/scottsdale.pdf (accessed February 5, 2011).
9. City of Scottsdale, Quick statistics (Scottsdale, AZ: Economic Vitality Department, June, 2009).
10. City of Scottsdale, 2001 general plan (Scottsdale, AZ: City of Scottsdale, 2001), p. 33.
11. City of Scottsdale, 2001 general plan (Scottsdale, AZ: City of Scottsdale, 2001), p. 32.
12. City of Scottsdale, 2001 general plan (Scottsdale, AZ: City of Scottsdale, 2001), p. 35.
13. City of Scottsdale, Scottsdale citizen (Scottsdale, AZ: City of Scottsdale, Winter, 2000).
14. City of Scottsdale, 2001 general plan (Scottsdale, AZ: City of Scottsdale, 2001), pp. 11–12.
15. City of Scottsdale, 2001 general plan (Scottsdale, AZ: City of Scottsdale, 2001), p. 1.
16. City of Scottsdale, City of Scottsdale General Plan, http://www.scottsdaleaz.gov/generalplan (accessed June 21, 2011).
17. City of Scottsdale, City of Scottsdale General Plan, http://www.scottsdaleaz.gov/planning/general/sensitivedesign (accessed June 21, 2011).
18. City of Scottsdale, Resolution No. 6644, http://www.scottsdaleaz.gov/greenbuilding/leed.asp (accessed June 21, 2011).
19. City of Scottsdale, Citizen's guide to the environmental sensitive lands (Scottsdale, AZ: City of Scottsdale, 2004), p. 2.
20. City of Scottsdale, Citizen's guide to the environmental sensitive lands (Scottsdale, AZ: City of Scottsdale, 2004), p. 7.
21. City of Scottsdale, Native Plant Ordinance—City Code, Chapter 46, Article V, 1981, see the following link for more information on native plant policies in Scottsdale at http://www.scottsdaleaz.gov/codes/nativeplant.asp (accessed June 21, 2011).
22. Meehan, P. J., *Truth Against the World* (Washington, DC: The Preservation Press, 1992).
23. Stein, B. and J.S. Reynolds, *Mechanical and Electrical Equipment for Buildings* (New York: John Wiley & Sons, 2000).

# 29

## *Sustainable Energy Alternatives for the Southwest*

**David Berry**

**CONTENTS**



### 29.1 Introduction

Electricity is necessary for modern life. The problem is how to produce and consume it in a manner that is environmentally and economically sustainable over the long haul.

The desert Southwest includes some of the country's largest metropolitan areas—Los Angeles, San Diego, Riverside-San Bernardino-Ontario, Tucson, Phoenix, Las Vegas, Albuquerque, and El Paso, as well as several smaller metropolitan areas. The 30 million people living in these metropolitan areas consume large amounts of electricity that is generated using a system whose basic design was developed decades ago. That system has provided reliable electric service but it faces enormous environmental conflicts and economic risks.

This chapter describes the emerging transformation of the electric supply and demand system in the desert Southwest from one dominated by central station fossil-fueled power plants to a cleaner energy future that relies much more on renewable energy and energy efficiency. This transformation depends on the development of institutional capabilities in the public and private sectors to plan for and implement a more sustainable energy system. Institutional capability refers to the competence of organizations to influence conduct, cultivate new paradigms, innovate, mobilize resources, and attract broad support to effectively design and execute clean energy programs in an environment of changing conditions.*

We begin with an overview of the energy system today. We then review the environmental and economic risks of the current system and describe how those risks can be managed through sustainable energy alternatives. In the next section, we highlight the institutional bases for a sustainable energy system. The prospects for establishing a sustainable power system are assessed in the last section.

To start, a little electric jargon. Energy is measured in kilowatt hours (kWh). Because huge amounts of energy are generated and sold, it is useful to measure energy in megawatt hours (MWh). 1 MWh equals 1000 kWh. 1000 MWh equals 1 gigawatt hour (GWh). Power generating capacity and instantaneous power consumption are measured in kilowatts (kW). 1000 kW is 1 megawatt (MW) and 1000 MW equals 1 gigawatt (GW).

## 29.2 Energy System Today

The demand for electricity in the desert Southwest has grown rapidly (Figure 29.1).† Between 1990 and 2008 the average annual growth rate for electricity sales in Arizona, Nevada, and New Mexico was about 3.5%, although after 2007 the recession reduced the level of sales.[2]

The intense desert heat shapes the demand for electricity. Peak demand is well above demand at other hours because consumers maximize air conditioner use during the hottest part of the day during the summer—in the late afternoon and early evening hours.

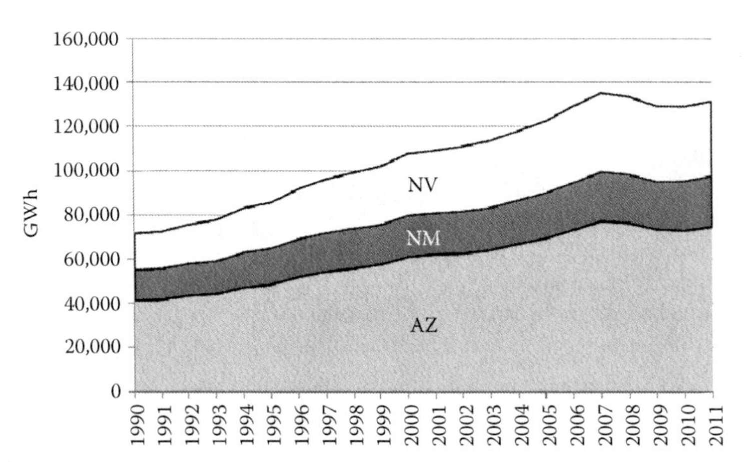


**FIGURE 29.1**
Retail electrical sales: AZ, NM, NV.

* On the role of institutions, see North.[1]

† The study region does not correspond to political boundaries or electric supply regions. Consequently, to make use of readily available data, we must sometimes present information for states. Because much of California and Texas lie outside the desert Southwest, this chapter presents state level data only for Nevada, Arizona, and New Mexico. However, policies for all five states are discussed.

The demand for electricity is served by a network of central station power generators. These include the Palo Verde Nuclear Generating Station west of Phoenix, plus numerous coal-fired power plants, gas-fired power plants, and hydropower plants located primarily along the Colorado River. Utilities typically classify their power plants as

a. Baseload plants, which run most of the time close to their generating capacities
b. Intermediate plants, which generate electricity some of the time, usually during periods of high demand
c. Peaking plants, which run only a few hours a year and usually serve very high loads in the late afternoon and early evening in the summer

Most baseload plants are located far from the major cities and burn coal to generate electricity. The Palo Verde nuclear plant is also a baseload plant. Intermediate and peaking plants are often located in or near load centers and typically burn natural gas.

Figure 29.2 summarizes the generation mix (in terms of MWh of electricity produced) of resources located in Nevada, New Mexico, and Arizona as of 2011. Coal- and gas-fired generation dominate; hydro and nuclear power are also important, but renewable energy (other than hydro) provided only about 3% of the electricity produced in the three state region. The system of power plants is linked together by a series of high-voltage transmission lines that deliver power to load centers.

The general design of the current power supply system has been in place for decades* and the utility industry and its regulators have worked out a complex set of rules and procedures for meeting load growth, operating the system, and pricing electric services. These rules and customary procedures constitute the institutional framework and

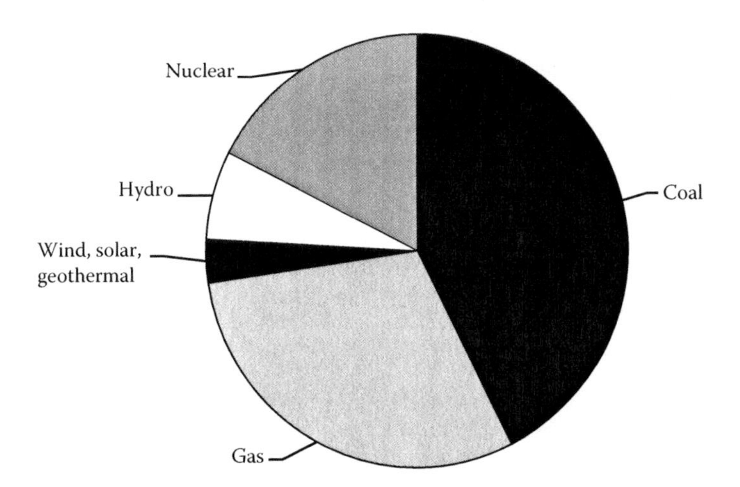


**FIGURE 29.2**
2011 Generation mix (GWh): AZ, NM, NV.

* The spatial design of the power supply system stems from concepts developed in the 1920s and 1930s. To increase the availability and benefits of electricity, Robert Bruère and others advocated a system of power plants located at coal mines delivering power to consumers through an interconnected transmission system. This arrangement was intended to supersede the location of power plants in cities and to bring electricity to rural areas. See Bruère.[3]

capabilities in place today, but that framework is being tested and new capabilities are being developed.

## 29.3 Risks and Risk Management

The current power generation infrastructure will be difficult to sustain and replicate in the future, for both environmental and economic reasons. In this section we look at environmental and economic risks and then summarize how these risks can be managed through a sustainable energy system.

### 29.3.1 Environmental Risks of Conventional Power Generation

Coal, and to a lesser extent, gas-fired power generation come with a lot of environmental baggage. Air emissions from these plants include the following:

- Carbon dioxide. In 2008, coal-fired power plants in Nevada, Arizona, and New Mexico emitted about 78 million metric tons of carbon dioxide into the atmosphere and gas-fired power plants emitted about 30 million metric tons.[4] Carbon dioxide contributes to climate change; emissions are regulated in California and some other states and may be further regulated in the future.
- Sulfur dioxide and nitrogen oxides. Power plants in Nevada, Arizona, and New Mexico emitted about 73,000 MT of sulfur dioxide in 2008 and about 159,000 MT of nitrogen oxides.[4] Sulfur dioxide and nitrogen oxides contribute to acid rain and react with other chemicals in the atmosphere to produce fine particulate matter. The fine particulate matter is associated with several types of health impacts, including premature mortality, bronchitis, hospital admissions, asthma, and heart attacks.[5]
- Mercury. Coal-fired power plants in Nevada, Arizona, and New Mexico emitted about 4400 lb of mercury in 2005.[6] The Environmental Protection Agency indicates that mercury exposure at high levels can harm the brain, heart, kidneys, lungs, and immune system of people of all ages and that high levels of methylmercury can harm unborn babies, young children, and wildlife.[7]

Emissions of sulfur dioxide, nitrogen oxides, particulate matter, and other pollutants from power plants directly and indirectly impair visibility by absorbing or scattering light.[8] Degradation of visibility is especially important in national parks and wilderness areas such as the Grand Canyon or Mesa Verde National Park.

In an arid environment water is a scarce resource, often with multiple competing uses. Power generation can be a locally major water consumer. Most, but not all, electricity produced in the Southwest uses steam to spin a turbine that, in turn, powers a generator. To complete the energy cycle, it is necessary to condense the steam back into water. With a few exceptions, this cooling process uses water in Southwestern power plants. The result is a fairly hefty demand for water—500–700 gal/MWh generated at a coal, gas, or nuclear steam plants and about 180 gal/MWh at a natural gas-fired combined cycle power plant.[9] A few Southwestern power plants use dry cooling (air cooling) or hybrid (combination of wet and dry) cooling which greatly reduces water consumption. The Silverhawk combined cycle plant in Nevada is an example of dry cooling.

### 29.3.2 Economic Risks

Conventional power plants are subject to several economic risks.

- Fuel price risk. The fuel for natural gas-fired power generation has exhibited wild fluctuations in price since the early 1970s (Figure 29.3).[10] Consequently, use of gas-fired power generation exposes power plant owners, utilities, and electricity consumers to periods of potentially high costs.
- Environmental regulation compliance cost risk. Coal-fired power plants face potentially significant costs to reduce emissions to meet environmental regulations. This is especially the case for complying with potential limitations on carbon dioxide emissions; future compliance costs may exceed fuel costs.[11] In addition, the Environmental Protection Agency is considering Best Available Retrofit Technology for power plants in Arizona and New Mexico to reduce emissions of nitrogen oxides and particulate matter so as to improve visibility and is considering further regulation to reduce other environmental impacts of power generation. Compliance could be expensive, costing hundreds of millions of dollars.[12]
- Financial risk. Obtaining financing for an investment in a new power plant is difficult today, even under favorable circumstance. If investors calculate that a power plant developer or utility may not fully recover its costs, then financing may not be possible. Impediments to full cost recovery include potential regulatory disallowances for imprudent investments that, for example, did not take account of the risks of environmental regulatory costs or fuel price volatility. In addition, future sales of electricity are subject to uncertainty, and new power plants might represent excess generating capacity whose costs might not be fully recovered.

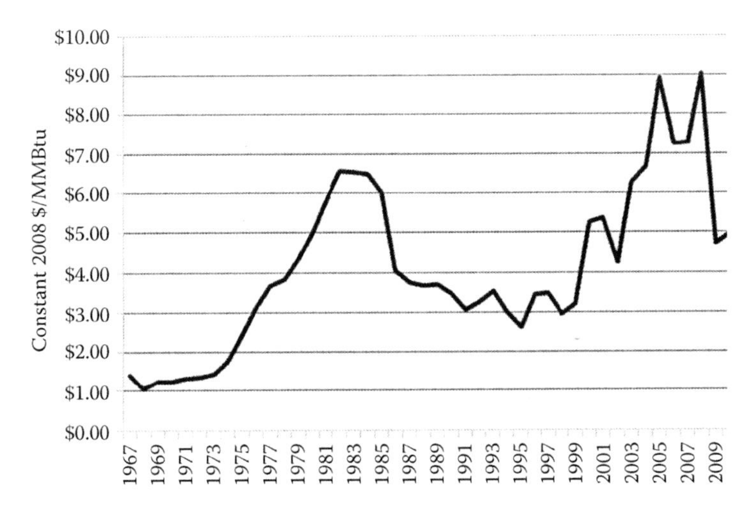


**FIGURE 29.3**
Natural gas prices paid by electric power sector.

## 29.4 Alternative Energy Resources

How can utilities and the public manage the environmental and economic risks of continued reliance on conventional power plants? A portfolio of renewable energy and energy efficiency can, to a large extent, provide a hedge (Figure 29.4).

The first resource alternative is energy efficiency. Using energy more efficiently means using less energy to attain the same outcome such as lighting, space cooling, water heating, motor power, and so forth. Greater efficiency can be achieved by substituting more advanced technology for older technology, by changing physical designs, and by changing behavior. In general, energy efficiency is less costly than generating electricity. We will discuss energy efficiency in more detail later.

Second, many renewable energy technologies can help utilities manage environmental and price risks. Renewable resources available in the desert Southwest include solar energy, wind energy, geothermal energy (which produces electricity using heat from the earth's crust), and some biomass resources such as landfill gas, gas from treatment of wastewater, and wood or agricultural waste. Solar energy includes a variety of technologies—only the major ones currently in use are discussed here. Photovoltaic (PV) technologies generate electricity directly from sunlight. Concentrating solar power (CSP) plants concentrate sunlight using mirrors to heat a transfer fluid which, in turn, makes steam for a turbine and generator. CSP plants may use parabolic troughs or a central receiver (power tower). Solar hot water for residential or business use is also a solar energy resource. In general, these renewable resources have little or no air emissions and so their costs would not be affected by future environmental regulations. Further, most use no fuel (biomass being the exception), so they are not subject to potentially high prices for fuel. For most renewable energy technologies, the major cost component is the fixed capital cost of the plant. Consequently, renewable energy is a stably priced resource that serves as a hedge against high fuel and environmental regulation compliance costs of conventional generation.

Renewable energy is not a perfect substitute for current power generation technologies, however. Some renewable resources are intermittent (wind and to some extent solar) and the power supply system has to accommodate rapid changes in output of these renewable energy generation resources. The costs of integrating intermittent resources

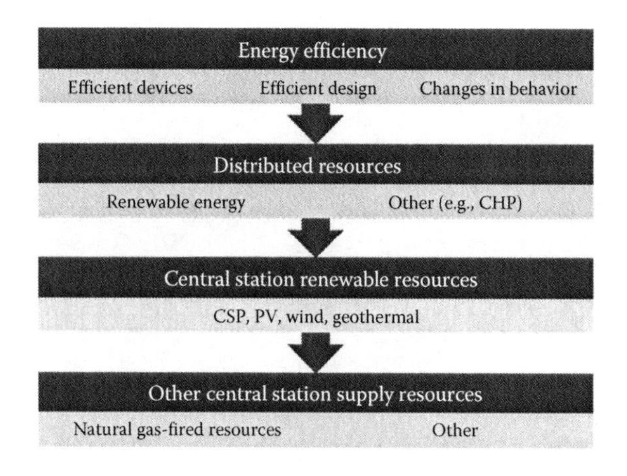


**FIGURE 29.4**
Loading order.

into the electric grid are generally low, at least until the proportion of power generated by these resources reaches roughly 20% of total power generation.[13] However, changes to the way the electric grid is operated could reduce the costs of integrating higher levels of intermittent renewable energy.[14]

Also, solar power is not available when the sun is not shining unless the generation facility can store energy. Some solar technologies do store energy—for example, CSP using parabolic troughs or central receivers can be designed to store heat to make steam to power a generator in the evening or other hours when the sun is not shining. Solar hot water systems store hot water in a tank for use when the consumer demands hot water.

Solar power plants based on a steam technology with conventional cooling consume large amounts of water—perhaps over 700 gal/MWh. Dry cooling could be used, but at additional cost and with lowered power production efficiency.[15]

Large scale renewable energy projects in the desert may interfere with wildlife and disrupt local habitat. For instance, wind turbines, if not carefully sited, can kill birds or bats.[16] Large solar energy projects can take up hundreds or even thousands of acres and may disturb desert tortoise or other habitat. However, some solar projects could be located on what is currently farmland or industrial or mining land, thereby avoiding undisturbed wildlife habitat (and which might come with water rights).

Some renewable energy technologies are more costly than their conventional counterparts. Various incentives, such as production tax credits, investment tax credits, and utility incentives for distributed energy projects, often lower the effective cost to the project owner. Broadly speaking, wind, geothermal, and some biomass plants are currently cost competitive with conventional generation, but most solar technologies are more expensive. Costs for PV facilities have decreased recently and may be on a path toward a competitive price.

New large hydropower projects are unlikely because there are few sites left that could accommodate these projects, because of their environmental impacts, and because long-term drought will reduce their power output. Some small hydro projects are possible, such as Salt River Project's 750 kW plant on the Arizona Canal in Phoenix.

Table 29.1 provides a summary comparison of coal- and gas-fired generation and some major clean energy resources that are likely to be developed over the next 10–20 years.*

## 29.5 Diminishing Role of Coal-Fired Power Plants

California's emissions performance standard is an important policy affecting existing and new coal plants serving California customers. The standard for new baseload power generation is 1100 lb of carbon dioxide per MWh. The standard applies to long-term contracts or ownership of baseload power plants, whether the power plant is located in California or elsewhere. The impact of the standard is to eliminate use of new conventional coal-fired power plants or new long-term contracts for purchases from conventional coal-fired power

* This chapter does not address the role of nuclear power in the future. While nuclear power has its advocates and some utilities are actively exploring additional nuclear power, its role is very unclear. There is no recent experience with building nuclear power plants in the U.S. Construction costs are likely to be high, but the absence of recent experience means that reliable projections of cost and of performance characteristics during early years of operation are not available. And there remains the problem of safe long-term storage of radioactive waste—a political controversy that apparently is unsolvable.

**TABLE 29.1**

Summary of Environmental and Economic Features of Major Southwestern Resources

| Factor | | Fossil-Fueled Generation | | Major Renewable Energy Technologies Likely to Be Developed over Next 10–20 Years | | | | | |
|---|---|---|---|---|---|---|---|---|---|
| | | Coal-Fired Generation | Gas-Fired Generation | Distributed PV[a] | Central Station PV | Wind | CSP with Thermal Storage | Geothermal | Energy Efficiency |
| Environmental compatibility | Carbon dioxide emissions | ● | ⊙ | ○ | ○ | ○ | ○ | ○ | ○ |
| | Health impacts | ⊙ to ● | ○ to ⊙[b] | ○ | ○ | ○ | ○ | ○ | ○ |
| | Impairment of visibility or visual intrusion | ⊙ to ● | ○ | ○ | ○ | ○ to ⊙ | ○ | ○ | ○ |
| | Impacts on wildlife | ⊙ to ● | ○ to ●[b] | ○ | ○ to ⊙ | ○ to ⊙ | ○ to ● | ○ | ○ |
| | Water use | ● | ⊙[c] | ○ | ○ | ○ | ○ to ● | ⊙[d] | ○ |
| Economic factors | Uncertainty of fuel prices | Moderate | High | Not applicable—no fuel required | | | | | |
| | Current economic costs | Low fuel cost, high capital cost | Low capital cost, uncertain fuel cost | High but declining | Moderate to high, declining | Moderate | Moderate to high | Moderate | Low |
| | Availability of power | Whenever needed | | Subject to availability of wind or sunlight | | | Whenever needed | | |

*Key:* Intensity of environmental incompatibilities: ●, high incompatibility; ⊙, moderate incompatibility; ○, low incompatibility.

[a] Distributed PV projects are located at or near the point of electricity consumption, such as on a consumer's rooftop or shade structure.

[b] Impacts occur at well sites; for example, there may be health impacts from obtaining natural gas from shale and there are impacts on wildlife from exploration, drilling, and extraction of natural gas.

[c] For a combined cycle unit; a combustion turbine would consume very small amounts of water.

[d] Assuming a binary plant with hybrid cooling.

plants, and to prohibit life-extending investments in existing coal-fired power plants.[17] Coal-fired power plants which capture and sequester carbon dioxide emissions may meet the standard, but such plants are not commercially available today.

With regard to coal plant retirements, the 1580 MW Mohave Generating Station, which started service in 1971, was shut down on December 31, 2005. The shutdown occurred after the power plant owners failed to install air pollution controls pursuant to a federal consent decree. In 2010, Southern California Edison (SCE) entered into an agreement with Arizona Public Service Company (APS) to sell SCE's share of Four Corners Units 4 and 5 to APS, and APS would, in turn, retire Four Corners Units 1–3 (560 MW).[18] APS could thereby avoid costly retrofits at Units 1–3 to reduce air emissions in response to new environmental regulations.

In addition, current low natural gas prices make operating older, less efficient coal plants relatively uncompetitive and make investments in new coal plants riskier. Thus, gas generation is expected to increase while coal generation may decrease.

## 29.6 Institutional Bases for a Sustainable Power System

Establishing a sustainable power system requires that numerous organizations develop and maintain capabilities to design and implement that system over a long period of time. These organizations include state utility regulators, utilities, local governments, community-based organizations, suppliers of renewable energy facilities, suppliers of energy efficiency measures, and the finance sector. This section briefly summarizes how some organizations are developing their capacities for a sustainable power system that serves the desert Southwest.*

### 29.6.1 State Capabilities

States in the southwestern United States are, in fits and starts, developing policies to promote clean energy. Key state policies are renewable portfolio standards, energy efficiency standards, and appliance standards and building codes. California has been the leader on these policies and, as a result, it has far lower electricity use per person or per dollar of state gross domestic product than the nation as a whole,[20] and it has developed a variety of wind, solar, geothermal, and biomass resources. Table 29.2 summarizes major state policies as of Spring 2010 for the five states that encompass the desert Southwest.[21]

* This section is concerned with developing clean energy resources to serve the Southwest. However, the vast solar resource of the Southwest could generate power for a large portion of the rest of the country. Several analyses have looked at using concentrating solar power and photovoltaics as major electricity resources that, when combined with other renewable energy, could provide most of the nation's electricity by 2050.[19] This export strategy would require thousands of square miles of land for solar energy projects and many thousands of miles of new transmission facilities. Energy storage technologies would have to advance significantly from current capabilities to provide electricity when consumers demand it. For example, molten salt storage for CSP projects may be improved to allow electricity to be generated around the clock. Compressed air or other storage of energy associated with photovoltaics could also be used, but this technology is not very far along at present. Planning for and implementing a large-scale export strategy would require massive investments, technological improvements, a high degree of coordination, and careful reviews to minimize environmental conflicts.

**TABLE 29.2**

Summary of State Energy Efficiency and Renewable Energy Standards[a]

| | Energy Efficiency Standard | | Renewable Energy Standard | |
|---|---|---|---|---|
| | Standard | Remarks | Standard | Remarks |
| California | 2010–2012 program goal for investor owned utilities = 7,000 GWh (2.6% of retail sales) | California has adopted appliance standards, building codes, and utility programs to reduce energy use | 33% of retail sales from eligible renewable energy resources by 2020 | |
| Nevada | Up to 25% of renewable energy standard can be met with efficiency savings | See renewable energy standard column; extra credit multipliers apply to efficiency | 25% of retail sales from renewable energy resources by 2025; solar carve-out | Extra credit multipliers for PV; see energy efficiency standard column |
| Arizona | Cumulative reduction in retail sales by 2020 of 22% of 2019 retail sales | Arizona Corporation Commission regulation | 15% of retail sales must be met with eligible renewable energy resources by 2025; carve-out for distributed energy | Arizona Corporation Commission regulation |
| New Mexico | Savings of 10% of 2005 total retail kilowatt-hour sales in 2020 as a result of energy efficiency and load management programs implemented starting in 2007 | State policy is to include all cost-effective energy efficiency and load management programs in utility energy resource portfolios | 20% by 2020 (10% by 2020 for cooperatives); minimum requirements for various eligible technologies and for distributed energy | Subject to cost limits (limits vary by technology); overall rate impact limit |
| Texas | 20% of load growth to be met with energy efficiency savings by 2010 | | 5,880 MW by 2015; 10,000 MW by 2025 (of which 500 MW is for non-wind technologies; extra credit multiplier) | |

[a] Standards may not apply to municipal utilities or cooperatives.

Energy efficiency standards require electric utilities to reduce kWh sales on a specified schedule by offering or subsidizing more efficient devices for space cooling or heating, water heating, motor drives, lighting, insulation, and so forth, and by adopting programs to change behavior. Energy efficiency may also be achieved by better design of buildings and sites as will be discussed later.

Renewable energy standards establish a minimum level of energy obtained for retail sales that must come from eligible renewable energy technologies. These standards are intended to

- Reliably serve load growth while replacing conventional resources with clean resources
- Diversify utility resource portfolios
- Encourage utilities to learn how to integrate renewable energy technologies into a conventional energy supply system

- Create greater market certainty for renewable energy suppliers
- Develop the market for renewable energy so that proposed projects are technically feasible and can obtain financing
- Induce improvements and cost reductions in renewable energy technologies
- Create niches for renewable energy technologies whose costs are above conventional energy resource costs
- Establish a means for utilities to recover the costs of acquiring renewable resources

State renewable energy policies coupled with federal tax incentives and other subsidies have, directly or indirectly, resulted in deployment of a variety of renewable energy projects throughout the desert Southwest. Several examples are listed below.

- Geothermal energy. Over 500 MW of geothermal generating capacity have been installed in the Salton Sea area of southern California.[22]
- Concentrating solar power. Nine solar electric generating stations built in the 1980s and early 1990s in the Mojave Desert produce about 350 MW of power combined. They were the first in the United States to use parabolic troughs on a large scale to make steam for generating electricity and are still in operation. The Nevada Solar One project near Las Vegas, which was completed in 2007, generates 64 MW of power using parabolic troughs, and a 100 MW power tower with thermal storage in Nevada is scheduled to start construction in 2011. In Arizona, Abengoa has begun construction of the 250 MW Solana CSP plant with thermal storage; the electricity will be sold to Arizona Public Service Company.
- Large photovoltaic projects. Sempra Energy completed a 10 MW and a 48 MW PV plant near Las Vegas, Nevada and sells the output to a utility. The 30 MW Cimarron PV facility in New Mexico provides electricity to Tri-State Generation and Transmission Cooperative.
- Wind energy. East central New Mexico and West Texas have good wind resources and numerous large scale wind energy projects. As of 2010, New Mexico had about 700 MW of wind generation and Texas had about 10,000 MW of wind generation.[23]

As additional utility-scale renewable energy projects are planned, more transmission capacity will be needed. For instance, if Arizona utilities want to add significant amounts of New Mexico wind energy to their portfolios, more transfer capability from New Mexico into Arizona will be needed. Similarly, if more solar energy facilities were to be installed in west central Arizona to serve the load in Phoenix or southern California, more transmission capacity would likely be needed. States are examining renewable energy zones where power plants would be sited and utilities are trying to coordinate investments in transmission to bring power from those zones into the grid. Transmission planning must also address the environmental impacts of proposed transmission corridors so that conflicts with scenic landscapes and with wildlife are avoided, mitigated, or minimized.

### 29.6.2 Local Capabilities

Local governments, entrepreneurs, and nongovernmental organizations are becoming established players in distributed energy and energy efficiency. Distributed energy refers to electric generation or other energy production (such as hot water) close to the place where the energy is demanded. For example, a 390 kW rooftop PV system produces electricity for Coronado High School in Scottsdale, Arizona.

Many local governments have prepared sustainability plans that review options for sustainable energy, identify specific sustainable energy projects or goals, and outline how projects can be built or how goals can be reached. For example, Pima County, Arizona, set a goal of obtaining 15% of the energy used by county facilities from renewable resources by 2025. Long Beach, California set as a goal the facilitation of development of at least 2 MW of solar energy on city facilities by 2020. And Sparks, Nevada, set a goal of generating at least 5% of energy for new city buildings with on-site renewable resources and increasing the capacity of a biogas combined heat and power (CHP) plant at a water reclamation facility.[24] CHP produces both electricity and heat or hot water or steam for an industrial, commercial, or agricultural purpose from a single heat source. It is, therefore, very efficient because the fuel, typically natural gas, is used to produce both electricity and useful heat. CHP facilities are designed to serve a particular "host" needing hot water, steam, or heat and may use the electricity on-site or sell the electricity to a utility. Table 29.3 summarizes common distributed energy projects.

Energy efficient design is part of a sustainable infrastructure. For building design, Leadership in Energy and Environmental Design™ (LEED™) provides a template. The City of Scottsdale, Arizona, requires all new, occupied city buildings to be designed, contracted and built to achieve the LEED Gold certification level, and to strive for the highest level of

**TABLE 29.3**
Overview of Common Distributed Energy Projects

| Project Type | Typical Scale | Typical Location | Energy Use | Examples in the Southwest |
|---|---|---|---|---|
| Small stand-alone energy systems (not connected to grid) | Up to 10 kW | Where warning signs or lighting needed | On-site | Park lighting in Santa Fe, NM |
| Rooftop PV or solar hot water | One to several hundred kW or equivalent | Residential or commercial rooftops | On-site | Residential rooftop PV, PV on school roofs or commercial building roofs, PV on buildings in ecological restoration areas and public gardens |
| Other small or moderate size PV | Up to several hundred kW | Built into or added onto individual structures | On-site or sold into grid | PV on covered parking structures (e.g., Riverside, CA, Utilities Operation Center Solar Carport) |
| Larger distributed energy projects | Several hundred kW or larger | Rooftops, ground mounted, inside buildings, or as separate complex | On-site or sold into grid | Combined heat and power projects at factories, greenhouses, hotels, hospitals, etc., biogas at wastewater treatment plants, PV at military bases, PV in water supply systems to provide pumping power, PV at airports |

certification whenever project resources and conditions permit (see Chapter 28).[25] Projects include a new fire house and a senior center. The New Buildings Institute found that, on average, energy use in LEED buildings was about 25%–30% lower per square foot than national average energy use for similar buildings.[26] Another analysis found that, on average, LEED buildings used 18%–39% less energy per square foot than similar conventional buildings, but that 28%–35% of LEED buildings used more energy per square foot than similar conventional buildings.[27]

A landscape design solution for new or existing homes is shade trees.[28] The tree canopy in Southwestern cities is often meager—in Las Vegas it is about 10%, and in Phoenix about 13% of the land area has plant cover.[29] In the low desert regions of Arizona and California, a mature tree casting shade on the west, east, or south side of a house can reduce electricity use for air conditioning by about 200 kWh/year; more trees would lower electricity use even more.* Shade tree programs are often carried out by community organizations as described in the next section.

### 29.6.3 Role of Community Organizations

Community organizations may lead the effort to deliver or install energy efficiency measures or to inform community members about those measures.† A common approach is for a community organization—a neighborhood association or environmental organization, for example—to educate the public about the benefits of energy efficiency and about how to obtain energy efficient measures. Participants are often recruited through social networks and personal contacts, and trained volunteers might conduct workshops or go door-to-door to provide information. The organization may obtain shade trees or other measures for distribution to participants or may give away or install low cost energy efficiency measures. Some community organizations have delivered hundreds of thousands of shade trees or compact fluorescent lamps (CFLs), for example.‡ Often, these programs are funded by utilities.

Thus, energy efficiency and, to a lesser extent, distributed energy present opportunities for civic engagement to promote sustainable cities. A community organization may serve as a catalyst for civic engagement, encouraging public participation in program design and implementation, often through volunteers. The community organization would typically build up trust in the community, serve as a conduit for expertise on technical matters, and engage in outreach to the community through social networks.

### 29.6.4 Utility Capabilities

Utilities in the Southwest are commonly the principal means of implementing state clean energy policies. Some utilities have become experts in designing and implementing large scale renewable energy and energy efficiency programs. They are learning how to use market processes to acquire large renewable energy facilities and how to stimulate markets for distributed energy and energy efficiency through incentives and education programs. As a result, they are able to achieve far more aggressive renewable energy and energy efficiency goals than was thought possible in 1990 or 2000.

An evolving regulatory process has helped foster utility capabilities. Regulators have established processes for preparing clean energy plans, reporting on progress, and

* For the desert, drought tolerant trees, especially those native to the Southwest, are most appropriate.[28,30]
† For a review of examples, see Berry.[31]
‡ For a review of shade tree programs, see Western Resource Advocates.[28]

reviewing and modifying plans. In addition, regulators are modernizing the way they set rates so that utilities are not faced with financial disincentives if they reduce kWh sales because of successful energy efficiency programs.[32] The trick will be to encourage innovation in utility program design and implementation, assure the public that utilities are spending money wisely, and do so without smothering the utilities in ponderous regulatory reviews.

## 29.7 Assessment

Development of a more sustainable energy system requires institutional capabilities to plan and implement a transformation of the power system into one that relies much more on energy efficiency and renewable energy. With that transformation, mainstream supply and demand decisions will primarily involve clean energy resources.

Public and private sector capabilities to achieve a more sustainable energy system are expanding in the desert Southwest.

- Utilities and regulators are recognizing the risk management benefits of transforming the power supply and demand system—stably priced, non-polluting clean energy resources serve as a hedge against uncertain fossil fuel prices and uncertain environmental regulation compliance costs of operating conventional power plants.
- State legislators and regulators in the desert Southwest have crafted and adopted efficiency standards and renewable energy standards, key public policies for creating a sustainable energy system. Implementing these policies has been accelerated through financial incentives or subsidies to induce individual consumers and utilities to acquire clean energy resources and through policies to allow utilities to recover, in rates, the costs of renewable energy and of utility-sponsored efficiency programs.
- Utilities are learning how to evaluate, acquire, and integrate renewable energy into their power supply systems and learning how to identify credible power plant developers.
- Some cities, building owners, distributed energy providers, consumers, and utilities are looking for opportunities to install PV and other distributed energy projects.
- Community organizations are developing the capability to deliver large quantities of energy efficiency measures, such as shade trees or CFLs, and to educate members of their communities. These programs are often implemented via partnerships with utilities or other organizations or government agencies.

Nonetheless, there are significant challenges. For example, state governments have been active leaders in the desert Southwest, but that leadership role may diminish as political forces change. A second challenge is short-term thinking that focuses on immediate cost minimization and underestimates long-term risks. And a third challenge is difficulty imagining how conventional coal-fired generation can be replaced by a portfolio of renewable energy and gas-fired generation on a large scale.

If the transformation process continues, in 20–40 years the power system will consist of a portfolio of renewable energy, natural gas-fired power plants, much more distributed energy resources, much more energy efficient consumption, and far less conventional coal-fired power generation. If not, we will live in a more polluted world with a vastly altered ecology, powered by old technology, and subject to price uncertainty.

## Acknowledgment

The author wishes to thank Amanda Ormond, Lowrey Brown, and Sandy Bahr for helpful comments on earlier versions.

## References

1. North, D., *Institutions, Institutional Change and Economic Performance* (Cambridge, UK: Cambridge University Press, 1990); Gertler, M., Rules of the game: The place of institutions in regional economic change, *Regional Studies* 44 (2010): 1–15.
2. U.S. Energy Information Administration, *Electric Power Annual*, State Historical Tables for 2010 (Excel file); U.S. Energy Information Administration, *Electric Power Monthly*, February 2012.
3. Bruère, R., Giant power–region-builder, *Survey Graphic* 7(1925): 161–164, 188; Cooke, M. L., Electricity goes to the country, *Survey Graphic*, http://xroads.virginia.edu/~MA01/davis/survey/articles/energy/energy_sep36_1.html, 1936.
4. U.S. Energy Information Administration, State electricity profiles 2008 (Washington, DC, March 2010).
5. Pope, C. A., M. Ezzati, and D. Dockery, Fine-particulate air pollution and life expectancy in the United States, *New England Journal of Medicine* 360 (2009): 376–386.
6. U.S. Environmental Protection Agency, eGRID 2007, Version 1.0 plant file (year 2005 data), Excel file.
7. U.S. Environmental Protection Agency, Mercury: Basic information, http://www.epa.gov/hg/about.htm (accessed August 20, 2011).
8. Committee on Haze in National Parks and Wilderness Areas, National Research Council, Protecting visibility in national parks and wilderness areas (Washington, DC, 1993); Malm, W., *Introduction to Visibility* (Fort Collins, CO: Cooperative Institute for Research in the Atmosphere, Colorado State University, 1999).
9. Western Resource Advocates, *A Sustainable Path: Meeting Nevada's Water and Energy Demands* (Boulder, CO: Western Resource Advocates, 2008).
10. Price data from U.S. Energy Information Administration, *Annual Energy Review 2008* (Washington, DC), *Monthly Energy Review*, March 2010, and *Short-Term Energy Outlook*, April 2010. Prices translated to constant dollars using the GDP Implicit Price Deflator, Bureau of Economic Analysis, National Income and Product Accounts, Table 1.1.9.
11. Western Resource Advocates, *Investment Risk of New Coal-Fired Power Plants* (Boulder, CO: Western Resource Advocates, 2008).
12. U.S. Environmental Protection Agency, Assessment of anticipated visibility improvements at surrounding Class I areas and cost effectiveness of best available retrofit technology for Four Corners power plant and Navajo Generating Station: Advanced notice of proposed rulemaking, Washington, DC, *Federal Register* 74(166) (August 2009): 44313–44334.
13. Smith, J. C., M. Milligan, E. de Meo, and B. Parsons, Utility wind integration and operating impacts state of the art, *IEEE Transactions on Power Systems* 22 (2007): 900–908.
14. GE Energy, Western wind and solar integration study (Schenectady, NY: Report to National Renewable Energy Laboratory, 2010).

15. U.S. Department of Energy, Concentrating solar power commercial application study: Reducing water consumption of concentrating solar power electricity generation (Washington, DC, 2009).
16. Lynn, J. and W. Auberle, *Guidelines for Assessing the Potential Impacts to Birds and Bats from Wind Energy Development in Northern Arizona and the Southern Colorado Plateau* (Flagstaff, AZ: Northern Arizona University, 2009).
17. California Senate Bill No. 1368 (2007), http://www.energy.ca.gov/emission_standards/documents/sb_1368_bill_20060929_chaptered.pdf (accessed August 20, 2011); NRDC, California takes on power plant emissions, 2007, http://www.nrdc.org/globalWarming/files/sb1368.pdf (accessed August 20, 2011).
18. Arizona Public Service Company, Application for authorization for the purchase of generating assets from Southern California Edison and for an accounting order, Arizona Corporation Commission Docket No. E-01345A-10-0474, filed November 22, 2010.
19. Zweibel, K., J. Mason, and V. Fthenakis, A solar grand plan, *Scientific American* 298(1) (2008): 64–73; Shinnar, R. and F. Citro, De-carbonization: Achieving near-total energy independence and near-total elimination of greenhouse emissions with available technologies, *Technology in Society* 30 (2008): 1–16.
20. Horowitz, M., Changes in electricity demand in the United States from the 1970s to 2003, *Energy Journal* 28 (2007): 93–119.
21. American Council for an Energy-Efficient Economy, State energy efficiency resource standard activity, April 2010 (Washington, DC, April 2010). Database of State Incentives for Renewables and Efficiency, http://www.dsireusa.org/ (accessed August 20, 2011). Arizona Administrative Code R14-2-2401 *et seq*. Arizona Administrative Code, R14-2-1801 *et seq*. New Mexico Statutes, NMSA 62-17-3 and 62-17-5(G). Public Utility Commission of Texas Substantive Rule §25.181(e).
22. Sison-Lebrilla, E. and V. Tiangco, *California Geothermal Resources* (Sacramento, CA: California Energy Commission, CEC-500-2005-070, 2005).
23. American Wind Energy Association, State fact sheets, http://www.awea.org/learnabout/publications/factsheets/factsheets_state.cfm (accessed August 20, 2011).
24. Pima County Board of Supervisors, Sustainable action plan for county operations (2008). Long Beach Office of Sustainability, Sustainable city action plan, 2009; City of Sparks, Sustainability action plan 2009, draft dated September 15, 2008.
25. Scottsdale Resolution of the Mayor and City Council No. 6644 (2005).
26. New Buildings Institute, Energy performance of LEED® for new construction buildings (White Salmon, WA, 2008). http://www.usgbc.org/ShowFile.aspx?DocumentID=3930 (accessed August 20, 2011).
27. Newsham, G., S. Mancini, and B. Birt, Do LEED-certified buildings save energy? Yes, but ..., *Energy and Buildings* 41 (2009): 897–905.
28. Western Resource Advocates, Phoenix green: Designing a community tree planting program for Phoenix, Arizona (Boulder, CO: 2009). See also City of Phoenix, Tree and shade master plan, http://phoenix.gov/PARKS/shade52010.pdf (accessed August 20, 2011).
29. City of Las Vegas, Nevada, Urban forestry initiative, 2008. Memo from Dale Larsen, Acting Director, Phoenix Parks and Recreation Department, to Rick Naimark, Deputy City Manager, Tree and Shade Task Force Overview, prepared for Phoenix City Council Work Study Session, May 26, 2009.
30. Simpson, J. and E. G. McPherson, Potential of tree shade for reducing residential energy use in California, *Journal of Arboriculture* 22(1) (1996): 10–18.
31. Berry, D., Delivering energy savings through community-based organizations, *Electricity Journal* 23 November (2010): 65–74.
32. Shirley, W., J. Lazar, and F. Weston, Revenue decoupling standards and criteria, Regulatory Assistance Project Report to the Minnesota Public Utilities Commission, 2008.

# 30

## *Search for a Lean Alternative*

**Paolo Soleri**

**CONTENTS**



### 30.1 Suburbia and Hyperconsumption

Industrial societies are demonstrating how powerful is the production–consumption cycle is when sustained by self-interest and technology. The exponential growth of industrialization is not matched by an exponential growth of intelligence and wisdom. Trivialization of society is an evident consequence. Too much power, not enough wisdom to use it, therefore misuse results from the use of this power. The strength of our virtues, dignity, and equity are not measuring up to our opportunism, and so our proclaimed "free enterprise" is emulating the law of the jungle, the survival of the fittest. In the developing wealth of society, greed is iconized as a virtue. The nation asks us to be dedicated consumers, which is "Americanese" for hyperconsumption.

Single-family homes, insulated and isolated, a patchwork of hermitages endlessly extending into the desert, is what we want and what the market says we must have. The ever-enlarging square footage of the single-family hermitage feeds the production–consumption engine. With the Internet's magic, every living room, playroom, computer room, guestroom, or garage is a personal marketing center on 24 h alert. That keeps the engine humming—the desert be damned! The resultant equation is: exurbia = the demise of nature.

Unfortunately, there is no such thing as virtual logistics, so gridlock is real, paid for in real dollars, time, frustration, land degradation, pollution, and social segregation. Nor is there any virtue in the case of all the umbilical cords necessary to keep the personal hermitage functioning: water, power, garbage, sewage systems, the road above, the utility below, and then the household supplies, the array of appliances, furnishings, groceries, and so on. We exhibit great cunning in improving the wrong things, exurbia being the case here. This is what I call the "better kind of wrongness" domain: improving the wrong thing only adds more layers of wrongness—a dismal domain. So again, *say goodbye to the desert* and the nature we have grown to love.

Now "natural capitalism" is giving pause to our consciences. We may be shifting onto the right track, but planetary equity stands frozen on another track. By proposing a lean alternative, we are trying to dematerialize the triumphal march of the social-Darwinian bent of the capitalist track: the explicit, indeed arrogant, hyperconsumerism we seem to have fallen into. Consider the projected global human population—8–10 billion people—and the biosphere's carrying capacity. In this context, the hyperconsumption frenzy (ignoring more than half the human population, which is a dereliction) seems ungainly and unrealistic. In addition, hyperconsumption might be detrimental to the aim of potential transcendence, in which *Homo sapiens* is and ought to be engaged.

## 30.2 Lean Alternative

There may not be many alternatives to materialism's often violent void. One way is the quest for a lean society that pursues equity and excellence. What I call the "lean alternative" offers an alternative option in face of the hyperconsumption engaging American society as it pursues and submits to limitless wealth. We are consumers because we exist, but we are also producers of that which, if we are wise enough to listen to our consciences, transcends consumption. Leanness—both physical and mental—might give us a more inspiring and cogent position with which to start. More life is the alternative to more wealth. Segregation of wealth and power is actually a diminution of life: the pursuit of a "better kind of wrongness." A sort of destiny gene seems to be confirming itself, advancing toward consumerist materialism's probable breakdown. More and more, pursuit of happiness on such an incline seems to move the hyperconsumer into the corner of dissatisfaction, redundancy, shallowness, guilt, and loss of meaning. Society has to come to terms with the three-headed dragon of ignorance, greed, and hypocrisy.

The lean-alternative imperative is as significant in consumerist countries as it is in have-not countries. In consumerist countries, the lean alternative is needed to abate consumerist aggression. That is, the have countries should spare the have-nots from hyperconsumption's ills and excesses, which are the mirages of the have countries. The planet at large would be the winner.

The inefficient logistical reticulum of our industrialized nation needs to be reformulated from the ground up. As the logistical infrastructure now in disrepair is obsolete anyway, we need a serious conceptual reformulation of the whole system along realistic guidelines, not futilely fighting the ever-increasing congestion of roads, highways, and parking areas by expanding roadways to accommodate ever increasing traffic, but reformulating the damaging patterns of our communities, especially our anti-cultural, anti-environmental, anti-social promulgation of one- to two-story single-family homes. One house or mansion per family requires a logistical landscape horrendously wasteful and brutally anti-environmental—the antithesis of greenness.

In the desert, especially, the pursuit of leanness in water use and consumption is crucial. In their competition with agriculture, suburban and exurban developments present a dilemma because they are the thirstiest of all human settlements, with private courtyards, lawns, swimming pools, golf courses, and so forth. Anticipating a doubled or quadrupled state population presents Arizona with possibly insoluble water problems. Suburban private property manifests a typical case of water waste. Millions of private "hermitage" houses result in rivers and lakes of underused and misused water. Even considering only the backyard, driveway and car washing, lawn and landscaping greenery, and similar suburban accessories, when fed into the million-times multiplier these collectively speak of a staggering misuse of water. By contrast, every organism is a circumscribed plumbing system—and this has been true of billions upon billions of them, working at their success for many billions of years. Which model is more convincing? The marvel of organism is directly dependent on the stupendously efficient logistical flow of fluids throughout its makeup, millions of trillions of molecular "water bags."

The task we face is one of total reformulation—a daunting task, and an inevitable one. The radicalism of reformulation implies a gradualist approach: laboratory-like institutes working on urban problems step-by-step and functioning as testing grounds for the nonsegregational attack that the problems demand. A reformulation of our self-creational *élan* requires the quarantine of the suburban "hermitage" and a total reformulation of our "greenness," compromised as it is now by *Homo faber** materialism. If it is impossible to redirect the tide by means of reform, it is possible and necessary to propose total reformulation. Here is where the hope of true green resides. We need an alternative to this excessive consumption engine that provides a more sustainable way for societies to use the resources of nature and live a life without adverse harm to the world. This is one of the key principles of the lean alternative theory and the urban proposals that are presented in this chapter.

## 30.3 Arcology Concept

To address the growing concerns of modern urbanization, I propose a new model for living that has the potential to show that humankind and nature can coexist in a form that promises a rich and abundant life for all. This model is the basis for the evolution of the arcology proposal as an alternative to modern suburban and exurban incoherence. Arcology is my concept of cities that embody the cooperation of architecture and ecology.† The arcology concept proposes a highly integrated and compact three-dimensional urban form that is the opposite of urban sprawl with its inherently wasteful consumption of land, energy, and time, tending to isolate people from each other and the community. The complexification and miniaturization of the city enables radical conservation of land, energy, and resources.

An arcology would need about 10% as much land as a typical city of similar population. Today's typical city devotes more than 60% of its land to roads and automobile services. Arcology eliminates the car from within the city. The multiuse nature of arcological design would put living, working, and public spaces within easy reach of each other, and walking and cycling would be the main form of transportation within the city.

* The Latin term *Homo faber* translates as "man the maker" or "man who fabricates," referring to the capability of humans to create artifacts, tools, and technologies.

† This concept is developed in my book *Arcology: The City in the Image of Man*. For introductory reading on arcology, see McCullough and Lima.[1]

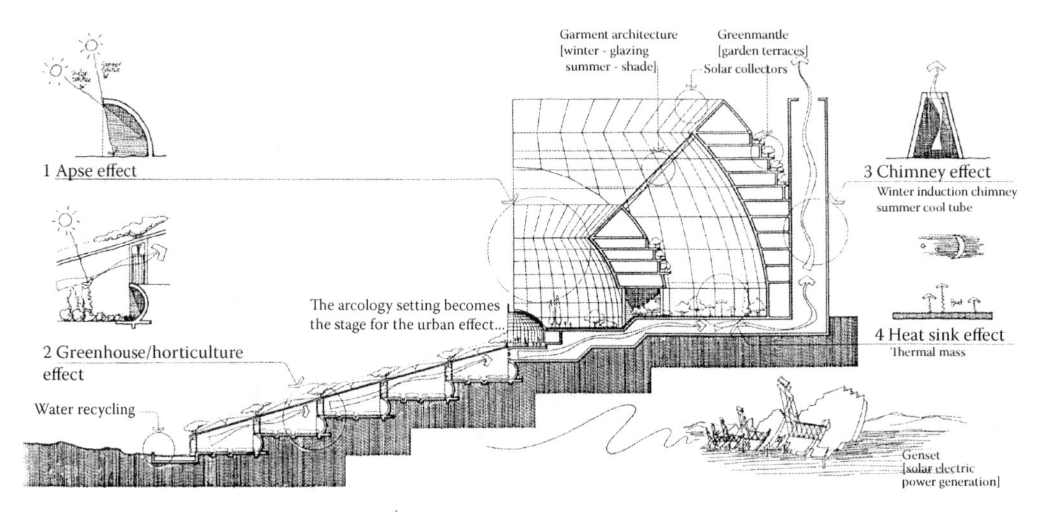


**FIGURE 30.1**
Two Suns arcology: using solar orientation as a resource to benefit urban life.

An arcology's direct proximity to uninhabited land would provide the city dweller with constant immediate and low-impact access to rural space, as well as allowing agriculture to be situated near the city, maximizing the logistical efficiency of food distribution systems. Arcology would use passive solar architectural techniques such as the apse effect, greenhouse architecture, and garment architecture to reduce the energy usage of the city, especially in terms of heating, lighting, and cooling (Figure 30.1). Overall, an arcology seeks to embody a lean alternative to hyperconsumption and wastefulness through more frugal, efficient, and intelligent city design.

Arcology theory holds that this leanness is obtainable only via the miniaturization intrinsic to the "urban effect," that is, the complex interaction between diverse entities and organisms, which mark healthy systems both in the natural world and in every successful and culturally significant city in history.

## 30.4 Arcosanti: A Lean Alternative Laboratory

In 1970, the Cosanti Foundation purchased land to begin the building of Arcosanti, the first arcology and experiential city in the high desert of Arizona, 70 miles north of metropolitan Phoenix.* When complete, Arcosanti will house 5000 people, demonstrating ways to improve urban conditions and lessen our destructive impact on the earth. Its large, compact structures and large-scale solar greenhouses will occupy only 25 ac of a 4060 ac land preserve, keeping the natural countryside in close proximity to urban dwellers.

Arcosanti constitutes an instrument and process for applying the lean alternative. It is a construction site, a process-architecture development in what I call the "lean-habitat mode." The structures and materials we use are the expression of a brain that developed, not in a nondescript and depressing environ, but in the Italian–Mediterranean physical and cultural landscape (Figure 30.2).

* http://www.arcosanti.org/ (accessed August 20, 2011).

**FIGURE 30.2**
Since 1970, construction of Arcosanti has occurred through the support of windbell sales, donations, and voluntary efforts.

Developing a small town to work as an urban laboratory was, and still is, daunting. No public or private resources were provided with the exception of tax exemption, and this remains so after almost 40 years of development. We ourselves subsidized the main task of constructing Arcosanti by gradually developing Cosanti Originals' production in ceramics, bronze, and aluminum. The production and marketing of our windbells has become a small national and international enterprise. The project's income has been steady and reliable for over four decades; modest, but enough to enable very gradual construction and development.

The purpose of Arcosanti, qua laboratory, is to explore an urban alternative, actively demonstrating ways to improve conditions of urban life while at the same time lessening our destructive impact on the earth. Suburban and exurban sprawl, by the nature of their demand on people, resources, and the biosphere, are counterproductive. The "better" they become, the worse the physical and cultural consequences are. The result might turn out to be not a global village, but a global hermitage, in which each habitat unit (the home) is virtually plugged into the whole world, but only in the abstract domain of brain–computer–brain relationships. This is an immensely dangerous, segregating, environmentally and humanly costly situation.

Arcosanti is an urban laboratory guided by what I call "lean minds" seeking to develop a lean habitat, keeping in mind that agriculture (food) and habitat (shelter) are the two indispensable ingredients of human life. Its principal intent remains a quest for the alternative of a lean and more equitable society. One consequence of a successful lean alternative would be a first step toward the reconciliation of the haves and the have-nots. Half the world's population is frugal by necessity. We think it desirable that the whole world become frugal by virtue.

The Arcosanti project, in its own way, wants to be an incarnation of the aforementioned preoccupations and not an absurd, empty, utopian world for a "chosen few." The project represents a viable and positive solution to population growth, environmentally appropriate living, frugality, miniaturization, efficiency, urban evolution, pollution, conservation, transportation, net energy utilization, social interchange, privacy, food production, preservation of natural habitats, aesthetics, affordable housing, global warming, ultimate recycling, education, and world awareness.

Arcosanti builds on little conquests in the routine of life that might come its way, well aware of our dependence on other people and things. An aversion to social, cultural, and economic inequities is paramount to the project. Men and women share responsibility at all levels. Very modest wages are common to all. Participation from many countries and different economic backgrounds guarantees interaction conducive to rich intercultural friendships and experiences among participants.

As an arcology, Arcosanti is designed such that the built and the living would interact and transcend into a highly evolved being. Many systems work together, with efficient circulation of people and resources, multiuse buildings, and solar orientation for lighting, heating, and cooling. At Arcosanti, apartments, businesses, production, technology, open space, studios, and educational and cultural events are all accessible, while privacy is paramount in the overall design. Greenhouses provide gardening space for public and private use, and act as solar collectors for winter heat.

Most of the buildings are oriented toward the south to capture the sun's light and heat, but with roof designs that admit the maximum amount of sun in the winter and a minimal amount during the summer. For example, the bronze-casting apse is built in the form of a quarter sphere (Figure 30.3). The layout of the buildings is intricate and organic, rather than a typical city grid, with a goal of maximum accessibility to all of the elements, increased social interaction and bonding, and a sense of privacy for the residents (Figure 30.4).

Existing structures at Arcosanti have a variety of purposes in order to provide for the complete needs of the community. They include a five-story visitors' center, café, and gift shop (Figure 30.5), a bronze-casting apse, a ceramics apse, two large barrel vaults, a ring of apartment residences and storefronts around an outdoor amphitheater, a community swimming pool, an office complex, and Soleri's suite. A two-bedroom Sky Suite occupies

**FIGURE 30.3**
Ceramics aspe. Arcosanti Foundation is partially funded by the sale of artisanal bronze and ceramic bells, which are sold worldwide.

**FIGURE 30.4**
Southern view of Arcosanti located in Cordes Junction, Arizona.

**FIGURE 30.5**
Crafts III building contains visitor's center, gallery, bakery, café, and residences.

the highest point in the complex and is available for overnight guests. Most of the buildings have accessible roofs.

As an urban laboratory, Arcosanti has sought to incorporate the practice of cooperation and cooperative use rather than ownership, longevity rather than obsolescence, containment rather than diaspora, integration, self-responsibility, spirituality and transcendence against materialism, faith in sensible technology, the struggle against homogenization, authority rather than power, and universalism rather than nationalism. Following are some concrete examples:

- Self-containment of the habitat and adherence to the paradigm of complexity-miniaturization duration
- Efficient use of resources (not self-sufficiency, but self-reliance)
- Reciprocal synergy between the urban and the rural
- Integration of dwelling, learning, working in a structure designed with walkable distances
- Preservation of the emotional, sensorial, and environmental sensibility
- Rejection of prepackaged information

The necessary frugality of the project makes residents aware of the site's limited resources. Prime among them is the value of water and the care required to limit its use. Arcosanti's response to this issue has centered on the concept of the public garden oasis as part of a miniaturized, collective urban habitat. The water saving is enormous when we compare it to the water used by single homes. The use of greenhouse culture is also a strategy for saving water at Arcosanti. Sitting as it does on marginal land, we have excluded any construction on the acreage of farmland below the habitat. But the main saving is defined by the habitat's structure: There are no single homes with their attendant waste of water in gardens, lawns, swimming pools, car washes, and so on. The immediate proximity of wilderness, ravine, canyon, mesas, and cultivatable land is a powerful reminder of the richness of the natural landscape, engendering intrinsic respect and care. Dusk and dawn are particularly beautiful, reinforcing this immediacy of response.

Since Arcosanti is the place for the alternative, the marketplace must adapt to it and not vice versa. The lean alternative is a challenge to the marketplace. Given the relative size and power of Arcosanti, the challenge is minuscule, ephemeral. That is one reason why being absorbed by the market now is unwise. One tenet is that prototyping is not profit-oriented. It ceases to be prototyping the moment it enters the straightjacket of profitability, with its strict rules of "I'll do it if it pays." The lean alternative would be no longer open to the nimbleness of novelty; it would have skirted the risk of challenging the market.

The market would have never accepted, endorsed, or financed the Old Town (the existing structures). The mesa would be there, pristine and untouched if we persevered in trying to bring in the market. None of the more than 6000 people who built the Arcosanti Old Town asked for stocks, ownership, guarantees, public debate, or a decision-making voice. They came, worked, and left, and I think none of the workshop participants came out the worse for it. Now we have a working fragment of a lean alternative. It should not yet be drowned in the oceanic marketplace.

The Cosanti and Arcosanti projects have been a modest attempt to become a testing ground for lean society. We had to generate resources for constructing Arcosanti, while residents—averaging 70 people at a time—carried on with dignity and some rewarding experiences. The awareness of being in a domain of excellence vis-à-vis waste, pollution, environmental disruption, and disavowing the segregation of people and things has been the constant reward. We would like to submit a proposal for accelerating Arcosanti's construction to corporate and noncorporate America. Speeding up Arcosanti's construction would produce more persuasive alternatives to the problems of hyperconsumption and materialism, which are inseparable twins. Extending the experience of lean culture from 100 or so people to 1000 or so people is the main goal; 1000 or so people would achieve Arcosanti critical mass.

Arcosanti, by intent and design, has had a 40 year experience of lean-minded persons living in a lean habitat. We are researching an alternative way of sheltering individuals and society by seeking *optimal* shelter. Hyperconsumption's natural home is the exurban waste camp: an opulent but humanly sterile place. What mainstream society has elected to be the optimal sheltering—suburbia, now mutating into exurbia—is rapidly destroying our physical and mental resources. The self-perpetuating labor of technology, inventiveness, and our innate greed (opportunism) are conjuring the planetary hermitage that we might discover, too late, to be lethal. Self-isolation, only virtually broken by the microchip, is what the gregarious, convivial body-brains that we are, will be unable to cope with. The environmental impact generated by the most extraordinary waste intrinsic to the one-house, one-family formula will overwhelm people, animals, and the green life on which all other forms of life depend.

## 30.5 Lean Linear Arterial City

In recent years I have developed Lean Linear Arterial Arcology as an elongation of the arcology principle intended to perform well with respect to the main logistical reticulum now so indispensable to urban life.* The Lean Linear Arterial proposes the study of an urban environ along main logistical systems existent or anticipated. The main promise of leanness consists in an unflagging thrust toward a recoordination of cultures within and along intense broad-ranging experiences available, as history tells us, only in urban coordinations. As we are definers of spaces and are defined by space, the natural environ

* For more on this proposal, see Soleri et al.[2]

and the manmade environs, the lean linear arterial city is a conjecture willing to test and improve different geometries of space.

"Mobility" is one of the virtues we prize and iconize in the present way of life. It might be smart to recognize: first, the stupefying ways organisms achieve complexity when the arterial–venous logistic is applied, and second, the no less stupefying leanness of those logistics, as life is channeled by arteries. The residuals of life processes are channeled by a virtually symmetrical set of veins. The artery–vein system is a world of fluidity, flexibility, responsiveness, emergence, reliability. And the versatile pump, the heart, is another performing singularity serving zillions of tenants. Our cumbersome, uncoordinated, segregative, materialistic, idolized, hyperconsumptive systems desperately need the best logistical network we can concoct. They must emulate the self-reliant, self-disciplined, nonsegregational lean reticulum, proven by the success of evolution.

Lean Linear Arterial Arcology proposes a dense and continuous urban ribbon consisting of interlinked city modules designed to take advantage of regional wind patterns and solar radiation, both photovoltaic and greenhouse. The habitat "coincides" with logistical channels by incorporating the means of transit within the societal presence; that is, hyperlogistics are embedded within hyperurban structures.

The Lean Linear Arterial habitat is designed to respond to some of the critical situations now taking form in China, and soon in India. Three major points of the project are as follows:

1. Food and habitat are mandatory (necessities), a priority, and a universal imperative.
2. A continent as populated as China at the edge of hyperconsumerism cannot afford to engulf its farmland into parkways, highways, roads, parking lots, garages, and dumpsites. These are consequences of the unchecked metastasis of the city into suburbia and exurbia.
3. A child separated from nature—as even the most opulent exurbia imposes—will be a deprived persona.

Lean Linear proposes a continuous urban ribbon of 20 or more stories high, extending for many kilometers. According to preliminary projections, each "module" of the city measuring 200 m in length accommodates about 3000 residents and the spaces for productive, commercial, institutional, cultural, recreational, and health activities. In a matter of a few minutes the pedestrian can reach most of the locations in his or her daily routine. In a matter of a few more minutes walking, cycling, or using public conveyors such as trains, he or she can reach the adjacent "town," or urban module, to the left or the right.

The continuous urban ribbon is designed to intercept wind patterns of the region. It will also be sensitized to the sun's radiation, both photovoltaic and greenhouse. Briefly stated, its main characteristics are (Figure 30.7) as follows:

1. Two main parallel structures of 30 or more stories extending several kilometers to hundreds of kilometers
2. A climate-controlled volume constituting the inner park defined by the two structures delineated previously, featuring greenhouse in winter and parasol in summer
3. Two wind generator continua
4. Two photovoltaic continua
5. One greenhouse apron continuum

6. One orchard apron continuum
7. Several logistical bands for local, regional, and continental trains (rail and maglev)
8. Moving walkways, shuttles, and "fast down and out" slides at appropriate locations
9. One water "stream" for the enjoyment of residents and travelers and for recreation
10. Two delivery and pick up networks
11. Two liquid and solid-waste networks

In the arterial city, 5 min on the train plus a 5 min walk takes you where you choose or need to be (daily cycles). In 5 min on the train you could traverse ten "mini provinces" (modules), each with its own distinct flavor akin to New York's ethnic neighborhoods. The modular characteristics could not be mandated; rather they would have to come about as the city started to click as a lean, continuous human habitat. Although a single module inhabited by 3000 residents is a relatively modest urban enterprise, a fully developed lean urban ribbon (tens or hundreds of kilometers) would be able to employ a very large, skilled, and diverse labor pool. As an infrastructural system advancing across the whole continent, LLAC could advance in parallel, coupling with each other within areas of highly concentrated population.

Lean Linear Arterial can be seen as a modular three-dimensional landscape that advances into two-dimensional topographies (Figure 30.6). The latter are carriers of local, physical, biological, and human characteristics. Lean Linear offers its own predisposed energy patterns and volumes that can be given the garments and interiors best suited to local needs. Like people holding hands, each brings its own "personality" to the continuum in ways occasionally jolting, but ultimately (physically and transphysically) indicative of the resilience and optimism of the inhabitant. It is a sort of coherent positivism focused on personal and collective intent and how such can penetrate the vast geophysical landscape.

The arterial modularity asks "local" designers to achieve the right fit by and for the "residents." To illustrate: In the first act, a moving machine rolls out the skeletal frame of Lean Linear, one module after another. In the second act, local and regional interests enter the three-dimensional frame and bring the modules to life by designing and building according to specific, local needs (Figures 30.7 and 30.8). (This is what "developers" do: first

**FIGURE 30.6**
Cross-section perspective of Lean Linear City with open land on either side.

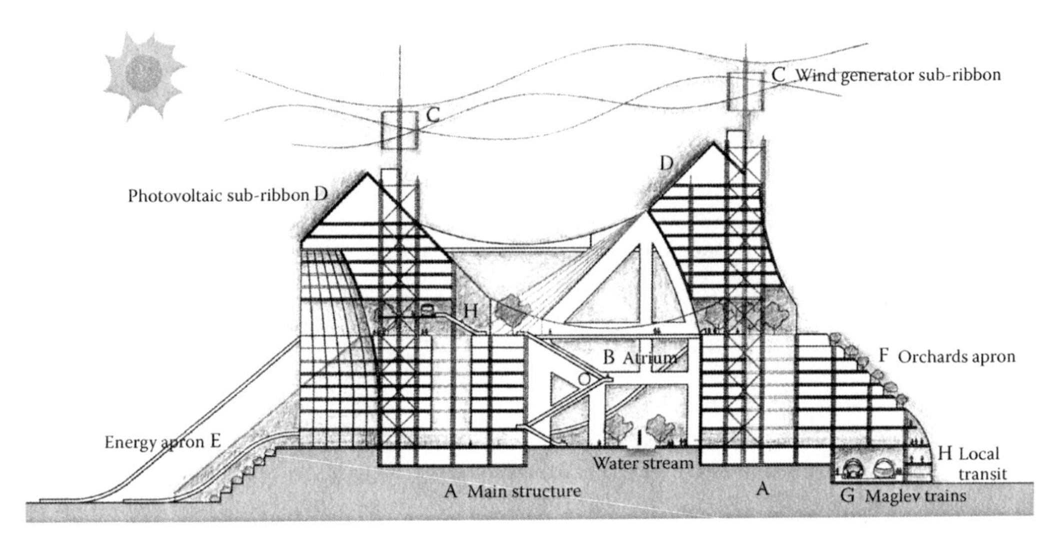


**FIGURE 30.7**
Cross-section of Linear City.

**FIGURE 30.8**
Linear living: atrium space and walkways provide community interaction.

formulating and formalizing the container, then allowing insertion of the content.) The process can be capricious and short lived, or it can be coherently honed for the long term. Most of the time human practicality imposes its bias and reality is thrown out, with consequences that are easy to see, feel—and pay for.

## 30.6 Conclusion

Lean society is more evolutionarily coherent than hyperconsumption society. It is a realistic proposition that could embrace have and have-not alike. That is why a true concern for desirability—even at feasibility's expense—is gaining an urgency never known before, if and where survival's constraints marginalize the human condition. Now we are reaching the threshold of food, shelter, and education for all people, but we are far from equity's threshold. Lean society could be lethal for greed and materialism by decontaminating individual people and groups of people. Decontamination fights the overstuffing that has besieged our environments and our minds. Leanness is agile, almost mercurial, sensitized, and alert about the always precarious condition of people that gets buried in materialism's obesity.

Automobiles colonizing Asia, Africa, and the Americas are sure catastrophes. A car per two persons means five or so billion automobiles for the globe to be produced every five or so years. Alternative logistics to the car could promote a lean way of life. Structuring and restructuring the habitat—for humankind, not for cars—of a projected 10 billion people is an immense task; however, the task is unavoidable if we want to move into more promising landscapes.

Circumstances on a crowded planet are demanding urban systems of all sizes and originality that coordinate in continental hyperorganisms, producing a homospherical network of arterial cities. Time to get planners and architects to ponder their responsibilities in comprehensively reformulating the landscape. The moment is unequaled in view of the transformative power of the production and marketing avalanches that *Homo faber* is generating. Evolution might well be poised for an unparalleled acceleration, courtesy of learning and doing's new technologies.

## Acknowledgment

The author would like to thank Lissa McCullough for providing thoughtful comments and input during the development and organization of this chapter.

## References

1. Soleri, P., *Arcology: The City in the Image of Man* (Cambridge: MIT Press, 1969; reprint, Cosanti Press, 2006); McCullough, L. (Ed.), *Conversations with Paolo Soleri* (New York: Princeton Architectural Press, 2012); Lima, A. L., *Soleri: Architecture as Human Ecology* (New York: Monacelli Press, 2003).
2. Soleri, P., T. Tamura, Y. Kim, C. Anderson, A. Nordfors, and S. Riley, *Lean Linear City: Arterial Arcology* (Paradise Valley, AZ: Cosanti Press, 2012).

## Bibliography

Lima, A. I., *Soleri: Architecture as Human Ecology* (New York: Monacelli Press, 2003).
McCullough, L., (Ed.), *Conversations with Paolo Soleri* (New York: Princeton Architectural Press, 2012).
Soleri, P., *The Bridge between Matter and Spirit Is Matter Becoming Spirit: The Arcology of Paolo Soleri* (New York: Anchor Doubleday, 1973).

Soleri, P., *The Omega Seed: An Eschatological Hypothesis* (New York: Anchor Doubleday, 1981).
Soleri, P., *Arcosanti: An Urban Laboratory?* 3rd edn. (Paradise Valley, AZ: Cosanti Press, 1993).
Soleri, P., *The Urban Ideal: Conversations with Paolo Soleri* (Berkeley, CA: Berkeley Hills Books, 2001).
Soleri, P., *What If? Collected Writings, 1986–2000* (Berkeley, CA: Berkeley Hills Books, 2003).
Soleri, P., *Arcology: The City in the Image of Man* (Cambridge, MA: MIT Press, 1969; reprint, 4th edn., Paradise Valley, AZ: Cosanti Press, 2006).
Soleri, P., *What If? Quaderno 13: Lean Linear Arterial City* (Paradise Valley, AZ: Cosanti Press, 2012).
Soleri, P., T. Tamura, Y. Kim, C. Anderson, A. Nordfors, and S. Riley, *Lean Linear City: Arterial Arcology* (Paradise Valley, AZ: Cosanti Press, 2012).

# 31

## *Creating Sustainable Futures for Southwestern Cities: The ProtoCity™ Approach in the Ciudad Juarez, Mexico/El Paso, Texas Metroplex*

**Pliny Fisk III**

**CONTENTS**



> The role of art is to make the commonplace special and that of science to make the special commonplace.[1]

## 31.1 Introduction

My professional experience with cities began in 1971 when I worked for Ian McHarg's office as the coordinator of engineering and ecology for New Orleans East, a new town of 100,000 on the Mississippi Delta. My role was to develop a system to filter the town's organic waste and waste from barges on the Mississippi River through a constructed wetland that in turn provided feed to the largest fishery in North America. A secondary mission was to support the establishment of industrial conversion processes that rendered the other parts of the waste flow into productive industries for the new town. This metabolic redesign was to set up economic development scenarios to divert toxic waste coming down the Mississippi away from the vulnerable Gulf of Mexico ecosystem by using the wastes as feedstock for eco-industrial parks. In hindsight, this type of approach may have helped prevent some of the ecological disasters caused by hurricanes Katrina and Rita, the massive release of oil from the Deep Water Horizon oil drilling rig explosion in 2010, and the 5800 mile$^2$ Gulf dead zone that now characterizes this formerly abundant, vital ecosystem.

Today, we call this approach ecoBalance™, a process in which decisions related to the built environment are informed by the full lifecycle of resources and in which by-products—intentional or unintentional, benign or toxic—are viewed as resources.[2] Through the work with McHarg, I learned lessons that would guide the rest of my career: the city, including the public, private, and non-profit sectors, requires considerable coordination to support measurable and sustained regeneration.

## 31.2 Center for Maximum Potential Building Systems

In the 40 years subsequent to this McHarg-inspired initiative, the Center for Maximum Potential Building Systems (CMPBS) has worked with communities along the U.S.–Mexico border, indigenous populations along Nicaragua's Atlantic Coast, ecovillages, American Indian Nations, university campuses, prominent federal building greening initiatives, a rural Chinese village and farm, and the preliminary design and plan for a 50,000 person EcoCity in Morocco. The experience of implementing a spectrum of tailored solutions in a kaleidoscope of cultures and ecologies helped prepare us for undertaking a project in a region as economically, ecologically, and socially complex as the border cities of El Paso and Juarez.

Key to understanding and regenerating the ecology of a place is establishing a framework to understand the conditions that influence how decisions are made and how an altered ecology of the city region is evolving. Cities are also producing unique ecosystems evolved from their surroundings. Cities are undergoing rapid transformation, as evidenced by the concepts presented elsewhere in this book.

This realization grew out of our National Input–Output/Life Cycle Assessment/Geographic Information Systems model development in the mid-1990s, supported by a Cooperative Agreement from the U.S. EPA. This approach spatially overlaid the 12.5 million businesses tracked by the U.S. Bureau of Economic Analysis with environmental impact data from the U.S. EPA, including greenhouse gases, criteria air pollutants, and toxic releases. An algorithmic correlation of dollar equivalency and human impact revealed a cell-by-cell accounting with the greatest NAICS (North American Industry Classification System) impact located along the urban edges. The pattern was consistent: Urban areas of higher population tended to mediate impacts while rural areas took on the brunt of the pollution (Figure 31.1). If we are to "save our planet," a very different planning paradigm is needed: one that uses the city, combined with its rural partners, to trigger systemic planetary health and well-being.

By the early 1990s CMPBS was well-established in Austin, Texas, and collaborated with City of Austin staff to establish the first municipal green building program in the world: the Austin Green Building Program.[3] The green building movement quickly spread from city to state and now, just 22 years later, the United States has metrics for green at almost every scale including home, commercial, school, neighborhood,

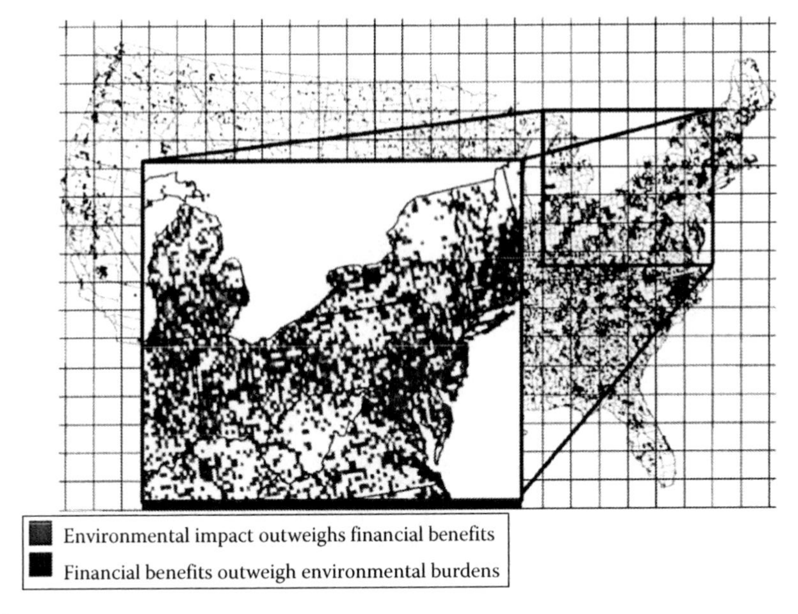


**FIGURE 31.1**
Cell by cell accounting of financial benefits versus environmental impacts revealed higher impacts on the outskirts of urbanization.

and soon the city itself. With more than 160,000 LEED® (Leadership in Energy and Environmental Design) accredited professionals, the U.S. Green Building Council[4] helps establish the green building movement as a success story across for-profit, non-profit and public sectors. Former London Mayor Ken Livingstone and the Clinton Climate Initiative partnered to create the Cities Climate Leadership Group (C40) which targets 40 large cities for strategic action.[5] This is a fitting approach considering, since 2007, the majority of the world's population lives in cities.

A recent research study at our Austin office illustrates the need to understand the city as a system of public, non-profit and private sector actors, where with the use of creative intervention, we can move the system to a regenerative mode. CMPBS and associates recognized the land use/environmental challenge created by the ubiquitous big box store/warehouse typology as an opportunity for Austin to continue its steps toward a model of the green city of the future (Figure 31.2). The plan that emerged involved retrofitting these buildings with combinations of commercially available roof top systems for rainwater harvesting, ecological waste water treatment, high yield organic food production, high efficiency organic fertilizer production, solar photovoltaic panels and even algal-based liquid fuel systems. The plan incorporated the same ecoBalance guiding principles of previous work by balancing local needs with life cycle procedures that involved, as much as possible, local sourcing, transport, processing, and re-sourcing. Without adding a single residential photovoltaic panel or LEED certified building, this system as proposed was estimated to supply almost 20% of Austin's electric needs, more than 15% of Austin's water needs, exceed Austin's fresh vegetable food demand, and meet close to 20% of Austin's biofuel needs (Figure 31.3).

This study and our prior experience with establishing Austin's Green Building Program suggested the key to city scale intervention is the establishment of partnerships with

**FIGURE 31.2**
Map of Austin, Texas, warehouses reveals an opportunity to further advance Austin as a model green city of the future.

strategic people and organizations. In the development of Austin's Green Building Program, we enjoyed the support of a progressive mayor and City Council, 42 citizen's commissions and many environmental groups. Furthermore we established a futuristic, "maximum potential" model. Its ultimate success was the process itself, that capitalized on existing checks and balances within Austin's urban system. The program's merits were recognized at the 1992 UN Earth Summit receiving one of 12 awards for local governmental environmental initiatives and the only one from the US.

## 31.3 Lessons Learned

What was learned that could help us understand and work with the even more complex conditions confronted by the El Paso/Juarez metroplex?

1. The need to identify and put into place a city/regional planning process so that all actors (i.e., the public, private, and non-profit sectors; citizen commissions; the mayor and city council) understand their interrelated responsibilities with a system for monitoring and measuring.
2. An understanding of four essential flows (information, currency, energy, and material) and how they organize society and resources in general.
3. Flow intervention must be handled strategically. For example, when capital investment is not available to support new technologies, the local business sector must

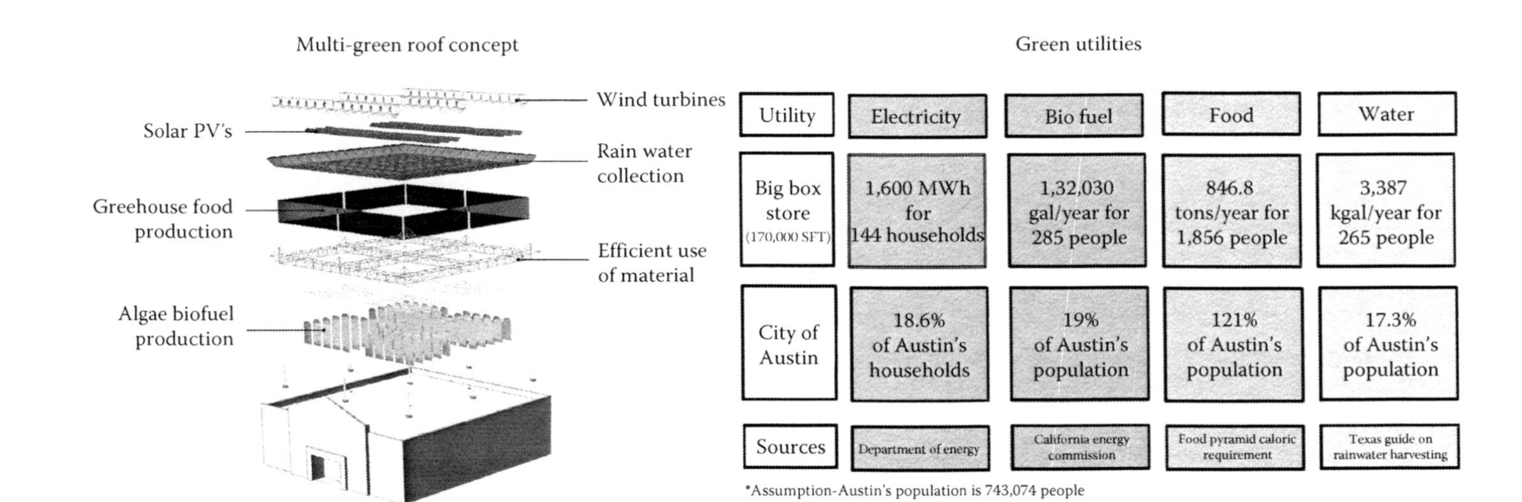


| Utility | Electricity | Bio fuel | Food | Water |
|---|---|---|---|---|
| Big box store (170,000 SFT) | 1,600 MWh for 144 households | 1,32,030 gal/year for 285 people | 846.8 tons/year for 1,856 people | 3,387 kgal/year for 265 people |
| City of Austin | 18.6% of Austin's households | 19% of Austin's population | 121% of Austin's population | 17.3% of Austin's population |
| Sources | Department of energy | California energy commission | Food pyramid caloric requirement | Texas guide on rainwater harvesting |

*Assumption-Austin's population is 743,074 people

**FIGURE 31.3**
CMPBS's plan for an ecoBalanced retrofit of Austin's warehouses would provide a green source of electricity, water, and food for the city.

be organized to step in. Similarly, if the information flow is incorrect, media must be leveraged to disseminate factual information or the energy flow may need decentralization with incentives.

4. The importance of assessing the state of balance within infrastructure (i.e., air balance, water balance, food balance, energy balance, and material balance).
5. The need to find and apply appropriate information regarding *place*. For example, its climate, hydrology, soils, and ecology trends of improvement or decline.
6. Cultivation of influential and effective partners in the private, public, and non-profit sectors is essential to facilitate the process.
7. The value of working within a national and international bio-metric network to help keep us abreast of developments in other similar programs around the world.
8. The need to integrate an information dissemination network to reach practitioners and the general public simultaneously.

Two conceptualization tools—one contextual and the other operational—have helped us understand how Austin, or other city/regions, could learn from and build on lessons to shape future programs.

- The *contextual* tool is the Development Ladder; it addresses the city/region's state of development at varying stages of its evolution.
- The *operational* tool, ProtoScope, provides a systemic representation of how the city could potentially function; it has evolved from our experiences with several cultures throughout the world. This idealized, systemic view of the city/region became what is now called a *ProtoCity*™, or an idealized place-based prototype city of the future.

## 31.4 Development Ladder

The Development Ladder establishes the current status of a place and identifies effective action steps. It is structured around four basic stages: surviving, maturing, anticipatory, and worldly. The first step in effecting positive development at any scale is to determine the city/region's position on the Development Ladder. To state that any of these stages are superior or inferior is not the point; indeed, one could cluster all within any stage and recognize key attributes. The Development Ladder can apply to the city/region as a whole or can assess a city/region's position in terms of specific issues such as public health, education, governance, employment, environmental sustainability, or superstructure (Figure 31.4).

Four essential flows determine a city/region's position on and movement along the Development Ladder:

1. *Information*: The most fluid and most useful as well as the most easily disrupted of all the flows. We use it in several ways such as locating global partners with success in similar issues in similar conditions. Embedding measurement and feedback mechanism is essential (progress in information flow means evolving to the smart grid).

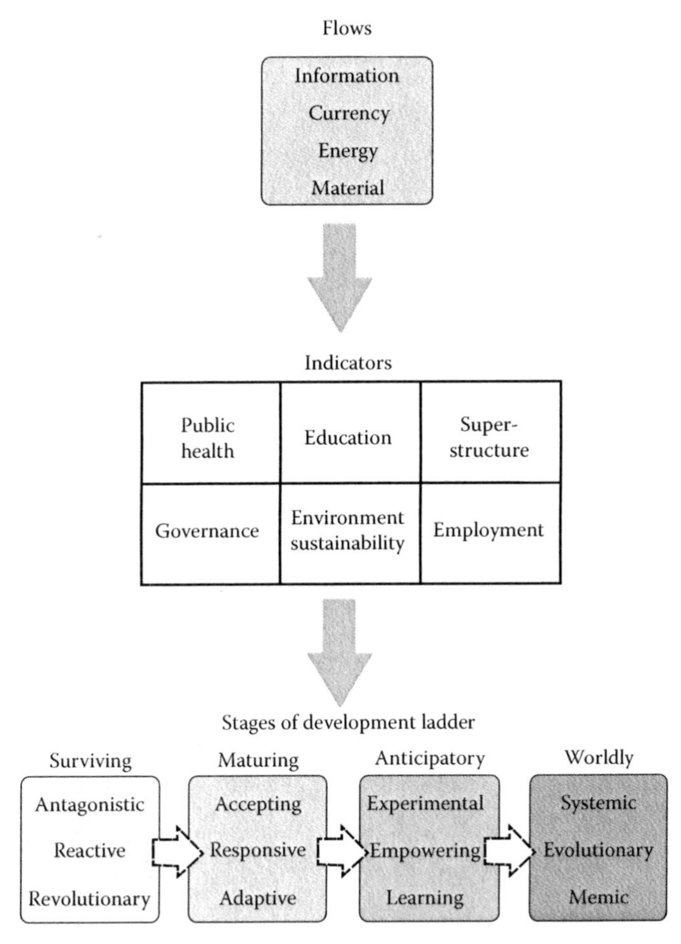


**FIGURE 31.4**
Development Ladder helps assess the current status of a community and its potential for development by tracking four essential flows in one or more indicator categories.

2. *Currency:* The strategic flow of money through the city/region. This flow includes the strategic placement of available dollars to improve specific triggers for change. Innovative financing that supports ecosystem services, community health, and life cycle connected businesses are particularly important.
3. *Energy:* The energy flow, like the material flow, needs to be understood from an ecoBalancing™ standpoint through localized sourcing, processing, use and re-sourcing. Energy flow must be worked with at every level of society so that codes, investment, design, and engineering become fail safe due to scalar life cycle redundancy from home to neighborhood to region.
4. *Materials:* Similar to the flow of energy, the flow of materials (physical, biological, mineral, chemical, etc.) becomes a significant area of a localized creativity so that it is not only sourced within the region but is also combined into sophisticated chemical processes. For example, aluminum and magnesium compounds derived through solar electrolysis from the briny ground water. The extracted compounds not only have metallic and cementitious physical properties but when combined

with other ingredients become the basis for hydrogen energy production with far better energetic and environmental efficiency than coal burning.

Austin's Green Building Program is an example of ambitiously advancing a city into an anticipatory learning environment where the idea of a holistic understanding of the city was not only possible but also desired by many. As a result, Austin continued along the Development Ladder into an almost evolutionary position pushed by explicitly embracing music and the digital arts, underpinned by a strong environmental ethic. Our continued work with the community yielded a sharper image of Austin's evolving place on the Development Ladder and a more refined definition of the Development Ladder itself. As we analyze El Paso/Juarez as a generic system condition, where would we place it?

## 31.5 Development of the ProtoCity

In 1960, the English psychiatrist and cybernetics pioneer Ross Ashby developed a model to conceptualize the brain's process of adaptation.[6] He described the brain as "goal seeking" and in constant pursuit of equilibrium. In his model the organism interacts with its environment and settles on an equilibrium defined within certain limits at a given point in time. Thirty years later, in 1990 while developing Austin's Green Building Program, CMPBS imagined the city as a brain and adapted Ashby's conceptual model as a tool to understand the interaction of the City's public and private sectors. The model that evolved placed public bodies in the role of the environment (better described as the keepers of the commons) and the private sector as the organism trying to respond to the environment but also effecting and helping to develop policy. Monitoring occurred via the commissions that kept close ties with the city reporting to them if and when environmental problems arose (Figure 31.5).

We have since determined an inherent problem with these early, seminal iterations of our ProtoCity methodology. The original model considers the city/region in homeostasis without accounting for change; we have since determined that a dynamic representation is required as shown in our 2008 rendition.

For example, urban vegetation may increase in extent and diversify with increased urbanization, just as locally produced, organic food may become more available. Additionally, a damaged ecosystem such as is present in Juarez and El Paso not only needs to be repaired but the city must also proactively intervene to the point of revitalizing and regenerating at a system level well beyond its present condition. A further adaptation of Ashby's diagram provides opportunity to create dynamism in the limits of a system. This is accomplished through introducing ProtoScope.

### 31.5.1 ProtoScope

Any type of development benefits from understanding where you are coming from in terms of trends and where you are trying to go in terms of best practices. ProtoScope provides the basis for triggering a city/region to move to the next stage of the Development Ladder by increasing the capacity to create the context for change. It may be introduced at any stage of the Development Ladder and is appropriate for any community on the edge of change. ProtoScope entails the following three phases (Figure 31.6).

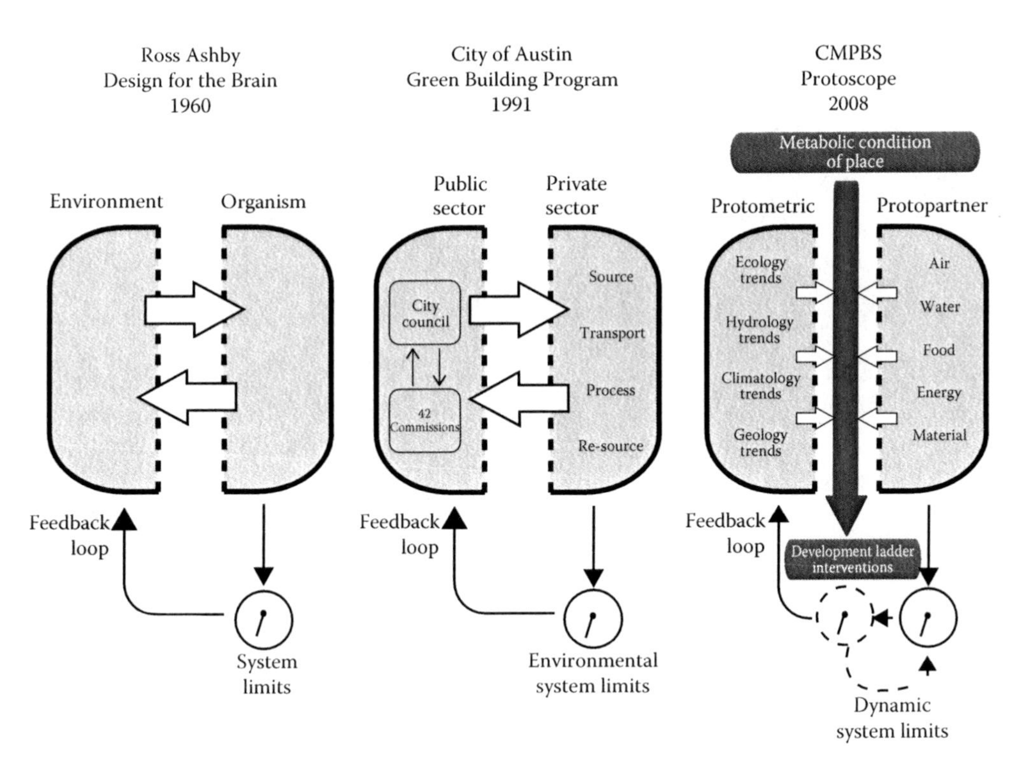


**FIGURE 31.5**
Ross Ashby's Design for the Brain diagram from 1960 inspired the approach used to develop Austin's Green Building Program and most recently CMPBS's ProtoScope.

#### *31.5.1.1 Phase 1: Information Gathering*

1. *ProtoMetrics*™ provides a location-specific assessment incorporating a globally oriented biophysical pattern finding procedure.
2. *ProtoSpace*™ locates similar biophysical zones.
3. *ProtoPartners*™ identifies organizations in ProtoSpaces that have relevant information to share in those same or similar biophysical zones.

#### *31.5.1.2 Phase 2: Assessment/Learning*

1. Establish flow indicators derived from assessing the local metabolism in light of environmental patterns and human experience.
2. Establish the position on the Development Ladder.
3. Locate the physical and digital ProtoSite or a place to prototype some or all of the solution sets.

#### *31.5.1.3 Phase 3: Implementation*

1. Establish the tool kit for intervention according to the flow indicators.
2. Identify ecoBalance (resource balancing) procedures at several scales including the home, neighborhood, and regional levels through an understanding of air, water, food, energy, and material balancing.

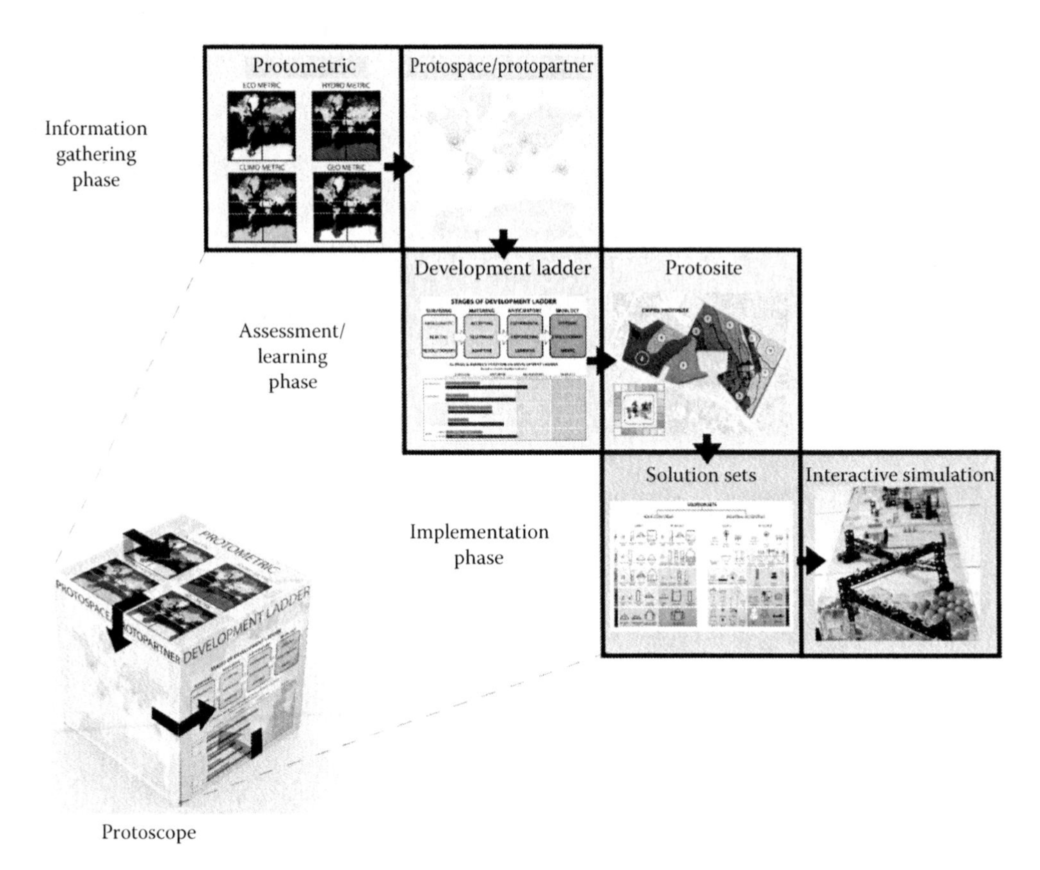


**FIGURE 31.6**
Three phases of ProtoScope.

3. Select strategic local partners oriented around indicators.
4. Set up immersive and interactive simulation experiences, from experimental gaming (informal) to demonstrations with focus groups (formal).
5. Create, leverage or acquire creative financing that bridges issues such as ecoBalance, health and economic development.
6. Assess using built-in feedback mechanisms and replicate successes.

The lessons of our previous work and the phases explained earlier gradually evolved into a far richer understanding of the city region partially spawned by the complexity of recent experiences gleaned, from hurricanes Katrina, Rita, and Ike. Although they were in many ways a crisis/reactive condition they were also occurring within societies and cultures that if they choose could use these crises as opportunities. The following describes the process as it involves El Paso and Juarez.

*31.5.1.3.1 Information Gathering*

- *ProtoMetric* represent the characteristics of a place, based on the biophysical scalar patterns, identified through trends from world to city/site (Table 31.1). ProtoMetric utilizes online databases covering ecology, hydrology, climate, and geology/soils (Figure 31.7).

**TABLE 31.1**

Proto-Scoping the City/Region of El Paso and Juarez Using ProtoMetric Trends

| | Trend | World | Country | State | Site |
|---|---|---|---|---|---|
| | | Global | United States/Mexico | Texas/Chihuahua | El Paso/Juarez |
| | Vulnerable, endangered, critically endangered, or extinct species | More than 17,000 species globally | 2,258 species in United States and Mexico | 32 in Texas, 0 in Chihuahua | 2 in El Paso and Juarez |
| Ecological ProtoCode | Status of species and habitat | Over 50% of all species are effected in some way by climate change | Half of N. America's ecoregions are now severely degraded with at least 235 threatened species | Texas ranks second in biodiversity after California, but threats to animal species have increased dramatically since the turn of the century | Sneed Pincushion cactus (*Coryphantha sneedii* var. sneedii) only occurs in El Paso country and two counties in New Mexico |
| Hydrological ProtoCode | Change in frequency and intensity of precipitation | Higher rate of evaporation, increase in frequency and intensity of storms, lower soil moisture in some regions | Much of the American southwest to become more arid with increases in seasonal flash flooding | Precipitation projected to decrease by 5%–30% in winter and increase 10% in other seasons | Drier climate could decrease stream flow by up to 35% with more intense storms and flooding in wet season |
| Climatological ProtoCode | Change in average temperature | Global surface temperature to increase by average of 1.6°F–6.3°F by 2100 | United States temperatures projected to increase more than the global average in warm areas | By 2100 temperatures in Texas could increase by an average of 3°F in spring and 4°F in other seasons | Temperatures increasing more and at a faster rate than rest of the world |
| Geological ProtoCode | Desertification | 27.5 million acres considered at high or very high risk of human induced desertification | Nearly 90% of N. American arid lands are moderately or severely desertified | Three of four areas of severe desertification are in Texas, Chihuahua, and New Mexico | Desertifying as rapidly as the worst areas in Africa and Asia |

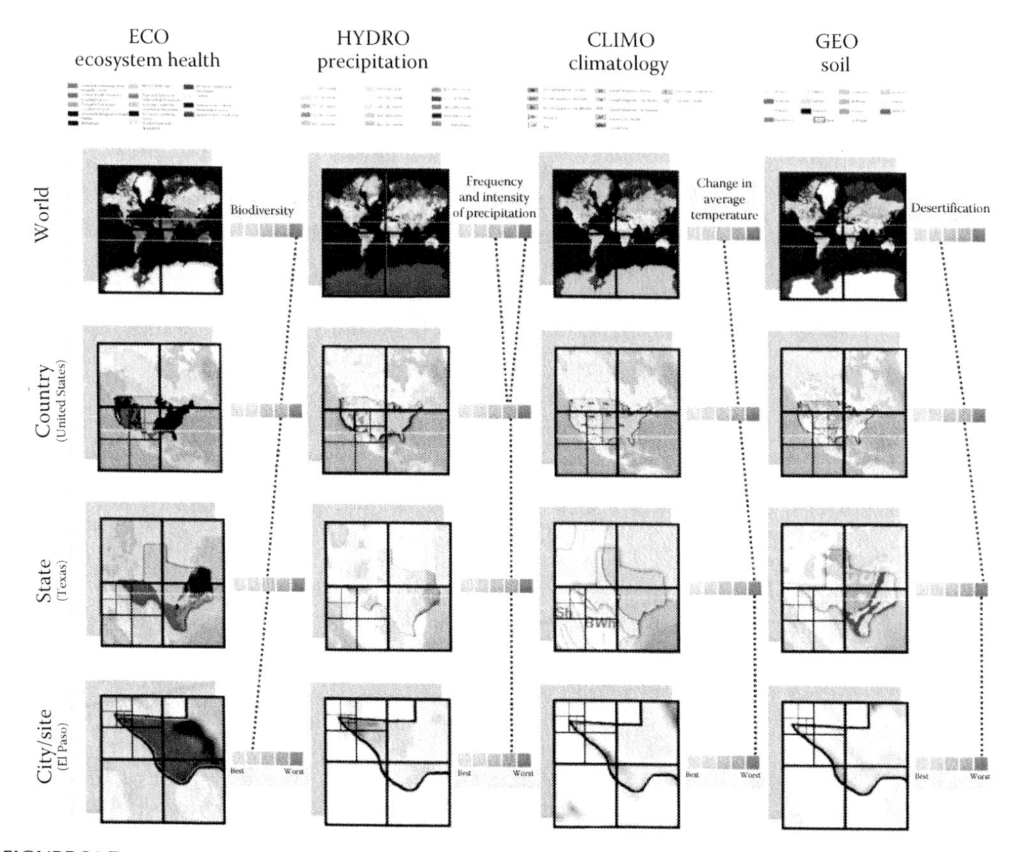


**FIGURE 31.7**
ProtoMetrics of El Paso/Juarez identify ecological, hydrological, climatological, and geological trends from world to site to better understand the specific physical characteristics of the place.

- *ProtoSpace* is the identity of all locations across the globe with statistically similar ProtoMetric characteristics (Table 31.2). ProtoSpace forms the basis for a global learning and organizing network linked by similar experience and expertise (Figure 31.8).

- The *ProtoPartner* work-net is a network of individuals, organizations, and other bodies working in the five ecoBalance impact categories (air, water, food, energy, materials) with the same essential biophysical conditions. See the following ProtoPartner chart identifying ProtoSpaces or cities/regions of similar biophysical zones divided according to the ecoBalance categories.
- *ProtoPartners and ProtoCities* for El Paso/Juarez.
- Key ProtoPartners exist for all four major topic areas and at all scales from world to site (Table 31.3). The ProtoCities also appear (Figure 31.9) in chart form as detail.

*31.5.1.3.2 Assessment/Learning Phase*

The Development Ladder is applied to El Paso and Juarez in terms of public health indicators.

**TABLE 31.2**

Detail of Figure 31.8, ProtoSpaces of El Paso and Juarez

| S. No. | Name | Country | Desert | Location | Elevation (ft) | Area (Mile²) | Population | Density | History | Geographic Features |
|---|---|---|---|---|---|---|---|---|---|---|
| 1 | El Paso | United States | Chihuahuan desert | 31°47′ 25″N 106°25′ 24″W | 3740 | 250.5 | 736,310 (2006) | 2,446/mile² | 1680 | Rio Grande river, Franklin Mountain, Heuco Bolson |
| | Ciudad Juarez | Mexico | Chihuahuan desert | 31°43′ 4″N 106°25′ 17W | 3675 | 72.6 | 1,400,891 (2005) | 12,167/mile² | El Paso del Norte 1659 | Rio Grande river, Juarez Mountain, Hueco Bolson |
| 2 | Yazd | Iran | Dasht-e Kavir desert | 31°30′ 0 N 54°30′ 0 E | 3990 | 6177 | 505,037 (2006) | 81.8/mile² | AD 226–640 | Shir Kuh mountain (4075 m) |
| 3 | Jaipur | India | Thar desert | 26.92°N 75.82°E | 1417 | 77 | 3,324,319 (2005) | 42,963/mile² | 1727 | Major rivers—Banas and Banganga |
| 4 | Alice Springs | Australia | Simpson desert | 23°41′ 60S 133°52′ 60E | 1998 | 57.1 | 26,486 (2006) | 461.0/mile² | 1872 | Todd River, MacDonnell Ranges |
| 5 | San Pedro de Atacama | Chile | Atacama desert (aridest desert) | 22°55′S 68°15′W | 7900 | 9,049 | 4,969 (2002) | 0.5/mile² | 1450 | Arid High Plateau, Oasis |
| 6 | Ghanzi | Bostwana | Kalahari desert | 21–42S 021–39E | 3710 | 45,525 (district) | 36,675 district | 0.8/mile² | Bushman; first white settlers came in 1874 | Terrain is flat and featureless |

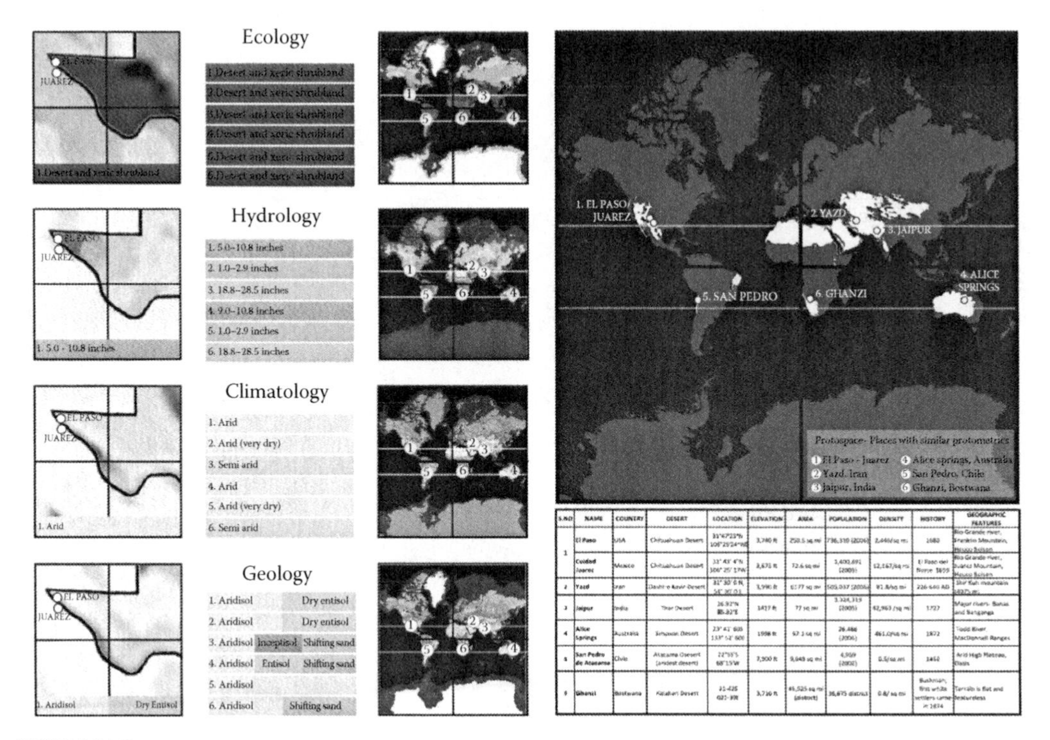


**FIGURE 31.8**
ProtoSpaces of El Paso/Juarez are locations across the world that share statistically similar environmental conditions in one of the four ProtoMetric categories. ProtoSpaces of El Paso/Juarez face similar concerns about water, desertification, access to food, and air quality.

The Development Ladder is based on assessing a location's four resource flows—information, currency, energy, and materials—and subjecting these to both the biophysical and human trends of ProtoMetrics and ProtoPartners (Figure 31.10). Indicators such as health are chosen due to their high coincidence in both categories (i.e., health of environment and health of people are defining characteristics of Juarez and El Paso).

By assessing the four flows we can identify the development stage of a place: in this case, *maturing* for Juarez and *anticipatory* for El Paso. The coincidence of flows was most pronounced when we tested relative to health, environmental sustainability, and employment (Table 31.4). Indicators were identified in the following sequence of importance:

1. Health
2. Environmental sustainability
3. Employment

A *ProtoSite* is a neutral location where public, private, or non-profit groups can digitally and/or physically prototype solution sets supported by an array of computer simulations, models, web gaming, and physical construction. Digital methods can range from basic climatic simulation on the building scale that prototypes performance before construction to more complex planning procedures to see how the viral evolution of a particular set of intervention strategies affects ground conditions. ProtoSite is an intervention laboratory to test ideas.

**TABLE 31.3**

ProtoPartners of El Paso and Juarez Are Organizations, Individuals, and Groups Working in ProtoSpaces That Deal with Air, Water, Food, Energy, or Material Issues Specific to Desert Climates

| **Some ProtoPartners of El Paso and Juarez** |
|---|
| *World ProtoPartners* (Corresponds to ProtoSpaces listed in Table 31.2) |
| Center for Maximum Potential Building Systems, Austin, Texas |
| CENESTA (Centre for Sustainable Development), Iran |
| TERI (Tata Energy Research Institute), India |
| Desert Knowledge Australia, Australia |
| Acción, Chile |
| EASD (Empowerment for African Sustainable Development), South Africa |
| *Country ProtoPartners* |
| Air: Clean Air Council, Pennsylvania |
| Water: Watershed Preservation Network, California |
| Food: Agricultural Council of America, Kansas; Environmental Research Lab, Tucson, Arizona |
| Energy: Institute for Energy and Environmental Research, Maryland |
| Material: U.S. Green Building Council, Washington DC |
| *State ProtoPartners* |
| Air: Environment Texas |
| Water: American Water Works Association, Desert Mountain, Texas Chapter |
| Food: Sustainable Agriculture Research and Education |
| Energy: Sustainable Energy and Economic Development Coalition |
| Material: Straw Bale Association, Texas |
| *Site ProtoPartners* |
| Air: Texas Commission on Environmental Quality |
| Water: El Paso County Water Improvement District |
| Food: El Paso County Department of Public Health |
| Energy: El Paso Solar Energy Association |
| Material: El Paso Boy and Girl Scouts |

The Austin site of CMPBS, a 35 year old 501(c)3, is developed as the original ProtoSite representing Texas' 10 ecozones. ProtoSites can be located anywhere in the world according to project needs, NGO involvement, and community support (Figure 31.11).

#### 31.5.1.3.3 *Implementation Phase*

The implementation stage is made up of two primary actions:

- Identification of specific collaborators that have evolved out of our ProtoPartner research.
- Development of a planning framework for a series of on-the-ground prototypes that test financial models. Eventually, these models gain financial support and creation of an interactive simulation tool that represents the city from home to neighborhood as well as the entire region. The entire effort creates a living laboratory for a convivial exchange of creative solutions engaging all citizenry.

With the support and collaboration of ProtoPartners, specifically selected for their germane experience in similar ecological and social contexts, the work in El Paso and Juarez

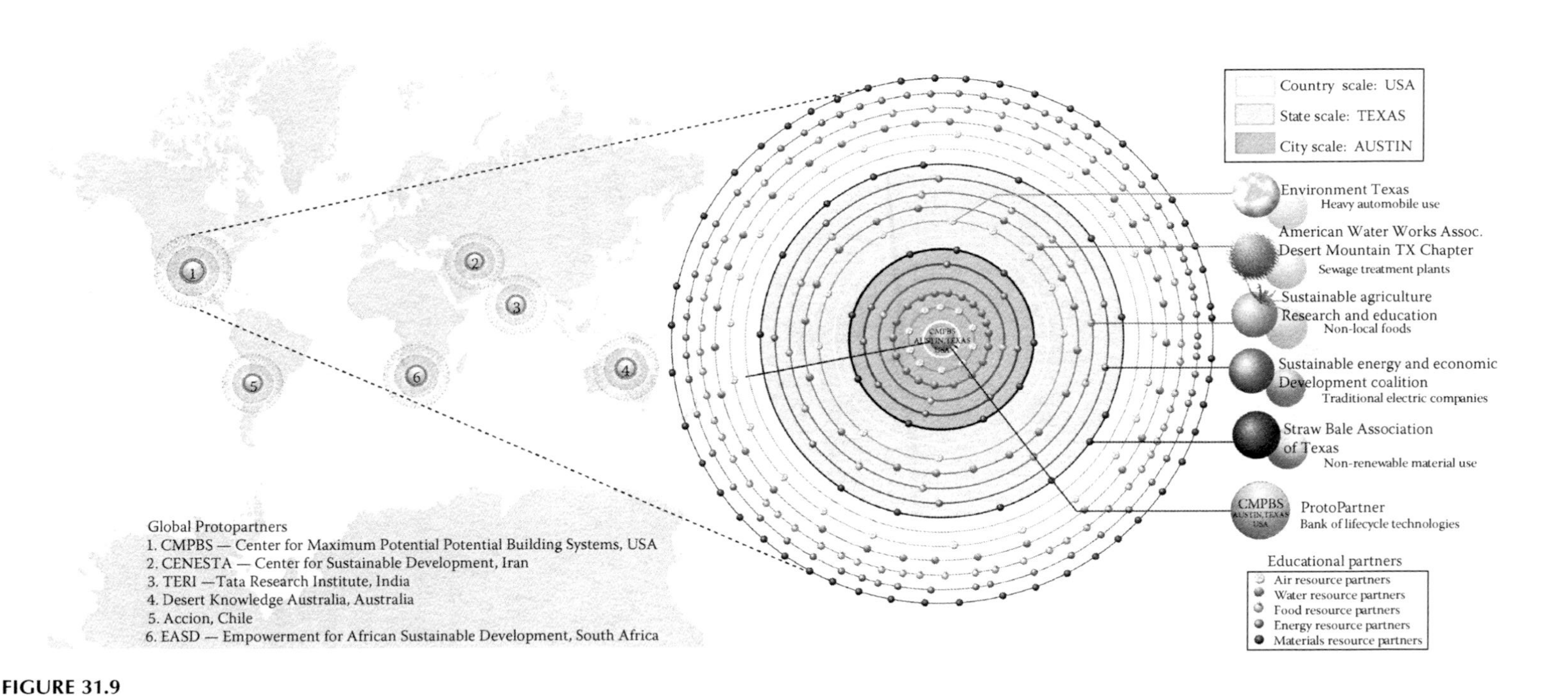


**FIGURE 31.9**
ProtoPartners of El Paso/Juarez are NGOs, research organizations, and universities across the world that share statistically similar environmental conditions in one of the four ProtoMetric categories. They can be correlated to concerns about water, desertification, access to food, and air quality.

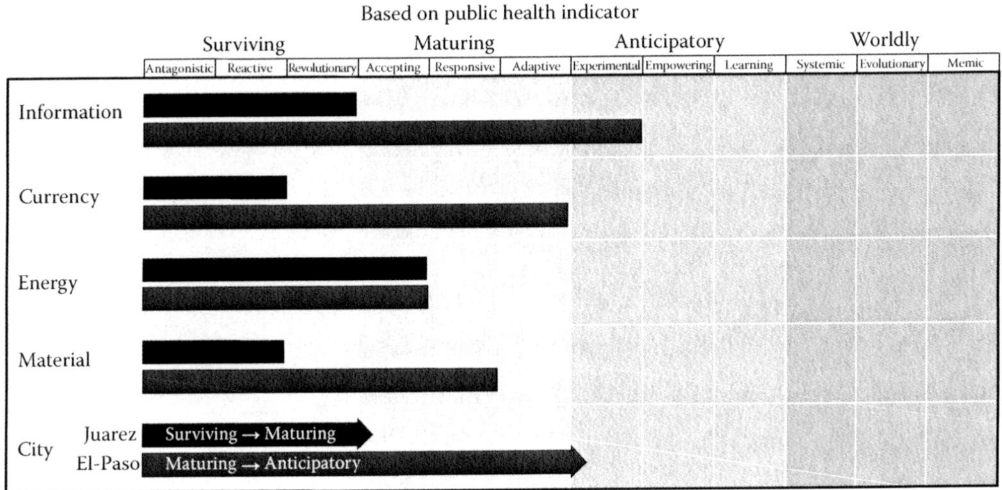


**FIGURE 31.10**
By assessing flows of information, currency, energy, and materials related to public health in El Paso and Juarez it becomes clear that Juarez is moving toward a maturing city/region, whereas El Paso is verging on an anticipatory city/region.

transitions into the on the ground testing of solutions. The development of a master planning framework for a given city or region is based on an interactive simulation procedure that brings various actors representing the environment and the community to the table. Such a system is conceptually supported by the previous phases of the ProtoScope methodology so that in-depth questions can be addressed.

The interactive simulation procedure draws upon the early vision of Will Wright, creator of the Sim City game series. Wright originally envisioned an urban game oriented to a real city that used remote sensors throughout the urban environment to collect information about the effects of community actions to improve the quality of air, water, and energy use systems. The intent was to create a game that provided a flow of real-time information to both planning departments and directly to the individuals in the community as they themselves simulated alternative futures. The game concept enables the citizenry, as well as city and regional officials, to be equally aware of results to better inform urban decisions at all scales of the urban environment. The resulting constant feedback creates a new type of information technology that becomes a vehicle for an ecodemocracy. Although Wright's concept was sophisticated and prohibitively expensive at the time, modern information and communication systems are rapidly approaching the potential to support such a system. Inexpensive and powerful sensors can easily function on the home and city-regional scale to providing immediate and long-term feedback to inform models of future scenarios.*

Our game is a life cycle based procedure that is won by the person, business, community, or government body that can complete whole life support needs within the boundary of their sphere of influence (i.e., home, neighborhood, city). Digital game board components are used to build a city combining horizontal development blocks for neighborhoods and vertical industry blocks for more densely populated urban areas where land value is more prohibitive (see Figure 31.12 through 31.14). The latter can become the basis for spanning

* GreenGoose is an example of an emerging interactive sensoring technology. See GreenGoose.[7]

**TABLE 31.4**

Phases of Development Ladder Defined as Resource Flows Related to Public Health

| | Surviving | Maturing | Anticipatory | Compassionate/Worldly |
|---|---|---|---|---|
| Information | No information about formalized medicine. Traditional medicine may be available but no meter for misinformation. Disease spread due to lack of information about diseases or nutrition | Some knowledge of formalized medicine. Population learns about health, wellness and disease through word of mouth and limited public service announcements. Public is aware of major health issues though not knowledgeable about cause or treatment | Formalized medical education system. Reliable information available to most of population. Some interest in alternative methods of healthcare | Medical knowledge drawn from global sources to establish the best practices. Information universally available and consumed by population. Increased knowledge encourages preventative medicine and healthier lifestyles. Information and currency begin to merge |
| Currency | No formal investment in community health. Poverty is a major contributor to community health issues | Community scale medical economy. Capital available for basic health services and education but still focused on short-term solutions rather than preventative measures. Highest classes can pay for high quality medical care from external sources | Capital available for healthcare and preventative programs. Functional local health economy. General population can afford healthcare. Health issues less linked to poverty | Universal free healthcare. Cash flows sustainable in long term. International investment in healthcare programs. Healthy business and health economy begin to merge |
| Energy | Energy produced at home scale often with fuel sources that contribute to air pollution and related health concerns | Community is linked to energy grid but service is unreliable at times. Furthermore, fuel sources often contribute to air and water quality problems | Reliable and regulated electrical grid. Fuel sources are mostly nonrenewables with integration of renewable technology in certain sectors | Wholly renewable energy economy. Green energy technologies enjoy full financial and regulatory support from populations and government |
| Material | Use of locally available material that can be foraged at no cost regardless of quality or environmental impact of materials. Low quality structures contribute to accidents, vector borne disease and other health concerns | Combinations of cheap and salvaged building materials contribute to unhealthful living conditions. Indoor air quality is not a consideration, and this contributes to upper respiratory disease and cancer | Sustainable and healthy material options are increasingly available. Older structures still largely constructed of harmful materials. Medical facilities slower to make transition due to traditional protocol. Limited air, water, food, energy, and material balance, usually at partial life cycle level | Use of by-products with mature technology approaches. Existing and new structures completely rehabilitated or reconstructed of integrated life cycle. ecoBalanced life supports air, water, food, energy, and materials. Medical facilities select materials that contribute to the patients' recovery |

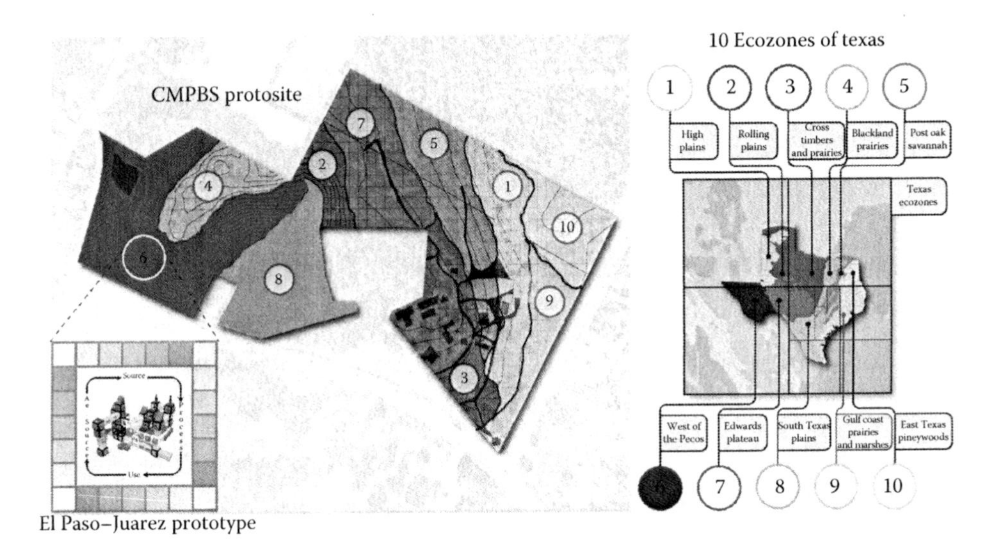


**FIGURE 31.11**
Center for Maximum Potential Building Systems in Austin, Texas, is the primary ProtoSite for El Paso/Juarez. At CMPBS, teams can digitally and physically prototype solution sets that are best suited for the needs of El Paso/Juarez.

large areas with space frames that together become robust frameworks within which a plug-and-play spatial need can evolve over existing overtaxed infrastructure in downtown El Paso. These components were chosen with inherent health and economic benefits in mind.

The fully integrated game is incentivized through a fully functional alternative currency. The interrelated issues of helping create the conditions for a healthy population and a healthy economy in the Juarez/El Paso border region requires transitioning the underground economy of illicit drug trafficking toward a more life-enhancing, health-based economy emanating from life cycle balancing. As we have experienced before—overcoming a crisis requires not only a leap of faith but also a leap of technique. In essence, the money and power derived from smuggling and trading of illegal substances must be replaced by a financial reward system based on good work at the individual and community level that improves living standards while creating surplus for external economic gains. To do this we borrow from highly successful alternative currency systems that strengthen individuals as well as local businesses, already in nascent phases in Asia. The approach is somewhat comparable to many other existing informal currency systems where acquired credit can be converted into useful community benefits, services, or products. The alternative currency being proposed becomes interchangeable between one micro-enterprise and the next with incentives rewarded for the completion of cycles within the abstracted city/region game. The game helps to prompt on-the-ground life cycle entities that function in a manner that completes multiple overlapping cycles amongst a wide spectrum of enterprises. We propose a kind of "wearable currency" that is connected to the simulation game and also records accrued credit in the real world. The individual or group can gain exponential credit with each successful scalar completion of a life support cycle all the way to the entire city system. City and regional planning facilities or private investment can then in turn convert these credits into on the ground development of crowd-sourced needs. The collective reward of ecoBalance directs this system beyond sustainability toward regeneration. Following is an example of how all cycles have a personal activity or micro-enterprise monitor that compiles cyclical results from person through to neighborhood

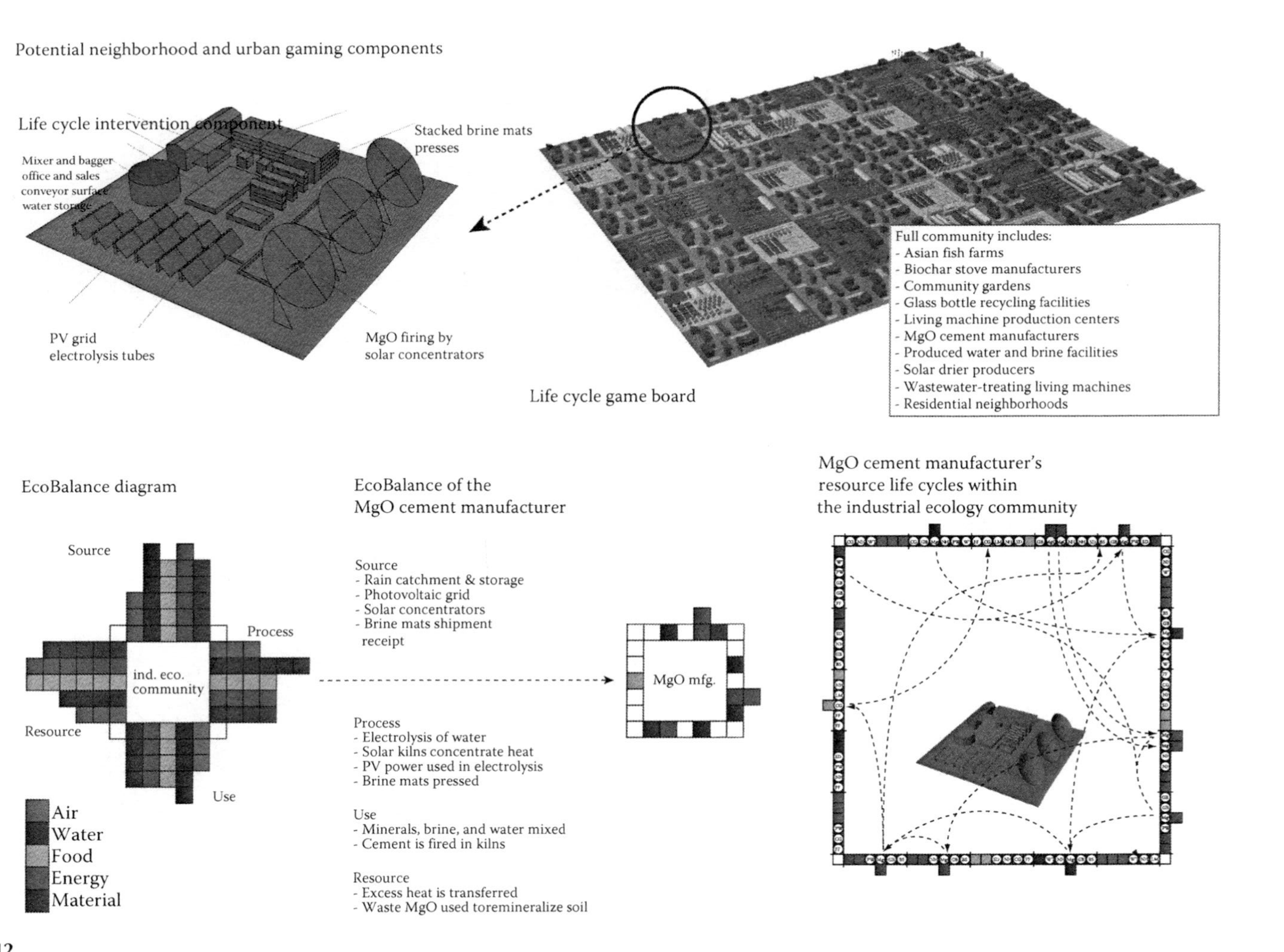


**FIGURE 31.12**
Representative portion of the proposed neighborhood scale game applied to Juarez with the ultimate objective of ecoBalancing all life support cycles.

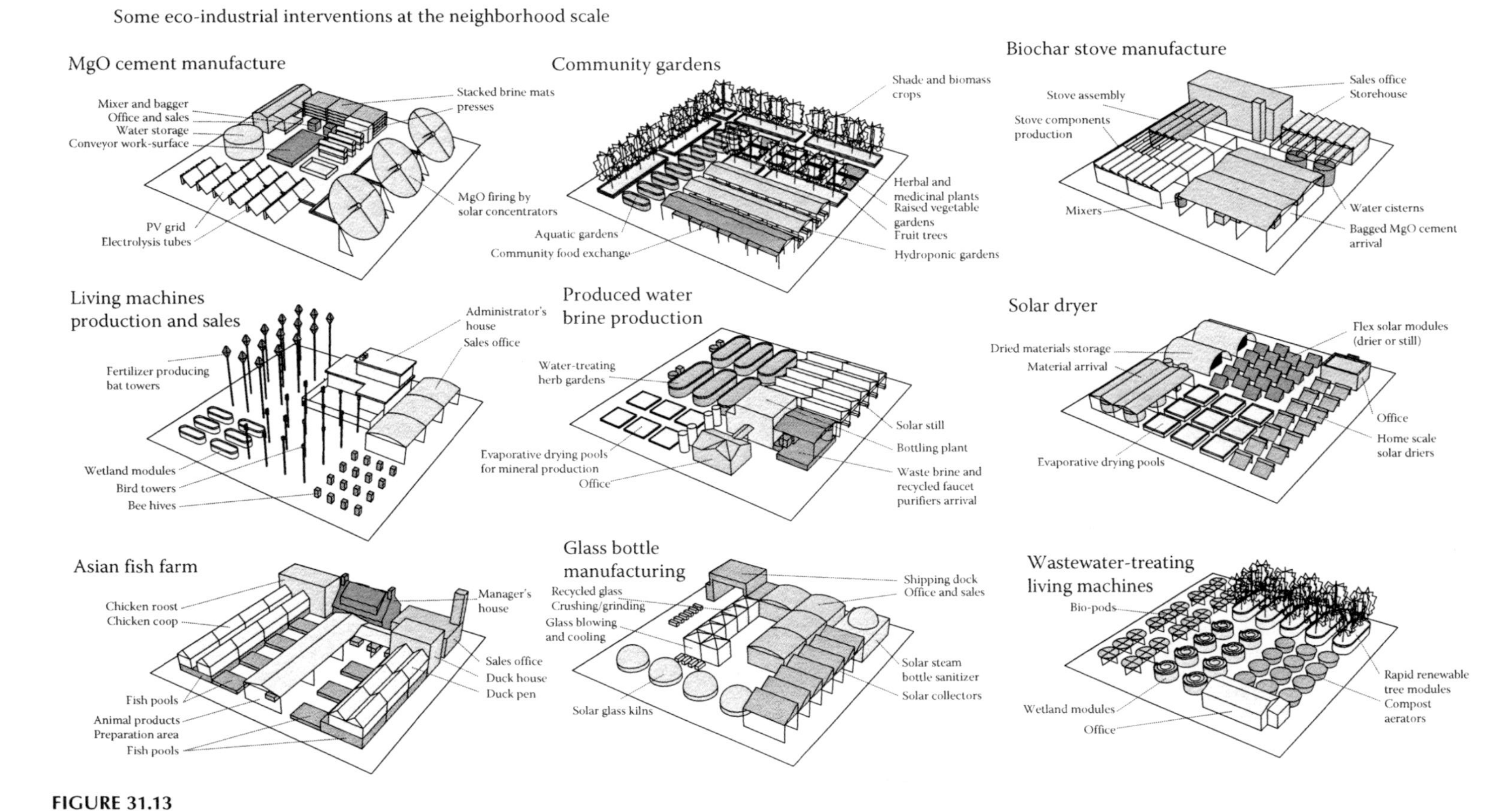


**FIGURE 31.13**
Examples of micro-ecoindustrial interventions at the neighborhood block scale.

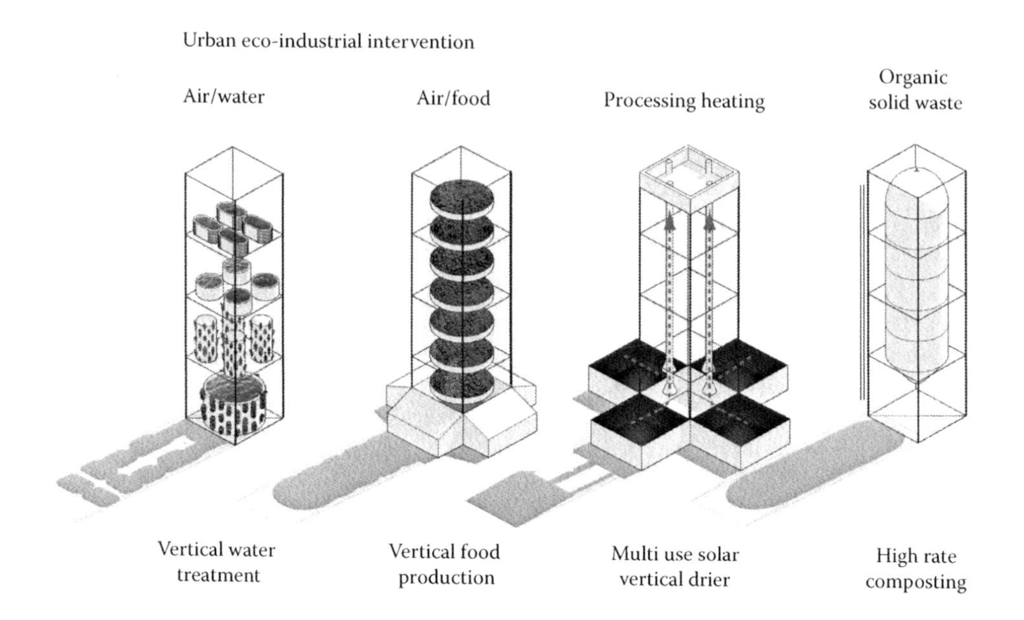


**FIGURE 31.14**
Ecoindustrial interventions at the intensive urban scale using vertically integrated, metabolically connected, industrial clusters used to economically support large space frame columns for the proposed ecoBalanced urban portions of El Paso.

currency. At the personal level, wearable data retrieving enables the individual to receive positive or negative feedback resulting in many levels of conscious behavior (Figure 31.15).

With "Similar to other activity we reward behavior that facilitates disease prevention with the understanding that responsible completion of cycles, such as the water cycle, prevents water born disease (see Figure 31.8). Figure 31.16 places this into the same sort of cybernetic feedback diagrams that we used for the City of Austin.

We find compelling precedent using our protoScoping survey strategy for the concept of alternative currency systems in operating economic systems at somewhat the same scale as our own city. Examples abound such as one developed by Ralph Borsodi in the 1930s called BerkShares in Great Barrington, Massachusetts, now available in five denominations, accepted at more than 350 area businesses, and now reaching nearly 2.7 million bills in circulation since 2006 making it the largest alternative currency network in the United States.[8] Other domestic alternative currency systems include Ithaca Hours of Ithaca, New York; Detroit Cheers in Michigan; and Humboldt Community Currency and Dillo Dollar in Austin, Texas. In Curitiba, Brazil the mass transit use is rewarded through a token-based currency connecting organic food producers to waste recyclers to mass transit.

E. F. Schumacher's *Small is Beautiful* suggests that these currency systems create "good work" by supporting businesses that cater to local services, organically grown food, a range of natural healing and health care methods and recycling.[9] As Mara Ortenburger explains,

> It is legal to print and circulate an alternative currency as long as it looks different than a U.S. dollar and using it is voluntary. Doing so can kick-start a sluggish regional economy by boosting sales of local goods and services. Because it cannot be spent at chain stores or online shops, it stays in the community instead of disappearing to out-of-area banks and corporate coffers. In addition, the nonprofit that manages the currency can issue loans and grants to community groups.[10]

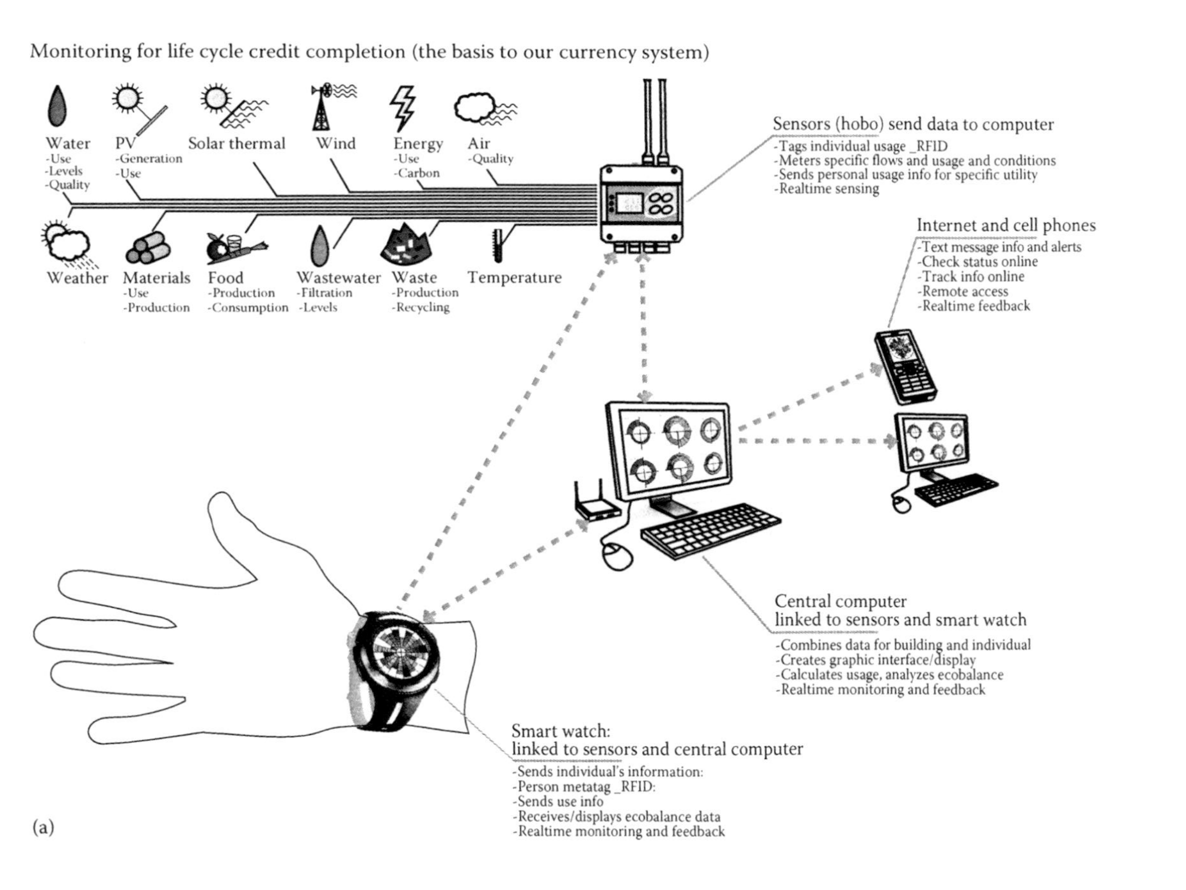


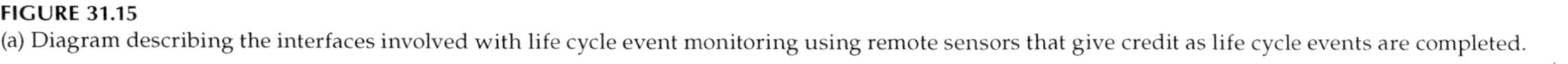
**FIGURE 31.15**
(a) Diagram describing the interfaces involved with life cycle event monitoring using remote sensors that give credit as life cycle events are completed.

(*continued*)

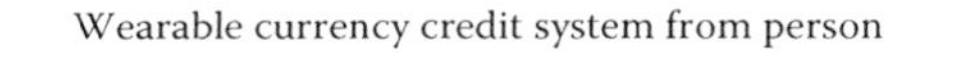


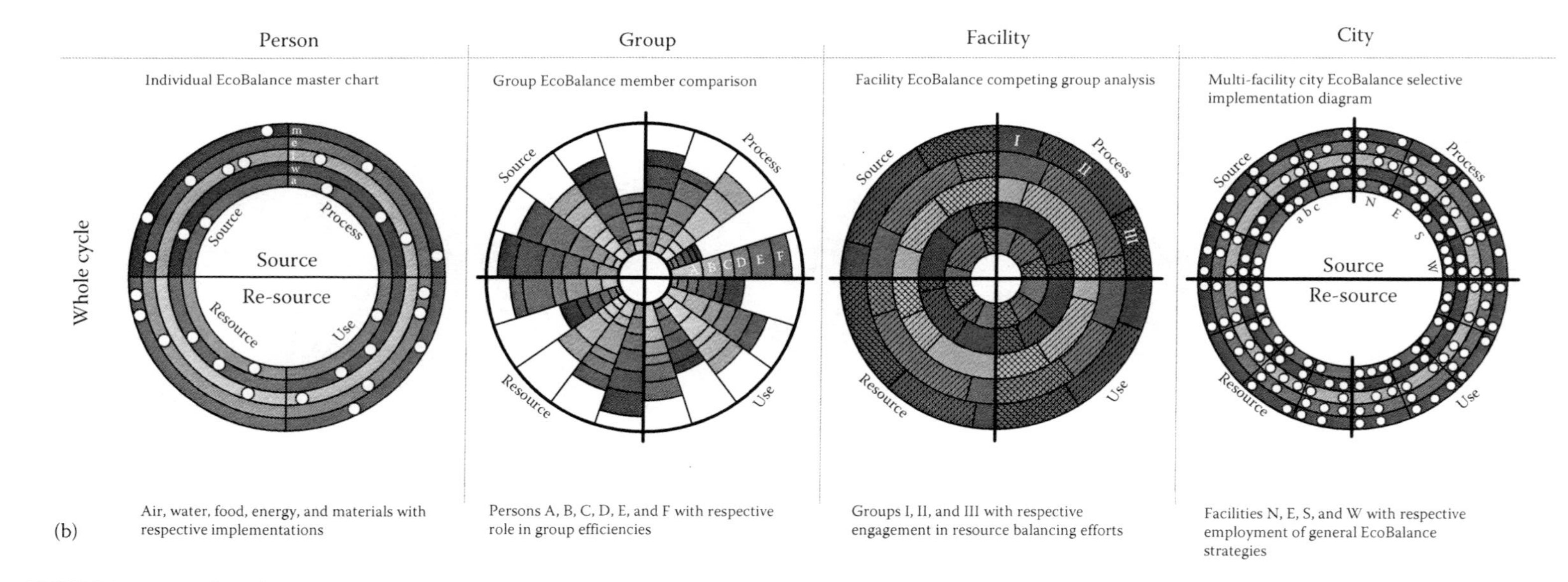


**FIGURE 31.15 (continued)**
(b) Life cycle balancing (ecoBalance) from individual to the city; one scale at a time with credits that accrue as group cooperation creates success at each subsequent scale.

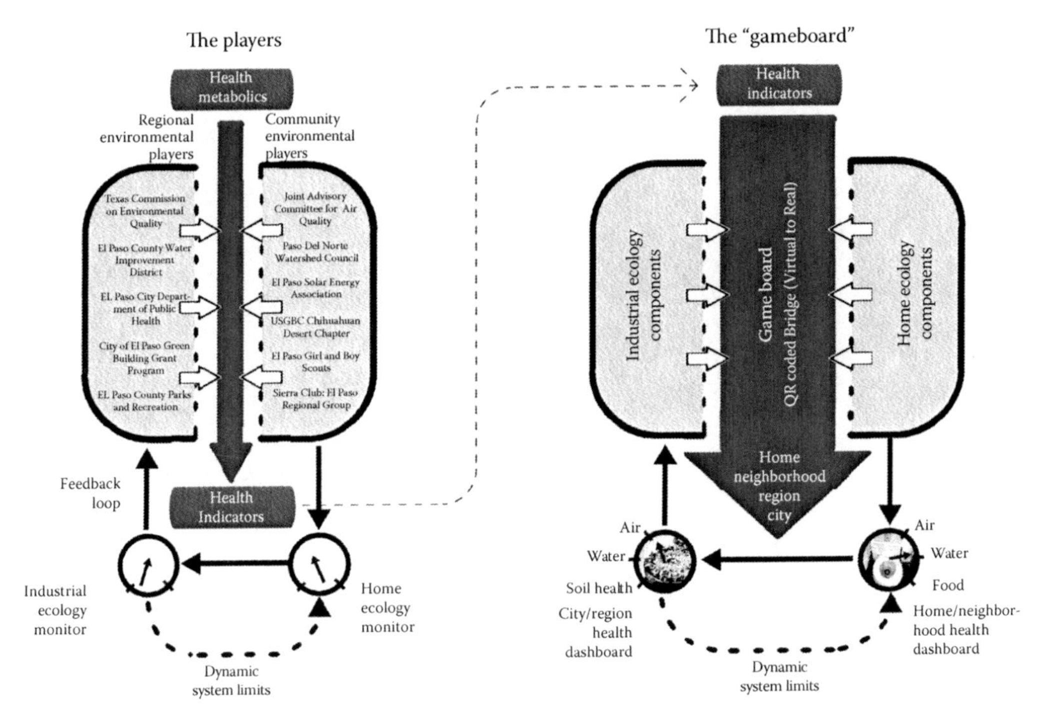


**FIGURE 31.16**
This health to currency feedback system picks up on the use of the Ashby diagram at two different levels within the proposed game, one at the player level and the other at a game board representing the city region.

The true test of a currency system is the building of a credit system that rewards the completion of air, water, food, energy, and material lifecycles at all scales of the city region while also functioning with other systems of alternative currency such as carbon trading.

The new world of incentives, based on life cycle balancing, will be strange at first to those of us entrenched in traditional currency systems that are not connected to an ecology of all human endeavors. Again, this new frontier of concentric, interrelated cyclesstarts with life cycle completion within the functioning of the household, then expands to the neighborhood scale as a set of micro-businesses, and finally to the city—region. The latter occurs in the form of large necklaces of industrial and non-industrial points of conversion rewarded by reaching a state of plenitude and the resulting freedom of time. See Pearltrees (Figure 31.17) an online tool used to explore linkages across ranges of topics.[11]

### 31.5.2 Solution Sets

Solution sets evolve from two types of investigation:

- Tracking ProtoPartners that relate to a given topic in a given ProtoSpace.
- Incorporating ideas and information cultivated from wide scale gaming into built solutions. Orienting the solution sets around the identified indicator of health ensures solution sets will make the greatest impact for our particular community.

Major global health issues as identified in the global health survey of the UN's Millennium Report can be directly correlated to life cycle stages of the five ecoBalance

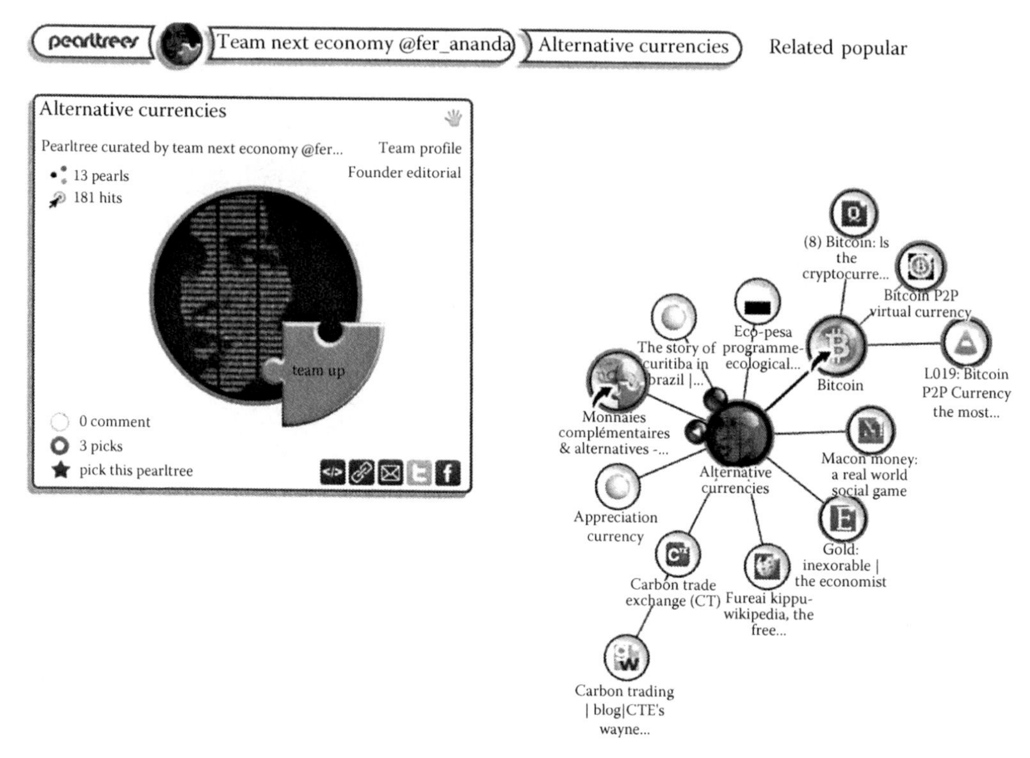


**FIGURE 31.17**
Pearltrees is a tool used to crowd source through a sharable mind mapping process. In this case the users are exploring alternative currencies and their inter-relationship with each other.

flows (see Figures 31.18 and 31.19). By analyzing health effects using ecoBalance we find a wide range of diseases that we tie into life cycle phases. These life cycles are then analyzed at all levels of community from those affected by individual actions to those at a city wide scale.

We bridge health and economic development as series of regenerative actions such as micro enterprises at the neighborhood scale and progress to the entire city scale depending on need and the investment available.

Both the neighborhood horizontal scale and the vertical city scale of regenerative interventions serve the purpose of balancing air, water, food, energy, and material at different intensities of flow relative to energy, material, information, and currency. Together they form a mosaic of economic development supporting an urban ecosystem.

Although the Juarez/El Paso metroplex is a highly industrialized city, due to a generally lower individual capital investment capacity the neighborhood ecology is a more reasonable scale for change in Juarez. Whereas city-wide industrial ecology implanted as vertically integrated ecoindustry functions better within El Paso's more highly developed financial structure, especially given El Paso's high land value relative to its neighbor. However, our presentation of these technologies in the eventual interactive decision-making process represents an ideal mix of technology and neighborhood collaboration.

Neighborhood and urban ecology interventions are directly correlated to major health issues afflicting the region (Table 31.5).

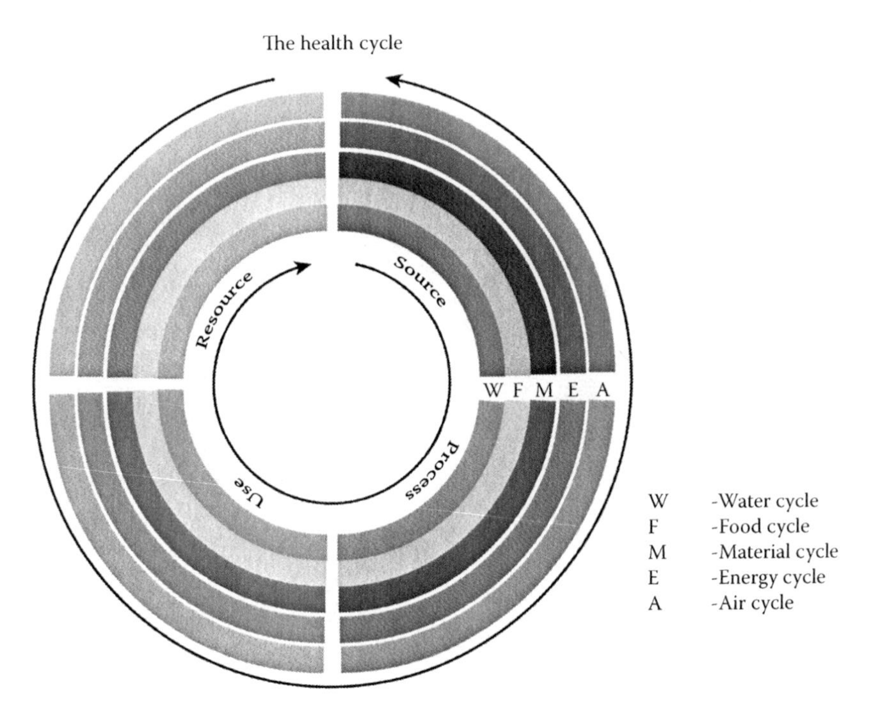


**FIGURE 31.18**
Human health is our strategic focal point in this ProtoCity. Health is intimately involved with all cycles of life and completion of cyclical events functions as the first order of prevention. (Note that the cycle of daily use sequence of life cycle events is opposite to that of the proposed alternative currency credits cycle, i.e., the reward system.)

## 31.6 Planning the Juarez El Paso Metroplex

> People don't change when they see the light,
> They change when they feel the heat.
> People don't believe it when they see it,
> They see it when they believe it.[12]

Human's responsibility to restore, rehabilitate, preserve, or even rebirth the cultural and biological importance of the place where we reside is perhaps the most important planetary function we may perform. Our ability to regenerate and repair through an openness and resilience is how we might intervene to fulfill this lofty responsibility and assesses a community's ability progress to the next stage in their development. If *people* do not see the light, the essence of place might forever disappear.

So a reorientation of the mind is the key; the ability of our everyday existence to be knowingly embedded with ecological thinking is imperative. Even the simplest routines in the home or within other micro-economies, built around the life cycle management of water, food, and energy, enables the continued expansion of the new economy by supplying the products and processes that elevate life cycle balancing to commonplace practice. There are obstacles; humans want to affect their habitat in a manner that is seemingly expedient and self-gratifying in the short term with little regard for the future. The mind, due to its evolutionary development, tries to outpace the very thing that preserves it: the steady, gradual, constant sustaining capability of a stable ecology

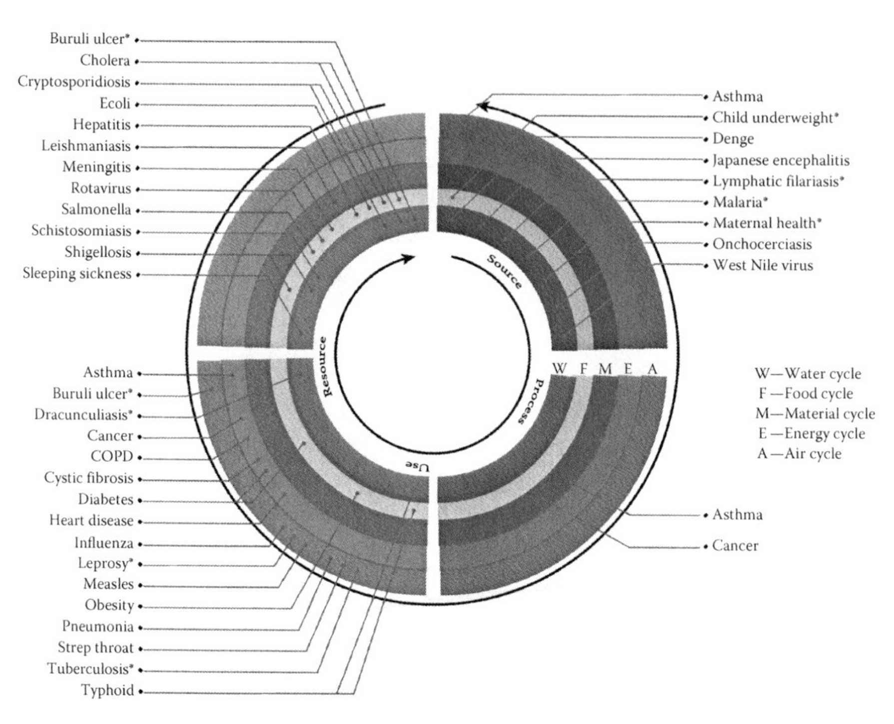


**FIGURE 31.19**
Life cycle coincidence with all diseases listed within the U.N. Millennium Report. The completion of life cycles is the key to preventitive health care. The completion of life cycles is simultaneously the key to earning life cycle credits. *These medical conditions are specifically concentrated upon by the UN under the Millenium Development Goal program.

of place. Refocusing the mind to crave the nature of place might be the supreme purpose of the design sciences, a process of persistently progressing toward some new ecology of mind that brings the at times overly creative neocortex in line with these cycles in our ultimate goal (Figure 31.20).

First there must be an understanding of where the community mind is and how it is increasingly embedded into a virtual world that needs a reorientation to place. Fortunately, our fast paced, technologically connected society is perhaps more receptive to the intervention required to establish this dialogue.[13] A world where mind expansion technologies such as the cell phone, the Internet, and computer gaming are the new reality. It is a world where the crowd is gathering and change is happening and hopefully a place to start to focus on a new ecology of mind and to reorient our goals to live within our planet means. It involves the following:

- Recognize the immensity of the problem through the use of tools that can actually mine the best of human experience while reaching the masses with the uniqueness of place-based knowledge.
- Embrace technology as a tool to build peer-to-peer relationships and facilitate crowd sourcing. Modern communication and information sharing is required to perform the task of steering toward a better ecological future.

**TABLE 31.5**

Causes and Example Solution Sets for 10 Primary Health Issues in El Paso and Juarez

| Home/Neighborhood Ecology Examples | Cause | Health Indicator | Cause | Neighborhood/Urban Ecology Examples |
|---|---|---|---|---|
| Tree/platform shade system | Poor indoor air quality | Asthma | Air pollution from burning of fossil fuels and agrochemicals | Brine-based solar/hydrogen electrical energy, solar pond, dust alleviation using sulfur/tire paving for sidewalks and roads |
| Living machine and cistern sales and use | Inadequate sanitation | Shigellosis (diarrhea) | Water contaminated by sewage | Megaflora/wetland wastewater plantation, reverse osmosis, desert plant biofuel production |
| Bird/bat living machines, community solar drying, bottled water biomass | Mosquitoes, rats, etc.—lack of screens, breeding sites | Vector borne disease (malaria, dengue fever, West Nile virus) | Pests breed in stagnant water in illegal dump sites | Megaflora/wetland wastewater plantation, tire processing, construction waste/debris pulverizer, desert plant BioFuel production, MgO, chloride, phosphate or silica cement |
| Intensive micro and neighborhood food production | Non-nutritious diet due to poverty | Diabetes/obesity | Lack of access to healthful foods, lack of recreational facilities | High rate vertical food production, integrated aquaculture system |
| Active life cycle work, healthy food farming and preparation | Lack of access to healthful foods aggravated by poor air quality and hypertension | Heart disease | Lack of access to healthful foods, lack of recreational facilities | High rate organic vertical composter for organic food, biochar energy, integrated aquaculture system, open space recreation corridors as biofilters |
| Rehabilitate neighborhood for better air scrubbing | Poor indoor air quality | Communicable disease (common cold, flu, pneumonia) | Crowded living conditions | High rate organic composter, megaflora wastewater/wetland plantation, MgO cement |
| Ventilation, vegetation, hydroponics, aquaponic, energy, water treatment | Poor indoor air quality/ exposure to agrochemicals and industrial by-products | Cancer | Exposure to air and water pollution from agrochemicals and industry | Megaflora/wetland wastewater plantation, dust alleviation using sulfur paving, methane from landfills, terra preta landfills, megaflora gasifier boiler generating electric energy, desert plant biofuel production, MgO, chloride, phosphate, or silica cement |
| Recreation (balcony) | Children playing in illegal dump sites | Accidents | Crowded living conditions | Dust alleviation using sulfur/tire paving |

- Create an applied educational approach to learning where classroom teachings are oriented around actually helping build and plan a healthy community so that classroom extends beyond four walls and is embedded more within the community.
- Make everyone a learner/teacher: parents and children, teachers and practitioners, planners and builders.
- Use public observation technologies that are currently relegated to traffic signal violators, shop lifters, identification surveillance, etc. into transformative community informative technologies. For example, a community's established good food growing practices can be "uploaded" and distributed to communities around the globe. Other practical best management practices experiences that can be learned through interactive procedures include water treatment, nontoxic paints, and other health connectable interventions.
- The use of environmental sensors (air, water, land) to communicate feedback progress on small and large scales using remote sensing technologies and identifying key points where significant information is attainable or needed (the drainage inlet, the toilet, the street corner).
- Encourage participatory planning by creatively using the gaming community to simulate planning procedures with recognizable elements that relate to community triggers (in this case environmental and human health interventions) via redirecting processes in the physical environment.
- Work interoperatively. The community simulation planning game, the physical board game, and the real-world working ecotechnologies need to all be connected with a tagging procedure such as QR codes (Figure 31.21).
- Implement the real and the virtual into the community through a multifaceted approach that creates demonstrations locally, regionally and at the state level using the ProtoSite concept where all players are included as equals.
- The gradual process of redirecting flows must start with a focus on the interventions that the community together has decided upon.
- Redirect currency into supporting these new flows at all levels—through legislation, codes, incentives, and the use of alternative monetary systems or "sustain-a-bills"—to support good work across the board.

## 31.7 Conclusion

1968 was an exciting time, I was in Paris at the Beaux-Arts Institute where I had been asked as a student to lecture on what American architecture and planning schools were doing from the stand point of addressing pressing issues of the day. I described how architecture, environment, and cities were trying to come together in some fairly big ways in our school. In the midst of my discussion, the doors of the building flung open and I came within moments of spending my summer traveling fellowship in prison.

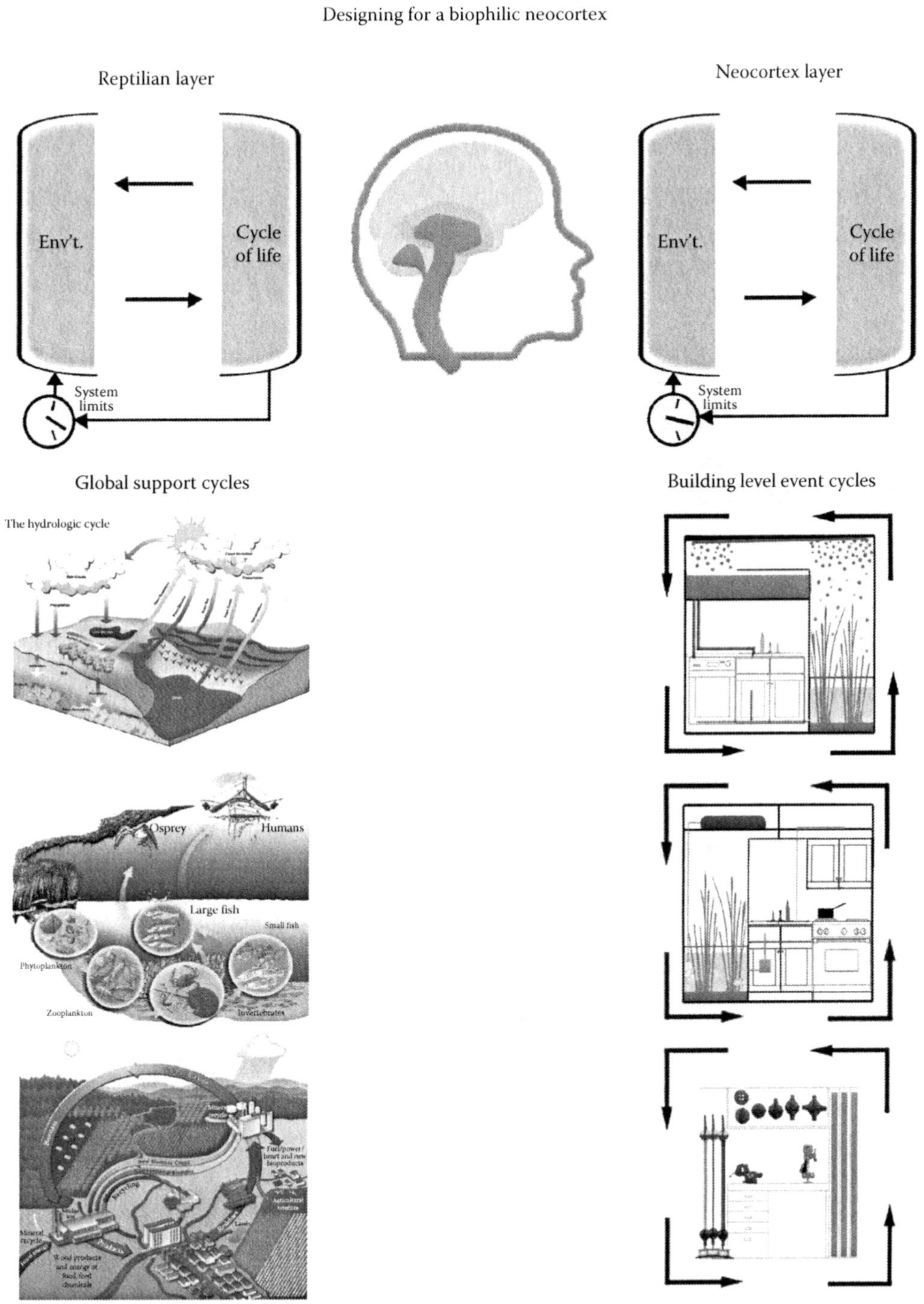

Designing for a biophilic neocortex

Reptilian layer

Env't.

Cycle of life

System limits

Neocortex layer

Env't.

Cycle of life

System limits

Global support cycles

The hydrologic cycle

Osprey

Humans

Large fish

Small fish

Phytoplankton

Zooplankton

Invertebrates

Building level event cycles

**FIGURE 31.20**
In order to bring the brain into synchronization with the cycles of nature we miniaturize the air, water, and food cycles that occur at the global-regional level so that human can make contact with each cycle phase.

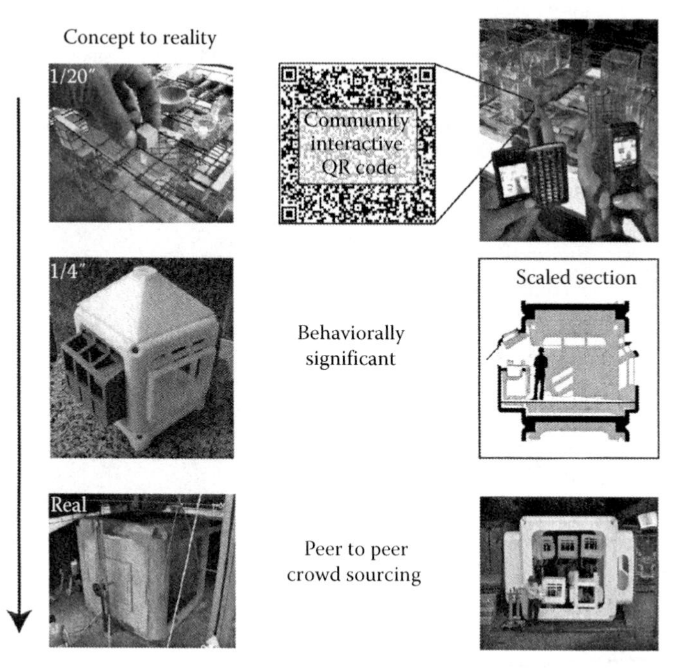


**FIGURE 31.21**
Diagram at three scales of user participation and brain functioning from gaming to model to actual scale prototype.

For a while I thought I had to be careful simply asking big questions but as I re-entered my university setting that fall and looked at options for my final year, the issue of relevance stuck in my head. My previous mindset would have preempted me from landing the position with working on that new town with McHarg a year and a half later. I also realized that even if our specific New Orleans project never happened it still would have been a key moment of understanding what ecological design could accomplish in such a key position in our country, a site that possessed both great potential and great risk.

What enabled me to be hired in such a prestigious position with Wallace, McHarg, Roberts and Todd, arguably the most reknown ecological planning and design firm in the country at the time, grew out of my ambition to complete two masters degrees simultaneously. Architecture studio projects were exciting enough at Penn but I remember hearing that a Professor Ian McHarg was not averse to trying the impossible. As a bright-eyed, eager student this piqued my interest, so I decided to join a studio well beyond my capabilities in the regional planning department.

The class objective was to save the Hackensack Meadows in New Jersey. The project, paid for by the state of New Jersey, was to sustain this gigantic salt marsh and preserve it as McHarg would say as an "Urban Oasis" in the midst of the largest metropolitan area in the world at the time. The studio quickly realized a major problem: The McHargian Method was not meant for an already transformed environment. As a result, we were only able to save 2 acres out of the 18,000 acre site. The natural wetland had been destroyed by development and it was not feasible to tear out Interstate 95 that connected New York to all points west in order to reinstate a natural drainage pattern to save the wetland.

The method was fundamentally flawed and I was disheartened because I had put years of effort into something that was not working. Instead of giving up I took an unexpected step, I decided to do the entire project over again—my way. My new work on the Hackensack Meadows became my thesis for a combined master's degree in architecture and landscape architecture. Luckily Penn was a haven for good thinking and I was exposed to the systems thinking through the Systems Sciences Program at the Wharton Business School. I was soon infected with the systems thinking bug and I decided to apply it to my thesis. System dynamics were the words of the day, so I decided to map-change by working slowly backward through history, recording change, to discover how it got to its current state. I figured out the control points from where I could best leverage and manage the system. If I was lucky I could use that knowledge to steer the system into what potentially could be a model for urban ecology whereupon both ecological and human systems could flourish.

The impairment of natural water flow by roads and railroads was unavoidable, but I noticed that the invasive *Phragmites communis*, a common reed, thrived in the polluted water. Through an early attempt at what I now call proto-Scoping and proto-Partnering I connected with Dr. Kathy Seidel, an underfunded and disregarded researcher at the Max Planck Institute in Germany, who had used a similar approach in a similar ecosystem.[14] Years later I realized that she was the godmother of wastewater treatment using wetland species when she placed levies around highly polluted ponds next to the Rhine, then purposely planted the same invasive reed on the banks to treat the water. The polluted water was filtered through the plants' roots and the levy sand filter and returned, greatly improved, to the urban system. Her work gave me hope and strangely connected me to those roads and rail roads in the Hackensack Meadows. Unfortunately, although I knew now I had a methodology and solution for the salt marsh, it was too late, the studio work failed to saved the wetland and as a consequence the state refused to pay for the studio's previous work and funding for my work.

I was looking for the maximum potential of the system by understanding where the system had been, where it was coming from and where it could potentially go. In other words I sought to capture the possibilities and use them because they most likely represented the way of the least effort. Sadly without funding the prototype that may have been the basis for the largest wastewater wetland system in the country was never constructed. Fortunately the journey was worth the disappointment, as the process has influenced my work for the subsequent 40 years. Although I discovered that my way of thinking about planning differed from my mentor, my experience with McHarg helped bridge the ecology and engineering of an entire new town in one of the most densely populated and polluted regions of the world.

The story is poignant because the stakes are infinitely higher now and the need for creative solutions and McHarg's willingness to take on the impossible is needed more than ever. The El Paso/Juarez metroplex is just as important as New Orleans East or New York City in establishing a new discipline referred to by many names (ecoindustrialization, sustainable urbanism, ecocities) but essential for our collective future nonetheless.

The issues are ever more complex and the traditional resource and funding structure to support positive work are diminishing. The new world requires a systems thinking approach which Donnella Meadows referred to as "discovering the key points in how we enter the system." It is the discovery of the pressure points that take the least physical and monetary resources and incorporate the creative power within all of us to accomplish what seems to be the impossible (Figures 31.22 and 31.23).

**FIGURE 31.22**
Physical model showing the hypothetical revitalization of El Paso and Juarez with a new protoScoped planning system.

**FIGURE 31.23**
Section-elevation of proposed protoCities of Juarez and El Paso.

## Acknowledgments

The author would like to acknowledge the following staff and interns at the Center for Maximum Potential Building Systems, in Austin, Texas, who helped in the development of this chapter including, writing, research, and editing assistance from Cassidy Ellis;

and writing, research and design from Lovleen Gill-Aulakh. CMPBS interns Charlie Candler, Youngjae Chung, Janis Fowler, Sebastijan Jemec, Nels Long and Jesse Miller also contributed to graphics for the chapter.

## References

1. Ackoff, R., An introduction to operations research (Philadelphia, PA: The Wharton School of Business, The University of Pennsylvania, Lecture, 1968).
2. Fisk, P. *Ecobalance: A Land Use Planning and Design Methodology* (Austin, TX: Center for Maximum Potential Building Systems, 2010).
3. Austin Energy Green Building—About AEGB, https://my.austinenergy.com/wps/portal/aegb/aegb (accessed August 11, 2011).
4. Katz, A. The LEED Green Building Certification System, http://archityperreview.com/19-sustainability/editorials/14-the-leed-green-building-certification-system (accessed Sepetmber 13, 2012).
5. Climate Leadership Group. C40 Cities: Climate Leadership Group: About, http://www.c40cities.org/about (accessed September 13, 2012).
6. Ashby, W. R., *Design for a Brain; the Origin of Adaptive Behavior,* 2nd edn. (New York: John Wiley & Sons, 1960).
7. GreenGoose—Play Real Life, http://www.greengoose.com/ (accessed August 23, 2011).
8. The Borsodi constant AKA "The Exeter experiment," inflation free currency, *Community Currency Magazine,* March 2009, p. 22.
9. Schumacher, E. F., *Small Is Beautiful: Economics as if People Mattered,* 2nd edn. (New York: Harper Perennial, 1989).
10. Ortenburger, M., Santa Cruz Cash, *The Santa Cruz Good Times,* December 8, 2008.
11. Pearltrees—Next Economy @fer_ananda—Alternative Currencies, http://www.pearltrees.com/#/N-f=1_2713621&N-fa=2292772&N-p=19755917&N-play=0&N-s=1_2713621&N-u=1_108097 (accessed August 23, 2011).
12. Murphy, M., Landscape Architecture, Texas A&M University, Personal communication.
13. Fisk, P., The greening of the brain, *in* S. Kellert, J. Heerwagen, M. Mador, Eds., *Biophilic Design* (New York: Wiley, 2008), pp. 307–312.
14. Matsch, W. J., J. G. Gosselink, *Wetlands,* 2nd edn. (New York: Wiley, 2000).

# Index

**D**

## I

## K



## L



## M



## N

**T**

## X